THE COMPLETE GUIDE TO CLIMATE CHANGE

'Climate change is a wide-ranging and complex subject – a global challenge involving governments, companies, professionals and public action. This book provides a comprehensive coverage of the key scientific, technical, social and political aspects of climate change in an easily referenced and reader-friendly style. I would recommend this book to students and professionals new to the field that wish to expand their knowledge.'

Tom Baumann, *CEO ClimateCHECK and Co-Founder Greenhouse Gas Management Institute*

For anyone trying to separate the fact from the fiction, *The Complete Guide to Climate Change* is an indispensable resource. Taking you through the A to Z of the key scientific, geographical, and sociopolitical issues involved in the study of the environment and the implications of mankind's effect upon it, topics covered include:

- Environmental science – the carbon cycle and the "greenhouse gases"
- The impacts of climate change on life, land, and sea
- Mitigation strategies from carbon capture to carbon taxes
- The Kyoto Protocol and UNFCCC
- Renewable fuel sources, from wind to solar power

Including guides to the latest scientific and governmental thinking on climate change, this book will tell you all you need to know about perhaps the biggest issue facing mankind today.

Brian Dawson is Climate Change and Energy Adviser for the Australian International Development Assistance Agency.

Matt Spannagle is Technical Manager of the MDG Carbon Facility of the United Nations Development Programme.

The Complete Guide To Climate Change

Brian Dawson and
Matt Spannagle

Routledge
Taylor & Francis Group

LONDON AND NEW YORK

First published 2009
by Routledge
2 Park Square, Milton Park, Oxon OX14 4RN

Simultaneously published in the USA and Canada
by Routledge
270 Madison Ave, New York, NY 10016

Routledge is an imprint of the Taylor & Francis Group, an informa business

© 2009 Brian Dawson and Matthew Spannagle

Typeset in Times New Roman by
Book Now Ltd, London
Printed and bound in Great Britain by
TJ International Ltd, Padstow, Cornwall

British Library Cataloguing in Publication Data
A catalogue record for this book is available from the British Library

Library of Congress Cataloging in Publication Data
A catalog record for this book has been requested

ISBN10: 0–415–47789–1 (hbk)
ISBN10: 0–415–47790–5 (pbk)
ISBN10: 0–203–88846–4 (ebk)

ISBN13: 978–0–415–47789–5 (hbk)
ISBN13: 978–0–415–47790–1 (pbk)
ISBN13: 978–0–203–88846–9 (ebk)

CONTENTS

CONTENTS

ILLUSTRATIONS

Tables

Figures

PREFACE

Climate change has emerged as one of the defining political and socioeconomic issues of the twenty-first century. Although it has been part of the scientific agenda since the 1970s, it only really began to attract widespread international attention during the 1990s. It is now a mainstream political issue that people need to understand and respond to. Climate change is a complex issue that covers the full spectrum of scientific, economic, social, and political disciplines, and few people have the opportunity to attain a comprehensive and in-depth understanding of all facets of climate change. The purpose of this book is to assist readers to gain a better understanding by distilling current knowledge and key issues in an easily understandable form.

Over the past two decades, enormous progress has been made in the understanding of climate science, the likely repercussions of a changing climate on human and natural systems, and what options are available to reduce the extent of future climate change. Thousands of research papers, reports, articles, and opinion pieces have been published on climate change, and these have presented a wide range of divergent, and often opposing, views. Climate change is an issue that has been the subject of ongoing, and at times heated, scientific and political debate. Although uncertainty remains in respect to the magnitude, timing, and extent of climate change and its impacts, the vast body of scientific evidence presents a compelling case that human activities have already changed the earth's climate and, if present emission trends persist, the changes that are projected to occur over the course of this century are likely to be very dangerous indeed.

The book consists of 64 separate entries, each of which covers a specific topic or issue relevant to climate change. Each entry is, in effect, a stand-alone essay that provides the reader with a summary overview of the current knowledge and understanding of the topic. The information contained in each entry is sourced from the most recent and authoritative literature on the subject. Each entry contains cross references to other entries, so that readers can find more detailed information on related issues. At the end of each entry, readers are directed to other entries most relevant to the topic under discussion and to further information sources that cover the topic in more detail. The book covers all the key concepts of climate change, including climate science, projected impacts, social and political drivers underlying future emissions, finance and carbon markets, the potential contribution of different mitigation options, and the international political and administrative infrastructure that guides the international community's response to climate change.

As authors, and practitioners in the field of climate change, we have attempted to present an objective and balanced account of the differing views and levels of uncertainty surrounding various climate change issues. We have generally refrained from drawing definitive conclusions on most topics, unless the empirical evidence is overwhelming, or it covers our particular areas of professional expertise.

Brian Dawson and Matt Spannagle,
Canberra and New York,
July 2008

Acknowledgments

To compile a book of this nature requires considerable time and effort on our part as authors. However, it also requires considerable understanding and tolerance on the part of our respective partners and families. For this we wish to devote a special note of thanks to Jan and Jits and to our children, Ella, Olivia, and Inka, who have made many sacrifices to enable us to compile this book.

The breadth and complexity of the climate change issue makes it difficult for any single person to have an in-depth understanding of all facets of the debate. During the compilation of this book, we have sought advice, guidance, and feedback from many people on key issues across a range of subject areas, and we thank them for opinions and inputs. In particular, we extend a special thanks to Rob Kelly, who provided insightful guidance and editorial feedback on many sections of this book, and Jamie Rogers for his specialist knowledge of international finance, risk markets, and the dynamics of the global economy.

ADAPTATION

Adaptation refers to all responses, adjustments, or actions by humans and natural systems to accommodate and/or reduce their vulnerability to the impacts of climate change. There are many different definitions of adaptation in the climate change literature: some relate purely to humans, while others take a broader perspective and include adaptive responses of natural systems.[1] Adaptation measures will not reduce climate change, but they can reduce vulnerability to the impacts of climate change.

Global temperatures have already risen by 0.76°C since 1850, and due to inertia in the climate system, a further 0.5–1.0°C of warming is expected over the next 50 years, even if atmospheric greenhouse gas concentrations could be held at today's levels. Emissions to date have already committed the earth to a global mean temperature increase of approximately 1.5°C above preindustrial levels, and possibly 2.0°C (see **global warming**). It is anticipated that humans will be able to adapt to a global temperature increase of this magnitude, although there will be adverse impacts. The ability of natural systems to adapt is much more problematic. Even low levels of warming can cause considerable stress on natural ecosystems and, over the course of this century, climate change is likely to result in further ecosystem degradation and species extinctions (see **biodiversity impacts** and **marine impacts**).

While adapting to the impacts associated with 1–2°C seems manageable, it is the additional warming expected over the next few decades that will present a much greater adaptation challenge. If current emission trends persist, it is likely that concentrations could reach double preindustrial levels by 2040 (see **future emissions trends**). This is expected to increase global temperatures by around 3°C, but possibly 5°C (see **climate sensitivity**). Temperature increases of this magnitude over such a short period will have significant, and potentially catastrophic, impacts on both humans and natural ecosystems (see **climate change impacts**). The further greenhouse gas concentrations rise, the greater the amount of global warming, and the greater the risk of strong positive **climate change feedbacks** that could amplify warming even further.

Since the **United Nations Framework Convention on Climate Change (UNFCCC)** was signed in 1992, the focus of international climate change negotiations has been on reducing greenhouse gas emissions and how responsibility for emission reductions are allocated amongst countries (see **Kyoto Protocol**). Adaptation issues have been part of the negotiations but have generally attracted less attention, though this has begun to change. Several factors have contributed to the increased focus on adaptation. First, the commitment to future warming continues to grow as, to date, the international community has done very little to reduce greenhouse gas emissions. The annual increase in concentrations has actually accelerated since the UNFCCC was signed.[2] Second, there is mounting evidence that climate change impacts are beginning to manifest themselves more quickly than initially anticipated, which may necessitate earlier adaptation responses. The third contributing factor is that the developing countries recognize that the burden of climate change will fall most heavily on them, even though they have contributed least to the problem. As a result, they are exerting increased political pressure on the high-income developed countries to provide the necessary financial and technical resources to help them adapt to climate change.

Human adaptation to climate change

Humans have been adapting to changes in climatic conditions for thousands of years, either in response to natural climate variations (such as those that have occurred since the end of the last ice age around 15,000 years ago) or to human-induced changes to local environmental conditions (such as reduced rainfall resulting from excessive deforestation and land degradation).[3] However, the pace and magnitude of climate change over the course of this century is likely to exceed, in both scale and geographic extent, that experienced in the human past and will require a much more significant adaptation response.

The extent and severity of **climate change impacts** rise nonlinearly with global temperature increases. At relatively low levels of warming (less than 2°C), the adaptation task appears manageable. In temperate regions, there may even be some benefits at low levels of warming. For example, agricultural productivity may increase in parts of Canada, Russia, and northern Europe. However, once temperatures exceed 2°C above preindustrial levels, and more so above 3°C, **climate change impacts** are expected to be overwhelmingly negative across all regions and the adaptation task will grow substantially. Beyond 4°C, the human cost of climate change is likely to be very large indeed, and some areas in the mid- and low latitudes may be incapable of supporting their current populations. Given that humans have not previously experienced such large and rapid rises in global mean temperatures, it is not at all certain how communities will cope.

Regional variation in adaptation needs

There is expected to be significant regional variation in the types and severity of climate change impacts. Some regions will face greater exposure to **extreme weather events**, impacts on crop yields will vary (see **agriculture and food supply impacts**), and changes in annual and seasonal variations in precipitation and water availability will have serious implications for some countries and regions but less so in others (see **water impacts**). Regionally different adaptation responses will be needed.

The uncertainty surrounding the timing and severity of **climate change impacts** makes the task of identifying and implementing appropriate adaptation strategies difficult. Adaptation responses will depend on where people live, the structure of their economic systems, and how they derive their livelihoods – adaptation largely entails a localized response. Some adaptation measures will be reactive, responding to the impacts as they occur, while others will be proactive and implemented in advance to reduce future climate risk and vulnerability.

Climate change is, and will continue to be, most rapid in higher latitudes. Some high-latitude regions have already experienced temperature increases greater than 2°C (see **polar impacts**). Extensive thawing of the permafrost will have major implications for existing infrastructure. Large investments to replace and/or strengthen infrastructure will be a necessary adaptation. Indigenous communities, such as the Inuits, are already facing a significant adaptation task, and those whose livelihoods are dependent on traditional hunting and fishing will be particularly vulnerable. However, as these regions are sparsely populated, the overall human impact cost will be relatively modest in a global context. Furthermore, the prospects for alternative livelihoods appear somewhat brighter for the higher latitudes than for the lower latitudes as less harsh climatic conditions will, over time, enable these regions to support more forestry, agriculture, and resource extraction activities and also support higher population densities than today.

The temperate regions, where most developed countries are located, are likely to be least vulnerable to climate change, at least for low to moderate levels of warming. People living in these regions are generally less dependent on natural ecosystems to support their livelihoods and have far greater technical and financial resources at their disposal to support necessary changes (see **socioeconomic impacts**). Nonetheless, they will still need to put in place adaptation measures to accommodate changes in rainfall and temperature patterns (especially for impacts on agriculture); invest in improved water security; and accommodate increased risk of severe droughts, floods, heat waves, and intense storm events. They will also need to minimize the impacts of changes in the prevalence and geographic extent of pests and diseases on agriculture and human health (see **health impacts**). In the longer term, **sea level rise** and higher storm surge peaks will place many coastal regions at risk of temporary or permanent inundation. The temperate and mid-latitude regions account for the majority of the world's existing capital infrastructure (industry, bridges, roads, ports, buildings), much of it located in vulnerable coastal areas. Effective adaptation will involve substantial investments to protect, rehabilitate, and/or relocate critical infrastructure (see **coastal zone impacts**).

It is the tropical and subtropical regions that are most vulnerable to climate change and where humans face the greatest adaptation challenges. It is also where the majority of the world's poorest people live. These regions are expected to experience the most pronounced falls in agricultural yields, suffer from potentially acute water stress (see **water impacts**), and be more vulnerable to **extreme weather events**. Most of the inhabitants of these regions have a relatively high dependency on natural ecosystems to support their livelihoods (such as agriculture, hunting and gathering, fishing and forestry activities), all of which are highly vulnerable to climate change. The low-income developing countries are likely to experience greater relative economic losses from climate change than the wealthier nations but have much more limited financial and technical resources available to implement adaptation responses.

The primary focus of low-income developing country adaptation strategies will be to maintain water and food security and reduce vulnerability to extreme weather events. Response measures could include improved irrigation and water storage; introducing more drought- and heat-resistant crop varieties, change planting times and/or cropping patterns; disaster risk reduction measures (such as improved building standards, infrastructure protection, and land use zoning); enhanced emergency response capabilities (such as early warning and evacuation/relocation measures); and strengthening health system capabilities to manage the increased vulnerability to diseases and heat stress.

If **global warming** cannot be kept to relatively moderate levels, these measures are, in many cases, only temporary solutions. There are limits to how far adaptation can maintain current socioeconomic systems. For example, in the case of agriculture, adaptation measures such as irrigation, increased water storage, and end-use water efficiency can be important adaptation responses to accommodate changes in annual or seasonal water availability. However, permanent long-term declines in annual precipitation, or permanent changes in seasonal water flows, may eventually require the total transformation of existing agricultural systems, or abandonment of agriculture altogether in some areas. Food and water scarcity, and in the longer term, sea level rise, may force people to migrate to other areas as their only remaining adaptation option. This may lead to political tension or conflict in areas receiving climate change refugees, particularly if resources are already stretched. Arrangements will need to be put in place to accommodate these people.

Reducing the vulnerability of the poor

There are a range of different views on how best to reduce vulnerability to climate change. The most common approach, and one that currently underpins most official development assistance support for adaptation, is to conduct vulnerability assessments, devise adaptation strategies, and implement adaptation projects. Others take the view that unless the underlying factors that make people more vulnerable to climate change in the first place are tackled, namely poverty and limited economic development, stand-alone adaptation initiatives will not be sufficient.[4] Economic growth may increase the ability of a country to fund adaptation measures, but growth alone is unlikely to be sufficient to meet the adaptation challenge, particularly if this growth is based on **fossil fuels**. In this case it would exacerbate climate change by adding even more **greenhouse gases** to the atmosphere and require yet further adaptation.

Irrespective of the merits of different vulnerability-reduction approaches, it is becoming evident that climate change is likely to result in substantial human and economic cost to the developing world. It may undermine the ability of the world's poorer nations to achieve their development aspirations and may also place in jeopardy many of the development gains achieved over recent decades.[5] Climate change adaptation is very much a development issue, whose political dimension will grow significantly in the years ahead.

Mechanisms have already been established to provide financial and technical support to developing countries for adaptation activities. The Global Environment Facility (GEF), World Bank, the United Nations Development Programme, the United Nations Environment Program, bilateral development assistance programs,

4

and other organizations are all actively supporting adaptation initiatives in developing countries. In particular, through the **UNFCCC**'s Least Developed Country Fund (LDCF), assistance has been provided to the world's poorest countries to develop National Adaptation Plans of Action (NAPA). These aim to identify priority adaptation needs and formulate adaptation plans. The NAPAs have proven useful in raising awareness of the need to adapt to climate change and in prioritizing possible adaptation responses. However, the key issue facing many developing countries is not whether they have suitable adaptation plans but how they can harness the necessary resources to implement these plans.

Financing adaptation

The estimated costs of adapting to climate change vary widely and are subject to much uncertainty. For the developed economies, estimated adaptation costs by 2030 range from a low of $ 15 billion per year to as high as $ 150 billion per year, depending on the level of warming and assumptions about the cost and effectiveness of different adaptation measures.[6] Estimates for the developing world are generally lower, and the **UNFCCC** estimates that by 2030 the required finance and investment flows to fund adaptation could be in the range of $ 28–67 billion per year.[7] However, given the uncertainty surrounding the timing and magnitude of **climate change impacts** and difficulties in estimating the costs of adapting to these impacts, these projections remain speculative. As yet no reliable comprehensive global estimate of future adaptation costs exists. Nonetheless, it is evident that adaptation costs are likely to be substantial and that they will grow significantly over the coming decades as more warming occurs.

The international community recognizes that significant resources will be required to finance adaptation responses and that, for the developing countries, a substantial amount of external sources of financial and technical assistance will be required to implement adaptation responses. So far three UNFCCC financial instruments have been established to assist with adaptation funding: the LDCF, the Special Climate Change Fund (SCCF), only a portion of which will be devoted to adaptation activities, and the Adaptation Fund. As of early 2008, these funds had limited resources: the LDCF had funds totalling US$ 86 million, the SCCF US$ 73 million, and the Adaptation Fund has yet to become operational.[8] The Adaptation Fund will be administered by the GEF and financed primarily through a 2% levy on credits issued under the **Clean Development Mechanism (CDM)** and possibly extra budgetary contributions from individual donors. The World Bank estimates that by 2012 the CDM could provide US$ 100–500 million of revenue to the Adaptation Fund. The World Bank administered Climate Investment Funds, estimated to total more than US$ 5 billion, will include a Pilot Program for Climate Resilience and significantly boost the amount of funds available to finance adaptation in developing countries.

Though these international funding mechanisms are likely to make a valuable contribution to adaptation financing needs in the short term, it is unlikely that donor-funded financial flows alone will be sufficient to meet the adaptation task in the longer term when climate change impacts become more pronounced. The private sector, and particularly the finance and insurance industries, will also need to play an important role through raising the capital to finance adaptation measures and managing the risks of climate change (see **finance and insurance**).

See also: climate change feedbacks, climate change impacts, climate sensitivity, global warming, socioeconomic impacts.

Notes

1 Schipper 2007
2 IPCC 2007
3 Smithers and Smit 1997
4 Schipper 2007
5 UNDP 2007
6 Stern 2006
7 International Development Association 2007
8 Global Environment Facility 2008

Further reading

Stern 2006; IPCC 2007; Schipper 2007.

AEROSOLS

Aerosols are solid or liquid particles between 0.01 and 10 μm in size that are airborne in the atmosphere for at least several hours. Aerosol emissions to the atmosphere have a negative effect on atmospheric **radiative forcing** (a cooling effect): offsetting as much as 40% of the **global warming** effect of **anthropogenic greenhouse gas emissions**.

Aerosols are a collection of various microscopic airborne solid (e.g. soot) or liquid (e.g. mist) particles. These particles may be of natural (dust storms, sea spray, forest fires, volcanic eruptions) or anthropogenic origin (such as combustion of **fossil fuels**, construction/mining dust, agriculture, biomass burning). Some sources, such as soot and sulfates from fossil fuel combustion (the largest source of aerosols) are clearly anthropogenic, while others, such as wind-blown dust from land altered for agriculture are difficult to distinguish between natural and anthropogenic origin. This uncertainty of origin makes determining the impact of aerosols on climate change difficult as natural aerosols sources are part of the preindustrial climatic equilibrium.[1]

Historic levels of aerosols are difficult to determine, as no pristine air conditions remain on earth for reliable comparisons to be made. Studies conclude that current aerosol loads in remote temperate regions are now up to 300% higher than in preindustrial times. Heavily industrialized regions, particularly in East and South Asia, have aerosol levels much higher again. In short, anthropogenic sources of atmospheric aerosols now clearly dominate, at more than 10 times the concentrations of natural sources in continental and most oceanic air sheds.[2]

Since 1950, aerosol emissions have increased substantially. The rapid expansion in industrial activity and fossil fuel combustion in North America, Europe, the Soviet

Union, and Japan in the decades following World War II not only resulted in large increases in greenhouse gas emissions but also substantially increased aerosol emissions. This caused major air pollution problems in many of the world's major cities and industrial areas, with numerous adverse side effects (such as acid rain, increased respiratory disease, and elevated levels of lead, mercury, and other heavy metals).

To improve urban air quality and curb acid rain, many industrialized country governments progressively introduced strict emission control regulations during the 1970s and 1980s. This led to the development and widespread deployment of technologies, such as catalytic converters in cars and flue gas scrubbers in power stations. By the 1990s, these measures had led to a significant reduction in aerosol emissions from North America, Europe, and Japan, with further reductions from these regions expected over the coming years.[3] However, these gains have been progressively offset by increased aerosol emissions from rapidly industrializing developing countries, particularly China and India.

Aerosols and climate

Determining the influence of aerosols on climate is complicated by numerous aerosol types and their associated properties (size, atmospheric lifetime, optical and hygroscopic properties, chemical reactivity), as well as the nonlinear cloud formation response to atmospheric aerosol loading.[4] Nonetheless, satellite observations, combined with ground observations and computer modeling, have been able to estimate the **radiative forcing** of the total aerosol load.

Aerosols influence the climate in three primary ways: (1) by directly warming or cooling the atmosphere; (2) indirectly by changing cloud properties and earth's **albedo** (reflectivity); and (3) through their precipitation and influence on surface **albedo** properties. The aerosol direct and indirect effects have reduced the amount of energy from the sun that reaches the earth's surface (see **global dimming**), and combined, these three primary mechanisms result in a net cooling effect. Aerosols reduce **radiative forcing** by -1.1 Wm^{-2}, offsetting around 40% of the 2.64 Wm^{-2} of warming from **anthropogenic greenhouse gas emissions**.[5]

Direct effect

Dark particles, such as black carbon (soot), can have a very low **albedo**, absorbing more incoming solar radiation and warming the atmosphere. Their direct effect is particularly strong when these aerosols are aloft over a bright surface such as desert or snow since they absorb more solar radiation than the reflective surfaces below. Conversely, high **albedo** aerosols (such as cement dust and sulphates) scatter and reflect incoming radiation and have a cooling effect, particularly over forests or tropical oceans that would otherwise absorb most of the incoming energy. When aerosols are at very high altitude, such as those from aircraft emissions, they effectively reflect energy before it enters the atmosphere. Thus, high and low **albedo** aerosols result in decreased warming when at high altitude. The direct effect of aerosols reduces **radiative forcing** by 0.5 Wm^{-2}, nearly 20% of the warming effect of **anthropogenic greenhouse gas emissions**.[6]

Indirect effects

More significant than the direct effect are the two indirect cloud formation effects of aerosols – both of which increase earth's **albedo**. First, aerosols provide a larger number of nucleation points for water vapor condensation, increasing cloud albedo (Twomey effect). Second, aerosols increase cloud thicknesses and lifetimes (Albrecht effect). Combined, they reduce **radiative forcing** by -0.7 Wm^{-2}, offsetting 27% of the warming caused by **anthropogenic greenhouse gas emissions**.[7]

Aerosol precipitation

The atmospheric residence time of aerosols can vary from as little as several hours to as long as several years. Typically most remain in the atmosphere for less than 10 days.[8] Eventually aerosols are either dissolved in water droplets and/or directly precipitated out of the atmosphere. When particles land on surfaces in sufficient density, they can change surface **albedo**. In particular, when black carbon (soot) lands on snow or ice, it decreases surface **albedo** – absorbing more incoming solar radiation. This adds 0.1 Wm^{-2} to **radiative forcing** (a warming effect): equivalent to 4% of the warming from **anthropogenic greenhouse gas emissions**.[9]

Future impact of aerosols on global warming

Atmospheric aerosol concentrations are closely correlated with fossil fuel combustion. As fossil fuel consumption is expected to increase substantially over the next few decades (see **fossil fuels**) so too will aerosol emissions. This is expected to increase the net aerosol cooling effect, at least in the medium term.

Public pressure to reduce chronic air pollution and improve air quality may, over time, lead to stricter emission controls in the rapidly industrializing countries (such as China, India, and Brazil), though these measures may take several decades to substantially reduce aerosol emissions. Aerosol emissions could decline as **mitigation** efforts lead to a decrease in aggregate fossil fuel consumption, although this is unlikely before 2030. Eventually consumption of fossil fuels must fall as they are a finite resource.

Due to the high levels of uncertainty surrounding projections of future fossil fuel combustion and efficiencies, estimates of when aerosol concentrations might fall to closer to natural levels remain speculative. Nonetheless, atmospheric aerosol loadings are expected to decline by the second half of this century and so too will their cooling effect. This may accelerate **global warming**, and temperature increases may approach the upper extreme of the range projected by the **Intergovernmental Panel on Climate Change (IPCC)**[10] (see **dangerous climate change** and **climate change impacts**).

See also: albedo, anthropogenic greenhouse gas emissions, climate change feedbacks, fossil fuels, global dimming, radiative forcing.

Notes

1 IPCC 2001
2 Andraea 2007
3 Breon 2006
4 Kaufman and Koren 2006
5 Ibid.
6 IPCC 2007
7 Ibid.
8 Andraea 2007
9 IPCC 2007
10 Andraea *et al.* 2005

Further reading

IPCC 2007; Andraea *et al.* 2005; Andraea 2007.

AGRICULTURE AND FOOD SUPPLY IMPACTS

The impacts of climate change on agriculture must be viewed in the context of ever-increasing demand for food and agricultural products. Agriculture currently accounts for nearly a quarter of world economic output and employment.[1] The world's population is growing rapidly, rising from 2.5 billion in 1950 to 6.5 billion in 2005 and to an expected 9 billion by 2050.[2] As per capita incomes rise in industrializing countries such as China and India, so too will the demand for food and, in particular, meat – which will in turn necessitate increases in fodder crop supplies. If **biofuels** production also expands, so too will the area devoted to source crops.

Global crop production will have to increase significantly over the coming decades to meet these demands.[3] Under normal circumstances, ongoing improvements in crop yields (due to technical advances and irrigation) and expansion of the amount of land under crop might be sufficient to meet the growing demand. However, "normal circumstances" are unlikely to prevail over the coming decades since climate change is expected to affect the environmental factors that determine global crop production.

Crop yields and agricultural production are primarily determined by soil moisture, temperature, sunlight, and soil fertility. Over the course of this century, climate change is expected to lead to higher average global temperatures, changes in annual and seasonal precipitation patterns (see **water impacts**), and increases in the frequency and intensity of **extreme weather events**. Elevated concentrations of atmospheric **carbon dioxide (CO$_2$)** could also influence crop yields through the *CO$_2$ fertilization effect*. These factors will fundamentally alter crop yields and the distribution of agricultural production. In some regions, changes in climatic conditions might improve crop yields, as is expected in the higher latitudes of Europe and North America. However, in most other regions, changes in temperature and water availability are expected to result in reduced yields, particularly in tropical regions, where temperatures are already at, or exceeding, optimal conditions for plant growth.

Climate-induced changes in biodiversity and ecosystems – such as reductions in the abundance of essential crop pollinators (such as bees) or natural insect control species (such as insect-eating birds) could also affect yields, as could expansions in the geographical ranges of pests and diseases.

Temperature and rainfall impacts

Different crops require different temperature and soil moisture regimes for optimum yields. Water is generally the major limiting factor with regard to crop production. If there is insufficient soil moisture, as occurs during droughts, or there is too much, as occurs during floods or periods of excessive precipitation, yields decline and can even result in total crop loss. Most crops are very sensitive to temperatures above 35–40°C as photosynthetic functioning can be impaired. Elevated temperatures tend also to reduce the growing period between sowing and harvest, thereby reducing the amount of light captured and the resulting biomass production.

The prevailing rainfall and temperature regime is a key determinant of the types of crops farmers choose to grow. The world's principal wheat, oat, and barley production regions, for example, are located in mid-latitude regions such as Australia and central USA, where winter and spring rains are reliable and where summers are warm and dry. Wheat is not suited to warm, wet conditions, such as those found in the humid tropics and subtropics, or to the cold, high-latitude regions where the growing season is too short. Rice, on the other hand, generally requires warm and wet conditions during the growing cycle to obtain optimum yields and is well suited to tropical regions and areas that have warm monsoonal regimes. As climate regimes evolve and migrate, so too will the crops that can be grown.

Within any specific climate regime, the impacts of short-duration changes, such as heat waves, cold snaps, and flooding, can be very significant. At critical stages of the crop growing cycle, such as crop flowering times, high and low temperature extremes can significantly reduce the formation of grains, fruits, and seeds. Threshold temperatures beyond which grain set is reduced vary from 31 to 37°C, depending on the crop.[4] In the case of cool-region wheat varieties, for example, a 2°C increase in average maximum air temperatures above 30°C during grain set can potentially halve yields.[5] As extreme weather events are expected to become more frequent and intense as the planet warms, their impact on yields will also become more pronounced.

The range of plant pests and pathogens is constrained primarily by temperature.[6] As temperatures increase, many pest species will thrive as they can often produce more eggs, live longer, and extend their range. It is estimated that, on average, up to half of the world's food production is currently lost each year to pests and diseases, even though applications of pesticides and fungicides has grown enormously.[7] There have been observed changes in the ranges of several major crop diseases since the 1970s, including those of corn grey leaf blight, wheat fungal diseases and soybean rust and charcoal rot: their general expansion toward the poles is highly suggestive of an underlying **global warming** cause.[8] Increased climate variability has led to the outbreak of several fungal diseases in the major soybean producing regions of both North and South America since the late 1990s. Extended periods of unusually dry (charcoal rot) and wet (soybean rust) conditions have not only reduced production in some regions but have also led to large increases in the quantity of fungicides applied.[9]

Our knowledge of how future climate change will alter the incidence of different agricultural pests and diseases is still limited and, as a result, considerable uncertainty surrounds their potential impact on global food production. However, a warmer and wetter world is likely to favor many pests and diseases and lead to an extension in their geographic ranges.

CO₂ fertilization effect

Rising CO_2 concentrations will increase the efficiency of photosynthesis in many plants and will also decrease water requirements (by adjusting the size of leaf pores). The response of different crops to elevated levels of CO_2 largely depends on plant physiology but is also conditioned by other key determinants of photosynthetic production, such as temperature, water, and nutrient availability. The vast majority of plants – approximately 95% of terrestrial plant biomass, including wheat, rice, and barley – possess what is termed the C3 photosynthetic pathway: such plants experience enhanced photosynthesis under elevated CO_2 levels. However, some plants, including maize, millet, and sugar cane, possess a C4 pathway and do not experience photosynthetic efficiency gains. Increased CO_2 concentrations do, however, improve their water use efficiency.

Scientific experiments conducted in laboratories and greenhouses indicate that crop yields can increase by 20–30% above today's level at CO_2 concentrations of 550 ppm. If this relationship holds true "in the field," then, in some regions at least, the negative impact of temperature and precipitation changes might be largely offset by enhanced yields from the CO_2 fertilization effect.[10] However, considerable uncertainty surrounds the strength of the CO_2 fertilization effect, and different crop models produce different results. Estimates of the impact of a doubling of CO_2 levels on rice production in Bangladesh, for example, gave a reduced yield of −2% in one study and −35% in another.[11] Recent field studies have concluded that actual performance in the field may be only half that achieved in controlled greenhouse experiments.[12]

Crop yields are also expected to be affected by levels of surface **ozone**, which can be detrimental to plant growth even at very low concentrations.[13] Ozone creates reactive molecules that destroy rubisco, an enzyme crucial for photosynthesis. As fossil fuel use increases, so will surface ozone concentrations, and in major crop growing regions of China and the United States, surface ozone levels are expected to rise by 25% by 2050.[14] A 2006 study on the impacts of CO_2 and ozone concentrations on soybean yields (simulating expected 2050 CO_2 and ozone levels) concluded that crop yields could fall by 10%, rather than increase as would be expected to occur with no ozone concentration change.[15] The study also indicated delayed crop maturity, which could increase crop susceptibility to early frosts. Most crop models do not take into account the potential impact of surface ozone changes and, as a result, may tend to overestimate future crop production for some regions.

Implications for future food production

It is difficult to predict with any accuracy the precise impacts of climate change on global food production. Uncertainties arise from our limited understanding of future temperature and precipitation patterns, the sensitivity of crops to such changes, and the strength of the CO_2 fertilization effect. Projecting global food production over the next 50–100 years is a complex task and requires detailed forecasts of a host of factors, many of which are nonclimate related: including population growth, land degradation trends, changes in agricultural subsidies, and market prices. Nonetheless, a broad scientific consensus is beginning to emerge on the likely impacts of climate change.

As the planet warms, the areas suitable for producing crops such as wheat, barley, and oats will tend to shift toward the poles and retreat from the lower latitudes. A warmer climate is expected to facilitate a significant expansion in the grain-producing areas of the mid-to-higher latitudes, mainly due to a longer growing season. For example, Canada is expected to experience increases in both yields and the spatial extent of suitable grain growing regions over the next half century, and this might also occur in central and northern regions of Europe and the United States, and possibly southern regions of Australia and Argentina.[16] There will, however, be significant regional variation due to changes in precipitation patterns. The Mediterranean, the Middle East, west Asia, southwest USA, and parts of southern Russia may experience declines in production, due to both temperature and precipitation changes. Africa is expected to experience significant adverse impacts on crop production under all future climate change scenarios, even at relatively low levels of warming. Crops in lower latitudes, especially the tropics, are already close to exceeding their optimum yield temperature thresholds and will thus be more susceptible to temperature increases.

If the CO_2 fertilization effect is strong, most mid-to-high latitudes regions are expected to benefit, with strong yield and production gains in Canada, Europe, parts of Australia, Argentina, Russia, and northern United States associated with moderate warming (2–3°C).[17] However, if the CO_2 fertilization effect is weak, then no region is expected to show significant yield gains, though Canada, Europe, and Argentina are expected to be the least affected. India, Russia, China, the Mediterranean, west Asia, and the Middle East regions are expected to experience declines in yields by 2050.[18] Again, Africa is expected to fair badly, regardless of whether the CO_2 fertilization effect is weak or strong.

Due to their C4 physiology, crops such as maize, millet, and sugar cane are likely to experience greater yield reductions than C3 crops such as wheat, barley, and rice. Declines in maize yields will have significant economic and livelihood effects in Africa and the tropical and subtropical regions of Latin America due to the prominence of maize in aggregate cereal production. There is also general agreement that the more global temperatures rise, the greater the likelihood of significant reductions in crop yields and aggregate food production. At temperature changes above 4°C, the negative impacts are expected to become widespread and very pronounced, and some areas may cease to be able to support agriculture.[19]

The **Intergovernmental Panel on Climate Change (IPCC)** emission scenarios are commonly employed (see **future emissions trends**) to determine the effects of different climate parameters in model simulations. Most studies use cereal production as the main benchmark indicator of global food production as cereals account for 80% of global food supply. The models used to derive food production projections usually rely on crop simulation models that are driven by average weather data generated by global climate models and then downscaled to local and regional levels.

Our inability to predict daily and weekly climatic variations means that models struggle to take into account the impacts of changes at critical times of the growing cycle or the impact of extreme events such as floods, droughts, and heat waves. As a result, the models are only able to provide a general indication of likely trends in food production and are subject to uncertainty. Different global climate models produce different climate results, as do different crop models in relation to yields and production estimates. Impact assessments generally include estimates derived from a range of models and scenarios as no single model prediction can be given a high degree of confidence.

Most studies forecast a net decline in overall aggregate global food supply over this century. The magnitude and timing of this decline is highly dependent on assumptions of future climate, population, and the global economy. Factors such as the magnitude and rate of change in atmospheric concentrations of greenhouse gases, **climate sensitivity**, the strength of the CO_2 fertilization effect, and our adaptive responses will all have an important bearing on the outcome.

Recent impact assessment studies have produced estimates of aggregate food production over the next 50–80 years under different scenarios.[20] Assessments are presented for specific atmospheric CO_2 concentration levels (usually a doubling of concentrations above preindustrial levels) and/or for different temperature increases. They also usually produce estimates that assume a strong and a weak CO_2 fertlization effect and take into account possible **adaptation** responses such as changing planting times, planting different crops, and increased access to irrigation.

Mid-range **IPCC** projections, with a strong CO_2 fertilization effect, indicate a decline in global food production of less than 5% by 2050.[21] If the CO_2 fertilization effect turns out to be weak, global production could fall by 10–15%,[22] representing a major threat to global food security and placing substantial upward pressure on agricultural commodity prices.

However, aggregate impacts tend to mask underlying shifts in the geographical distribution of food production. Even assuming strong CO_2 fertilization, most gains are expected to be in the developed world, where aggregate production is expected to increase by 5–15%, while most losses will be felt in the developing world, where losses could be of the order of 10–15%.[23]

Reductions in global food supply of 5–15% would have potentially devastating consequences for a majority of the world's population. Africa, in particular, is expected to be especially hard hit. Agriculture accounts for one-third of African Gross Domestic Product and employs three-quarters of its workforce. The majority of the rural poor depend on rain-fed agriculture. Yields of many of Africa's staple crops are likely to fall 5–10% by 2050 and, when viewed in the context of rapid population growth, the food security threat becomes severe.[24] Those at "risk of hunger" (defined as people with insufficient income to purchase the base level of cereal to support normal dietary requirements) are likely to increase. At present, an estimated 800 million people are at risk of hunger, especially in Africa.[25] Some studies suggest that temperature increases of 2–3°C could lead to an additional 30–200 million being considered at risk of hunger.[26] Other studies give higher and lower estimates, but no study predicts a significant reduction in the number at risk.

Overall, it appears that there may be some agricultural benefits with low to moderate climate change, but predominantly in the higher latitude developed countries. However, beyond warming of 2–3°C, global agricultural output is likely to fall, with the greatest falls expected in the developing countries.

See also: biodiversity impacts, biofuels, extreme weather events, global warming, socioeconomic impacts, water impacts.

Notes

1 Stern 2006
2 Cohen 2005
3 Parry *et al.* 2004
4 Challinor *et al.* 2006a
5 Based on data from Wheeler *et al.* 1996
6 Rosenzweig and Hillel 1998
7 Epstein 2005
8 Rosenzweig *et al.* 2001
9 Epstein 2005
10 Parry *et al.* 2004
11 Challinor *et al.* 2006a
12 Long *et al.* 2006
13 Ibid.
14 *Nature News* 2005
15 Long *et al.* 2006
16 Parry *et al.* 2004
17 Ibid.
18 Ibid.
19 Hare 2006
20 See Stern 2006 and Parry *et al.* 2004
21 Ibid.
22 Parry *et al.* 2004
23 Stern 2006
24 Nyong and Niang-Diop 2006
25 Stern 2006
26 Ibid.

Further reading

Parry *et al.* 2004; Stern 2006; Hare 2006; Rosenzweig and Hillel 1998

ALBEDO

Albedo is a measure of the reflectivity of a surface. It is a scale from zero to one, where zero represents perfect absorption of light and one represents perfect reflection.

Table 1 Albedo of various surfaces[1]

Surface	Albedo
Equatorial oceans at noon	0.05
Dense forests	0.05–0.10
Forests	0.14–0.20
Cities	0.14–0.18
Green crops	0.15–0.25
Grassland	0.16–0.20
Sand	0.18–0.28
Polar oceans with sea ice	0.60
Old snow	0.40–0.60
Fresh snow	0.75–0.95
Clouds	0.40–0.90
Spherical water droplet with low angle of incidence	0.99

In practice, the scale is somewhat more limited since an object with an albedo of zero would be invisible and an object with an albedo of one would be indistinguishable from the original source of light.

Surface albedo values are largely determined by color, texture, and the angle at which the sun hits the surface (the angle of incidence). Surfaces that are light colored, such as fresh snow, can reflect up to 95% of solar radiation back into space as visible light: albedo = 0.95 (see **greenhouse effect**). In contrast, dense forests absorb up to 95% of solar radiation (albedo = 0.05). Table 1 lists the albedo values of different surfaces.

The lower the sun's angle of incidence, the greater the reflectivity and higher the albedo, especially for transparent surfaces such as the ocean. Consider a window pane: when looking from an angle, the pane reflects like a mirror, but when looking directly through the window, there is no discernable reflection. Similarly, sunlight falling near the poles has a low angle of incidence, enabling greater reflection, while equatorial oceans are consequently significantly lower than polar oceans in terms of their reflectivity. When bright (white) sea ice covers polar oceans, albedo can exceed 0.5, meaning that more than half of the energy reaching the surface is reflected back to space.[2]

Albedo and climate

Albedo is important in relation to the earth's climate as the amount of energy reflected or absorbed by the planet's surface strongly influences global temperatures. The temperature of the earth's atmosphere is fundamentally governed by the balance of incoming solar energy of 342 Wm^{-2}, outgoing energy of 107 Wm^{-2} (31%) as reflected solar radiation, and 235 Wm^{-2} (69%) as longwave (infrared) radiation (see **greenhouse effect**). The 31% of energy that is reflected back to space does not warm the planet. Any change in albedo changes this energy balance. If albedo increases, there is more energy reflected back to space and the earth cools. If albedo decreases, less energy is reflected to space and the earth warms. If global albedo were to decrease by 1% (i.e. to decrease from the present 0.31 to 0.30), it would represent an increase of 3.4 Wm^{-2} absorbed by earth, an equivalent warming effect as doubling the atmospheric **carbon dioxide (CO_2)** concentrations.

On a geological timescale, the earth's albedo changes between glacial and inter-glacial periods as ice cover advances and retreats from the poles in response to **Milankovich cycles**, amplifying their impact on earth's climate.[3] These changes occur over millennia, but on timescales of hundreds of years relevant to humans, naturally occurring changes to the earth's albedo are very minor and largely undetectable.

Recent changes in the earth's albedo

The albedo of landscapes and of the earth as a whole can change through direct human impact such as deforestation, or through earth-system responses to **global warming** (see **climate change feedbacks**) such as decreasing snow cover. Over the past two centuries, the earth's albedo has increased (i.e. become more reflective) by approximately 0.4%. This anthropogenic albedo increase has counteracted more than a third of **global warming** due to **anthropogenic greenhouse gas emissions**, by far the largest negative (cooling) **radiative forcing** component of anthropogenic influence on climate.

The primary anthropogenic albedo changes relate to cloud formation (linked to **aerosols**), aerosol direct effects, and land use changes.

Clouds

The most important factor influencing earth's albedo is clouds: 77 Wm^{-2} (72%) of the total 107 Wm^{-2} reflected back to space is due to clouds (approximately 22% of incoming solar energy). Clouds are effective reflectors of energy because their high altitude reflects insolation before it can be absorbed by **greenhouse gases** in the lower atmosphere and because clouds have such an enormous geographical spread over the planet.

For water vapor (see **greenhouse gases**) in the atmosphere to condense into droplets and form clouds, a nucleus such as pollen, dust, dimethyl sulfide (formed by marine plankton), or man-made particulates must be present. As initial droplets attract more water vapor, they become bigger and eventually grow to a size where their weight causes them to fall as precipitation (rain, hail, snow).[4] Significant increases in fossil fuel combustion by humans have greatly increased the amount of **aerosols** in the atmosphere in the form of sulfates, organic carbon, and black carbon (see **global dimming**). By increasing the amount of **aerosols** in the atmosphere, there are more nucleation points for water vapor condensation. This results in more water droplets of a smaller size since the same amount of available water vapor is distributed across a larger number of condensation sites.[5] Clouds that contain a larger number of smaller droplets are more reflective than those comprised of larger droplets. Smaller droplets also take longer to reach the weight of precipitation, meaning clouds persist for longer and maintain higher altitudes. These changes in cloud properties have increased albedo, cooling the planet.[6]

Even in a stable climate, the physics of cloud genesis, formation, movement, and decay is not well understood, and modeling the changes to cloudiness with changing aerosol loads and greenhouse gas concentrations is complex. Fundamentally, an increase in atmospheric temperature will result in greater evaporation, and an atmospheric column with more water vapor will increase cloudiness, all other conditions being equal (a negative **climate change feedback**). In addition, higher atmospheric concentrations of aerosols will increase cloud formation and persistence (though there is considerable variation depending on aerosol composition and formation conditions).

Although satellite and ground based observations have greatly improved since 2000, considerable scientific uncertainty remains in relation to the cloud albedo effect. What is known is that cloudiness has increased over the past century and so too has earth's albedo. Estimates show that increased cloudiness has resulted in a rise in earth's albedo and a negative **radiative forcing** of -0.7 Wm^{-2} (a cooling effect).[7] While this represents only about a 1% increase over naturally occurring cloud albedo, it is the single largest negative component of the total anthropogenic **radiative forcing** and represents a 27% reduction (or cooling) from the 2.64 Wm^{-2} of direct warming from **anthropogenic greenhouse gas emissions**.

Aerosols

In addition to altering cloud formation, **aerosols** in the atmosphere affect albedo through two other mechanisms. First, aerosols have their own albedo and, overall, reflect more solar energy back to space than would be the case without anthropogenic aerosols. The direct aerosol albedo effect on **radiative forcing** is estimated at -0.5 Wm^{-2}, a net global cooling that offsets approximately 20% of the warming associated with **anthropogenic greenhouse gas emissions** (see **aerosols** and **global dimming**).[8]

Second, aerosols, especially heavier particulates, eventually precipitate out of the atmosphere. When they land on surfaces, they can change surface albedo. This is not significant when black carbon lands on a forest or the ocean, both of which have low albedo values. However, when black carbon lands on snow or ice, it can lower albedo by more than 0.13.[9] Furthermore, maximum atmospheric black carbon peaks in late Northern Hemisphere spring (after fuel burning during winter), coinciding with snow melt onset, and can trigger strong local feedbacks. Black carbon deposition accelerates snow melt, exposing rocks, vegetation, and soil, further lowering albedo, increased warming, and yet further snow melt (see **polar impacts**).[10] The decrease in albedo due to black carbon on snow is, however, relatively small, contributing $+0.1$ Wm^{-2} to **radiative forcing** (approximately 4% of total anthropogenic warming).

Land-use change

Human agriculture (cropland or pasture) has expanded from around 8.5 million km^2 (approximately 6.5% of the global land surface) in 1750 to around 50 million km^2 (nearly 40%) today. Prior to 1950, most of the expansion was due to land clearance in North America and Australia, but since 1950, large-scale deforestation in South America, and more recently Asia and Africa, has been the major contributor to the total area under agriculture.[11] Changes in land use can influence local albedo, though not always significantly: a shift from grassland to green crops will result in little change, but conversion from dense forest to green crops can increase local albedo by as much as 0.15 (see Table 1). Generally, a decrease in vegetative cover results in an increase in albedo – a negative **climate change feedback**. This is compounded in high latitudes when snow exposure during winter is higher after boreal forests have been cleared. In spite of the spatial variability of albedo, and the difficulty of establishing historic global albedo figures, land-use change since 1750 is estimated to have resulted in a negative **radiative forcing**[12] (cooling) of -0.2 Wm^{-2}.

Ice and snow cover

Collapsing glaciers and melting snow are common in imagery of climate change. Loss of snow cover, particularly in spring, has substantial albedo impacts locally, with earlier thaws and higher local temperatures (see **polar impacts**). However, due to the relatively limited geographic extent, the impact on global albedo from loss of snow and ice cover to date has been negligible compared with the 1.3 Wm^{-2} reduction in **radiative forcing** from clouds, aerosols, and land-use change.

While the earth's albedo has only increased from 0.313 to 0.317 since 1750 (a 1.2% increase in reflectiveness), this is equivalent to a 40% reduction in the warming impact of **anthropogenic greenhouse gas emissions** to date.

Future albedo changes

The uncertainty around future changes in albedo is very high as there will be factors both increasing and decreasing albedo. Increasing cloudiness and aerosol loadings and ongoing deforestation and desertification will continue to increase albedo over the coming decades, which will add to the current cooling effect from these factors. However, in the longer term, reduced ice and snow cover and an eventual reduction in anthropogenic aerosols will decrease earth's albedo, which could amplify **global warming**.

Aerosols

Atmospheric **aerosol** concentrations are expected to increase in the short to medium term in line with fossil fuel consumption (increasing albedo, a cooling effect) but in longer term, especially beyond 2030, could fall if fossil fuel combustion efficiency improves and stricter emission controls are introduced in rapidly growing developing economies (see **aerosols**). Eventually, aerosol emissions will fall and the cooling effect of aerosols will diminish (see **global dimming**). It is uncertain when aerosol loadings might return to near natural levels, but when they do, the net warming effect of anthropogenic emissions will increase.

Clouds

The projected increase in global temperatures this century will increase atmospheric water vapor concentrations, with consequent increases in cloudiness. This will increase albedo and moderate **global warming**, regardless of aerosol effects. The extent to which cloudiness will continue to increase remains uncertain as broader changes in climate, such as temperature gradients and ocean and wind circulation patterns, may alter cloud formation and circulation patterns.

Land-use change

Although tropical rainforest deforestation rates have slowed in recent years (particularly in Brazil), net deforestation continues. Climate change-induced drying of continental landmasses, particularly in central Africa, Southwestern USA, and Australia, will accelerate desertification. Both these effects will tend to increase albedo – a cooling effect. These effects may be counterbalanced to some extent by the poleward expansion of boreal forest cover (since forested land has a lower albedo – a warming effect; see **polar impacts**) and the extent to which deforestation rates are reduced and large-scale reforestation is undertaken. There is much uncertainty surrounding projections of future land-use change, but the increased albedo effect of land-use change is expected to also diminish over the course of this century.

Ice and snow cover

While decreases in the geographic extent of snow and ice cover have already been observed, it has, to date, only had a minor impact on albedo. However, the rate of ice and snow cover loss is expected to accelerate significantly over the coming decades, particularly summer Arctic sea ice, which could disappear completely by mid-century (see **polar impacts**). These changes will reduce albedo and amplify **global warming** (a positive **climate change feedback**).

Given the considerable uncertainty surrounding the relative strength of different factors influencing earth's albedo, projecting future albedo values is somewhat speculative. If the global community took significant measures to reduce deforestation, promote reforestation, and reduce fossil fuel consumption (and thus aerosol emissions), it is likely that the net albedo cooling effect observed to date could begin to decline and possibly amplify, rather than reduce, **global warming**.

See also: aerosols, climate change feedbacks, global dimming, global warming, greenhouse effect, greenhouse gases, Milankovich cycles, polar impacts, radiative forcing.

Notes

1 Ricklefs 1993
2 Hardy 2003
3 Labeyrie 2000
4 Charlson *et al.* 2001
5 Geresdi *et al.* 2006
6 IPCC 2007
7 Ibid.
8 Ibid.
9 Flanner *et al.* 2007
10 Hansen and Nazarenko 2003

11 IPCC 2007
12 Wielicki *et al.* 2005

Further reading

IPCC 2007.

ANTHROPOGENIC GREENHOUSE GAS EMISSIONS

Greenhouse gases released into the atmosphere as a result of human activities are termed *anthropogenic* (human created) greenhouse gas emissions. Although anthropogenic emissions from industry, agriculture, and deforestation remained relatively modest from the beginning of the Industrial Revolution (generally taken to be 1750) until the mid-twentieth century, they have since grown rapidly, increasing at an average rate of 2.5% per year since 1950.[1] Between 1970 and 2004, global emissions increased 70%, and carbon dioxide (CO_2) emissions alone increased by 80%.[2] By 2004, global emissions, in **carbon dioxide equivalence (CO_2e)**, had reached 49 Gt[3] per year (compared with 28 Gt CO_2e in 1970).[4]

The two principal factors driving the increase in emissions have been the significant expansion in the global economy, which grew by 77% between 1970 and 2004, and the growth in world population, which increased by 69% over the same period.[5] These two factors have led to a substantial rise in emissions from fossil fuel consumption (the world's main source of primary energy) and, to a lesser extent, land-use change (primarily deforestation). Over the past two centuries, these two sources alone have released an estimated 2,300 Gt of CO_2 into the atmosphere, more than half of which has been released in just the past 30 years.[6] While half of these emissions have been reabsorbed through the **carbon cycle**, the rest remain in the atmosphere, increasing atmospheric greenhouse gas concentrations appreciably.

Compared with their preindustrial levels, concentrations of the three principal greenhouse gases have grown substantially: **carbon dioxide (CO_2)** (from 280 to 382 ppm), **methane (CH_4)** (from 715 to 1,774 ppb[7]) and **nitrous oxide (N_2O)** (from 270 to 320 ppb).[8] When the six **greenhouse gases** covered by the **Kyoto Protocol** (the "Kyoto gases"), plus those covered by the Montreal Protocol (see **synthetic gases**), are converted to their **CO_2e**, aggregate atmospheric greenhouse gas concentration had reached 455 ppm CO_2e by 2007 – more than 40% higher than in 1750.[9]

This rise in concentration is the main cause, via the enhanced **greenhouse effect**, of the observed 0.76°C increase in global mean temperature over the past century.[10] In the past two decades, global mean temperature has been rising at a rate of 0.2°C per decade, and based on current emission trends, this rate is likely to accelerate. The rise in global mean temperature over the coming century is projected to be between 2°C and 5°C, and possibly more, depending on emission levels and **climate sensitivity**. Increases of this magnitude will have adverse impacts on human and natural ecosystems (see **climate change impacts** and **dangerous climate change**).

Given the rapid rise in atmospheric greenhouse gas concentrations, the global community is, understandably, becoming increasingly concerned about climate change and is currently grappling with how to reduce emissions to a level that will stabilize concentrations below dangerous levels (see **stabilization targets**). For decision makers to formulate an effective global action plan, it is essential that they have accurate information on the major greenhouse gas emission sources, the underlying emission drivers, and likely future trends.

Emissions data reliability

Global emissions data are predominantly derived by aggregating national **greenhouse gas inventories** and are subject to some uncertainty. For most developed countries, emissions data are reasonably accurate and up-to-date, but this is not the case for many developing countries where emissions data can be less reliable and often quite dated (the data for many countries are often 5 or more years old). For this reason, estimates of current global emissions levels are generally based on extrapolations from the most recent national inventories available. The comparative emissions data used in this section refer mainly to emissions in the year 2000, which is the most recent comprehensive data set available.

Of course, things have changed since 2000. For example, China has experienced rapid economic growth and, as a result, CO_2 emissions from fossil fuel consumption have increased substantially. China's energy consumption grew by 14% in 2004, 15% in 2005, and 11% in 2006, with most of the new energy sourced from coal (for electricity production) and oil (primarily for road transport).[11] Energy sector emissions have also grown considerably in the United States, Canada, India, Australia, and several European countries. Nonetheless, the underlying trends, sources, and relative contributions of each country to total global emissions tend to change relatively slowly, usually over decades rather than annually, and the relative shares of the major greenhouse gas emitting countries have not changed significantly since 2000.

The quality of emissions data also varies considerably between different gases and sources. Data for energy-sector CO_2 emissions are generally the most accurate and up-to-date. In contrast, data from land-use change activities face a range of measurement and other data-gathering constraints and are, as a consequence, subject to considerable uncertainty. In general, the quality of data from industrialized countries is more robust than data from many developing countries, and care needs to be taken when comparing data across countries. Nonetheless, the quality of emissions data from those countries that account for a large proportion of global emissions is generally sufficiently robust to enable meaningful comparisons.

Anthropogenic emissions

Humans release approximately 50 different gases into the atmosphere that can contribute to the **greenhouse effect**. Many of these gases are released only in small quantities and, at this stage, do not contribute significantly to **global warming**. Fewer than 10 gases account for nearly all the atmospheric **radiative forcing** that can be attributed to humans. These gases can be further subdivided into two main

Table 2 Greenhouse gas emissions and shares 2004[12]

Greenhouse gas	Billion tonnes CO_2e	% share of each gas in 2004 global emissions (based on GWP)	Radiative forcing (W/m^2)	% share of radiative forcing (2005*)
Carbon dioxide	37.5	77	1.66	63
Methane	7	14	0.48	18
Nitrous oxide		8	0.16	6
HFC, PFC, SF_6	0.5	1	0.34**	13
Total	49	100	2.64	100

*Radiative forcing in 2005 based on projections from available data and measured surface energy balance.
**High radiative forcing figure due to the inclusion of the Montreal Protocol gases (CFCs and halons).

categories: (1) those covered under the **Kyoto Protocol – carbon dioxide, methane, nitrous oxide**, hydroflourocarbons (HFCs), perfluorocarbons (PFCs), and sulphur hexafluoride (SF_6); (2) the **ozone** depleting substances covered under the Montreal Protocol – chloroflourocarbons (CFCs), halons, and several other gases (see **synthetic gases**). Most published national and global greenhouse gas emissions data refer only to the Kyoto gases, even though the Montreal Protocol gases and other emissions are also important contributors to **global warming** (see **synthetic gases** and **ozone**). The following discussion refers mainly to those gases covered under the **Kyoto Protocol**.

Carbon dioxide is by far the most important Kyoto gas (accounting for 77% of 2004 emissions), followed by methane (14%), nitrous oxide (8%), and the synthetic gases (1%). The relative contribution of these gases to global warming has changed in the past 50 years, with carbon dioxide growing in relative importance. Though methane and nitrous oxide emissions have grown considerably since 1970, it has been at about half the rate of CO_2, and their share of total emissions has declined marginally. Table 2 indicates the emission volumes, the relative contributions of the six Kyoto gases, and their contribution to atmospheric **radiative forcing**.

Principal emission sources

Greenhouse gas emissions emanate from a wide variety of sources and are driven by a range of different factors.

Carbon dioxide

The most significant source of CO_2 emissions is the production and consumption of fossil fuels in the energy sector, accounting for just over 70% of global CO_2 emissions. Just two activities account for more than two-thirds of energy-sector CO_2 emissions: electricity and heat production (43%) and transport (25%). The remainder is attributable to fossil fuel consumption in industry, agriculture, and the commercial and residential building sectors (mainly for space and water heating).[13] After fossil fuels, the next most important source of CO_2 emissions is from land-use change (24%), notably deforestation. The remaining CO_2 emissions emanate from industrial processes, principally cement production.

Methane

The four major sources of methane emissions are: (1) livestock and manure management (33%), (2) energy-sector emissions (29%), (3) waste management (23%) and (4) rice cultivation (11%). The remainder is attributable primarily to biomass combustion.[14]

Nitrous oxide

The bulk of N_2O emissions derive from agricultural soils (mainly related to fertilizer use). Small quantities of N_2O also emanate from industry, fossil fuel combustion, and wastes.

Synthetic gases

PFC emissions are attributable to aluminium and semiconductor production, HFCs derive principally from refrigeration, and SF_6 originates primarily from high-voltage electricity transformers.

Emission trends

Emissions growth has been largely driven by increases in global economic output (particularly, per capita income) and by population growth. This has been tempered to some degree by long-term declines in the **emissions intensity** of economic production, largely through reductions in energy intensity per unit of output.[15]

The main contributors to elevated **carbon dioxide** emissions over the 1970–2004 period were electricity production, which increased by 145% (two-thirds of electricity is generated from fossil fuels), and transport energy demand (almost totally reliant on oil), which increased by 120%.[16] The growth in per capita income was one of the principal drivers of the heightened demand for electricity and transport (largely road and air transport). Nearly one-third of global transport emissions emanate from just one country, the United States.[17]

Other sources of CO_2 emissions have grown, but at a much slower pace. Industrial emission sources have grown by 65% since 1970, with the lower growth rate partly due to **energy efficiency** gains and fuel substitution (mainly increased use of natural gas and reduced dependence on coal and oil). Emissions from land-use change increased significantly during the 1970s and 1980s, due to large-scale deforestation in tropical regions, but since the mid-1990s, deforestation rates have moderated but remain significant – overall emissions from land-use change have increased 40% since 1970.[18] At present, land-use change emissions are approximately 7.5 $GtCO_2$ per year.[19]

The growth in **methane** emissions has been largely driven by increases in the size of the global livestock herd and growth in fossil fuel use (particularly coal mining). Emissions from the waste sector and rice cultivation have risen much less quickly. Methane emissions data are much less reliable than for CO_2, and some uncertainty remains. Emissions largely stabilized during the 1990s, increasing by only 3% (from 5.7 to 5.9 GtCO$_2$e between 1990 and 2000) and remained largely unchanged in the period 2000–2004.[20] However, emissions are expected to begin rising again over the next two decades and could possibly reach 7.9 GtCO$_2$e by 2020, unless additional mitigation measures are introduced (see **future emissions trends**).

Nitrous oxide emissions have maintained relatively constant growth rate over the past few decades, driven primarily by increased food demand arising from population growth and from the intensification of agriculture and consequent increases in fertilizer use in the agricultural sector. Nitrous oxide emissions reached 3.5 GtCO$_2$e by 2000 (the latest reliable global figure) and are expected to continue to increase over the coming decades.[21]

Major emitting countries

The size of each national **greenhouse gas inventory** relates to the **emissions per capita** and the size of a country's population. Emissions per capita are largely determined by the level of income per capita (usually measured as Gross Domestic Product [GDP]/capita) and the **emissions intensity** of the economic activities that generate national income. As a general rule, the larger a country's population and the higher the level of GDP/capita, the higher the level of greenhouse gas emissions. For example, the United States, Japan, and Germany are large emitters of greenhouse gases (together accounting for 27% of global emissions) as they have relatively large populations (8% of global population) and high levels of GDP/capita (all greater than US$ 26,000/capita).[22] Nonetheless, the sheer population size of some countries (e.g. China and India, which together account for 38% of global population) means that they are also large emitters of greenhouse gases (combined accounting for approximately 20% of global emissions), even though they have low emissions per capita and incomes per capita (less than US$ 5,000/capita). There is considerable variation in greenhouse gas emissions between countries, even for those that have similar populations and per capita income levels, largely as a result of variations in the **emissions intensity** of their economies.

In total, just 25 countries accounted for 83% of global emissions in 2000 and just five accounted for half (United States, China, Russia, India, and Japan).[23] Of the 192 countries that have signed the UN climate change convention (see **UNFCCC**), 140 account for less than 10% of global emissions.[24] Table 3 compares the relative contributions to global emissions of 15 selected countries (accounting for two-thirds of global emissions), together with their respective emissions and income per capita. While emissions data from land-use change are often excluded from international emission comparisons (mainly due to the lack of accurate data), these activities are, nonetheless, important sources of emissions and need to be considered when comparing the relative contributions of different countries: their inclusion changes the relative contributions of some countries significantly, particularly in the case of Brazil and Indonesia (combined, these two countries account for half of global emissions from deforestation). For this reason, Table 3 includes data on the share of global emissions, both including and excluding emissions from land-use change.

Table 3 Greenhouse gas emissions of selected countries in 2000[25]

Country (t/yr)*	Emissions (Mt CO_2e/ year)	% world (excluding. LUCF)	% world (including LUCF)	Population (m)	$GDP/capita (2002 PPP)**	CO_2e/ capita
United States	6,928	20.6	15.8	293	34,557	24.5
China	4,938	14.7	11.9	1,280	4,379	3.9
Russia	1,915	5.7	4.8	144	7,993	13.2
India	1,884	5.6	4.5	1,049	2,572	1.9
Japan	1,317	3.9	3.2	127	25,788	10.4
Germany	1,009	3.0	2.5	82	26,141	12.3
Brazil	851	2.5	5.4	174	7,480	5.0
United Kingdom	694	1.9	1.6	59	25,139	11.1
South Korea	521	1.5	1.3	49	16,570	11.1
France	513	1.5	1.2	59	26,090	8.7
Mexico	512	1.5	1.5	101	8,662	5.2
Indonesia	503	1.5	7.4	212	3,057	2.4
Australia	491	1.5	1.2	20	27,256	25.6
Turkey	355	1.1	0.9	70	6,145	5.3
Argentina	289	0.9	0.8	36	10,664	8.1
Pakistan	285	0.8	0.8	145	1,941	2.1

*Emissions per capita data excludes emissions from forestry and land use change.
**Income data is based on purchasing power parity (PPP).

Although some countries have experienced significant growth in emissions over the period since 2000, particularly in China and India and to a lesser extent the United States, Australia, and South Korea, the overall relative shares and rankings have not changed significantly. Preliminary 2007 data indicates that China has surpassed the United States as the largest emitter, though in terms of cumulative historic emissions, China has so far contributed only one-third as much to the present atmospheric greenhouse gas concentrations as the United States (see below).

Cumulative emissions

Greenhouse gases can reside in the atmosphere for long periods of time (more than 3,000 years in the case of SF_6). When the total quantity of greenhouse gases released by each country since 1750 (referred to as cumulative emissions) is calculated, it is clear that some countries have historically contributed much more than others to the present level of atmospheric greenhouse gas concentrations. In the process of determining the relative responsibilities of countries for reducing emissions over coming decades, it is important to consider both past and present emissions, as well as future emissions trends.

There are three different ways of assessing the cumulative impact of past emissions on global warming: (1) aggregate past emissions (which weights all emissions equally no matter when they occur), (2) the contribution of cumulative emissions to the present atmospheric concentration levels (which takes into account the decay

Table 4 Cumulative greenhouse gas emissions for selected countries in 2002[26]

Country	% global CO_2 emissions (2002)	% cumulative CO_2 emissions 1850–2002	% contribution to 2002 concentrations	% contribution to temperature increase
United States	24.0	29.2	27.8	29.0
China	14.4	7.6	9.0	7.5
Russia	6.4	8.1	8.3	8.5
India	4.4	2.2	2.5	2.1
Germany	3.6	7.3	6.4	7.1
United Kingdom	2.3	6.3	5.0	5.9
Indonesia	1.2	0.5	0.6	0.5
Pakistan	0.4	0.2	0.2	0.2

rate of different gases), and (3) the contribution of past emissions to the increase in global mean temperatures observed to date. These three approaches produce different results, but whichever approach is used, they all indicate that the older industrialized countries (particularly the United States, United Kingdom, and Germany) have contributed more to current greenhouse gas concentrations than their present emission levels would suggest. Table 4 presents the results of these three approaches for four industrialized and four developing countries in terms of cumulative CO_2 emissions since 1850 (the data are for CO_2 emissions from fossil fuels only, as reliable long-term data for other sources and other gases are not available).

It is evident that humans have released large quantities of greenhouse gases into the atmosphere, especially in the last 50 years, increasing atmospheric greenhouse gas concentrations appreciably. Most of the increase in concentrations has, historically, been due to emissions from industrialized countries, but the contribution from developing countries is expected to increase over the coming decades, particularly from the major emerging economies of China, India, Mexico, and Brazil (see **future emissions trends**).

See also: carbon dioxide (CO_2), emissions intensity, emissions per capita, future emissions trends, greenhouse gases, methane (CH_4), nitrous oxide (N_2O), ozone, radiative forcing, stabilization targets, synthetic gases.

Notes

1 IPCC 2007
2 Ibid.
3 A Gigatonne (Gt) is 1,000,000,000 tonnes; 1,000 Megatonnes (Mt) or a thousand million tonnes.
4 IPCC 2007
5 Ibid.
6 WRI 2005
7 ppm (parts per million) or ppb (parts per billion) is by mass. That is, the ratio of the number of greenhouse gas molecules to the total number of molecules of dry air. For current concentrations of CO_2 of 382 ppm, this means there are 382 g of CO_2 per tonne of dry air.

8 NOAA 2007
9 Stern 2006
10 IPCC 2007
11 Fung 2006
12 Derived from data contained in IPCC, 2007 and WRI, 2005
13 WRI 2005
14 Scheehle and Kruger 2007
15 IPCC 2007
16 Ibid.
17 WRI 2005
18 IPCC 2007
19 Houghton 2005
20 Scheehle and Kruger 2007
21 Ibid.
22 WRI 2005
23 Ibid.
24 Ibid.
25 WRI 2005
26 Ibid.

Further reading

IPCC 2007; WRI 2005; Stern 2006.

BIODIVERSITY IMPACTS

Biodiversity is a term used to describe the variability that exists between organisms, species, and ecosystems. Although 1.6 million species have been documented to date, it is estimated that the total number of species on earth could be as many as 30 million.[1] The earth's biodiversity changes constantly as new species emerge and others disappear. Biodiversity is also concentrated in specific areas, particularly tropical rainforests. Scientists have identified 32 biodiversity hotspots that contain over half the world's known species, but covering only 2% of the earth's surface.[2]

Based on studies of the earth's biological history, it is estimated that the natural average rate of emergence and disappearance of species is around three per year.[3] This rate has not been constant, with some periods being characterized by rapid increases in biodiversity and others featuring mass extinctions. Changes in prevailing climatic conditions due to fluctuations in solar activity and long-term variations in the earth's orbit (see **Milankovich cycles**), periods of intense volcanic activity, and cataclysmic meteorite strikes have been important determinants of biodiversity change in the past.

The current rate of species extinction is well above the natural background rate.[4] Currently, 25% of the world's mammals and 12% of birds are at risk of extinction.[5] This is largely due to human activities such as deforestation, urban development, pollution, overgrazing, desertification, damming of rivers, overexploitation of species, and a host of other major disturbances. Land-use change, particularly tropical deforestation and habitat fragmentation, was the single most important contributor to biodiversity loss during the twentieth century and is expected to continue to be a major cause this century.[6]

The **Intergovernmental Panel on Climate Change (IPCC)**[7] and a host of other recent studies[8] concluded in 2001 that global mean surface temperature increases exceeding 2°C are likely to result in significant adverse impacts on natural ecosystems and biodiversity and that these impacts would become progressively more severe as temperatures rise. **Global warming** is expected to substantially accelerate biodiversity loss. Based on IPCC warming projections and ecosystem impact assessments, climate change may become the single largest contributor to biodiversity loss this century. Combined with other human impacts, it is estimated that more than half of the species on earth may cease to exist by 2100.[9] This clearly represents a major species mass extinction event.

How does climate change affect biodiversity?

Ecosystems are influenced by an array of factors, and it is often difficult to isolate the changes attributable to climatic change from those attributable to other variables, such as habitat destruction through deforestation or the impacts of pollutants. Nonetheless, there are certain changes that can be directly attributed to **global warming**. For example, climate-induced variations in maximum and minimum air and water temperatures can be linked to observed changes in the timing of major life-cycle events. These life-cycle changes, or changes in *phenology*, include the timing of spring flowering, the emergence of insects, spawning episodes, and migration times. Reliable records of major life-cycle stages (such as the first appearance of a particular species of bird or flower) can stretch back many centuries and can provide a valuable benchmark against which scientists can base their climate change impact assessments.

The most obvious impact of global warming on a species or ecosystem is mediated through changes in climatic conditions within a specific geographical area (the "climate niche"). Changes in variables such as temperature, water availability and seasonal flows, or the increased occurrence of more **extreme weather events** can have major impacts on the health of ecosystems. This is particularly the case during vulnerable life-cycle periods (e.g. flowering, birthing, or hatching times), when sudden variations in climatic conditions can significantly influence mortality rates, breeding success, or seed set. Changes in variability and extremes often cause more damage than underlying changes in average climatic conditions.

Some species rely on a very narrow range of climatic conditions for their survival and are more susceptible to changes in climatic conditions than others. This is particularly the case for those species that live in geographically confined ecosystems, such as mountain tops and isolated islands. Once their climate niche is displaced, such species are often left with nowhere to go. For example, the Pigmy Possum in Australia, which depends on winter snow cover for its survival, seems doomed to extinction this century as snow cover disappears from the Australian Alps – the Possum will simply run out of altitude.[10]

Forest clearance for agriculture, timber production, and urban development has created a patchwork landscape in many regions, containing remnant, and often

28

totally isolated, pockets of natural vegetation. This has reduced the migration options for the remaining plants and animals in these areas. Many national parks, some of which were originally established to protect unique and endangered ecosystems, could be particularly badly affected. For example, the unique plant species of the succulent Karoo in South Africa will be unable to migrate south as the climate changes because they are blocked by the unsuitable soil and topography of the Cape Fold Mountains – on current trends, the Karoo is likely to completely vanish soon after 2050.[11]

Many marine and freshwater species are sensitive to small changes in water temperature. Such changes can influence mortality rates, breeding success, and migratory patterns. Coral reefs are particularly susceptible, and even a variation of 1°C can lead to widespread bleaching and coral mortality – these impacts are already being observed across many major reef systems (see **marine impacts**). Marine species, such as fish and plankton, will tend to migrate toward the poles as sea temperatures increase, a phenomenon already observed over recent decades. However, for many native freshwater fish species, particularly those found in the upper reaches of mountain streams or those that are endemic to specific lakes, migration opportunities are limited or nonexistent.

Reptiles and amphibians are also very sensitive to temperature change. The abundance and range of many amphibians have declined since the 1970s, and some of these changes can be directly attributed to climate change. Gender, and hence breeding success, of many reptile species is dependent on the maintenance of specific temperatures during the incubation of eggs. For example, American alligators only produce males if the eggs are incubated at average temperatures less than 31°C and only females when incubated above 32°C.[12] The same sensitivity applies to the painted turtle where a higher percentage of female than male offspring has already been observed.[13]

Climate change can also have a significant impact on the structure of food chains. Ecosystems are characterized by sets of complex interactions between species: the livelihoods of many species depend on the timing of life-cycle events in other species. However, not all species within an ecosystem respond to climate change at the same rate, and this can cause mismatches in the life-cycles of species. For example, the Pied Flycatcher in Europe has advanced its egg-laying date by approximately one week over the past two decades, but the major food source for its offspring, the Winter Moth caterpillar, has been appearing two weeks earlier. This mismatch in timing has significantly reduced the breeding success of the Pied Flycatcher.[14] Many sea bird species are experiencing similar difficulties (see **marine impacts**). Even though individual species may be able to accommodate increases in temperature, they may have difficulty adjusting to changes in the timing and abundance of their main sources of food. Such mismatches in life-cycle events will act to amplify the effects of climate change on biodiversity.

The effects of climate change will not be adverse for all species. Indeed, many organisms have already benefited from climate change. Many insects breed and mature more rapidly in higher temperatures. The Mountain Pine Beetle of North America, for example, is one species that has clearly prospered. As a result of warmer and shorter winters in northwest United States and Canada, the Beetle's population size and range has expanded considerably in recent decades, albeit at the cost of increased mortality for millions of hectares of native pine trees.[15] Many other insect species, like mosquitoes and ticks, also benefit from a warmer climate (see **health impacts**). Opportunistic species that can disperse quickly or that have relatively short life cycles (e.g. weeds) also tend to prosper during ecosystem disruptions and rapid change.

Elevated temperatures and **carbon dioxide (CO_2)** concentrations are expected to lead to a general increase in biological productivity (via the *CO_2 fertilization effect* – see **agriculture and food supply impacts**), at least for moderate levels of warming (less than 2°C), even though biodiversity may decrease. Fast-growing trees will tend to outcompete slow-growing trees, changing the composition of forest ecosystems, and hence biodiversity. Although plants may grow faster, their leaves are often less nutritious. This can reduce the productivity of herbivores and, in turn, undermine the capacity of an ecosystem to support megafauna.

The impact of climate change on biodiversity clearly involves a complex set of interactions between many different variables, some of which we do not yet fully understand. Much depends on how far and how rapidly global temperatures rise. Evidence from earth's history shows that when the planet warms, species tend to migrate toward the poles or to higher altitudes; when the planet cools, they tend to move toward the equator. Although there have been sudden climate shifts in the past, the rate of past climate change has generally been more gradual than that expected this century. If the rate of climate change is sufficiently slow, species and ecosystems can adapt and evolve. Some scientists consider that many ecosystems can adjust to temperature changes of 0.05°C per decade (a quarter of the current global average rate of 0.2°C per decade and one-eighth the current rate of change in the Arctic).[16] Others suggest that up to 0.1°C per decade may be tolerable for half of the world's species. Perhaps one-third of ecosystems may be able to adapt to rates of change up to 0.3°C per decade, but few ecosystems are able to accommodate rates of change higher than this.[17] However, the ability of ecosystems to adapt today differs markedly from past climate change episodes due to the migration barriers (such as vast tracts of grain fields and grazing lands) that humans have created. It seems likely that many species and ecosystems will eventually simply cease to exist.

Observed changes to date

The past decade has witnessed significant scientific advances in understanding of the impacts of climate change on biodiversity, and there is now a much more extensive body of empirical evidence available to scientists for assessing present and future climate change impacts.[18]

Even the relatively limited global warming we have experienced so far (0.76°C since 1900) has resulted in considerable phenological change across all continents. Warmer autumn and spring conditions have affected the emergence, reproduction, growth, and migratory cycles of many species, particularly in the higher latitudes. Over 60% of Europe's butterfly species have shifted northward by 35–240 km over the twentieth century, and 20 of the 60 bird species studied in the United Kingdom have advanced their egg-laying dates.[19] A comprehensive analysis of over 140 studies covering nearly 1,500 species found that more than 80% of the changes in phenology, abundance, and distribution were in a direction consistent with climate change.[20] The study found that the timing of spring events, such as egg-laying and flowering, had shifted, on average, by 5 days per decade since the 1970s and had resulted in adverse consequences for many species.[21]

There have been noticeable impacts on both freshwater and marine ecosystems due to increased water temperatures. In freshwater aquatic systems, temperature increases have affected fish breeding and mortality levels. Summer stream temperature increases have risen to levels lethal to fish in some areas, particularly in the southern and central regions of North America and in Europe.[22] Higher freshwater lake surface temperatures have also resulted in thermal stratification (differences in temperatures at different depths) of many freshwater lakes, which can impede the upwelling of nutrients from deeper lake water and reduce biological productivity of lakes and also lead to increased incidence of anoxia (oxygen depletion).[23] Changes in the hydrological cycle, such as shifts in the distribution and timing of rainfall, have also placed many fluvial and wetland ecosystems under increased stress (some have completely dried out in extended drought periods).

One group of species, the amphibians, is very sensitive to changes in environmental conditions and is often viewed as an indicator of environmental change. While chemical pollution, such as acid rain and pesticides, and artificial changes in hydrology, such as the damming of rivers and the drainage of wetlands, have adversely affected amphibians, there is also compelling evidence that climate change is contributing to a decline in amphibian populations and habitat ranges. It is estimated that one-third of frog species are already threatened with extinction from current climate change.[24]

One amphibian, the Golden Toad of Costa Rica, became the first documented species extinction attributable to climate change (the last reported sighting was in 1989). The Golden Toad's breeding was dependent on the existence of pools of water on the forest floor, which were maintained by a constant mist that shrouded the mountains where the Toad lived. As the atmosphere warmed, the mist formation zone migrated to a higher altitude, leaving behind pools that evaporated too quickly to enable the tadpoles to survive to maturity.[25] Other frogs, such as the Day Frog and Gastric Brooding Frog in the rainforests of Queensland, Australia, have also vanished and are also believed to be victims of climate change.[26]

The impact of climate change has been most noticeable at high latitudes and the poles, where temperatures have increased at more than twice the rate of the rest of the planet. This has had significant impacts on terrestrial and marine ecosystems, particularly in the Arctic. Polar bears, seals, and several other large mammal species have experienced a significant decline in numbers and breeding success since the 1980s, mainly due to loss of habitat (such as reduced sea ice) and changes in marine food chains (see **polar impacts**).

There have also been observed impacts on marine and coastal ecosystems that can be directly attributable to climate change, notably in the form of warmer surface oceans and salinity changes in coastal wetlands. This has had a significant impact on the location and migration patterns of fish stocks, plankton productivity, coral reefs, and sea bird populations (see **marine impacts** and **coastal zone impacts**).

Projected future impacts

The severity of future impacts will be highly dependent on the magnitude and the rate of climate change, and this will vary between regions. Different areas of the planet are warming at different rates, with the higher latitudes warming at twice the rate of the low latitudes. Some areas will become drier and others wetter. While it is possible to identify contemporary impacts, it is much more difficult to predict how species and ecosystems will respond to future climate change. Scientists must assess current and future species distribution and responses against a range of different variables, not all of which are related to climate change.

Nonetheless, sufficient scientific knowledge and empirical evidence now exist to enable scientists to draw some reasonably robust conclusions about the likely future impacts of different levels of global warming. To assist with future impact assessments, scientists have constructed models (often termed *bioclimatic models*) that simulate ecological responses to changes in key climatic variables. These models incorporate our knowledge of species sensitivity to changes in ambient climatic conditions, the interactions and interdependencies between species, and evidence of biological responses to past climate change.

There is general agreement that those species that are already classified as vulnerable or endangered face a greater risk of extinction than those that are currently abundant across a large geographical range. It is also generally agreed that biodiversity impacts will increase significantly and will become more widespread if global mean temperatures exceed 2°C above preindustrial levels.

The migration of species toward the poles seems likely to continue as the planet warms and climatic zones shift to higher latitudes. Climatic zones suitable for boreal forest, for example, are likely to shift a further 100–150 km northward for each degree of warming, which is five times faster than ecosystems have had to adapt in the past according to paleoclimatic records.[27] Rising sea levels will also have significant impacts on biodiversity in coastal zones, particularly in tidal estuaries, coastal wetlands, and deltas (see **coastal zone impacts**).

Ecosystems that have very limited opportunity to migrate to enable them to maintain their climate niche are particularly vulnerable to extinction.[28] Some ecosystems, like the alpine and arctic systems, coral reefs, the highland mist forests of tropical regions, the succulent Karoo and Fynbos regions in South Africa, and the Dryandra forests of southwest Australia, are at high risk of degradation or total collapse at even relatively low levels of warming (less than 1.5°C). Most southern Indian Ocean coral reefs, for example, are unlikely to be able to tolerate temperature increases above 1.5°C.[29] Some countries and regions are also more vulnerable than others. Australia, which has the highest proportion of endemic species of any continent and also the highest contemporary rate of extinctions, contains a number of unique biodiverse ecosystems that are highly vulnerable to climate change.[30] Even low levels of global warming will have significant impacts on Australian alpine areas, the north Queensland rainforests, and the Kakadu tropical wetlands of the Northern Territory. One study of 60 species living in the north Queensland rainforests concluded that 40% would lose their habitat range with a 1°C rise in temperature and that this would increase to a 90% loss of habitat range for warming of 3.5°C.[31]

Above a 3°C global temperature rise, the impacts become even more severe and widespread: essentially, all ecosystems on the planet suffer major impacts. By this stage, the Arctic and Antarctic regions would be experiencing temperature increases of possibly 5°C or more, with devastating impacts on polar ecosystems.

Overall, the impacts of climate change on biodiversity are expected to be significant, even at relatively low levels of warming. A major multi-continent study concluded that 15–37% of all species studied faced extinction if the mid-range IPCC warming scenario to 2050 eventuated (see **future emissions trends**).[32] If these results are scalable to a global level, it would mean that once the earth's average temperature rises by 3°C or more human-induced climate change could initiate the largest mass extinction event the earth has experienced for millions of years. Once temperatures increase above 4°C, the impacts on the earth's natural ecosystems and biodiversity are expected to be profound and a majority of the planet's species would be doomed to extinction.[33]

Based on these projected impacts, many scientists have concluded that the increases in global mean temperature beyond 2°C are unacceptable for ecosystems and biodiversity.[34] An increasing number of scientists have concluded that to avoid large-scale biodiversity and ecosystem impacts, global temperature increase should be kept under 1.5°C and that the rate of change should not exceed 0.5°C per century.[35] Considering that global temperatures have already risen 0.76°C in the last century, that there is probably at least another 0.5–1°C of warming commitment from past emissions (see **climate sensitivity**), and that there will be a reduction in the cooling effect of aerosols if particulate emissions fall (see **global dimming**), it is possible that exceeding the 1.5°C temperature threshold may already be locked in, even if atmospheric greenhouse gas concentrations can be held at current levels. Based on present emission trends, it appears likely that global temperature will rise by at least 2°C this century, and possibly as much as 5°C (see **future emissions trends**).

Article 2 of the **United Nations Framework Convention on Climate Change (UNFCCC)** states that concentrations of greenhouse gases should be stabilized "within a time frame sufficient to allow species to adapt naturally to climate change." It appears unlikely that the global community will be able to honor this Convention commitment. What seems inevitable is that the complexity of organization, interaction, and diversity of species will decline substantially over this century.[36] Climate change is likely to be the single largest contributor to this decline.

Some options are available for assisting the climate adaptation responses of some species. These include the following: reducing deforestation and pollution; connecting isolated pockets of remnant forests and ecosystems to provide migration corridors; physical relocation of species to areas that have more suitable climate niches; better-managed harvesting of certain species; and ecosystem regeneration activities. However, most species will need to rely on their own adaptation responses for survival.

The impacts of biodiversity loss on humans are also likely to be significant. There are direct impacts on the livelihoods of many indigenous people and on the key economic sectors of agriculture, forestry, and tourism (see **agriculture and food supply impacts**, **coastal zone impacts**, and **socioeconomic impacts**). There are also other noneconomic considerations, such as the impacts on spiritual and social well-being. People value the existence of different species merely because they are there, irrespective of any direct economic value to themselves (often referred to as existence value or intrinsic value). These values are rarely included in the climate change impact assessment cost estimates, even though we know that the value is not zero.

See also: agriculture and food supply impacts, climate sensitivity, coastal zone impacts, dangerous climate change, extreme weather events, global warming, health impacts, marine impacts, polar impacts, socioeconomic impacts.

Notes

1 Wilson and Perlman 2000
2 Pheonix *et al.* 2006
3 IPCC 2002
4 Ibid.
5 Ibid.
6 Ibid.
7 IPCC 2001, 2007
8 Lanchbery 2006
9 Meyer 2006
10 Steffen *et al.* 2006
11 Flannery 2005
12 Cogger and Zweifel 1992
13 Flannery 2005
14 Van Vliet and Leemans 2006
15 Ibid.
16 Ibid.
17 Leemans and Eickhout 2004
18 Van Vliet and Leemans 2006
19 IPCC 2002
20 Parmeson and Yohe 2003
21 See IPCC 2002 and Parmeson and Yohe 2003
22 IPCC 2002
23 Ibid.
24 Stokstad 2004
25 Flannery 2005
26 Ibid.
27 IPCC 2002
28 Ibid.
29 Sheppard 2003
30 Steffen *et al.* 2006
31 Williams *et al.* 2003
32 Thomas *et al.* 2004
33 Hare 2006
34 CBD 2003
35 Van Vliet and Leemans 2006
36 Meyer 2006

Further reading

IPCC 2002; UNCBD Secretariat 2003; Hare 2006; Van Vliet and Leemans 2006.

BIOFUELS

Biofuels is a term used to describe all liquid, solid, or gas fuels derived from recently living organisms or their metabolic waste. As a **renewable energy** source, biofuels are a potential greenhouse gas **mitigation** option.

Solid biofuels (biomass), such as wood, manure, and crop residues account for nearly all global bioenergy supplies and are the primary energy source of 2.5 billion people. Biogas, derived from the anaerobic (without oxygen) digestion of municipal wastes (landfill gas) or animal and human wastes to produce **methane (CH_4)**, is also a biofuel, and production has grown considerably in recent years, in part due to financial incentives provided by the **Clean Development Mechanism (CDM)**. There has also been wider application of small-scale biogas generators using human and animal wastes to provide cooking gas, but together these energy sources are too small to even register in accounting of global energy use.[1] Virtually all commercial use of biofuels comes from the liquids bioethanol and biodiesel. This section focuses on liquid biofuels as a transport sector **mitigation** option.

The transport sector accounts for 14% of **anthropogenic greenhouse gas emissions**, and emissions from the sector are expected to increase substantially over the coming decades – the global demand for transport fuels is expected to grow by as much 50–60% by 2030.[2] Reducing transport emissions (from road transport in particular) is particularly vexing for policy makers. Policies that increase fuel prices "at the pump" are met with strong resistance by consumers, so introducing a cost penalty on emissions (see **carbon tax**) is politically difficult. Biofuels, along with more fuel-efficient vehicles, electric and hybrid-electric vehicles, natural gas, and modal shift (from private car to public transport and bicycles) are all expected to contribute to reducing transport sector emissions. In the longer term, beyond 2030, fuel cells and hydrogen could also make a contribution.

Liquid biofuels' contribution to energy supplies

There are two primary commercial biofuels: (1) *bioethanol*, a substitute for gasoline derived from fermented plant sugars and starches and (2) *biodiesel*, a substitute for mineral diesel produced by processing plant and animal oils. Both have been used as transport fuels for more than a century. Rudolf Diesel, the inventor of the diesel engine, ran his 1892 demonstration model on peanut oil, and the Model T Ford, first produced in 1908, ran on ethanol, which Henry Ford described as "the fuel of the future." However, for most of the twentieth century, biofuels have been more expensive than **fossil fuels**.

The 1973 and 1979 oil price shocks generated increased interest in biofuels, which soon languished as oil prices fell during the 1980s. Brazil was a notable exception and has actively promoted the use of biofuels and maintained strong support for bioethanol production from sugarcane. Biofuels now provide over 40% of Brazil's road transport fuel requirements.[3] This stands in stark contrast to the less than 2% share of biofuels in global transport fuel consumption.[4]

Bioethanol use increased rapidly during the 1990s, with production doubling over the decade to reach 20 billion liters by 2000, primarily driven by corn-based ethanol production in the United States. By 2000, the US share of global production reached

40%, though Brazil remained the largest producer.[5] Production more than doubled again, reaching 52 billion liters in 2006, mainly from continued strong growth in Brazil and the United States, but also an increasing contribution from the European Union, Canada, and China.[6] Bioethanol's proportion of global gasoline consumption (approximately 1,250 billion liters in 2006) remains small at just 4% but, nonetheless, more than doubling its 1.5% share in 2001.

Biodiesel production growth has been over a longer time period, beginning during World War II, to reach 6.5 billion liters by 2006. More than 80% of biodiesel is produced from rapeseed (canola), with the remainder largely derived from soybean and palm oil, with small amounts from sunflower seeds and other crops. Production is dominated by Germany (50%), with France (15%), the rest of the European Union (10%), and the United States (13%) also being significant producers. Biodiesel's proportion of global diesel consumption (approximately 1,100 billion liters in 2006) is much smaller than bioethanol at just 0.6%.[7]

Biofuel production increases have generally been underpinned by government subsidies and incentives to producers, particularly US support for corn-based ethanol production. Although some governments are supporting biofuel production for environmental reasons (greenhouse gas emissions and urban air quality), soaring oil prices and energy security concerns (dependency on imported petroleum) have been the major driving force. More than 40 governments have enacted biofuel consumption mandates that set annual targets, and most also provide tax incentives and subsidies to producers.[8] Together biofuels made up 2.4% of world transport fuel consumption in 2006, and this share is expected to continue strong growth in the near term. In 2007, there were 951 biofuel plants – 386 biodiesel and 565 bioethanol – operating or under construction in 56 countries, with a combined output capacity in excess of 163 billion liters.[9] If all these plants operated at full capacity by 2009, biofuels could supply as much as 5% of the road transport fuel market.

Advantages and disadvantages of biofuels

As a **mitigation** alternative to **fossil fuels**, biofuels have several advantages and disadvantages. The principal advantages are that they:

- are a **renewable energy** source which can be produced in virtually any country, potentially providing energy security to nations;
- unlike some other **renewable energy** sources, can be stored and hence provide reliable supplies;
- burn cleaner than fossil fuel alternatives, giving air-quality benefits and have limited health and safety risks;
- can result in a reduction of greenhouse gas emissions and, for some technologies, can result in a net removal of **carbon dioxide (CO_2)** from the atmosphere;
- are proven technologies that can be operated at large scale and opportunities exist to reduce production costs further; and
- can utilize existing fuel distribution infrastructure and, when blended with gasoline or diesel, can be used by existing vehicles without engine modification and as a replacement for fossil fuels in other applications.

The main disadvantages are the following:

- Potential land availability constraints if production is significantly scaled up and a range of other environmental and social issues that may constrain growth.
- Apart from sugarcane bioethanol, and possibly palm oil biodiesel, most biofuel production is only marginally economic (even at oil prices of US$100 per barrel) and is often reliant on government subsidies and incentives.
- Most crop-based biofuels are subject to variations in global food prices, particularly corn, wheat, and sugar, and food price increases may divert biofuel feedstocks to other markets, possibly resulting in fluctuating supplies and prices.
- Biofuels provide less energy per liter of fuel (in the case of bioethanol) which, in combination with other negative public perceptions, may slow uptake.
- They provide, with current production methods, only modest greenhouse gas mitigation benefits relative to fossil fuels and, in some cases (depending on how and where they are produced) may result in a net increase in greenhouse gas emissions.

Technology status

Bioethanol is produced by fermenting sugars with yeast to produce ethanol which is then distilled. In tropical regions, bioethanol is primarily derived from sugarcane while, in more temperate climates, is primarily based on sugar beet and starch rich crops, primarily corn, wheat, and barley, with the starch first being converted to sugars using enzymes. These bioethanol technologies are mature and well proven. Future cost reductions will largely be derived from increases in production scale.

These "first generation" technologies for producing bioethanol are well proven and widely available. Ethanol is less dense, burns more cleanly and at a higher temperature than gasoline, but provides less energy per liter. A 100% substitution of bioethanol for gasoline in standard engines can result in engine damage and a 10–20% loss of power. Nonetheless, the difference between gasoline and bioethanol is not great, and blends of up to 10% bioethanol (so called E10) can be used without modification or noticeable loss of power. Blends of 85% (E85) require slightly modified engines and result in some loss of power that can be overcome with minor driver behavior changes. Flexible Fuel Vehicles (FFVs) are being mass produced in Brazil and the United States to enable wider uptake of E85, with only a nominally higher (less than 3% without subsidies) initial purchase price. In Brazil, FFVs now make up 55% of new vehicles sold, while in the United States, less than 10% of new vehicles are FFVs.

Several new bioethanol "second-generation technologies" that utilize cellulose or lignose (a glucose-generating carbohydrate found in plant cell walls) as the primary feedstock are being developed. Technologies include hydrolysis/fermentation, gasification, or pyrolysis. Feedstock supplies can be drawn from a wide range of grasses, herbaceous and woody plants (such as switchgrass, straw, wood, and seaweed), or crop by-products such as wood pulp, grain and sugar crop residues, citrus waste, or organic municipal wastes.

There are two primary advantages of second-generation bioethanols. First, they enable fuel to be made from nonfood crops, avoiding competition with food production and consequent upward pressure on food prices. Second, feedstocks can be grown on marginal lands such as rangelands and prairies and may even be planted on degraded lands to facilitate rehabilitation. When feedstocks are a mixture of native grassland perennials, they can be grown on a large scale without fertilizer or herbicides and with little land management (such as ploughing or irrigation) inputs. Such Low-Input High Diversity (LIHD) plantations can provide more usable energy per hectare at potentially lower cost. Natural grasslands typically have higher species diversity per hectare, which can yield higher cellulosic density per hectare. Yields of up to 50% more energy/ha can be achieved on degraded lands compared with corn bioethanol produced on fertile agricultural lands. Furthermore, when grown on degraded lands, LIHD plots result in net carbon uptake in the soil of up to 1.7 tCO_2 per year (see **carbon sinks** and **biosequestration**). Overall LIHD bioethanol replacing gasoline could result in a 110% reduction in emissions relative to a similar quantity of fossil fuels – as such it can be a "carbon negative" biofuel.[10]

The Iogen plant near Ottawa, Canada, has been producing second-generation bioethanol from straw and wood waste since 2005, but only at a rate of around 100,000 liters (about one tanker truck) per day. A much larger plant is planned in Germany for 2009. While the potential of second-generation bioethanol as a mitigation option is significant, there is currently no large-scale commercial production.

Biodiesel technologies are also well proven and involve the transesterifcation of animal and vegetable fats – alcohol is used to break down fats into glycerol and biodiesel. Unlike bioethanol, biodiesel can be used up to a 100% direct replacement for fossil fuel-derived diesel. No engine or infrastructure modifications are required, unless being used in extremely cold temperatures, as biodiesel has a slightly higher melting point and may begin to thicken at temperatures below around −15°C, in which case small amounts of antifreeze can be added. There are no technical barriers to biodiesel as a replacement for conventional diesel. Nonetheless, even at oil prices of US$100 per barrel, most biodiesel production still requires subsidies to be cost competitive.

Table 5 provides information on the estimated costs of production of different biofuels and an indication of the potential greenhouse gas emission reductions from each fuel type. The cost figures can be compared with the cost of producing gasoline and diesel of around US$ 0.70 per liter at oil prices of US$ 100 per barrel.

Greenhouse gas emission benefits

Biofuels produced from different feedstocks are identical at the time they are combusted in an engine, but estimates of biofuel greenhouse gas emission reduction benefits are subject to considerable debate as they vary widely according to how biofuels are produced. To assess emission reduction benefits, all **greenhouse gases** emitted over the full life cycle of production (seed to wheel) must be accounted for.

The largest variable for emissions from biofuels relates to the use of fertilizer, which releases the powerful greenhouse emission **nitrous oxide (N_2O)**. Emissions from fertilizer application are difficult to ascertain and subject to considerable uncertainty – these emissions are rarely included in estimates of biofuel production emissions. Recent research indicates that prior assumptions of N_2O emissions were

Table 5 Biofuel production costs and emission reduction potential[11]

Production method	Production cost range* in 2005 (US$/liter)	Projected production cost range* in 2030 (US$/liter)	Estimated emission reduction relative to fossil fuels (%)
Bioethanol			
Sugarcane	0.20–0.50	0.20–0.35	10–90
Corn	0.60–0.80	0.35–0.55	40–150
Sugar beet	0.62–0.82	0.40–0.60	10–240
Wheat	0.70–0.95	0.45–0.65	40–210
Lingo-cellulose	0.80–1.10	0.25–0.65	–10 to 30
Biodiesel			
Vegetable oil – canola	0.50–1.00	0.40–0.75	20–170
Palm oil	0.5–0.80	No estimates available	20–550

*Excluding any subsidies to biofuel production.

significant underestimates and that the use of fertilizers on feedstock crops may negate most of the benefits of substituting bioethanol for gasoline and can even result in a net increase in greenhouse gas emissions.[12] For example, bioethanol from sugar beet produced without fertilizers can deliver as much as a 90% reduction in emissions per liter, but when significant quantities of fertilizers are used, emissions intensity can be up to double that of regular gasoline.

Biofuel emission intensity is also highly dependent on the land-use type prior to bio-fuel production commencing. Where biofuel production uses the same or similar crops as previously produced (little land-use change), the change in greenhouse gas emissions per unit of biomass output is likely to be negligible. Where biofuel crops are planted on degraded or marginal land, there may even be a net increase in carbon stored in the soil – a net contribution to **land carbon sinks**. However, if feedstocks are grown on land made available by clearing areas that store large amounts of carbon (see **land carbon sinks**), large emissions of CO_2 will occur, negating the benefits of biofuels. This is particularly of concern for sugarcane, where tropical rainforest destruction is a common way to access new land, reducing the climate benefits of sugarcane bioethanol compared with gasoline from 90% to just 10%. By far the worst impact, however, is from clearing and draining tropical wetland forests for the plantation of palm oil (see **biosequestration**), which results in a net increase in greenhouse gas emissions from palm oil biodiesel of more than five times that from using conventional diesel.

Bioethanols produced from starch-based feedstocks (like corn and wheat) requires a two-step process which increases energy input requirements, reducing greenhouse gas mitigation benefits. Even without the increase in emissions from fertilizer application or land-use change, starch-based bioethanol at best achieves a 60% reduction in emissions from displacing gasoline.

Calculating the exact emissions benefits of a particular liter of biofuel use can be complex and, as a result, determining the net emissions benefit of increasing the use of biofuels is difficult. Most literature concludes that biofuels are presently delivering modest reductions in greenhouse gases, relative to fossil fuel alternatives, but they are clearly not a "carbon neutral" product.

Social and environmental impacts

Apart from greenhouse gas mitigation and energy security benefits, there are several other environmental, social, and economic issues that are important to the biofuels debate. Biofuels can provide urban air quality benefits as they burn more cleanly and produce less particulate **aerosols**, but they produce more nitrogen compounds (see **nitrous oxide**) which have other adverse environmental effects. Biofuels production is also likely to offer employment and income generation opportunities in rural areas.

Current biofuel production is almost entirely based on capital intensive monoculture plantations using fertilizers, pesticides, herbicides, and often irrigation. Such production systems can result in loss of biodiversity, water quality and quantity problems, and a buildup of persistent organic pollutants (POPs) in the biosphere with questionable greenhouse gas mitigation benefits. Genetically Modified Organisms (GMOs) have been proposed as a way to increase productivity and possibly reduce the use of pesticides and fertilizers, but these remain controversial issues and face considerable resistance from some groups. Environmentalists fear that a dramatic expansion in biofuel production to meet **mitigation** targets will exact a heavy toll on the environment.

However, the most significant issue facing a large scaling up of biofuels production is the competition for arable land and the related impact on global food supplies and prices. Already biofuels production uses around 1% of arable land. If biofuel production increases in line with projections to 2030, they will account for up to 4% of arable land area (equivalent to the area of Australia, Japan, and New Zealand combined). Other studies estimate that if biofuels were to substitute for 10% of global gasoline and diesel consumption (using existing production methods), they would require 9% of earth's existing agricultural land area.[13]

Given that global food demand is expected to grow significantly in coming decades, the combination of food and biofuel demand will inevitably increase pressure for expansion of agriculture into existing forest and grassland areas, with subsequent further biodiversity loss and increased greenhouse gas emissions. Moreover, local landholders, indigenous people and those with informal land tenure are at a high risk of losing their livelihoods (see **renewable energy**).

Biofuels produced from food crops (including corn, wheat, and barley) increases demand for grains and directly increases food prices (by up to 50% by 2016),[14] impacting most heavily on the poor. The doubling of biofuel production between 2000 and 2005 has already resulted in upward pressure on food prices (particularly corn and wheat) and led to additional deforestation.[15]

Second-generation technologies, particularly LIHD bioethanol production, may alleviate some of these issues and potentially provide land rehabilitation benefits, but LIHD may also put pressure on livestock grazing and nomadic peoples currently using marginal lands. LIHD biofuel production methods can also be applied to fertile agricultural lands, which would yield more biofuel per hectare than the same techniques on marginal lands, but would also potentially displace food production. New research into using algae, seaweed, or plankton as biofuel feedstock may offer hope of alleviating land conflict issues, but these technologies remain at the research scale.[16]

What contribution can biofuels make to greenhouse gas mitigation efforts?

Biofuels are already an important part of greenhouse gas mitigation policies, particularly in the European Union, United States, and China, and biofuels have been an important component of Brazil's energy policy framework for many years.

The International Energy Agency projects that biofuel production will grow by 6–8% per year over the period to 2030, representing between a four- to seven-fold increase in production.[17] If this eventuates, biofuels could potentially supply between 4% and 7% of road transport fuel requirements by 2030. Several other studies project higher growth rates and suggest that biofuels could meet up to 10% of transport fuel requirements by 2030.[18] However, a sustained increase in food prices (such as that seen since late 2006) will force up the cost of feedstocks and adversely affect the cost competitiveness of some biofuel producers. Even though most studies project that biofuel production growth rates are unlikely to sustain the rapid growth rates experienced over the 2000–2005 period, when production doubled, they all project a significant production increase over the next decade.

If it is assumed that biofuels are able to displace 7% of road transport fossil fuels by 2030 (a mid-range estimate), their contribution to greenhouse gas mitigation would be relatively small, around 0.5 $GtCO_2$ per year. At higher growth rates, it is plausible that biofuels could reduce global emissions by up to 1 $GtCO_2$ per year. This would provide 1–3% of the emission reductions required to stabilize atmospheric greenhouse gas concentrations at 550 ppm CO_2e – a very modest contribution (see **stabilization targets**).

Most projections do not, however, include potential biofuel supply from second-generation technology sources. Rapid recent technology advances have greatly improved the prospects of commercially available second-generation biofuels by 2020 or sooner. If by 2030 a quarter of biofuels are produced using second-generation technologies with feedstocks from LIHD sources produced on degraded or marginal land, the mitigation contribution of biofuels could be considerably higher.

Although biofuels offer some potential as a mitigation option, they face a range of political, economic, and physical constraints. The contribution of biofuels faces land availability constraints, marginal economics for some biofuel sources, and substantial political and social issues in terms of global food prices and food security, particularly for the world's poorer nations. At best they offer only a minor contribution to global-mitigation efforts. Second-generation technologies offer considerable potential, and potential exists, over the medium term, to scale up production on degraded or low-value lands with fewer implications for global food production or pressure on existing forests.

See also: biosequestration, fossil fuels, land carbon sinks, mitigation, renewable energy.

Notes

1 IEA 2006a
2 Ibid.
3 Isherwood 2006

4 Pearce 2006
5 WBCSD 2007
6 Ibid.
7 Derived from WBCSD 2007 and IEA 2006a
8 Hubbard Preston 2008
9 Ibid.
10 Tilman *et al.* 2006
11 Derived from WBCSD 2007, Crutzen *et al.* 2007, and IEA 2006a
12 Crutzen *et al.* 2007
13 Righelato and Spracklen 2007
14 WBCSD 2007
15 Hooijer *et al.* 2006
16 Nowak 2008
17 IEA 2006a
18 WBCSD 2007

Further reading

Hubbard Preston 2008; IEA 2006a; Pearce 2006; WBCSD 2007.

BIOSEQUESTRATION

Biosequestration refers to the process of removing **carbon dioxide (CO_2)** from the atmosphere and storing it in land or ocean reservoirs through biological processes managed, promoted, or facilitated by humans. Biosequestration is one of a wide range of **mitigation** options available to limit or reduce the buildup of greenhouse gases in the atmosphere.

The cycling of carbon between the land and oceans and atmosphere is a fundamental driver of life on earth, and each year large quantities of CO_2 are exchanged between these three active carbon reservoirs (see **carbon cycle**). By managing or manipulating the flows of carbon between these three reservoirs, either through enhancing the biological processes to sequester carbon from the atmosphere (such as planting trees) or through preventing carbon stored in the land and ocean reservoirs from entering the atmosphere (such as preventing trees from being cut down), humans can influence the amount of CO_2 stored in the atmosphere. While humans may be able to influence the biological uptake of CO_2 by oceans (e.g. through iron fertilization – see **ocean carbon sinks**), biosequestration is more commonly used to describe measures that influence the amount of carbon stored by the land system. The remainder of this section deals with measures that influence the **land carbon sink**.

Through the natural processes of the carbon cycle, the land system is currently absorbing more CO_2 from the atmosphere each year than it emits back to the atmosphere – it is, therefore, a net **carbon sink**. Uncertainty surrounds the magnitude of the land carbon sink effect, but it is estimated to be around 6–8 $GtCO_2$ per year, though the uncertainty range is between 4 and 9 $GtCO_2$ per year (see **land carbon sink**). Overall net CO_2 uptake by the land is currently helping to slow the buildup of **greenhouse gases** in the atmosphere. This is providing the global community with some breathing space in which to introduce the full range of emission **mitigation** measures to stabilize atmospheric greenhouse gas concentrations.

Biosequestration (including avoided deforestation) has the potential to be one of the more important greenhouse gas mitigation options in the short to medium term (out to 2030) and arguably offers greater potential as a mitigation option than **renewable energy** or **carbon capture and storage**, at least in the short to medium term. Nonetheless, a range of technical, political, economic, and social factors are expected to constrain the biosequestration mitigation contribution to well below the technical potential. Increasing demand for food, timber, cash crops (such as rubber and palm oil), and **biofuels**, plus the expansion of urban and industrial areas are all placing increasing demands on the existing arable land and also drive ongoing deforestation and land degradation. All these factors are expected to limit the type and extent of biosequestration activities in the coming decades.

Permanence and leakage

Before discussing the types of biosequestration and their greenhouse gas mitigation potential, it is important to understand two concepts that are pertinent to biosequestration: permanence and leakage. These are both issues that constrain the uptake of large-scale biosequestration measures.

Permanence relates to the length of time that CO_2 is removed or withheld from the atmosphere. To have a lasting effect on atmospheric greenhouse gas concentrations, and hence future **global warming**, CO_2 needs to be permanently withheld from the atmosphere. Some define this as a period of at least 100 years (the assumed CO_2 atmospheric residence time), though others argue that biosequestered CO_2 should be stored in perpetuity. Nonetheless, the objective of biosequestration is to increase, or at least maintain, the total stock of carbon in the land system. For example, if carbon dioxide is removed from the atmosphere by establishing a fuelwood plantation that is cut down and burned after 15 years, then the CO_2 is only in temporary, not permanent, storage. This does not mean that short cycle fuelwood plantations cannot form part of an effective biosequestration project, but that they cannot be credited with a full greenhouse gas reduction benefit unless there has been an overall increase in the total stock of carbon within the project boundaries. There are other uncontrolled events, such as fire and insect infestations, that may reduce the permanence of biosequestration activities, even if the intention was to permanently store the sequestered carbon. Though a temporary reduction in atmospheric CO_2 loadings is beneficial, as it temporarily reduces the **radiative forcing** below that which would otherwise have been the case, it will only delay, rather than avoid, future warming. This is a major issue confronting biosequestration activities: how to guarantee that the carbon removed or withheld from the atmosphere represents a permanent increase in the land carbon sink.

The risk of nonpermanence reduces the attractiveness of biosequestration as a mitigation option. Measures such as instituting legal covenants in land titles to ensure that the biosequestration gains are maintained in perpetuity; insuring the stored carbon against loss; establishing a large and diversified pool of biosequestration activities across different geographical regions to minimize the risks of fires, pests, and storms; and/or crediting only a smaller proportion of the CO_2 actually sequestered can reduce, but not eliminate, the risks of nonpermanence. Nonpermanence also makes it more difficult to ensure the environmental integrity of any tradeable carbon credits issued for use in meeting mandatory emission targets (see **emissions trading**).

Leakage refers to the risk that a specific biosequestration activity in one location causes a change (usually a reduction) in land carbon stocks in another location so that the net effect on atmospheric concentrations is either reduced or negated. For example, if a biosequestration project that involves the cessation of timber logging activities at a particular location to gain carbon credits merely results in an equivalent increase in logging in another area, or country, to fill the shortfall in timber supply, then there is no net benefit to the atmosphere.

Leakage can take several different forms, including *activity displacement*, where the cessation of a particular practice in one location is replaced by the same practice at another location; *demand displacement*, for example, where preventing access of fuelwood gatherers to a newly protected area merely forces them to use an alternative energy source, such as kerosene; or *supply displacement* where, for example, the loss of new agricultural land by preserving a wetland causes the intensification of agriculture elsewhere through applying more fertilizers (generating an increase **nitrous oxide (N_2O)** emissions); and *investment crowding*, where, for example, public demand for (and willingness to pay for) protected areas (such as national parks and nature reserves) diminishes as more areas are protected. Although leakage is nearly always negative, in some circumstances it may be positive. For example, regenerating degraded rangelands by planting native grasses, or adopting low till agriculture, can result in CO_2 sequestration as well as increase agricultural productivity. If other farmers adopt similar practices, for the purpose of increasing yields, a greater quantity of CO_2 may be sequestered than from just the original project activity.

Many of these leakage types overlap, and it is often not possible to foresee or account for unintended leakage outcomes. Leakage is particularly difficult to control across national boundaries. For example, efforts to enforce regulations against illegal logging in Malaysia may prove successful, but this may also drive up international tropical hardwood prices and thus create an increased financial incentive for illegal loggers in another country, where Malaysian regulations obviously have no jurisdiction.

A major technical greenhouse gas accounting constraint facing biosequestration mitigation projects is the limited ability to effectively quantify and monitor leakage. At present, the ability to accurately quantify and monitor changes in land carbon stocks at the national level is limited, and reliable land carbon accounting systems are confined to just a few countries. Establishing national and regional land carbon or forest baselines, that provide a reference point against which changes in carbon stocks can be measured, is an essential prerequisite for reliable biosequestration accounting frameworks, and several countries are in the process of developing national baselines. Rapid improvements in remote sensing technologies will increase the ability to track leakage at a national and international level, but at present there is no means of tracking leakage at a global level.

Leakage can, to some extent, be addressed at the local or subregional level, through such measures as the provision of alternative incomes and livelihoods for those that may be displaced as a result of a biosequestration project (e.g. by establishing fuelwood plantations and direct employment in the biosequestration project activities themselves). However, at a national or global level, the options for controlling leakage are much more limited.

Types of biosequestration

There are two main types of biosequestration: (1) those that prevent the release of CO_2 to the atmosphere and (2) those that remove CO_2 from the atmosphere. Measures that prevent the release of CO_2 can be further subdivided into avoided deforestation and wetland/peatland conservation, while those that remove CO_2 can be subdivided into afforestation/reforestation (A/R) activities and improved land management activities.

Avoided deforestation

Avoided deforestation is a term used to describe activities that preserve existing forests and woodlands from conversion for the purpose of agriculture, timber/fuel-wood production, or other purposes such as human settlement. The aim of avoided deforestation (now commonly referred to as Reduced Emissions from Deforestation and Degradation – REDD) is to retain the net stock of carbon and ensure that it is not released to the atmosphere.

Deforestation and forest degradation has been a source of anthropogenic emissions for centuries but has become a much more important source of emissions over the past 150 years. During the period 1800–1950, vast tracts of forests were cleared in the mid- and higher latitudes, particularly Europe, North America, and Australia (see **land carbon sinks**). Since the 1950s, the lower latitude tropical and subtropical regions have accounted for the majority of deforestation activity, particularly Brazil, Central America, Southeast Asia, and, more recently, West and Central Africa.

The average rate of deforestation remains at a staggering 13 million hectares (Mha) per year (an area about the size of Greece). Though lower than the 17 Mha per year average that prevailed in the 1980s, annual deforestation rates have persisted at or near current levels for nearly two decades and show few signs of abating.[1] Emissions from land-use change (deforestation, fuelwood extraction, agricultural land management, land degradation, and wetland drainage) currently accounts for nearly one-fifth of **anthropogenic greenhouse gas emissions**. There are a range of complex factors driving deforestation and include, among others, global timber demand, increasing demand for food and cash crop production, and fuelwood and charcoal supply. Although more than a third of the earth's original forest cover is now gone, 3,900 Mha still remains, divided roughly equally between the developed countries (1,900 Mha) and developing countries (1,970 Mha). Any measures that avoid further deforestation would clearly make a significant contribution to reducing global greenhouse gas emissions. This is particularly the case in tropical countries as mature tropical forests store about 240 tC/ha compared with 150 tC/ha for temperate forests, or around 60% more carbon per hectare.[2]

To date, avoided deforestation measures have not been a major component of the global mitigation response. This is partly due to the fact that avoided deforestation is not recognized as a creditable activity under the **Kyoto Protocol** carbon trading mechanisms of the **Clean Development Mechanism (CDM)** or **Joint Implementation (JI)**. It is also partly due to the technical constraints associated with measurement, verification, leakage monitoring, and the subsequent land use; calculating emission reductions from avoided deforestation remains a complex task. Emissions from deforestation are, however, included in national **greenhouse gas inventories** under the **United Nations Framework Convention on Climate Change (UNFCCC)** category of Land Use, Land-Use Change, and Forestry (LULUCF), and this provides an incentive for countries with binding emission targets under the Kyoto Protocol to reduce deforestation activities.

Avoided deforestation is an important topic in the post-2012 climate change negotiations, largely in relation to possible future developing country mitigation commitments, but remains politically sensitive. While uncertain, there may be some provision for the creation of some type of carbon credits in the post-2012 international climate agreement (see **Kyoto Protocol**). Many would view the inclusion of avoided deforestation as a vital element of any post-2012 climate agreement and an essential means of engaging the developing world in global mitigation efforts. Several governments and nongovernmental organizations are already supporting avoided deforestation initiatives. For example, the Australian Government is currently working with Indonesia and Papua New Guinea on avoided deforestation measures.[3] There are other initiatives, such as the Forest Carbon Fund proposed by the World Bank, and forest preservation projects funded by a range of nongovernment organizations (primarily directed at biodiversity and ecosystem service objectives) that also may assist in reducing deforestation rates.

Estimating the potential mitigation contribution from avoided deforestation is difficult. As yet there is no direct financial incentive, at least in a greenhouse gas mitigation sense, for developing countries to reduce deforestation rates. The lack of a global market incentive to preserve existing forests remains a major factor constraining avoided deforestation as a mitigation option and complicates the task of projecting its future contribution. The potential is, nonetheless, very large as even a 20% reduction in current deforestation could yield up to 1 Gt per year of emission reductions. Some modeling indicates that avoided deforestation and wetland conservation could deliver up to 2 $GtCO_2$ per year by 2030 for a $30/$tCO_2$ emission reduction price or 3.5 $GtCO_2$ per year by 2030 for $100/$tCO_2$ (see **mitigation**). Others models indicate that market prices for emission reductions make virtually no impact on deforestation rates until they reach $150/$tCO_2$ and do not halt deforestation worldwide until they reach over $1,000/$tCO_2$.[4] While a financial incentive would almost certainly assist in delivering emission reductions from avoided deforestation, the quantity of reductions that could be delivered on a global basis, and the price required to achieve these reductions, currently remains speculative.

Wetland conservation

Wetlands are areas in which the water table is near or at the surface, covering soils made up of thick layers (up to 20 m) of organic material accumulated over thousands of years. Wetlands include the subcategories of peatlands, peat swamp forests, and mangrove swamps. As the name suggests, wetlands are waterlogged and anoxic conditions (limited oxygen availability) prevail, which prevents, or substantially reduces, the oxidation of organic matter that accumulates over time. Wetlands represent only about 2.3% of the global land surface but contain nearly 10% of global land carbon stocks (only 5% of which is accounted for by above ground vegetation).[5] Wetlands store an average of 700 tC/ha, about triple that stored in tropical forests, and tropical peatland forests with deep peat-soil deposits can contain up to 5,800 tC/ha.[6] Approximately 80% of all wetlands are located in just three countries, Russia, Canada, and the United States, but these are mainly in the Arctic region and most are permanently frozen (see **polar impacts**). Of the active (unfrozen) wetlands, more than 60% are found in Southeast Asia, with a few significant areas in Africa and Central America.[7]

Wetlands are traditionally seen as low-value lands for agriculture or human habitation due to waterlogged soils. However, when cleared of vegetation and drained, they can support a range of agricultural activities. For example, many oil palm plantations in Southeast Asia have been established on drained tropical peatlands. Reliable data are limited, though it is estimated that 1 Mha/year of wetlands are currently being lost, most of the loss occurring in the carbon-rich tropical peatland forests of Southeast Asia. Since 1981 these peatland forests have been lost at an average rate of around 2% year – Indonesia and Malaysia alone have now lost almost half of their peatland forests over this period.[8] Wetland draining is akin to deforestation, though most of the CO_2 emissions come from the oxidizing soils and peat fires rather than the burning or decay of above ground biomass.

In total, wetland drainage is estimated to release approximately 2,000 MtCO$_2$ per year to the atmosphere – around 4% of total **anthropogenic greenhouse gas emissions**. This occurs through two main mechanisms: (1) oxidation of organic soil carbon (approximately 600 MtCO$_2$ per year) and (2) peat fires (approximately 1,400 MtCO$_2$ per year) – 70% of wetland emissions come from Southeast Asia (mainly Indonesia) and about 10% from North America and Europe.[9] The rate of soil oxidation depends on soil depth and occurs rapidly in shallow soils in the tropics, but oxidation may continue for 100 years or more for deep (greater than 4 m) soils. The dried, carbon-rich soils are extremely fire prone – in 2006 over 40,000 fires occurred in drained wetlands in Southeast Asia.[10]

If further wetland drainage were to cease today, the current 600 MtCO$_2$ per year of emissions from oxidizing soils would diminish gradually over time but may still be over 50 MtCO$_2$ per year by 2100. For this reason, the option of wetland rehabilitation, through reflooding and replanting cover vegetation are being evaluated and in some areas implemented. However, while this will prevent further oxidation (as anoxic conditions are recreated), it can initially result in a spike in **methane** emissions for several years after reflooding. It may take up to ten years before the benefits of reduced CO_2 emissions from oxidation outweigh the warming effect of the methane emissions. While there is likely to be a net emissions benefit in the long term, the short-term benefit is negative. In relation to investing in wetland rehabilitation projects for the generation of emission reduction credits (though these do not presently qualify under the Kyoto trading mechanisms), this lag in credit creation (credit payments would not commence for at least ten years after the initial investment) represents a significant disincentive.

As a biosequestration activity, wetland rehabilitation and/or protection (avoiding drainage) faces similar measurement and leakage issues as avoided deforestation. Furthermore, wetland drainage emissions are not required to be included in **greenhouse gas inventories** under the **UNFCCC** and, as such, have received little attention in global climate change discussions. If, for example, wetland emission sources were required to be accounted, Indonesia would jump from its current 21st place ranking to being the third largest emitting country after the USA and China. At present there is no financial incentive for countries to rehabilitate or protect wetlands for greenhouse gas emission reduction purposes – in fact, the current demand for cash crops such as palm oil (for edible oil and **biofuels** markets) provides a direct financial incentive to continue the conversion of tropical wetlands to agricultural uses.

While there is currently no financial incentive for wetland preservation or restoration in developing countries, and a range of technical accounting issues remain, the potential of this mitigation option remains significant. Until such time as wetland preservation is included as a creditable action under the international climate change regime, little of this mitigation potential is likely to be realized.

Afforestation and reforestation (A/R)

Establishment of forest is classified under the **Kyoto Protocol** as either reforestation, replanting trees on land that had been previously cleared, or afforestation, planting trees on land that was either not naturally forested or has not been forested for a period of at least 50 years. Afforestation and reforestation activities are the only measures that are recognized under the **CDM** and **JI**.

Around three-quarters of A/R activities are plantation forests for future wood and fiber harvesting for paper and construction timber or forest cash crops (such as palm oil and rubber). The remaining A/R activities largely involve revegetating areas for the purposes of soil and water conservation, combating desertification and habitat restoration. A/R activities can involve active tree planting and also include human-induced natural forest reestablishment on abandoned agricultural land (such as fencing to exclude livestock).

Lands currently supporting A/R activities are estimated to be removing between 500–1,000 MtCO$_2$ per year. The additional area being devoted to A/R activities each year is presently around 6 Mha, most of which is occurring in Europe (about 4 Mha per year) and China (which has average plantings of 1.3 Mha per year since 1990).[11] This is less than half the rate of deforestation, and the global net forest loss is still running at approximately 7 Mha per year, an area about the size of Panama. However, the estimated land area considered available for A/R is very large, about 1,000 Mha – an area the size of Canada. Most of this potential area consists of marginal, degraded, or unused agricultural lands, and more than half of this is located in Africa, South America, and the United States, but also other countries, such as Australia, have significant areas that could be available for A/R.

While some investments in A/R will occur for reasons other than emissions mitigation, to restore degraded watersheds, provide windbreaks, and erosion protection measures on agricultural land, or for future wood supplies, the existence of a market value for each tonne of CO$_2$ sequestered will drive large-scale plantings over the years ahead. Key constraints facing investments in A/R are the long periods required to achieve an attractive return on investment compared with other investment options (the investment opportunity cost) and the value of the land for other uses (the land-use opportunity cost).

As a greenhouse gas mitigation investment, A/R activities face a significant cash flow barrier. Most of the cost of establishing an A/R project is borne upfront at the start of the project (preparing land and planting, watering, and tending trees), whereas most of the carbon sequestration occurs many years, and in some cases decades, into the future. A/R projects can sequester CO_2 for 40–80 years, depending on the time trees take to reach maturity, and the total quantity of CO_2 sequestered can be large – thus, over the life of a project they generate substantial income flows. In the carbon markets, credits are only issued for CO_2 actually sequestered, and generally at least five years or more can elapse before the first credit payments are earned.[12] Maximum carbon sequestration per hectare, and hence annual carbon credit earnings, usually occurs between 10 and 25 years after planting (depending on trees species and growing conditions) – presenting discounted cash flow constraints for investors.

If investors can be assured of a long-term market for credits, and the price of credits is sufficiently high, investment returns can be attractive. However, this is not presently the case and A/R investors face considerable investment risk. For example, A/R is an eligible activity for **CDM** and **JI**, but the market for Kyoto credits is only assured up to 2012 and carbon credit prices in these markets have generally been below \$20 – this does not provide a sufficiently attractive investment horizon for A/R investments. As a result, A/R credits are expected to account for less than 1% of all carbon credits issued to 2012.[13] If, however, investors can be guaranteed a market for 30–50 years, A/R becomes a potentially attractive investment. It is very likely that a substantial market for carbon credits will exist after 2012, but what is not yet certain is how long into the future the market will be assured. Hence, the outcome of the current post-2012 climate negotiations will have a major bearing on the attractiveness of future A/R investments.

In terms of the land opportunity cost, there are a range of variables that affect the A/R investment outlook. Most of the available land for A/R activities is likely to be in the mid- to higher latitudes where significant amounts of marginal or surplus lands exist. At present the opportunity costs for these lands are relatively low, but over the coming decades this may change, particularly if global food prices continue to increase in real terms and the demand for **biofuels** increases as projected – both will increase the land-use opportunity cost. As A/R projects often require land to be locked up for 60 years or more, this could also represent a growing investment constraint for A/R projects. In the lower latitudes where most of the existing deforestation is occurring, the land-use opportunity costs are, at least in a relative sense, a significant barrier to A/R investments. Increasing demand for agricultural land, and attractive short-term returns from logging, are presently driving deforestation rates. These present a significant land-use opportunity cost barrier for A/R projects. In addition, higher investment risk levels in many developing countries necessitate a higher annual rate of return on investments, making investments in A/R projects relatively unattractive.[14]

Nonetheless, many studies suggest that A/R could make a reasonable contribution as a mitigation option over the coming decades. The maximum potential biosequestration from A/R is around 700 $GtCO_2$ by 2050, or 17 $GtCO_2$ per year, if linearly spread over the period – enough to offset about a third of all **anthropogenic greenhouse gas emissions** at current rates. Unfortunately, nothing like this level of biosequestration is, or is likely to be, achieved in the near future due to the long payback times and high capital investments required and other barriers such as competing land use (see **agriculture and food supply impacts**). Projections vary considerably from up to 1 G tCO_2 per year by 2030 using estimates of viable and institutionally available land areas to much lower levels using more detailed financial models.[15] Conceivable carbon prices in the range of $ 35–75/tCO_2 in 2010 deliver less than 0.5 G tCO_2 per year by 2030. Biosequestration of 1 G/tCO_2 per year by 2030 is only likely from A/R with costs of more than $ 300/tCO_2 in or soon after 2010 – an unrealistic price. Over the medium to long term A/R could potentially deliver substantial reductions using market mechanisms as market prices climb. Nonetheless, A/R is unlikely to deliver emission reductions of more than 3 $GtCO_2$ per year before 2050.[16]

Improved land management

Improved land management encompasses the remaining land biosequestration activities that increase land carbon stocks including improved land management such as reduced grazing and low till agriculture, and improved forest management. Approximately 50% of the global land carbon stock is stored in nonforest areas such as grasslands, tundra, scrub, and semiarid woodlands, and 80% of all land carbon is stored in the soil (see **land carbon sinks**). Any measures that retain or increase the quantity of carbon stored in soils and forests will contribute to reducing atmospheric greenhouse gas concentrations.

Over the past few centuries, vast quantities of carbon have been lost from the soil through vegetation clearance, overgrazing, and wind and water erosion. Ploughing and tilling accelerates the oxidation of soil organic matter and this is released to the atmosphere as CO_2. Soil carbon loss is generally more rapid than the slow process of soil carbon accumulation, and following vegetation clearance, some soils may lose much of their stored organic carbon in less than five years.[17]

Soil biosequestration manages land to maximize carbon accumulation and minimize oxidation. Activities that promote soil carbon retention or accumulation can include maintenance of continual plant cover, crop stubble retention, controlled irrigation and water management to avoid the loss of dissolved carbon in runoff, low or zero tillage, fencing off degraded grasslands to allow regeneration, establishing biodiversity conservation corridors, and many other practices. These activities also often contribute to increased soil fertility as organic carbon stocks increase, which can boost agricultural productivity.

Forest vegetation management can also be undertaken as a biosequestration activity. Traditionally, plantation forest management removed or suppressed the growth of other species and understory vegetation – often resulting in understory "deserts."

Tree branches are often removed up to 3–4 m above ground and debris and litter periodically removed or burned to facilitate machinery access. These practices result in substantial carbon loss from above ground vegetation and the underlying soil. However, forest management practices are changing and improving, tending to increase stored carbon per hectare. Improvements in satellite imagery and global information systems (for tracking forest growth) reduce the requirement for onground forest access; the introduction of less destructive logging machinery, reduced understory pruning and thinning, and selective logging have all reduced the levels of forest disturbance and increased carbon retention.

Wild fire suppression measures generally involve the periodic burning of forest debris to reduce fuel loads. It has been widely practiced as a means of reducing the risk to human settlements and valuable timber trees. However, more recently the value of traditional forest fire suppression measures has been questioned, and many now hold the view that allowing historic fire regimes to prevail actually improves forest health and productivity and, counterintuitively, carbon storage.[18] Fire suppression measures can result in forests dominated by younger, smaller trees and the loss of fire-tolerant species. When fires do take hold, they are more catastrophic "crown fires," damaging a forest such that it takes decades to recover, or may be permanently transformed to a different ecosystem, such as grassland, with a much lower carbon storage potential.[19]

While improved land management practices offer only small carbon sequestration benefits per hectare, if practiced over large areas, the aggregate biosequestration potential can be significant. Improved land management practices are included in **UNFCCC** national **greenhouse gas inventories**, and both Canada and the United States have reduced aggregate national emission levels significantly through land and forest management practices (see **land carbon sinks**). However, financial incentives for improved land management as a greenhouse gas mitigation measure are confined to Annex I countries that have ratified the **Kyoto Protocol** and are not an eligible activity under the **CDM**. This is due to similar technical constraints as those facing wetland management and avoided deforestation.

The potential for land management biosequestration activities is large, particularly in the developing world where significant tracts of degraded rangelands and degraded forests and woodlands exist. There are also a range of ancillary benefits that can accrue from improved land management, including increased agricultural productivity and rural incomes, water quality improvements, and biodiversity gains. The adoption of sustainable practices could deliver carbon sequestration gains over the coming years, irrespective of the existence of a financial incentive from the carbon market. However, the availability of carbon market finance would undoubtedly improve the prospects for land management biosequestration projects.

Future mitigation prospects

The medium-term mitigation potential from biosequestration measures is large. Some measures, especially avoided deforestation, offer immediate emission benefits, while others provide benefits over a longer period of time. However, issues such as permanence and carbon leakage, the general lack of financial incentives and long-term market assurance, and competing land-use factors means that only a small proportion of the biosequestration potential is likely to be realized in the short term.

In the medium term, the prospects are more promising. However, for biosequestration to substantially increase its mitigation contribution will require existing technical accounting and monitoring difficulties to be overcome, including the possible establishment of an agreed international land carbon baseline regime (supported by reliable remote sensing techniques); appropriate financial incentives and support to be made available through carbon markets and other measures; and for sustainable management practices to be more widely adopted. The outcomes of the post-2012 negotiations, particularly how avoided deforestation is handled, and how the underlying factors driving deforestation are managed, will have a major bearing on biosequestration mitigation. Grant funding mechanisms to keep forests and wetlands intact will be important if issues of monitoring, permanence, leakage as well as an equitable method of disbursement have yet to be overcome.[20] Nonetheless, many studies estimate that the contribution from biosequestration will increase over the medium term, though estimates vary from 1 GtCO$_2$ per year (less than 2% of current total **anthropogenic greenhouse gas emissions**) to as high as 5.5 GtCO$_2$ per year (about 11% of current emissions) by 2030.

In the longer term, the potential is much more speculative as key variables governing land carbon stocks could change significantly. A major issue that may emerge later this century is the impact of climate change on the land carbon sink. Climate change could cause the land system to transform from a net carbon sink to a net carbon source (see **land carbon sinks**).

The prospects of biosequestration making a significant contribution to "buy time" to avoid the worst **climate change impacts** are unlikely to be achieved under current policies and foreseeable carbon market prices. Valuing land purely on its carbon storage overlooks the enormous array of ecosystem services that are provided, uncosted, to society. These include direct, tangible benefits such as water and soil quality improvement; flood and drought attenuation through water storage; the provision of food, fiber, fuel and shelter; employment; and less tangible benefits such as biodiversity, recreational use, and a more robust environment that can facilitate **adaptation**. Some of these benefits are already included, and many are beginning to be costed and explicitly included in financial considerations of land management under the approach of "payment for ecosystem services."[21] Policy and market changes are needed to accelerate biosequestration rates. If payment for ecosystem services becomes a widely adopted policy, it could substantially change the economic attractiveness of biosequestration, particularly avoided deforestation and wetland conservation, and result in a much greater and much more rapid contribution to mitigating climate change.

Biosequestration offers considerable mitigation potential over the period to 2050 and, if the appropriate market signals and supporting frameworks are put in place, could plausibly contribute between 5% and 10% of the emission reductions required to stabilize atmospheric greenhouse gas concentrations (see **stabilization targets**).

See also: anthropogenic greenhouse gas emissions, biofuels, carbon capture and storage, carbon cycle, carbon sink, Clean Development Mechanism (CDM), Kyoto Protocol, land carbon sinks, mitigation.

Notes

1 FAO 2006
2 IPCC 2000a
3 Turnball 2007
4 Sathaye *et al.* 2006
5 IPCC 2000a
6 Hooijer *et al.* 2006
7 UNDP 2006a
8 Hooijer *et al.* 2006
9 Ibid.
10 Wetlands International 2007
11 FAO 2006
12 Schlamadinger *et al.* 2007
13 UNEP Risoe 2008
14 UNDP 2006a
15 IPCC 2001
16 Sohngen and Sedjo 2006
17 IPCC 2000a
18 Union of Concerned Scientists 2007
19 Kloor 2000
20 World Bank 2007a
21 UNEP/IUCN 2007

Further reading

IPCC 2000a; FAO 2006; Hooijer *et al.* 2006.

CARBON CAPTURE AND STORAGE

Carbon capture and storage (CCS) refers to the process of capturing **carbon dioxide (CO$_2$)** from large-scale emission point sources, for example, **fossil fuel**-powered electricity-generating stations or industrial plants, and depositing the gas in geological formations or the deep ocean for long-term storage. CCS can also include the conversion of CO$_2$ gas streams into stable mineral carbonate compounds by reacting CO$_2$ with magnesium or calcium oxides.

Global coal consumption is projected to increase by nearly 60% over the period 2005–2030, nearly all at large-scale stationary energy facilities.[1] Coal is plentiful and is generally the least-cost option for base-load electricity generation. Every week, on average, two new large coal power stations (each typically emitting two million tonnes of CO$_2$ a year) are commissioned around the globe, each with an operational life of 30–40 years. They are very large investments (of the order of US$ 1 billion), so once constructed they lock in a pattern of emissions that are difficult to avoid. CCS has been identified as a possible means of preventing most of these emissions from entering the atmosphere while enabling the continued use of coal and other **fossil fuels** for power generation.

The technology required to capture, transport, and store large quantities of CO_2 exists and is already commercially applied in the oil and gas industry. However, CCS technology has yet to be applied on a commercial basis to fossil fuel-based electricity-generating facilities, the main source of CO_2 emissions.

Advantages of CCS as a mitigation option are as follows:

- Fossil fuels could be consumed on a large scale while emitting far fewer **greenhouse gases** to the atmosphere.
- CCS technology can be integrated with existing electricity generation and distribution infrastructure.
- The technology has already been demonstrated to work in some applications.

The principal disadvantages of CCS are as follows:

- The technology has yet to be deployed on a commercial basis with fossil fuel power stations.
- It is likely to be relatively expensive and is unlikely to be adopted on a large scale unless a significant CO_2 cost penalty is introduced (see **carbon tax** and **emissions trading**), particularly in relation to retrofitting existing plants.
- It is a technology that has a relatively long time horizon before it could make a large contribution to mitigation efforts.
- The location of existing fossil fuel plants is often a considerable distance from suitable storage sites.
- The environmental risks of storing large quantities of CO_2 in geological and ocean reservoirs have yet to be fully assessed.

There are three principal stages involved in CCS: separating and capturing the CO_2, transporting it to a suitable storage location, and permanently storing it.

CO_2 capture

CO_2 capture and separation is a well-proven technology that has been used in the chemical industry for many years, mainly for the purpose of producing pure industrial gas streams (where CO_2 is considered an impurity). Anthropogenic CO_2 emissions are primarily generated through the combustion or transformation of fossil fuels and biomass. It can be removed either prior to combustion or by separation from the industrial gas stream after combustion. Precombustion removal is possible with some processes (for example, hydrogen production facilities) and with new, advanced coal-based generating technologies (e.g. Integrated Gasification Combined Cycle technologies – IGCC). In this process, coal is transformed into a synthetic gas that consists of CO_2 and hydrogen; the CO_2 is then removed from the synthetic gas stream prior to combustion. IGCC is a power generation technology that is well suited to CCS.

Postcombustion removal requires separation of CO_2 from other waste gases (mainly nitrogen), as CO_2 usually only accounts for 5–15% of the flue gas. Higher concentrations of CO_2 in flue gas streams can be generated by combusting the fossil fuel in a pure oxygen environment. Higher oxygen concentrations permit more efficient postcombustion CO_2 removal, which in turn substantially improves the economics of CCS, but such postcombustion systems are still at the research and development stage.

It is estimated that 85–95% of the CO_2 contained in either the pre- or postcombustion gas streams can be captured.[2] However, the capture and compression of CO_2 requires considerable energy. A standard pulverized coal power station fitted with CCS technology would consume 20–40% more energy per unit of electrical output than a similar plant without CCS (for combined-cycle natural gas and IGCC technology only 10–20% more energy is required).[3] Thus, a CCS plant produces more CO_2 for producing the same electricity output. The net effect is that CCS can reduce the amount of CO_2 per unit of electricity by approximately 80–90%. The additional fossil fuel use also increases other pollutants, such as sulphur and nitrogen compounds, and, in the case of coal, larger quantities of heavy metals and residual ash. Thus, whilst adoption of CCS technology can result in significant CO_2 mitigation benefits, facilities fitted with CCS would not be completely carbon neutral, nor would they reduce emissions of other environmental pollutants (except IGCC plants).

The cost of capturing CO_2 is dependent on the type of fossil fuel combustion technology used, fuel costs, the efficiency of the plant, and the cost of capital. Based on the most recent **IPCC** analysis, the cost per unit of electricity generated from standard pulverized coal plants fitted with CCS technology is expected to be 40–85% higher than the same plant without CCS. For other fossil fuel electricity-generating technologies, the cost penalty per unit is marginally lower (35–70% per unit for combined cycle gas technology and 20–50% per unit for IGCC). This equates to a cost of US$25–50 for each tonne of CO_2 avoided. Capture costs for steel and cement plants are estimated to be $25–115 per tonne CO_2. The least-cost option (at US$ 10–15/tonne) is associated with hydrogen gas plants fitted with CCS (which produce a high-concentration CO_2 gas stream), though these plants currently account for only 1% of stationary source emissions. Some suggest that further research and development could reduce capture costs by up to 50% over the next few decades.[4]

Transport

The second stage in the CCS cycle is the transportation of the captured CO_2 to the location where it is to be stored. For facilities that are located adjacent to a suitable storage site, the costs would be small, generally less than $1/tonne. However, for most facilities, the captured CO_2 would need to be transported some distance. For distances of up to 1,000–1,500 km, pipelines are expected to be the most cost-effective option, costing approximately $10–15/tonne of CO_2 transported. For long distances (greater than 1,500 km), the cheapest transport option is likely to be pressurized gas transport ships, at around $15–25/tonne CO_2. Rail or road transport is technically possible but would be much more expensive.

Storage

The third and final stage of the CCS cycle is long-term CO_2 storage. The three principal storage options are: (1) injection into suitable geological reservoirs, (2) storage in the deep ocean, and (3) conversion of CO_2 into stable mineral carbonates.

(1) Geological storage: Permanent geological storage sites potentially include deep saline aquifers, existing or depleted oil and gas reservoirs, suitable geological structures in sedimentary basins, and abandoned mines or coal seams that possess the right permeability and storage properties. Suitable gas injection technology is already in commercial use in the oil and gas industry. Estimates of the total geological storage capacity are very large. The capacity of depleted oil and gas reservoirs in close proximity (less than 250 km) to major emission sources is known to be at least 200 billion tonnes (gigatonnes – Gt) of CO_2, and possibly as much as 900 Gt. If deep saline aquifers (as currently used by the Sleipner gas project in Norway) are included, the potential storage capacity could exceed 2,000 Gt.[5] This would be sufficient to store all the CO_2 emissions produced by fossil fuel-powered electricity-generating stations (at current emission rates) for at least 150–200 years. The cost of injecting and storing CO_2 in geological structures is estimated to be US$1–8 (including monitoring and verification costs), depending on specific site characteristics.

(2) Ocean storage: The oceans are already a major CO_2 storage reservoir and they will eventually absorb most of the excess CO_2 in the atmosphere (see **ocean carbon sinks**). Deep ocean injection would, in effect, be a means of accelerating the natural **carbon cycle**, bypassing the atmospheric storage phase and thereby avoiding additional atmospheric warming through the **greenhouse effect**. If CO_2 is injected into the deep ocean (at depths greater than 1,000 m), prevailing temperature and pressure, combined with the slow turnover rate of the surface and deep ocean waters (see **thermohaline**), would mean that CO_2 could potentially be withheld from the atmosphere for many centuries. However, ocean storage is considered less permanent than geological storage.

The cost of ocean storage is dependent on the distance offshore, the depth of injection, and whether it is deposited through subsea pipelines or via ships. Cost estimates range from $6 to $30/tonne CO_2 but remain uncertain as it has yet to be implemented on a commercial scale.[6]

(3) Storage in mineral carbonates: The final storage option is to chemically react CO_2 with calcium oxide or magnesium oxide to form calcium carbonate ($CaCO_3$ – also known as limestone) or magnesium carbonate ($MgCO_3$). Carbonates are very stable over long periods of time and are considered to have a zero leakage rate.

There are sufficient supplies of extractable oxides in silicate rocks to "fix" all the CO_2 that could potentially be generated from the earth's total stock of fossil fuels. However, conversion to mineral carbonates is a much more energy-intensive process than geological or ocean storage, requiring some 60–180% more energy than non-CCS electricity generation. It would also produce 2.5–4 tonnes of carbonate material for each tonne of CO_2 captured. If implemented on a large scale, it would require the disposal of very large quantities of material. Disposal sites could include land fill or abandoned mines. Small quantities could also be used as inputs to other industrial processes, such as the production of construction materials.

Overall, the costs of conversion to mineral carbonates are estimated at US$ 50–100/tonne of CO_2 captured.[7] The high cost is attributable mainly to the significant amounts of additional energy required for the process. While mineral carbonation may be economically viable for certain niche applications, the higher cost suggests that it is unlikely to become a major CSS option.

Permanence

The risks of leakage from well-managed geological storage sites are considered low, probably less than 1% over periods of up to 1,000 years. There may be some risk associated with earthquakes or failures of capped injection holes, but existing oil and gas industry technologies for managing well blow-outs could be applied in these situations to minimize CO_2 leakage. Any CCS site would require stringent testing and analysis prior to storage approval and would also need long-term management, monitoring, and verification procedures. The costs associated with monitoring and verification are estimated to be less than US$ 1/tonne CO_2.

The risks of leakage from ocean storage are much more significant. Over time periods spanning several centuries, deep ocean waters mix with surface waters. If CO_2-rich deep-ocean water were to reach the surface, stored CO_2 would be released to the atmosphere. The IPCC has calculated that for CO_2 injected at a depth of 1,500 m, 25% of the stored CO_2 could subsequently be released to the atmosphere within two centuries, and as much as 50% in four centuries. At 3,000 m, the leakage rate is estimated to be 20% over four centuries.

Environmental risks

Carbon dioxide is not flammable and only becomes a health risk (from asphyxiation) when air concentrations exceed 10% by volume. However, there could be local environmental hazards associated with sudden large-scale releases, particularly from ruptured pipelines or well blow-outs. These risks would be similar to those associated with the transport and storage of other industrial gases.

With ocean storage, there are potential adverse ecological impacts. CO_2 is a slightly acidic gas, and injecting large volumes into the oceans would result in an increase in ocean acidity, particularly in the vicinity of injection sites. Ocean acidification is a potentially serious adverse side effect (see **marine impacts**). Deep ocean storage could accelerate ocean acidification. Although considerable uncertainties remain about the degree of acidification that might occur, and the associated ecological impacts, negative public perceptions may nonetheless limit the attractiveness of the ocean storage option.

What are the future prospects for CCS?

The principal factor constraining the rollout of CCS technology is cost. At present, under the best case scenario, the total estimated cost of CCS (including capture, transport, and storage) is in the range US$30–70/tonne CO_2 for standard pulverized coal plants, US$ 40–90/tonne for combined-cycle natural gas plants and US$25–50/tonne for plants utilizing IGCC technology. Retrofitting old plants is technically feasible but more expensive. The additional cost per kilowatt hour for electricity generated by CCS-equipped facilities ranges from US 2 to US 5 cents. This would raise fossil fuel electricity costs to levels equal to or greater than several other alternative electricity-generating options, such as **renewable energy**. However, if CCS is combined with enhanced oil recovery (EOR), its economic attractiveness could be considerably improved. EOR involves injecting gas or other liquids into oil and gas reservoirs to force out additional oil and gas and is a practice widely employed in the petroleum industry. Combined CSS/EOR projects could reduce net costs to below US\$10/tonne of CO_2 stored in some applications due to additional revenues from increased oil and gas production.[8]

For CCS to mitigate 10% (or 2.5 $GtCO_2$) of projected emissions from stationary sources in 2030, 1,250 CCS-equipped power stations would have to be built, each one capturing 2.0 million tonnes of CO_2 per year. Given the current state of technology and the costs of CCS technology, it is difficult to foresee such large-scale rollout in the next few decades.

Recent long-term energy modeling projections suggest that CCS will deliver as much as 20% of the required reduction in emissions to return emissions to 2004 levels by 2050.[9] For this to eventuate, there would need to be the introduction of a cost penalty on CO_2 emissions,[10] significant technology advances that reduce costs, and the adoption of proactive government policies. Studies[11, 12] conclude that without CO_2 cost penalties (of at least US\$25–50/tonne CO_2), or the introduction of regulatory requirements for CCS technology to be fitted, CCS is unlikely to be adopted on a large scale. There may be some uptake in certain niches (particularly when combined with EOR projects), but these are not expected to make a significant contribution to global emission reductions. With further research and development and "learning by doing," the costs of CCS are expected to fall. This will increase the cost competitiveness of CCS as a mitigation option over the coming decades. Until then, conventional pulverized coal technologies (without CCS technology) are likely to remain the technology of choice for new fossil-fueled power stations.

There are also legal issues (such as long-term storage liability provisions) and greenhouse gas accounting rules (e.g. relating to cross-border transport and storage) that need to be resolved prior to the large-scale adoption of CCS. While these issues and rules are unlikely to create insurmountable obstacles, they will need to be formulated and agreed upon before large-scale deployment can occur.

Overall, carbon capture and storage is a relatively expensive mitigation option with a medium- to long-term horizon in terms of making a substantial contribution to greenhouse gas mitigation. If fossil fuel power generation is to remain the primary means of supplying electricity over the next several decades, as is projected, then carbon capture and storage will most likely need to play an important part in the mitigation strategies of some countries.

See also: carbon dioxide (CO_2), fossil fuels, marine impacts, nuclear power, ocean carbon sinks, renewable energy, thermohaline.

Notes

1 IEA 2006a
2 IPCC 2005
3 Ibid.

4 IEA 2004
5 IPCC 2005
6 Ibid.
7 Ibid.
8 Ibid.
9 IEA 2006a
10 IEA 2004
11 Ibid.
12 IPCC 2005

Further reading

IPCC 2005; IEA 2004; *Scientific American* 2006.

CARBON CYCLE

The **carbon cycle** refers to the processes by which carbon moves between the atmosphere, the terrestrial (land) system, and the oceans. Carbon is constantly moving between the three active *reservoirs*, and these exchanges are called *carbon fluxes*. Understanding how the carbon cycle works, and how these fluxes influence **carbon dioxide (CO_2)** concentrations in the atmosphere, is essential to understanding how **anthropogenic greenhouse gas emissions** influence the global climate. Reservoirs that absorb more CO_2 from the atmosphere than they emit to the atmosphere are termed **carbon sinks**.

In total around 41,000 billion tonnes (or gigatonnes – Gt) of carbon (C) are available for exchange between the three principal reservoirs. The major reservoir of carbon is the ocean, which is estimated to contain around 38,000 GtC, or 93% of all exchangeable carbon. The ocean can be further subdivided into the surface ocean (down to about 100 m), which contains around 1,000 Gt C, and the deep ocean, which contains the remaining 37,000 GtC. The land carbon reservoir is estimated to contain just over 2,000 GtC, about 5% of exchangeable carbon. Of this, approximately 30% is stored in vegetation and other living organisms and the remainder in the soil and detritus. The atmosphere is the smallest of the three active reservoirs and is estimated, at present, to contain around 800 GtC, roughly 2% of exchangeable carbon.

There is also a vast reservoir of geological carbon (20,000,000 Gt) stored in the earth's crust, mainly as carbonate rocks. Of this, a small fraction (about 5,000 Gt) is stored as **fossil fuels** (coal, oil, and natural gas) and **methane hydrates** (5,000–10,000 Gt). Natural annual carbon fluxes into and out of the geological reservoir are tiny (less than 0.1% of the cycled carbon), so geological carbon is not generally considered part of the active carbon cycle.

Human impacts on the carbon cycle

Historical records show that the atmospheric carbon reservoir had been relatively stable at around 600 Gt and atmospheric CO_2 concentrations at around 280 parts per million (ppm), for at least the last 10,000 years. Until 1930, concentrations had not exceeded 300 ppm for at least 800,000 years. However, in the two centuries to 2000, combustion of fossil fuels has released around 285 Gt of geological carbon to the atmosphere.[1] Humans have also been responsible for the release of an additional 180–200 Gt from the land reservoir to the atmosphere through clearing forests and other land-use changes. These releases have created an imbalance between the three active carbon reservoirs, with much more contained in the atmosphere than when the carbon cycle is in equilibrium.

Given these large inputs of carbon to the atmosphere, basic arithmetic suggests that the atmospheric carbon reservoir should be much larger than it actually is. However, only about half of these emissions have remained in the atmosphere, the rest has been absorbed either by the oceans or the terrestrial land system through the action of the carbon cycle. The carbon cycle has been working to restore equilibrium between the reservoirs, but at only half the rate humans are adding carbon to the atmosphere. As a result, the amount of carbon in the atmosphere continues to grow, presently at a rate of around 4 GtC per year (15 $GtCO_2$), or 0.5%. By 2007, atmospheric CO_2 concentrations had reached 382 ppm, around one-third higher than preindustrial levels.[2]

The carbon fluxes

The ocean and land reservoirs both emit and absorb large quantities of CO_2. The difference between these opposing fluxes, plus emissions from human activities, determines the net annual addition of carbon (in the form of CO_2) to the atmosphere. Uncertainty still remains regarding the magnitude of annual carbon fluxes between the different reservoirs. This is due to natural variations in climatic conditions from year to year, gaps in scientific knowledge of the processes at work, and the physical constraints of actually measuring the different fluxes. For this reason, carbon flux estimates are usually presented as a range rather than exact figures. There is even greater uncertainty over how these fluxes will change over the coming decades in response to increased CO_2 concentrations and global warming.

The two dominant natural carbon fluxes are the land–atmosphere exchange (around 110–120 Gt per year) and the ocean–atmosphere exchange (around 90–100 Gt per year). There is also a small flux from the land to the oceans (around 1 Gt per year), from weathering, erosion, and the transport of dissolved organic carbon in rivers. These balanced natural fluxes are nearly 20 times greater than those associated with human activities. Although a molecule of CO_2, once emitted to the atmosphere, may stay there for a hundred years or more, CO_2 is cycled relatively quickly between the reservoirs. It is this constant recycling process that underpins the existence of living organisms. The carbon in our bodies has, over the years, been cycled through many different forms, both living and nonliving. The CO_2 molecule that was emitted from a car 50 years ago may actually have been absorbed by the apple tree (through photosynthesis) to produce an apple that was consumed today and released back to the atmosphere as CO_2 (through respiration). Alternatively it could have been absorbed by the ocean and be heading toward the deep oceans where it may remain for a thousand years or more. There is even a small chance that it may have been deposited in ocean sediments and be locked up for millions of years.

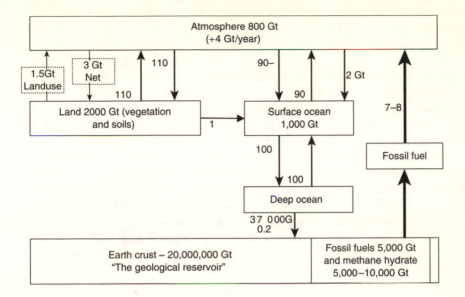

Figure 1 The carbon cycle

Role of the oceans

The oceans, as the largest reservoir of exchangeable carbon, play the most important role in regulating the carbon cycle and the earth's climate. It is the oceans that ultimately determine atmospheric CO_2 concentrations. The ocean carbon fluxes vary considerably over time and between different locations. The oceans are a major *source* of CO_2 to the atmosphere in some regions (e.g. the Equatorial Pacific) and in other areas a *sink* (e.g. the North Sea).

The ocean flux is dependent on the solubility of CO_2 in seawater, biological activity, kinetic forcing by the wind, and temperature. The main driver of CO_2 gas exchange between the ocean surface and the atmosphere is the difference between the partial pressure of CO_2 in seawater and that in the atmosphere. The partial pressure of CO_2 in seawater is influenced by temperature (it falls as it gets colder), the amount of dissolved inorganic carbon (it rises as the concentration of dissolved inorganic carbon increases), and alkalinity (it rises as alkalinity decreases).

There is considerable variability in partial pressure across the ocean at different times. At any one time, ocean partial pressure will be higher in some places and lower in others. When the partial pressure of CO_2 in the surface ocean is lower than that in the atmosphere, CO_2 is absorbed (sink). When the partial pressure in the ocean is higher, CO_2 is released (source). When averaged over the globe, the partial pressure of CO_2 in the surface ocean is presently less than the atmosphere. As a result, the oceans are a net carbon sink. The partial pressure of CO_2 in the atmosphere has increased since preindustrial times due to increased atmospheric concentrations.

Atmosphere–ocean fluxes are also influenced by the wind and physical disturbance and, most important, by biological activity through photosynthesis and respiration. The prevailing wind regime has a major influence on ocean–atmosphere CO_2 exchanges, and fluxes vary significantly across different regions. Changes in mean average wind speed of ± 2 meters per second can affect the CO_2 flux by $\pm 25\%$.[3] With improved remote sensing data and better wind speed statistics, the ability to estimate annual fluxes will also improve.

Estimates of the magnitude of the ocean carbon sink, or *negative carbon flux* to the atmosphere, range from 1.4 to 2.6 GtC per year,[4] and some estimates are even higher, but is generally considered to be in the order of 2 GtC per year (7 GtCO_2).

Nearly all (around 98%) of the exchangeable carbon in the ocean is stored as Dissolved Inorganic Carbon (DIC). A small but important fraction (2%) is stored as Dissolved Organic Carbon (DOC), or in other forms of organic matter. The size of the dissolved organic carbon reservoir is estimated to be around 680–700 Gt, or slightly less than the quantity of carbon in the atmosphere.[5]

DIC can be in three inorganic forms: dissolved CO_2 (carbonic acid), bicarbonate ions, or carbonate ions. Carbonic acid is relatively short lived and is quickly converted to bicarbonate and carbonate ions. The concentration of DIC increases with depth and is about 15% higher in the deep ocean waters than those at the surface. This gradient is maintained by the operation of the two major mechanisms in the ocean, namely, the "biological pump" and the "solubility pump." It is these two mechanisms that underpin the ocean carbon cycle and determine how much CO_2 is removed from the atmosphere.

The solubility pump

The solubility of CO_2 in water is related to the temperature of the water. CO_2 is twice as soluble in cold polar waters than it is in warm equatorial waters. As the ocean currents move warm waters from the equator to the poles they cool and the ability to absorb more CO_2 increases. As water becomes colder, it also becomes denser (heavier), and eventually it sinks into the deep oceans and carries the dissolved CO_2 with it. It is the difference in water temperature and salinity (see **thermohaline**) that drives the circulation of carbon through the ocean, operating like a giant carbon conveyor belt, playing an essential role in the carbon cycle. The **thermohaline** will eventually return this CO_2 to the atmosphere through the upwelling of deep ocean waters rich in DIC, but this can take a thousand years or more. The solubility pump accounts for around one-third of the transport of DIC to the deep ocean.

The biological pump

The biological pump has an important role in absorbing and distributing carbon throughout the ocean and is estimated to account for just over two-thirds of the export of carbon to the deep oceans. Through the biological pump, inorganic carbon is converted to organic carbon compounds through photosynthesis by marine organisms (phytoplankton), just as land plants use photosynthesis to convert CO_2 to organic compounds. The primary production from phytoplankton photosynthesis forms the basis of the marine food web, on which the marine ecosystem depends.

Phytoplankton is either consumed by zooplankton (microscopic marine organisms), which can also be consumed by larger zooplankton, and so on down the food chain, or sinks into the ocean as Dissolved Organic Matter (DOM) or other organic carbon solids. A significant proportion of these organic compounds (70–80%) are recycled in the surface ocean relatively quickly, and some are rereleased as CO_2 back into the atmosphere through respiration of marine organisms. Some of the carbon is also used to form the shells and bones of these organisms (which sink when they die) or is excreted in the wastes (which also sink). Only 20–30% of the organic carbon actually moves from the surface oceans into the deeper ocean, where most is remineralized back to inorganic forms by marine bacteria. Only a very small proportion (less than 1%) actually settles on the ocean floor to be locked away in sediments. Some of this carbon may eventually end up as fossil fuel over millions of years. It is through this process that deposits of oil and gas were originally formed.[6]

Ocean primary production

The level of primary production depends on the amount of sunlight, water temperature, availability of inorganic carbon, and other essential nutrients (mainly the macronutrients nitrogen and phosphorous, and also the micronutrient iron). Ocean primary production shows strong seasonal variation. For example, in the North Atlantic, there are large inflows of nutrient rich deep ocean water during the winter months, and when increased sunlight warms the waters in spring, there is a significant increase in biological activity (manifesting themselves as phytoplankton blooms). These intense periods of biological activity occur in many regions of the world's oceans. Net global production of dissolved organic carbon in the ocean is estimated to be about 1.2 GtC per year.[7] Estimates of the percentage of primary production that is exported to deeper waters range from 10% to 50%, and it appears to be more efficient in cold waters.[8]

The availability of nutrients has a major bearing on the level of biological activity in the oceans. Nitrogen and phosphorous are essential nutrients for biological processes, and their cycles in the ocean are also closely linked with the carbon cycle. In the middle ocean (200–1000 m), sometimes called the "twilight zone," inorganic nutrients (like nitrate and phosphate) are regenerated from organic matter sinking from above. Nitrogen fixing *diazotrophs* are important in this process and provide a valuable source of nitrates that feeds surface ocean biological activity. Scientists have found that the micronutrient iron (mainly supplied through atmospheric dust deposition), rather than just the macronutrients nitrogen and phosphorous, is also a major determinant of primary production. It is believed that primary production in up to half of the world's oceans could be limited by lack of iron (see **ocean carbon sinks**).[9]

The ocean carbon cycle is also influenced by prevailing climate conditions. The **El Nino Southern Oscillation (ENSO)** varies over multiyear cycles and can influence ocean biogeochemistry and the amount of CO_2 that is exchanged with the atmosphere. This, in turn, influences the rate of CO_2 uptake and mixing of nutrient rich deep ocean waters. In strong ENSO conditions, global ocean primary production is generally lower. When the ENSO phase weakens, there can be large increases in ocean primary production. For example, during the transition from the strong

1997–1998 ENSO, there was estimated to be a 10% increase in global ocean primary production.[10] Upwelling of deep ocean water rich in DIC also results in outgassing of CO_2 to the atmosphere. In the equatorial Pacific, about 3.7 $GtCO_2$ per year is released in this way.

CO_2 uptake and release zones

At different times of the year, the same area of the ocean can take up or release CO_2. For example, the northern temperate regions of the Atlantic and Pacific oceans release CO_2 during the northern winter and take it up in summer. The Equatorial Pacific and the Arabian Sea (particularly during the strong southwest monsoon season) are both major sources of CO_2 to the atmosphere.

The major uptake areas are the high-latitude areas of the northern oceans in summer and the Southern Ocean and south Atlantic in the southern hemisphere summer. This uptake is linked to biological activity. The changes in the partial pressure of CO_2 in the higher latitudes are influenced by deep water upwelling in winter and biological uptake in spring and summer, whereas in the subtropical and temperate oceans, it is largely influenced by water temperatures.[11]

The Atlantic Ocean accounts for more than 40% of CO_2 uptake but occupies only 23% of ocean surface.[12] The highest single uptake region is the North Atlantic (especially the North Sea), which accounts for 20% of the net ocean uptake of CO_2 but only 5% of the ocean surface. This is due to a set of specific characteristics in this region, such as surface water stratification and the intense biological primary production in the summer. The Southern oceans are also important uptake zones, particularly the Southern Indian Ocean, which accounts for around 15% of ocean uptake. In contrast, the Pacific Ocean as a whole takes up the smallest amount (18%), yet it accounts for 49% of the ocean surface.[13] This is because the uptake of CO_2 in the mid- to high latitudes is partly outweighed by CO_2 release in the equatorial Pacific.

The slow rate of circulation of the ocean waters means that the ability of the oceans to absorb atmospheric CO_2 is time constrained. If humans stopped emitting CO_2 to the atmosphere, eventually most (80–85%), but not all, of the anthropogenic CO_2 will end up in the oceans as bicarbonate ions. This would take at least 500 years after anthropogenic emissions ceased, but possibly much longer (see **ocean carbon sinks**).

Role of the land

The terrestrial system (the land) plays an essential role in the carbon cycle. The land absorbs CO_2 from the atmosphere through photosynthesis by plants, and the resulting primary production forms the basis of the land food web on which living organisms depend. Through the food chain, carbon is cycled and released back into the atmosphere directly through respiration (as CO_2) of the many different living organisms, or through respiration of other organisms as part of decay. The balance between these processes determines whether the terrestrial land system is a CO_2 sink or source.

There is some uncertainty over the magnitude of the land carbon sink. Estimates of net uptake range from 6 to 14 GtCO$_2$ per year, but normally half of this is offset by ongoing deforestation and land-use change. Each year an estimated 3–9 GtCO$_2$ is released to the atmosphere through these activities.[14] The *net flux* between the land and the atmosphere can vary considerably from year to year due to other climatic variables like the **ENSO**, temperature changes, and the distribution of rainfall. Plant growth has also been stimulated by anthropogenic emissions of nitrogen from industrial and agricultural activities. Fires can also result in sudden releases of CO$_2$ from the land to the atmosphere.

The land contains about three times as much carbon as the atmosphere, about one-third of which is contained in living organisms (biota) and the remainder in the soils and dead organic matter. The forests, and the soils in which they grow, account for about two-thirds of the land carbon reservoir, while the remaining third is stored in grasslands, wetlands, and tundra, mainly in the soils. Unlike the oceans, which respond relatively slowly to changes in temperature and atmospheric CO$_2$ concentrations, the land–atmosphere carbon cycle is much more dynamic. Some carbon can be locked up in soils and vegetation (like the stems of trees) for centuries, but generally most of the carbon is cycled relatively quickly through the land system.

Recent research on outgassing of CO$_2$ from Amazonian rivers suggests that most of the carbon originating from the Amazon Basin is less than five years old, which suggests that carbon is cycled through tropical systems more quickly than previously believed, possibly limiting their long-term storage capability.[15] Nonetheless, tropical forests remain an important sink. The uptake of CO$_2$ in the temperate and boreal forest regions has long been considered an important sink, but recent evidence suggests that the uptake of CO$_2$ in these regions may be less than previously thought (see **land carbon sinks**).[16]

Humans, through land-use change and forestry practices, have had a major influence on the land carbon cycle and the size of the land carbon reservoir. Over the period 1750–2000, land-use change and forestry activities accounted for around 40% of anthropogenic CO$_2$ emitted to the atmosphere.[17] Due to our ability to influence the uptake and storage of carbon by the land, for example, through revegetation activities and changed agricultural practices (see **biosequestration**), humans have greater ability to control the land carbon cycle than the ocean carbon cycle.

The land–atmosphere flux is known to be very sensitive to changes in climate and atmospheric concentrations of CO$_2$. Over the course of this century, as global temperatures and atmospheric concentrations increase, the net uptake of CO$_2$ by the land is expected to reduce. Eventually the land could transit from being a net CO$_2$ sink to a net source of CO$_2$, possibly before the end of this century (see **land carbon sinks**). If the land does make the transition from *sink* to *source*, this would magnify the current imbalance in the carbon cycle and accelerate, rather than reduce, the rate of climate change (see **climate change feedbacks**).

See also: anthropogenic greenhouse gas emissions, carbon dioxide (CO$_2$), carbon sinks, El Nino Southern Oscillation (ENSO), fossil fuels, land carbon sink, ocean carbon sink.

Notes

1 Prentice 2001
2 NOAA 2007
3 Feely *et al.* 2001
4 Ibid.
5 Hansell and Carlson 2001
6 Ducklow *et al.* 2001
7 Hansell and Carlson 2001
8 Doney *et al.* 2001
9 Ibid.
10 Behrenfeld *et al.* 2001
11 Feely *et al.* 2001
12 Ibid.
13 Ibid.
14 Prentice 2001
15 Magoya *et al.* 2005
16 Stephens et al. 2007
17 Houghton 2003

Further reading

IPCC 2007; Houghton 2003; Kasting 1998.

CARBON DIOXIDE (CO_2)

Carbon dioxide is a colorless, odorless, faintly acidic gas that consists of one carbon atom double bonded to two oxygen atoms. It is commonly referred to by its molecular formula, CO_2. At standard atmospheric temperature and pressure, it has a density 1.5 times that of air. CO_2 is a very stable, nonreactive, nonflammable gas.

CO_2, like other **greenhouse gases**, is only present in the atmosphere in tiny quantities. By volume, CO_2 accounts for less than 0.04% of all gases in the atmosphere. Relative to nitrogen (78%) and oxygen (21%), its share is so small that the standard measurement used is parts per million (ppm). By 2007, the concentration of CO_2 in the atmosphere had reached 382 ppm.[1]

CO_2 plays an essential role in the maintenance of the earth's biosphere. Living organisms absorb CO_2 from the atmosphere through photosynthesis and release it back to the atmosphere through respiration. The operation of the **carbon cycle** ensures that CO_2 is constantly cycled between the atmosphere, terrestrial land system, and the oceans (see **carbon cycle**).

CO_2 is an important greenhouse gas. It has a **global warming potential (GWP)** of one (GWP = 1) and is the standard unit against which the warming potential of all other greenhouse gases is measured. Although the least powerful of the three principal greenhouse gases (CO_2, **methane** and **nitrous oxide)**, it is present in much larger quantities than the other two gases. Consequently, it accounts for around two-thirds of the cumulative atmospheric **radiative forcing** experienced to date.

CO$_2$ is assumed to have an *atmospheric residence time* of approximately 100 years. In other words, a molecule of CO$_2$ emitted to the atmosphere would, on average, take 100 years to be recycled to either the land or the ocean carbon reservoirs (via the **carbon cycle**). While the IPCC uses 100 years as the standard benchmark residence time, the IPCC states CO$_2$ can reside in the atmosphere between 5 and 200 years.

Major CO$_2$ sources

CO$_2$ emissions to the atmosphere can emanate from natural or anthropogenic (man-made) sources. Natural emissions are 20 times greater than anthropogenic emissions but are largely counterbalanced by natural uptake by the oceans and the land (see **carbon cycle**). The main natural sources are from the respiration of living organisms, oxidation of organic carbon, outgassing of carbon dioxide from the oceans, and volcanic activity. Anthropogenic emissions emanate primarily from the production and consumption of **fossil fuels** (74%), land use, land-use change and forestry (LULUCF – 24%), and cement production (2%).[2] Scientists can trace emissions to anthropogenic sources through the analysis of carbon "signature" isotopes.

Anthropogenic emissions have resulted in atmospheric CO$_2$ concentrations rising by approximately one-third since preindustrial times and is the main contributor to the **global warming** experienced over the past century. It is the *concentration* of CO$_2$ in the atmosphere, rather than the volume of emissions per se, that is the primary factor of relevance to global warming.

Past concentrations

Continuous, precise measurements of CO$_2$ have only been recorded since 1958, when Charles Keeling commenced readings at the Mauna Loa observatory in Hawaii. While readings are taken in many different locations around the globe, the Mauna Loa measurements are the official reference indicator of average CO$_2$ atmospheric concentrations. Mauna Loa's altitude (over 3,000 m) and isolation from major CO$_2$ emission sources ensures a consistent supply of well-mixed air.

Estimates of atmospheric concentrations prior to 1958 rely on a range of different methods. These include the study of sediments, fossils, and ice cores, with the latter providing the most reliable long-term estimates. By drilling into the Antarctic and Greenland ice caps, scientists have been able to retrieve ice core samples that date back more than 800,000 years. Each year, new snow accumulates and traps tiny air bubbles in successive layers. By analyzing the air in these bubbles, scientists can determine the historical concentrations of CO$_2$, methane, and nitrous oxide with a reasonable degree of accuracy.

Ice core records suggest that CO$_2$ concentrations over the past 800,000 years have been as low as 180 ppm in glacial periods (ice ages) and as high as 300 ppm at the peak of interglacial periods. The historical average concentration over the entire period up to 1750 was 220 ppm. At the beginning of the industrial revolution, atmospheric concentrations of CO$_2$ stood at 280 ppm. This is because the earth has been in a relatively warm interglacial period over the past 10,000 years (referred to as the Holocene period). The highest preindustrial CO$_2$ concentration is estimated to have

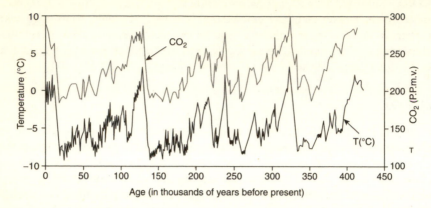

Figure 2 Climate and atmospheric history of the past 400,000 years from the Vostok ice core

Source: http://www.livescience.com/environment

been 300 ppm, some 330,000 years ago. This level was exceeded for the first time around 1930. CO_2 concentrations have since increased rapidly.

Figure 2 shows the evolution of atmospheric CO_2 concentration over the past 400,000 years – as suggested by an ice core extracted at the Vostok research station in Antarctica – and relates this to temperature changes over the same period. It is obvious from the chart that there is a close correlation between changes in CO_2 concentration and global temperature.

Recent concentration changes

The industrial revolution heralded the dawn of the fossil fuel era and is often considered to be the point at which humans first started to have a noticeable influence on the concentration of CO_2 in the atmosphere. Prior to 1750, CO_2 concentrations are estimated to have fluctuated between 270 and 290 ppm for the previous 10,000 years, with a midpoint of 280 ppm. By 2007, CO_2 concentrations had reached 382 ppm. This concentration level is the benchmark against which the impact of anthropogenic emissions is usually assessed. Table 6 indicates the change in concentrations over the period 1750–2007.

Table 6 Change in atmospheric CO_2 concentrations

Years	Concentration at end of period	Rate of increase (ppm/decade)
Pre-1750	280	—
1750–1900	290	0.7
1900–1930	300	3.33
1930–1958	315	5.44
1958–1975	331	9.4
1975–2000	369	15
2000–2007	382	25

At first, anthropogenic CO$_2$ emissions had only a very small influence on atmospheric concentrations, but during the twentieth century, the annual rate of increase grew dramatically. In just the first five years of this century, atmospheric concentrations increased by the same amount as the first 150 years of the Industrial Revolution – an astonishing rate of increase.[3]

Annual fluctuations in concentrations

It is only since 1958 that scientists had been able to observe changes in concentrations over the course of each year. Plotting changes over time yields what is known as the "Keeling Curve" (see Figure 3). The curve reveals three important observations.

First, every single year since 1958 has seen an increase in atmospheric concentrations over the previous year, though there has been some variation from year to year in the magnitude of the increase. Annual variations are often attributable to changes in **ENSO** cycles, the incidence of major fires, and volcanic activity.

Second, the slope of the curve measuring the annual average concentration values has steepened over time, indicating that the annual rate of increase is accelerating.

Third, the Keeling Curve is not a single monotonic trend but, instead, fluctuates up and down according to a regular cycle, varying by around 3% (or 7 ppm) each year. The reason for this annual fluctuation is the seasonal variation in the amount of photosynthesis by plants (which absorb CO$_2$ from the atmosphere) and in respiration by living organisms and decomposition of organic matter (which both release CO$_2$ into the atmosphere). The Mauna Loa observations relate to the northern hemisphere, but a similar, less pronounced, pattern is also found in the southern hemisphere. The oscillations are larger in the northern hemisphere because this is where most of the earth's soil and vegetation is located, and hence where most photosynthesis and respiration takes place.

In spring and summer, when plants are growing vigorously, CO$_2$ uptake is greater than that released through respiration, so atmospheric concentrations fall (with the minimum in October – the end of the growing season in the northern hemisphere). In autumn and winter, when many plants shed their leaves and cease growing, the release through respiration and decomposition of organic matter exceeds the uptake by plants, so concentrations increase (the maximum concentration is reached around May – the start of the northern hemisphere growing season). These variations are averaged out for the year to derive a mean annual CO$_2$ concentration. It is the trend in the mean that is important from a global warming perspective.

Future CO$_2$ concentrations

The three key determining variables of future CO$_2$ concentrations will be the quantity of anthropogenic CO$_2$ emissions (mainly the level of **fossil fuel** consumption and deforestation), the response of the **land carbon sink**, and the response of the **ocean carbon sink**.

If we assume that anthropogenic emissions remain constant at 2007 levels, and that the uptake of CO$_2$ by the land and ocean remains constant, then atmospheric concentrations would most likely increase, on average, by around 2.0–2.5 ppm per

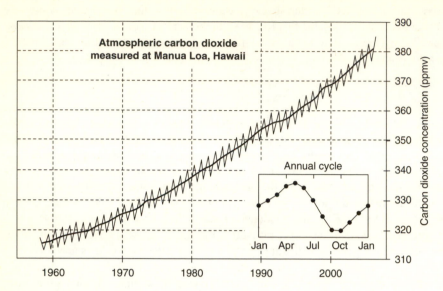

Figure 3 The Keeling curve

Source: http://www.globalwarmingart.com

year over the coming years (25ppm/decade). This would mean that concentrations would reach 475–500 ppm by mid-century and 575–650 ppm by 2100. However, given existing trends and projections of future fossil fuel consumption and defor-estation, this is an unrealistic assumption (see **future emissions trends**).

It is more likely that the annual rate of growth of CO$_2$ concentrations will continue to increase for at least the next few decades. This is due largely to the underlying momentum of the global economic system, driven primarily by population increase and economic growth (see **future emissions trends**). It also takes time to replace the existing stock of fossil fuel-consuming technologies (e.g. coal-fired power stations have a life of 30–40 years and a new station is commissioned every few days). One could liken it to trying to stop a giant oil tanker – once the decision to turn off the engine and bring it to a stop is made, it takes a very long time for this to actually happen. For this reason, the world is, to a large extent, locked into a given emissions trajectory for at least the next 10–20 years. Nonetheless, given time and appropriate action by governments, emissions could eventually stabilize, and possibly fall.

Based on the most recent International Energy Agency (IEA) projections, by 2030, CO$_2$ emissions from fossil fuels are expected to be 40–60% above 2005 emissions. These projections translate into an additional atmospheric CO$_2$ loading of 15–18 Gt/year above present levels by 2030. All other things being equal (no changes in land-use emissions and the uptake by the land and oceans remaining constant – see **carbon sinks**), this may result in an additional 1.0–1.5 ppm per year on top of the current 2.5 ppm annual growth already being experienced (3.5–4 ppm/year overall). This means that, without a fundamental shift in policies, CO_2 concentrations are likely to reach at least 500 ppm, and probably 550 ppm, by 2050.

Predicting what will happen beyond 2050 is much more difficult and open to conjecture. Based on IPCC emission scenarios, the concentration in 2100 could be as low as 500 ppm (low emissions case) to more than 1000 ppm (high emissions case). These estimated concentration levels do not take into account any major **climate change feedback** effects, such as a shutdown of the **thermohaline**, sudden release of methane from **methane hydrates**, or a transition of the land system from a net carbon sink to a net carbon source (see **land carbon sink**). Furthermore, when the other major greenhouse gases are included (converted to CO_2e), this adds another 50–70 ppm (see **anthropogenic greenhouse gas emissions**).

In the absence of very aggressive emission reduction measures, it appears that atmospheric CO_2 concentrations are likely to reach levels double to quadruple preindustrial concentration levels over the course of this century.

Is it possible to stabilize and then reduce concentrations?

To stabilize CO_2 concentrations emissions would have to fall to the natural rate of CO_2 uptake, estimated to be around 10–20 $GtCO_2$/year (see **stabilization targets**). If emissions fell below the level of natural uptake concentrations, or ceased altogether, CO_2 concentrations would begin to fall but it would take some time. While the assumed CO_2 atmospheric residence time is 100 years, it can take much longer for the actual CO_2 perturbation to fully work its way out of the atmosphere. Estimates of the long-term decay time are usually based on the findings of Shine in the first IPCC assessment report.[4] Under the Shine hypothesis, it is estimated to take 50 years or so for the first half of the excess CO_2 in the atmosphere to be absorbed (in other words, the first half-life of CO_2 is 50 years). The half-life of the remaining CO_2 is much longer and is estimated to be around 250 years. So, after 250–300 years, 75% of the original amount of excess CO_2 in the atmosphere would have been reabsorbed through the operation of the carbon cycle. Over successive half-lives, more and more excess CO_2 would be removed from the atmosphere, albeit at a diminishing rate: the Shine hypothesis suggests that in 500–1,000 years only 10–15% would remain.

Whilst this timeline remains widely accepted amongst the scientific community, some scientists believe that the rate of excess CO_2 decay is more protracted and that as much as 25% of the anthropogenic CO_2 loading could still remain after 10,000 years.[5] Whichever timeline is chosen, it is clear that CO_2 emitted today will continue to exert a warming effect in the atmosphere for a long period to come.

See also: anthropogenic greenhouse gas emissions, carbon dioxide equivalence (CO_2e), carbon sinks, El Nino Southern Oscillation (ENSO), fossil fuels, global warming potentials, greenhouse effect, greenhouse gases, land carbon sinks, methane (CH_4), methane hydrates, ocean carbon sinks, thermohaline.

Notes

1 NOAA 2007
2 WRI 2005
3 Shukman 2005
4 IPCC 1990
5 Archer 2005

Further reading

IPCC 2007; Pew Center 2007; Shukman 2005; Tschumi and Stauffer 2000.

CARBON DIOXIDE EQUIVALENT (CO$_2$e)

In order to make decisions regarding **mitigation** actions, it is important for policy makers to have a uniform basis of comparison between the climate impacts of different **greenhouse gases**. Carbon dioxide equivalent (CO$_2$e) is a metric that facilitates this comparison. It does so by normalizing the emissions from different **greenhouse gases** into the equivalent emissions of **carbon dioxide (CO$_2$)**, based on the **global warming potential (GWP)** of each gas.

The CO$_2$e of a greenhouse gas is derived by multiplying the number of tonnes of the gas emitted by its associated GWP.[1] For instance, if 1 tonne of **methane** (with a GWP of 21) is released into the atmosphere, 21 tonnes of CO$_2$ must be released to achieve the same level of **radiative forcing** (or warming effect). Thus, the emission of 1 tonne of **methane** is expressed as 21 tonnes CO$_2$e.

CO$_2$ is used as a base unit as it is the most abundant anthropogenic greenhouse gas, nonreactive under normal atmospheric conditions, and has a mean atmospheric residence time of around 100 years, making its atmospheric life comparable with the relevant time frames under consideration for **climate change impacts**.

CO$_2$e is used widely by climate change practitioners. It is almost used universally for the preparation of national greenhouse gas inventories, by the carbon markets, and as the basis for calculating and tracking emission reduction commitments under the **Kyoto Protocol**.[2]

See also: carbon dioxide (CO$_2$), global warming potentials, greenhouse gases, greenhouse gas inventories, Kyoto Protocol, radiative forcing.

Notes

1 IPCC 1996
2 UN Kyoto Protocol 2005

Further reading

IPCC 1996, 2007.

CARBON SINKS

A carbon sink is a carbon reservoir that absorbs more **carbon dioxide (CO_2)** than it emits. The amount of carbon contained in the three reservoirs of exchangeable carbon (the atmosphere, the land, and the ocean) has increased since preindustrial times due to the release of CO_2 (through burning **fossil fuels**) from the geological reservoir (the earth's crust). Although the atmosphere is technically a carbon sink (for anthropogenic emissions of CO_2), it is rarely ever referred to in this sense.

Through measuring the amount of CO_2 in the atmosphere, we know that only about half of anthropogenic CO_2 released to the atmosphere has actually remained there. The rest has been absorbed by the land and the oceans through the operation of the **carbon cycle** (see **land carbon sinks** and **ocean carbon sinks**). Both the land and oceans are *net carbon sinks*. The process of absorbing atmospheric CO_2 is termed *carbon sequestration*. If all anthropogenic CO_2 emissions over the past two centuries had remained in the atmosphere, the present CO_2 concentration would be around 480 ppm, rather than the 2008 recorded level of 382 ppm.

As the land and oceans are *net sinks* for CO_2, they act to reduce the rate of growth of atmospheric concentrations and are important natural buffers to increased anthropogenic emissions. Their ability to absorb CO_2 provides valuable breathing space for humans to reduce emissions to a level that stabilizes atmospheric CO_2 concentrations. At present, the best estimate is that the land and oceans absorb approximately 15 Gt of CO_2 each year from the atmosphere.[1] However, the precise amount varies from year to year as do the methods used for estimating carbon fluxes and estimates range from 10 to 20 $GtCO_2$/year.

Although the land is presently an important carbon sink (see **land carbon sinks**), it is the ocean that is the primary carbon sink. The oceans will determine the long-term equilibrium concentration of CO_2 in the atmosphere and the primary means by which the **carbon cycle** is brought back into balance, though this will take many centuries (see **ocean carbon sinks**).

Anthropogenic sinks

Apart from the natural sink responses (those not involving human intervention), there are also opportunities for humans to increase CO_2 sequestration rates through actions such as increasing the amount of land under forest and improved land-use management practices (see **biosequestration**). Other anthropogenic actions that may contribute to reducing atmospheric CO_2 concentrations include storing carbon in harvested wood products and the geological burial of carbon dioxide via **carbon capture and storage**.

The fact that carbon can be stored for extended time periods in wood products is well known. There are buildings and items of furniture still in existence that were constructed from wood cut centuries ago. Although some of this carbon can be stored for decades and even centuries, it is a temporary sink and nearly all of it eventually ends up back in the atmosphere through the process of decomposition or through burning.

See also: biosequestration, carbon capture and storage, carbon cycle, carbon dioxide (CO_2), fossil fuels, land carbon sink, ocean carbon sinks.

Note

1 Feely *et al.* 2001

Further reading

IPCC 2007; Feely *et al.* 2001; Lewis *et al.* 2006.

CARBON TAX

A carbon tax is a specific tax levied on each unit of **carbon dioxide (CO_2)**, or **carbon dioxide equivalent (CO_2e)** of other **greenhouse gases**, emitted to the atmosphere. It is normally levied at a rate per tonne – for example, $ 10/tonne CO_2. As with **emissions trading**, the primary objective of a carbon tax is to establish a cost penalty on greenhouse gas emissions (normally referred to as a *carbon price*). Introducing a cost penalty on emissions is one of a suite of options available to help reduce emissions.

Carbon taxes alter the relative prices of goods and services according to the embodied emissions associated with their production and/or consumption (see **emissions intensity**). In the case of electricity, a $ 10 carbon tax would increase the cost per kilowatt hour (kWh) of electricity from coal fired power plants by nearly 1 cent/kWh, efficient gas turbines by 0.5c/kWh, while the cost of electricity from **renewable energy** and **nuclear power** would remain largely unchanged. In an emissions intensive electricity system such price increases provide a financial incentive to reduce electricity consumption (e.g. through **energy efficiency**), substitute for less greenhouse gas intensive electricity products (e.g. electricity from renewable sources), or a combination of these – whichever way greenhouse gas emissions are reduced. If the carbon tax is low, consumers may just choose to pay the higher costs and demand may not change significantly. The higher the carbon tax, the greater the financial incentive to reduce for the consumption of emissions intensive goods.

In theory, the optimal carbon tax (carbon price) should equal the *social cost of carbon* – namely, the costs to society of **climate change impacts** resulting from **anthropogenic greenhouse gas emissions**.[1] Currently, the price of most goods and services that humans consume does not include the climate impact cost associated with their production and consumption – product prices understate their true economic cost (economists call this an unpriced negative *externality*).

However, it is difficult to determine the *social cost of carbon* as it is not known precisely when impacts will occur, the costs of these impacts, or the cost of **adaptation** to accommodate these impacts. Furthermore, even if the impacts are known, attaching a value to them can be problematic. For example, in the case of climate change-related **biodiversity impacts**, little direct market value will be attached to these losses, even though it may result in significant long-term cost to society. In general, the full climate impact costs are likely to be underestimated and, thus, not reflected in product prices (see **socioeconomic impacts**). In these cases, the social cost of carbon is likely to be higher than the carbon price.

Policy makers do not attempt to estimate the actual social cost of carbon but instead set taxes based on their perception of the cost impost required to deliver an adequate emission abatement incentive; the economic cost on industry; what is viewed as politically acceptable; and, possibly, government revenue implications.

Market-based versus nonmarket measures

Governments have an array of policy options they can employ to reduce emissions. These include market-based mechanisms (such as carbon taxes, **emissions trading**, the introduction and/or removal of subsidies, and investment tax credits) or nonmarket-based options (such as direct regulatory controls, public investment in research and development, education and information dissemination, and others). All instruments have advantages and disadvantages depending on where and how they are applied.

The use of market-based instruments to reduce greenhouse gas emissions has attracted considerable attention from policymakers as they should, at least in theory, provide emission reductions at lower economic cost than traditional regulatory approaches (often termed *command and control* measures). Market-based approaches aim to establish a market signal (through establishing a cost penalty on emissions and/or reducing the cost of low emissions technologies) and then leave the decision on where and how pollution is reduced to the market. Carbon taxes and **emissions trading** are the two market instruments policy makers are likely to adopt to control emissions, though specific subsidies to promote the adoption of low emission technologies, like solar water heaters or photovoltaic panels (see **solar power**), are also commonly deployed.

Command and control type measures have often been employed to control pollution. Examples of such measures include direct controls on the discharge of pollutants to waterways to achieve a specific water quality standard, the prohibition or strict controls on the releases of dangerous substances (such as emissions of dioxins), or mandatory pollution control equipment (such as catalytic converters on transport vehicles). These approaches are often criticized by economists as they only provide the polluting entity with an incentive to reduce emissions to meet the set minimum standard, even though it may still be technically possible to achieve superior environmental outcomes with little or no additional cost. They can also entail significant administrative costs for monitoring and enforcement, or are not enforced at all, and rarely enable burden sharing among polluters to achieve a specific objective at minimum cost.

In some circumstances, however, traditional regulatory measures can deliver greater emission reductions at lower cost than market-based instruments, particularly where significant barriers prevent the effective operation of the market (such as information constraints that limit awareness of cost-effective **mitigation** options) or where the demand for a product is relatively insensitive to price changes. For example, the introduction of minimum **energy efficiency** performance standards for household appliances has been demonstrated to be the most cost-effective means of reducing emissions associated with domestic appliances (where energy use is rarely a major consideration when purchasing the product), and often at negative cost. Carbon taxes and emissions trading are unlikely to deliver the same reductions at lower cost. The choice of whether to adopt market-based measures or other measures depends on the emission source under consideration and the market characteristics of the product or source in question. The optimal policy mix is likely to include a mix of market and nonmarket measures to achieve least cost emission reductions (see **mitigation**).

Economic theory suggests that market-based mechanisms will, in many circumstances, deliver greenhouse gas emission reductions at least cost. However, there is limited empirical evidence to substantiate this assertion. As yet no country has introduced a comprehensive economy wide carbon tax regime (one that applies a uniform tax to all sectors and activities) nor has any country introduced a full market-based emissions trading system (one based on full permit auctioning rather than free permit allocation). Nonetheless, market-based measures do offer considerable promise and are likely to feature strongly in the mitigation policy mixes governments adopt over the coming years.

Advantages and disadvantages of carbon taxes

The principal advantages of carbon taxes are that they:

- provide a clear market incentive (via a cost penalty) to reduce emissions and, like emissions trading, they allow the market to decide when and where emissions are reduced – thus utilizing the efficiency of the market mechanism;
- are relatively easy to establish and administer and can be readily incorporated into existing tax systems;
- can provide a significant source of revenue for governments, which can be used to offset other taxes (such as income and company taxes) with no net increase in the tax burden, potentially offering economic efficiency gains in other areas; and
- provide a known carbon price in advance for market participants and thus provide considerable price certainty and stability that can assist investment planning (unlike emissions trading where the carbon price is not known in advance and subject to much greater volatility).

The main disadvantages of carbon taxes are that:

- they do not guarantee a specific emission reduction outcome, unlike **emissions trading** where the quantity of emissions is fixed;
- they are not well suited to situations where the demand for emission intensive products is relatively insensitive to price;
- empirical evidence suggests that a relatively high carbon tax is required to deliver significant emission reductions, and high taxes usually face considerable political resistance; and
- polluters may be more willing to just pay the tax rather than reduce emissions, compared with emissions trading where noncompliance penalties are often several times greater than the expected permit price (the European Union's Emissions Trading Scheme has a €100/tonne CO_2e penalty, higher than what most governments would view as politically feasible for a carbon tax).

An important difference between carbon taxes and **emissions trading** is that carbon taxes aim to provide certainty on economic outcomes, while emissions trading aims to provide certainty on emissions outcomes.

Carbon taxes in operation

Some countries have already introduced carbon taxes (e.g. Norway, Sweden, and Denmark), although, as yet, no country has introduced a broad-based carbon tax that provides a common price signal across all sectors of the economy.

The Norwegian carbon tax experience does, however, provide some useful insights into the value of carbon taxes as a mitigation policy instrument. The tax was introduced in 1991 and set at varying levels for different sectors and activities. These varied from around $50/tonne CO_2 (gasoline and offshore petroleum production), $10–25/tonne CO_2 (for most industrial grade fuels), with a significant number of other sectors and emission sources exempt from the tax, many of which are emissions intensive (aviation, fishing, agriculture, land-based gas use, metals refining, cement, and several others). Overall, the economy-wide average carbon tax of $21/tonne CO_2 was applied to just over 60% of Norway's total CO_2 emissions. One economic study found that over the period 1991–2000, the carbon tax is estimated to have reduced Norway's aggregate CO_2 emissions by only 2.3% below what they would have been without the carbon tax, and only 1.5% for land-based sources – a relatively minor impact.[2] While a more broad-based tax applied equally to all sources may have resulted in greater emissions reductions, the evidence suggests that carbon taxes have to be levied at a relatively high level to deliver substantive emissions reductions. This appears to be especially the case in the transport sector where the relatively high carbon tax (around $50) that Norwegian motorists faced had only a negligible impact on road transport emissions.

While only a few countries have actually implemented a direct tax on emissions, many others have in place taxes that act as quasi carbon taxes (such as road transport fuel taxes), though these taxes are more often levied to raise general revenue, or to cover the costs of providing transport infrastructure and related services, and not as a greenhouse gas mitigation measure. Some countries have in place subsidies that act as negative carbon taxes, such as the significant subsidies applied to petroleum products (such as in Indonesia), or subsidized electricity provided to industry (as is the case for some aluminium production facilities in Australia). On a global basis, the total value of fossil fuel subsidies is currently in the order of US$200 billion/year, providing a market incentive to increase, rather than decrease, greenhouse gas emissions.[3] Removal of these subsidies would be an important first step in establishing a more appropriate market signal that reflects the true cost of emissions.

Carbon taxes (either direct or indirect) are likely to remain a potentially important policy option governments can adopt to control emissions, and in some situations may prove the most cost-effective instrument. However, their application is likely to be more limited than what was envisaged when the **United Nations Framework Convention on Climate Change (UNFCCC)** was signed in 1992. While in theory carbon taxes and emissions trading are both capable of delivering efficient economic emission abatement outcomes, most policy makers view **emissions trading** as the preferred means of establishing a carbon price.

See also: emissions trading, energy efficiency, mitigation, socioeconomic impacts.

Notes

1 See Stern 2006 for a discussion of the social cost of carbon.
2 Bruvoll and Larsen 2002
3 Stern 2006

Further reading

Stern 2006; Hepburn 2006; Ekins and Barker 2001.

CLEAN DEVELOPMENT MECHANISM (CDM)

The Clean Development Mechanism (CDM) is one of the three flexibility mechanisms defined by the **Kyoto Protocol**.[1] The CDM is a project-based "baseline and credit" **emissions trading** mechanism between Annex I Parties (developed countries with emission targets) and non-Annex I Parties (developing countries with no emission targets).

The CDM has two principal objectives:

- assist non-Annex I Parties "in achieving sustainable development and in contributing to the ultimate objective ..." of the **United Nations Framework Convention on Climate Change (UNFCCC)** and
- assist Annex I Parties "in achieving compliance with their quantified emission reduction commitments."[2]

Under the CDM, non-Annex I Parties host projects, which are specific activities designed to reduce greenhouse gas emissions or enhance greenhouse gas removals from the atmosphere. If a project satisfies the CDM project approval criteria, each tonne of **carbon dioxide equivalent (CO_2e)** mitigated by the project generates an emission reduction credit, called a Certified Emission Reduction (CER). CERs can be transferred, by purchase, to Annex I Parties to meet their **Kyoto Protocol** emission reduction commitments. One CER is equivalent to one Assigned Amount Unit (AAU), the emission allowance units allocated to Annex I parties.

How does the CDM operate?

The **Kyoto Protocol** has an aggregate emissions cap for Annex I Parties of 5.5% below 1990 levels. Under the Kyoto Protocol, Parties are allowed to trade AAUs between themselves to assist them meet their targets – the total stock of AAUs is fixed (capped) and trading operates under a "cap-and-trade" system (see **emissions trading**). In contrast, the CDM is based on projects in non-Annex I countries that do not have emissions reduction targets under the **Kyoto Protocol**: there is no fixed "cap" in the CDM. Rather, the CDM is a "baseline-and-credit" system in which new units (CERs) are created and traded, increasing the total number of AAU equivalent units in the system.

The CDM offers Annex I Parties the flexibility to deliver emissions reductions outside their domestic economies, potentially at lower cost. CDM may also involve the transfer of low emission technologies to non-Annex I Parties that contribute to promoting sustainable development. However, the CDM can only be (and was only ever intended to be) a supplementary measure to domestic actions by Annex I Parties. Furthermore, the CDM does not generate emission reductions that are additional to the emission reductions achieved by Annex I parties but merely transfers responsibility for emission reductions to a non-Annex I Party in exchange for payment. In essence CDM is a "zero sum" mechanism – aggregate global emission reductions remain the same; just the method of achieving these reductions changes.

Since CERs are generated outside the Annex I aggregate emissions cap, they are not reflected in the **greenhouse gas inventories** of Annex I countries. To ensure the environmental integrity of the **Kyoto Protocol** it is essential that CERs do not arise from emissions reductions that would have occurred anyway in the absence of the CDM, that is, they are "additional" to the "business as usual" case. For this reason, the Conference of the Parties (COP) agreed in 1997, during the negotiations of the **Kyoto Protocol**, that CERs arising from a CDM project must be certified as meeting the basic requirements of the following:

- voluntary participation by each Party involved;
- generating real, measurable, and long-term mitigation of emissions; and
- reductions in emissions must be additional to any that would occur in the absence of the CDM project.

Subsequent COPs, notably COP7 in Marrakech, have elaborated the rules of operation and have developed procedures for managing the CDM. These rules and procedures were formally adopted after the **Kyoto Protocol** came into force at the first Meeting of Parties (MOP) in Montreal, Canada, in 2005.

Ensuring "additionality"

Since it is in the interests of project proponents (the entity seeking project approval) to maximize the number of CERs (and hence revenue) generated from a CDM project, strict procedures have been put in place to ensure emission reductions are "additional" and correctly quantified. An important means of achieving this is to ensure consistency in the measurement and accounting of emission reductions. For this reason the CDM Executive Board (CDM EB) must approve the methodology a project proponent employs to quantify emission reductions.[3]

There are 15 general sectoral categories in which CDM projects can take place. The project proponent must use one of the existing approved CDM methodologies or develop a new methodology (and have it approved) if none already exists for that project type. All methodologies approved by the CDM EB (with the exception of "small-scale" methodologies) were initially put forward by a project proponent. Essentially, the CDM employs a "case law" approach: once a precedent is established with a methodology approved for one project, every similar subsequent project can use the same methodology. Any deviations from an approved methodology must be justified and are reviewed by the CDM EB.[4]

Under the CDM, the number of CERs that a project generates is calculated as the baseline emissions (emissions that would have occurred in the absence of the project) minus the project emissions. For example, a CDM project may involve the construction of a wind farm to supply **renewable energy** (a project with very few greenhouse gas emissions) to a national grid that is dominated by emission-intensive **fossil fuel** plants. It would be easy to assume that the wind farm is displacing electricity that would otherwise be generated by a fossil fuel-powered plant and therefore should be credited with displacing the fossil fuel plant emissions. However, it is often much more complex to ascertain how much a specific project reduces emissions across the system as there are a range of potential generating sources that could be displaced, some of which are also low emission sources.

Although it is relatively easy to quantify the emissions from a project, quantifying what the emissions would have been in the absence of the project (the hypothetical "business as usual" future) is more difficult since there are so many uncertainties in possible future scenarios. For example, for the wind power project, key questions to resolve include the following: Would the energy otherwise be produced by a fossil fuel-power plant, or would **energy efficiency** measures be introduced? Would the power plant be fueled by coal, oil, or gas (each with different emission intensities), or would it otherwise be produced by some other generating source such as nuclear power or hydroelectricity (which have virtually no emissions)? Most important, would the wind farm have been built regardless of whether or not the CDM existed? The answers to these and other types of project-specific questions have a significant impact on the amount of CERs that can be credited to a CDM project.

The need to demonstrate "additionality" is essential to the credibility of the CDM. Reasons why a project may be considered nonadditional include project implementation occurred prior to seeking approval from the CDM EB; the project is required by local, provincial, or national law; the project type is already common practice; or the project is an economically superior option relative to the project or activity it displaces (unless the project proponent can demonstrate that the project would not proceed without the CDM due to other barriers). The CDM EB has produced several guidance tools to assist project proponents grapple with the "additionality" issue.[5]

While a majority of projects so far approved under the CDM appear to satisfy the "additionality" criterion, some studies have questioned the additionality of some projects.[6] For example, a recent review of CDM projects in India (where more than 600 CDM projects had been registered by end 2007) found that up to one-third of the projects reviewed did not pass the additionality test.[7] While it must be recognized that determining what "would have happened" in absence of the CDM will always contain uncertainty, these studies, at least to some extent, call into question the credibility of at least some CDM emission reductions.

CDM participants

There are a range of actors that fulfill different functions in the CDM. These include formal government and UNFCCC administrative entities, project proponents, carbon brokers, and independent third parties that must validate and verify emission reductions.

At an international administrative level, the two principal entities are the **Kyoto Protocol**'s MOP, which has overall responsibility for decision making on CDM rules and procedures, and the CDM EB, which has been tasked by the MOP to supervise the administration of the CDM – all rules, procedures, and approval processes are administered by these two bodies.

At a national level, both non-Annex I host Parties and Annex I Parties have a role in the CDM project approval process. To enable non-Annex I host country governments to ensure that a CDM project contributes to and is consistent with their national development priorities, every CDM project must be approved by the host government's Designated National Authority (DNA). The host government DNA is an administrative unit whose primary responsibility is to screen and review all proposed CDM projects to ensure that they meet regulatory requirements and environmental guidelines and that the project will contribute to sustainable development in the country. If satisfied, the DNA issues a "letter of approval" to the project proponents. This letter of approval is required before the project can be approved by the CDM EB. This process ensures that the non-Annex I Party is voluntarily participating in the CDM.

Project proponents are of course a key player in the CDM – they are the actors responsible for designing the project, seeking approval, implementing the project, and delivering the emission reductions. CDM projects can be undertaken by private and/or public entities, such as state-owned utilities, companies, and joint ventures. Ownership of CERs typically rests with the project proponent rather than with national governments, although governments can also own a project and the CERs that the project generates.

Another key actor is the Designated Operational Entity (DOE), which is an independent party that reviews projects to provide assurance that the project meets the rules and procedures of the CDM and that the project proponent's estimate of emission reductions are calculated correctly – this process is termed "project validation." A DOE also needs to independently verify that the project has actually delivered the emission reductions claimed by the project proponent – this is termed "project verification."

The CDM project cycle

For a project to get to a point where it is issued CERs, a series of specific steps must be completed.[8] The CDM project cycle consists of five key stages:

Project Design Document (PDD)

The first step in the CDM project cycle is for the project proponent to prepare a Project Design Document (PDD). An initial stage (although not compulsory) in this process can be the preparation of a Project Idea Note (PIN), which can be used to attract or interest investors, and/or to gain preliminary approval from the host government DNA. Following this, the proponent prepares a formal PDD, using an existing approved methodology or developing a new methodology.

Project validation/host country DNA approval

This is the process by which the DOE systematically assesses the PDD in order to provide a statement on the reasonableness of the claims regarding future emission reduction performance. Essentially, the DOE checks that the project meets CDM rules and is likely to deliver the emission reductions stated in the PDD. This involves reviewing assumptions, calculations, methodologies, procedures, inputs, and, in particular, the justifications and assumptions employed regarding the baseline and the robustness of the planned monitoring and reporting measures. The DOE also makes the PDD publicly available on the Internet to ensure transparency and to enable other stakeholders to raise any issues that may have been deliberately concealed, or simply overlooked, by the project proponent. The validation statement (or "opinion") does not provide any assurance on the project's future performance but provides assurance to the project proponent and CDM EB that the PDD conforms to the rules and requirements of the CDM. The project proponent must also obtain a "letter of approval" from the host country DNA before submitting the PDD to the CDM EB for registration.

Registration

Once a PDD has been validated and has received a "letter of approval," it is then submitted to the CDM EB for approval. At this stage, the EB can approve or reject the project, seek additional information from the project proponent, or subject the project to a formal review process.

Project implementation and monitoring

Once the project receives EB approval, the project proponent can implement the project. There is provision for the project to be implemented prior to EB approval if it qualifies for "prompt start" exemption (applies to projects that were formulated prior to all the CDM rules, procedures, and administrative structures being established – pre-November 2004). During operation of the project, the project proponent monitors all the relevant parameters relating to greenhouse gas emissions as laid out in the PDD monitoring plan, and after a period of operation – typically, but not necessarily, one year – the proponent prepares a monitoring report.

Verification

The monitoring report must be verified by a different DOE than the one that undertook the initial project validation (except for small-scale projects). The intention of this rule is to avoid conflicts of interest between the project approval and project crediting processes. Verification is a means of confirming that the project proponent's emission reduction claim is accurately calculated and reported. That is, verification aims to confirm the truth of the report's assertions and to ensure that the project described in the approved PDD is the one that has actually been implemented. This typically involves rigorous collection, testing, and evaluation of evidence in order to present a complete audit trail.

Issuance of CERs

Once verified, the monitoring report is submitted to the CDM EB, which issues and distributes CERs to the project proponent. For the remainder of the CDM crediting period, the cycle repeats itself, with the project proponent operating the project, the DOE verifying emission reductions at regular intervals, and the CDM EB issuing the appropriate number of CERs. The CDM EB also withholds 2% of the issued CERs (via the Adaptation Levy) which are held in reserve, then sold, and the resulting revenues deposited in the Adaptation Fund (see **adaptation**). The Adaptation Levy is not applied to CDM projects in Least Developed Countries.

Crediting periods

A project can choose one of two periods in which credits can be generated: either a single 10-year crediting period (with no option for renewal) or a 7-year crediting period that can be renewed twice for a total of up to 21 years (the project must be reregistered with the CDM EB prior to the commencement of each new crediting period). Sequestration projects are an exception to this rule and can opt for a single 30-year crediting period or a 20-year crediting period that can be renewed twice for a maximum of 60 years of crediting (see **biosequestration**). The different treatment of sequestration projects reflects the length of time required for trees to grow to a point where they can sequester sufficient quantities of carbon dioxide from the atmosphere to be financially attractive as CDM projects.

The choice of crediting period is linked closely with the concept of additionality. Although a project initiated today may satisfy the "additionality" test, this may not remain true for future crediting periods. For instance, a proposed CDM project may have been deemed "additional" when first approved (e.g. it may have utilized a technology that was unproven or new to a particular industry – thereby not considered common practice). However, once the technology had been introduced and demonstrated its technical suitability and cost-effectiveness, it can be reasonably expected that the technology would, in future, be adopted by other industry participants – it would become common practice. On this basis it is unlikely that the project would qualify for more than one crediting period and the project proponent would most likely opt for a single 10-year crediting period. For other projects in which emission reductions are the primary reason for implementing the project, proponents generally opt for the renewable 7-year crediting period since it is usually straightforward to demonstrate that the project would cease, and emission reductions stop, if the project were not to have CER revenues (e.g. as is the case with the capture and destruction of landfill gas where there is no regulatory requirement, or financial incentive, to do so).

There is no penalty under the formal CDM process for projects failing to meet the emission reductions projected in their PDDs. If projects exceed their projected CER production, these CERs will be issued, provided the monitoring and verification approaches laid out in the PDD are adhered to.

The CDM is, however, a market mechanism involving contracts between private parties. In many of these contracts, the buyer requires the seller to guarantee delivery of all, or some proportion of, the CERs projected in the PDD. If a project proponent fails to deliver these CERs, they may be contractually required to purchase the shortfall of CERs from the open market, which will likely be considerably more expensive. This represents a considerable risk to project proponents, providing them with a strong incentive to ensure that the project performs as planned.

CDM projects and CERs issued

Initially the CDM started slowly, and only four projects were registered by early 2005. However, following ratification of the **Kyoto Protocol**, project activity increased rapidly and, by the end of 2005, 138 projects had been registered. This grew to 560 in 2006 and to 875 in 2007. By early 2008, more than 100 million CERs had been generated and, over the 2008–2012 First Commitment Period, the projects that were already registered at the start of 2008 were estimated to be capable of generating, on average, an additional 200 million CERs/year. In total, these projects could deliver more than 1 billion CERs by 2012. However, based on the average project delivery rates (around 90% of the CERs estimated from the PDDs), actual delivery may be slightly lower. Furthermore, as of early 2008, an additional 1,750 projects were at an advanced design stage but not yet registered.[9] If all these projects were also to be registered and deliver 90% of their estimated emission reductions, then an additional 1 billion CERs could be generated by the end of 2012 – in total, as much as 2 billion CERs could be available for purchase by Annex I Parties. If this number of CERs were to be available, it would make a significant contribution toward covering the expected Annex I emission allowance shortfalls.

An important consideration in determining the benefits CDM provides for non-Annex I countries is the type of projects that have been registered and the geographic distribution of projects between countries. It is evident from the CDM projects that have been registered to date that the project mix is reasonably diverse and employs a wide range of low emission technologies: projects have been registered in 11 of the 15 possible sector categories, cover more than 100 methodologies, and comprise a roughly even split between large- and small-scale projects.[10] Over two-thirds of projects utilize either **renewable energy** (mainly biomass, hydropower, and **wind power**) or **energy efficiency** technologies, which are usually considered as the types of technologies that contribute to sustainable development. The remainder consist mainly of end-of-pipe gas capture and destruction/transformation technologies (such as **methane** and industrial gas projects), which are generally considered to provide less discernible sustainable development benefits (although not necessarily devoid of development benefits).

The diversity of registered project types is a positive attribute of the CDM and reflects the wide range of cost-effective opportunities to reduce greenhouse gas emissions that are available in non-Annex I countries. It does, however, mask the fact that some projects (particularly those that involve reductions in emissions of **synthetic gases**, primarily HFC-23 destruction projects) account for more than one-third of projected CERs to 2012. These projects destroy greenhouse gases with very high **global warming potentials** (e.g. HFC-23 has a **global warming potential** of 11,700[11]) so that a small reduction in tonnage of these gases results in a large number of CERs generated by a CDM project. This does not make these emission reductions less valuable since each CER, regardless of the project, represents one tonne of **carbon dioxide equivalent** reduced. However, since synthetic gas projects are effectively "end-of-pipe" solutions for a small number of companies (often multinationals), the sustainable development benefits are generally considered to be very low.[12]

In terms of the geographical distribution of projects, the contribution of the CDM to sustainable development in non-Annex I countries is less clear. To date, only slightly more than one-third of the 140 countries that are eligible to host CDM projects are actually doing so, and just four countries (India, China, Brazil, and Mexico) host almost 80% of all CDM projects. Furthermore, very few projects are located in the Least Developed Countries (see **UNFCCC**).[13] Thus, on a geographical basis, the CDM project distribution is skewed toward just a handful of countries. This may change as the CDM matures and more countries gain experience with the mechanism.

In terms of the distribution of projects, among the five major UN developing country regions only two, Asia-Pacific, followed by Latin America and the Caribbean, dominate the CDM pipeline, and account for approximately 95% of projects and of expected CERs generated to 2012. It is, of course, to be expected that these two regions (especially Asia-Pacific) should host the largest number of projects since these regions account for a larger proportion of the developing country population and greenhouse gas emissions. However, when normalizing the expected CERs for each region by the population of each region, the CERs per capita in Asia-Pacific and Latin America and the Caribbean are substantially higher than in the other three regions of Africa, Arab States, and Europe/Commonwealth of Independent States.

The skewed geographical distribution of CDM projects has generated considerable criticism – notably from the Non-Governmental Organization (NGO) community – that the benefits of the CDM are unevenly distributed and only provide a development benefit to a limited number of countries.

The CER market

Since CERs are fully fungible (exchangeable) with other Kyoto emission trading units (AAUs and ERUs – see **emissions trading** and **Joint Implementation**), they should theoretically trade at the same price. In practice, however, this has not been the case and the average CER market price over the 2005–2007 period was around US$ 8/CER, compared with forward trades of €20–30/Phase II EU Allowance – the main representative benchmark emission credit price of the European Union Emissions Trading Scheme (EU-ETS).[14]

There are several reasons why this market price divergence has emerged. The signing of Emission Reduction Purchase Agreements (ERPAs) between buyers and sellers is usually done at the outset of a CDM project, when no CERs have yet been produced. As a result, buyers have many delivery uncertainties: the project may not be registered with the CDM EB; future market prices may evolve unfavorably; the project may not produce as many CERs as expected in the time frame; the project may not be properly managed; environmental and stakeholder opposition to the project may arise; and the host country may not provide a stable investment climate owing to natural disasters, political changes, and conflict.

These risks generally diminish as the project progresses through design, registration, construction, and operation: the further along this path a project is before it signs an ERPA, the more likely it will receive a better price. However, since implementing a project generally requires loans, a project proponent typically wants to settle with a buyer at an agreed price as soon as possible to help secure project financing. The buyer will lower the offer price for the CER according to the perceived risk. However, over time as the volume of CERs and carbon market liquidity increases, the divergence in market prices should, theoretically, converge to some degree.

In terms of which Annex I countries are purchasing CERs, it is evident that the European Union and Japan dominate, driven largely by their level of commitment to meeting their **Kyoto Protocol** targets. Initially, most buyers were governments, but since early 2006, approximately 90% of purchases were being made by the private sector.[15] The increased role of the private sector is largely as a result of purchases by entities covered under the EU-ETS (which permits the use of CERs to meet emission obligations).[16]

Although the market is dominated by the Kyoto and EU-ETS markets, there is a small but growing group of buyers composed of individuals and corporations that are seeking to offset their greenhouse gas emissions – namely, to be "carbon neutral." These "voluntary buyers" – so called because they are under no legal or treaty obligation to buy credits – account for less than 1% of CER purchases to date, but they are growing in number and importance. Where CERs are purchased by such buyers, and retired without being used to meet Annex I emission commitments, such voluntary purchases tend to enhance the environmental integrity of the CDM – by reducing its "zero-sum" characteristic (as they are not available to meet Annex I emission commitments).

The existence of the EU-ETS, and the coming into force of the **Kyoto Protocol** in 2005, has resulted in a significant assured market for CERs and increased their market value. Annex I Parties are likely to have a shortfall of 3–4 billion tonnes CO_2e by 2012 and CERs will be valuable in meeting individual country reduction targets.[17] Combined with a growing voluntary market, it is evident that the short-term market outlook for CERs looks robust.

The longer-term market outlook is less certain and depends to a large degree on what form a post-2012 climate change regime takes. The market faces a dilemma at the end of 2012 (when the **Kyoto Protocol** commitment period is due to end) as there is currently no agreed international agreement to follow the Protocol or that the CDM will continue to exist in its current form. While there is a generally held view that the CDM will continue operating after 2012, there is still some market uncertainty as to whether CERs will be worth anything after 2012. The decision by the European Union that CERs will continue to be accepted by the EU-ETS after 2012, regardless of whether or not there is a future international agreement has underpinned some market certainty. In addition, other potential markets that could emerge (e.g. the Regional Greenhouse Gas Initiative in the northeast states of the USA and the Australian emissions trading scheme) may also be available post-2012 and provide some future market surety. Nonetheless, the uncertainty surrounding the post-2012 market has an influence on the types of projects that investors will support and also this discontinuity is reflected in the much lower prices being offered for forward sales of CERs that are due to be generated after 2012.

Will the CDM be considered a success?

The CDM's success must be assessed in terms of the extent to which it meets its two objectives, namely, providing flexibility to Annex I countries in how they meet their Kyoto commitments and contributing to sustainable development in non-Annex I countries.

In terms of providing emission reduction flexibility to Annex I countries, it appears that the CDM will be a success and could potentially meet as much as half the expected shortfall in Annex I emission allowances.

With regard to the CDM's second objective, its contribution to sustainable development in non-Annex I countries, the success of the CDM is less clear-cut. The lack of clear definition of sustainable development adds considerable complexity to the assessment process, and it is, in many respects, a value judgment.

For the CDM, it is up to host governments to determine whether or not a project contributes to sustainable development. Most host government DNAs have established sustainable development screening criteria against which projects are assessed. The screening process should, in theory, enable host countries to weed out projects that do not make a contribution to sustainable development or at least those that would appear unsustainable. In practice, only a handful of projects have actually been rejected by host country DNAs on sustainability grounds. While this may mean that all the projects to date do in fact contribute to sustainable development, it may also indicate that non-Annex I countries are reluctant to reject projects that provide much-needed foreign capital investment.

In the early stages of its establishment the CDM attracted criticism from a number of organizations that the CDM was not delivering the sustainable development dividend that many had expected when it was first devised.[18] This assessment was largely due to the early dominance of **synthetic gas** projects (mainly HFC-23 destruction), that offered almost no sustainable development benefits to the host country, and the skewed geographical distribution of projects (just five countries accounted for nearly all projects and that the Least Developed Countries were almost entirely ignored by investors).[19] However, as the number and diversity of projects have grown, it is evident that the CDM is contributing at least some sustainable development benefits. Many projects have resulted in the transfer of low emission technologies. This may, over time, contribute to less emission-intensive development pathways. Furthermore, if the CDM results in a transformation of energy markets, where low emission technologies become the most common technology of choice by investors, then the CDM will prove to be a very useful instrument.

Several organizations have implemented initiatives that aim to increase the sustainable development benefits of the CDM. For example, the CDM Gold Standard,[20] endorsed by more than 50 NGOs worldwide, has promoted CDM projects in the **renewable energy** and **energy efficiency** sectors with greater local stakeholder engagement; the World Bank, through the Community Development Carbon Fund, has also focused on projects with a sustainable development dividend; and the United Nations Development Programme (UNDP) established the Millennium Development Goals Carbon Facility (MDG Carbon Facility), with the twin objectives of achieving greater geographical diversity in the CDM and generating greater sustainable development benefits for the CDM.[21]

While these efforts have increased the sustainable development focus of the CDM, the overall CDM portfolio remains dominated by a small number of countries, and a significant number of projects appear to have few discernible sustainable development benefits. As yet it is too early to judge the extent to which the CDM has delivered on this objective, but the evidence to date suggests that the contribution of the CDM to sustainable development may be less than what was envisaged when the Kyoto Protocol was negotiated.

See also: dangerous climate change, emissions trading, Joint Implementation, Kyoto Protocol.

Notes

1 UN Kyoto Protocol, Article 12
2 Ibid.
3 UNFCCC, COP8 Decision 21
4 See UNDP 2006a
5 See UNFCCC CDM Web site
6 Schneider 2007
7 Michaelowa and Ourohit 2007
8 See UNDP 2006a
9 UNEP Risoe 2008
10 See UNFCC CDM Web site
11 IPCC 1996
12 UNDP 2006a
13 UNEP Risoe 2008
14 World Bank 2007b
15 Ibid.
16 Ibid.
17 UNDP 2006a
18 See IISD 2006 and UNDP 2006a
19 UNDP 2006a
20 The Gold Standard 2007
21 See www.mdgcarbon.org

Further reading

UNEP Risoe Centre CDM guide books; UNDP 2006a; IISD 2006; World Bank 2007b.

CLIMATE CHANGE FEEDBACKS

Climate change feedbacks refers to **global warming**-induced changes to the earth's natural climate regulation mechanisms that either amplify (positive feedback effects) or subdue (negative feedback effects) the rate of warming. Climate change feedbacks can manifest themselves through two principal means:

1   changing the quantity of **greenhouse gases** residing in the atmosphere, through increasing (or reducing) emissions from natural sources (such as increased emissions of **greenhouse gases** from the thawing permafrost) or through altering the natural rate of removal of gases from the atmosphere (such as reducing the uptake of **carbon dioxide** by the land and oceans), or

2   varying the amount of warming associated with a given atmospheric concentration of **greenhouse gases** through such mechanisms as changing the earth's **albedo** (the amount of incoming solar energy reflected back into space), the amount of cloud cover, or several other mechanisms.

While positive and negative feedback mechanisms are generally associated with natural processes, climate change feedbacks can be expanded to include human-induced activities that may accelerate or reduce warming rates. For example, the increase in airborne particulate and sulphur emissions associated with **fossil fuel** consumption have had a net cooling effect on the planet and have kept global temperature increases below what would have occurred in their absence (see **global dimming** and **aerosols**).

Projections of mean global temperature change over the course of this century and beyond are based largely on expected values of **climate sensitivity** (the amount of warming associated with a given concentration of greenhouse gases in the atmosphere) and estimates of future **anthropogenic greenhouse gas emissions** (see **future emissions trends** and **stabilization targets**). The complexity of the earth's climate system, and gaps in the scientific understanding of the processes at work, means that the timing and strength of different feedback mechanisms remain an area of considerable uncertainty. For instance, relatively small temperature increases have increased the propensity and spread of low-intensity but widespread fires in the Amazon and Southeast Asia (a positive feedback), resulting in enormous areas of smoke haze, reflecting solar energy (a negative feedback). The overall feedback effect is estimated to be positive, but the complexity of the calculation results in much uncertainty.

Models have been developed to simulate the range of different climate feedback mechanisms, and these are generally termed coupled carbon-climate models. Each of these models is specified differently (they embody different assumptions about how the system works) and, as a result, they provide different estimates of feedback strengths. This is why there is such a wide range in the estimates of **climate sensitivity**. Uncertainty about the relative strengths of feedbacks remains, but, in general, the models tend to agree on whether a particular feedback is positive or negative.[1]

Scientists know that major, and abrupt, changes to the earth's climate have occurred in the past and that various climate feedback mechanisms have been significant contributors to the rate and extent of warming, and cooling, that occurred at these times. For example, a rapid period of warming 55 million years ago was in part due to the release of significant amounts of **methane** from subsea **methane hydrate** deposits. These methane releases caused atmospheric concentrations of greenhouse gases to rise, resulting in global warming, which led to even more releases of greenhouse gases from the methane hydrates and land system, creating a self-reinforcing positive feedback mechanism (see **methane hydrates**). Another example of a natural feedback mechanism occurred at the end of the last ice age (15,000 years ago) when rising global temperatures led to the release of significant quantities of green-house gases from the frozen permafrost as it melted, which accelerated the warming, thus accelerating deglaciation still further, which reduced the earth's **albedo**, creating another positive feedback mechanism that led to even more warming.

Likewise, global cooling has in the past been accelerated by negative **albedo** feedback effects where a cool period has lead to increased ice cover, which increased the earth's albedo, further cooling the planet, generating even more ice cover, and increased albedo, and so on – a negative climate change feedback effect. In most cases once a feedback mechanism is triggered, it often results in a self-reinforcing train of events that amplifies the initial warming or cooling. Eventually these run their natural course and, primarily through the operation of the **carbon cycle**, equilibrium is restored.

Scientists are also aware that the climate can remain relatively stable for long periods of time but then enters a phase of rapid climate change, once a certain temperature threshold is breached, that triggers these self-reinforcing feedback mechanisms. Instead of the earth's climate changing gradually in a linear fashion, it has, on occasions, experienced abrupt nonlinear changes. For example, the **thermohaline** (the major current responsible for mixing the waters of the surface and deep oceans) has shut down suddenly in the past and caused very rapid climate change (see **thermohaline**).

There is mounting evidence that the extent of global warming experienced over recent decades has already initiated some positive climate change feedback mechanisms, such as reduced albedo from accelerated ice loss and increased greenhouse gas emissions from permafrost thawing. However, it is the potential magnitude of human-induced temperature changes over the course of this century that are of most concern. It is likely that based on current trends, the earth's mean surface temperature will increase by at least 2°C, and possibly more than 5°C, by 2100. Mean global temperature increases of this magnitude may trigger a range of natural feedback mechanisms that could result in much more rapid warming, and possibly something humans will be unable to control. The mix and magnitude of different positive and negative feedback mechanisms are complex and subject to considerable uncertainty. Nonetheless, it is important to understand the possible feedback mechanisms that could be triggered, and the potential repercussions of these feedbacks.

Positive climate change feedbacks

Climate scientists have identified an array of positive feedback mechanisms that human-induced warming could initiate, or in some cases have already initiated, that are likely to accelerate global warming. The main positive feedback mechanisms include changes to the earth's **albedo**; reductions in the net uptake of **carbon dioxide (CO$_2$)** by the land and oceans; changes to ocean currents and, in particular, reduced flow of the **thermohaline**; increased emissions of **methane** and CO$_2$ from thawing permafrost and **methane hydrates**; and increased atmospheric water vapor concentrations.

Reduced albedo

As the earth's lower atmosphere warms, the proportion of the earth's surface under snow and ice cover will decline, and there will be changes to the relative shares of the earth's surface covered by forest, grasslands, deserts, and human settlements. Each of these changes alters the earth's **albedo**. Global warming, largely through its impact on surface ice cover, will tend to reduce the earth's albedo, retaining more of the sun's energy and resulting in additional warming – a positive feedback. This is expected to be counterbalanced to some extent by negative feedback mechanisms increasing albedo (particularly increased cloud cover, see below). The IPCC projects that the earth's albedo will decline this century, amplifying global warming. The current scientific consensus is that decreased surface albedo is likely to amplify surface temperature changes this century by around 10%.[2]

Reduced CO_2 uptake by the land

The land system currently removes (sequesters) on average approximately 3–9 billion tonnes more **carbon dioxide (CO_2)** from the atmosphere each year than it releases to the atmosphere – it is a net **carbon sink** (see **land carbon sinks**). The fact that land carbon stocks have grown over the past century (even when deforestation is taken into account) has played an important role in slowing the rate of increase of atmospheric CO_2 concentrations. The land has provided a valuable buffering mechanism by absorbing around one-quarter of the CO_2 emissions humans have released to the atmosphere.[3]

However, as the planet warms, the emissions of CO_2 from the land system are expected to increase, primarily through enhanced respiration of living organisms, particularly in the soil, and higher rates of oxidation of organic matter. If the earth's average temperature rises sufficiently, the quantity of greenhouse gas emissions from the land system could exceed the quantity being sequestered, resulting in a transition of the land from being a **carbon sink** to a carbon source. Soil carbon stocks in peatlands (especially in Southeast Asia) and the permafrost regions in the higher northern latitudes are particularly susceptible to increased temperatures, though soils in many other climatic zones are also susceptible. Increased emissions from the terrestrial land system could significantly accelerate the increase in atmospheric CO_2 concentrations and global warming – a very strong positive climate change feedback.

It is not yet known what the sink-to-source transition point is, but most models suggest it lies between 2°C and 4°C above preindustrial temperature levels (see **land carbon sinks**).[4] Nonetheless, net CO_2 uptake by the land is likely to decline over the course of this century as temperatures rise – in other words, a positive feedback effect will occur, even if the land remains a net carbon sink. The more global temperatures rise, the stronger this positive feedback will become. Humans may be able to arrest, at least to some extent, the rate of decline in land carbon stocks by halting deforestation and implementing major revegetation activities (see **biosequestration**). Whether this can counterbalance the expected temperature-induced decline in the land carbon sink remains an area of continuing debate.

Climate change is also expected to increase the spread of plant pests and diseases and the prevalence of fires, both of which are likely to affect land carbon stocks. For example, millions of trees in North America have already been lost due to Pine Bark Beetle infestations, whose survival range has in the past been controlled by night time temperatures falling below thresholds for larvae survival, but with warmer temperatures, larvae have survived the winter and devastated increasing areas of standing trees, reducing above ground carbon stocks in some regions and increasing the risk of fire due to increased fuel loads (see **biodiversity impacts**). Hotter, drier summers in some regions induced by global warming could increase fire risk and lead to sudden additional CO_2 loads entering the atmosphere – another positive feedback.

Reduced CO_2 uptake by the ocean

The ocean is an important **carbon sink**, currently similar in magnitude to the **land carbon sink** (see **ocean carbon sinks**). Unlike the land system, the oceans are likely to remain a net carbon sink for many centuries and will continue to absorb CO_2. Global warming will, however, reduce the oceans' capacity to absorb atmospheric CO_2 over time (see **ocean carbon sink**).[5]

Several feedback mechanisms will contribute to this reduction. First, as oceans warm, their ability to absorb CO_2 is reduced since the propensity of water to absorb CO_2 falls as water temperature increases. Second, as the oceans take up more and more CO_2, they will become progressively more CO_2 saturated, which reduces their ability to absorb additional CO_2.

Third, global warming is expected to alter the rate of mixing of ocean waters due to thermal stratification and changes to ocean currents (particularly the **thermohaline**), which will impact on ocean atmosphere CO_2 fluxes (see **ocean carbon sinks**). There is some evidence to suggest the thermohaline flow rate is already slowing (see **thermohaline**). As the thermohaline currently accounts for about a third of the transport of carbon to the deep ocean for long-term storage, any reduction in its rate of flow means more CO_2 remains in the atmosphere than would otherwise be the case: atmospheric CO_2 concentrations would increase more rapidly and accelerate global warming – the more the thermohaline slows, the stronger the positive climate change feedback.

Although the ocean and land remain important CO_2 sinks, the fraction of total anthropogenic CO_2 emissions remaining in the atmosphere has been increasing in recent decades, implying that the sink potential of the land and oceans has begun to moderate, and possibly decline.[6] Depending on which model is used, the increase in atmospheric CO_2 by 2100 from carbon cycle feedbacks could be as little as 20 ppm on top of anthropogenic emissions (a mild feedback) or as much as 200 ppm (a very significant amplifying effect).[7]

Increased methane emissions from permafrost and subsea hydrate deposits

Significant quantities of methane and organic carbon are stored in permafrost, and vast quantities are trapped in frozen marine sediments. In fact, the subsea **methane hydrates** are the single largest reservoir of organic carbon on the planet, storing as much as twice the quantity of organic carbon as **fossil fuels**.

As the planet warms, permafrost will continue to thaw and result in the release of methane and CO_2 and, thereby, contribute to additional warming. Increased emissions of methane from permafrost have already been observed, and annual emission levels appear to have accelerated considerably since the early 1990s (see **polar impacts**). If global temperatures continue to increase in line with IPCC projections, it is inevitable that a significant proportion of the organic carbon contained in the permafrost will be liberated to the atmosphere, and possibly a quarter of this carbon reservoir (equivalent to 350 billion tonnes of CO_2) could be released this century even if temperature increase can be limited to just 2°C above preindustrial levels.[8] This would be equivalent to around two-thirds of the current contribution of deforestation to anthropogenic CO_2 emissions (the next most important source after fossil fuel combustion).

While **methane** releases from the permafrost will represent a strong positive feedback, it is the potential large-scale releases of methane from subsea methane hydrate deposits that are of much greater concern. More than 95% of the stock of organic carbon stored as **methane hydrates** are contained in the ocean sediments. These deposits are much less sensitive to changes in global surface temperatures due to the time it takes for ocean temperatures to change – most of the methane hydrates are located several hundred meters below sea level and are considered to be quite well insulated from surface temperature changes. However, over the next few centuries, ocean temperatures will eventually equilibriate with surface temperatures and, as ocean temperatures rise, the risk of large-scale releases from the subsea methane hydrates will also increase. Once ocean temperatures increase by 3°C or more above preindustrial levels releases from the methane hydrates are inevitable (see **methane hydrates**).

Although scientists consider it unlikely that major releases from the hydrates will occur this century, the risks over the longer term are significant, and possibly much larger than the eventual contribution from **anthropogenic greenhouse gas emissions**, and could deliver potentially catastrophic results. In many respects methane hydrates are the wildcard in terms of positive climate change feedbacks.

Water vapor

As the atmosphere warms, the concentration of water vapor (a greenhouse gas) in the atmosphere increases producing a positive feedback. Some modeling studies indicate that the water vapor feedback effect could add as much as 1.8 Wm^{-2} in **radiative forcing** over the course of this century.[9] This significant additional radiative forcing is offset to some degree by the temperature lapse rate in the troposphere – the rate at which temperature decreases with height. Global warming is expected to result in a negative lapse rate feedback, estimated at approximately 0.8 Wm^{-2}. When the water vapor and lapse rate feedbacks are combined, the net effect is to amplify global surface temperatures by as much as 40–50%.[10]

Negative feedbacks

Global warming is also likely to trigger a range of negative climate change feedback mechanisms that could moderate the future rate of warming. The main negative feedbacks include possible increased cloud cover, temperature- and CO_2-enhanced photosynthetic biomass production, and a reduction in the atmospheric residence time of methane. In addition to these natural feedback processes, there are also several human activities, such as increased emissions of sulphur compounds and particulates to the atmosphere that could result in a planetary cooling effect.

Increased cloud cover

As the earth's lower atmosphere warms, it is able to hold more water vapor, which increases the conditions for cloud formation. Clouds reflect energy back into space increasing the earth's **albedo**. While increased cloud cover could help counterbalance the reductions in albedo due to retreating ice cover and the advance of boreal forests into tundra regions, considerable uncertainty surrounds the magnitude of this offset effect. While it is generally agreed that cloudiness is likely to increase as the globe warms, the extent to which this helps cool the planet depends on when and where the clouds form. Day time clouds help reflect energy back to space (cloudy days are generally cooler than clear days), while night cloud cover tends to prevent heat from escaping (cloudy nights tend to be warmer than clear nights).

Water vapor in the atmosphere that converts to water droplets reflects solar energy back into space and the availability of nucleation points such as dust or **aerosol** particles. With an increasing number of aerosol particles as a result of fossil fuel and biomass combustion (often related to deforestation), a larger number of smaller droplets form and the clouds persist for longer periods of time (see **aerosols**). While cloudiness is expected to increase and possibly provide a negative feedback, the effect is not well understood and the calculating of any net effect is complex and uncertain. Estimates of the strength of cloud feedbacks exhibit the widest range of uncertainty of all the feedbacks modeled. Best estimates conclude a negative radiative forcing of -0.7 Wm^{-2}, with a 90% confidence range from -0.3 to -1.8 Wm^{-2}. This compares with the total **radiative forcing** of long-lived **greenhouse gases** (excluding water vapor, stratospheric and tropospheric **ozone**) of 2.63 Wm^{-2}. Thus, increased cloudiness has had a negative feedback that has diminished the estimated direct radiative forcing of anthropogenic greenhouse gas emissions by more than a quarter, with a range of 11–68%.[11] In short, increased cloudiness may deliver an important negative feedback but remains an area of considerable uncertainty.

Enhanced photosynthetic production

Higher temperatures are likely to stimulate photosynthetic production in areas where growth is currently temperature constrained (especially through longer growing seasons in the higher latitudes). An upward trend in net primary production (NPP) of the terrestrial system has already been observed, and over the period 1981–2000, NPP increased by 6%.[12] Forest cover is also expected to extend toward the poles and encroach on existing tundra regions, thus increasing the stock of carbon contained in above ground vegetation (see **land carbon sink** and **polar impacts**). These represent negative climate change feedbacks. However, as forest replaces tundra, it also has a counterbalancing **albedo** effect, which is expected to outweigh the increased CO_2 sequestration benefits as forests are darker and absorb more solar energy than tundra (see **polar impacts**).

Most plants have increased growth rates (and hence carbon sequestration) in response to increased concentrations of CO_2. The strength of the *CO_2 fertilization effect* remains a subject of ongoing scientific debate but could increase plant photosynthetic production by at least 5%, and possibly as much as 10–15% (see **land carbon sinks** and **agriculture and food supply impacts**). This negative feedback is expected to moderate the rise in atmospheric CO_2 concentrations and the rate of global warming – assuming that plant growth is not significantly constrained by the availability of water and other essential nutrients.

There is also a potential negative feedback at work in the oceans. Higher temperatures are likely to accelerate geochemical reactions and the flow of organic and nutrient-rich sediments to the oceans, which in turn could stimulate the growth of algae, and increase the level of organic carbon sequestration. The extent to which climate change will stimulate biological activity in the oceans remains subject to considerable uncertainty, and nutrient limitations (such as the availability of soluble iron) could restrain biological activity. Furthermore, climate change is expected to reduce the rate of upwelling of nutrient-rich deep ocean waters (see **ocean carbon sinks**), which may counterbalance some of the increased productivity effects. If surface ocean algal growth is stimulated, it may also set in train another negative feedback mechanism, namely elevated levels of dimethyl sulphide emissions from the algae: a precursor to cloud formation and increased **albedo**.

Aerosols and particulates

The significant increase in fossil fuel consumption over the past century, increased condensation trails (contrails) from aircraft, and higher levels of airborne dust associated with human activities (like broad acre agriculture) have substantially increased atmospheric aerosol and particulate concentrations, which has reduced the amount of solar energy reaching the ground (see **global dimming**). These emissions have tended to cool the planet and keep global temperatures lower than they would otherwise have been in their absence. The projected increase in fossil fuel consumption over the next few decades (see **fossil fuels** and **future emissions trends**) is likely to maintain the strength of this negative feedback. However, this cooling is short term and tends to mask the underlying warming resulting from elevated atmospheric greenhouse gas concentrations. Once fossil fuel consumption begins to fall, or controls are introduced to improve local air quality, the associated reduction in atmospheric aerosols and particulate levels will lead to accelerated warming (see **global dimming**).

Reduction in atmospheric methane residence time

Surface warming has resulted in more convective activity (warm moist air ascending to form clouds) and, thereby, more lightning (which generates OH molecules). Increased atmospheric abundance of OH molecules accelerates the breakdown of methane. Already scientists have detected a reduced atmospheric residence time for **methane**, albeit by only 1–2%. Nonetheless, methane is 60 times more potent a greenhouse gas than CO_2 (molecule for molecule), and any reduction in the atmospheric residence time for methane will reduce its **global warming potential** and its contribution to atmospheric **radiative forcing** – a negative feedback effect (albeit expected to be relatively small).

Positive feedbacks outweigh negative feedbacks

Although uncertainty remains in relation to the strengths of different climate feedback effects, the general consensus is that the positive feedbacks will, over the course of this century, outweigh the negative feedbacks. This is expected to accelerate global warming relative to a situation in which global mean temperatures maintained a constant lineal relationship with atmospheric greenhouse gas concentrations.

It is also evident that the risks of triggering major positive feedbacks, particularly reductions in the land and ocean carbon sinks, and major releases of **methane** from the **methane hydrates**, increase substantially with global temperature rise. The risk of abrupt climate change also increases significantly once temperatures exceed 2–3°C above preindustrial levels. The only way to reduce the risk of initiating strong positive feedback effects is to limit atmospheric greenhouse gas concentrations to the lowest achievable level.

See also: albedo, agriculture and food supply impacts, carbon cycle, future emissions trends, global dimming, land carbon sinks, methane (CH$_4$), methane hydrates, polar impacts, ocean carbon sinks, thermohaline.

Notes

1 See Friedlingstein *et al.* 2006
2 Bony *et al.* 2006
3 Raupach and Canadell 2007
4 See Canadell *et al.* 2007
5 See Matear and Hirst 2001
6 Raupach *et al.* 2007
7 Raupach and Canadell 2007
8 Ibid.
9 Soden and Held 2006
10 Bony *et al.* 2006
11 IPCC 2007
12 Raupach and Canadell 2007

Further reading

IPCC 2007; Cox *et al.* 2000; Feely *et al.* 2001; Canadell *et al.* 2007; Raupach and Canadell 2007; Bony *et al.* 2006.

CLIMATE CHANGE IMPACTS

Average global temperatures have increased by 0.76°C since 1900 and have already resulted in changes to the earth's geophysical and biophysical systems.[1] There is general scientific consensus that the rapid rise in **anthropogenic greenhouse gas emissions** over the past two centuries, and particularly since 1950, has been a major contributor to the global warming that has occurred. The **Intergovernmental Panel on Climate Change (IPCC)** estimates that global temperatures could rise another 1–2°C by 2050 and 2–5°C by 2100, depending on the additional quantity of greenhouse gases humans emit to the atmosphere over the coming decades (see **global warming** and **future emissions trends**).[2] The current scientific consensus is that even if future warming can be limited to the lower end of this range (2°C), there will be significant impacts on natural and human systems.

Our understanding of climate change processes and impacts has improved considerably since the early 1990s as the body of scientific research and analysis has expanded. Recent major reports, including the IPCC Fourth Assessment Report (2007), Avoiding Dangerous Climate Change (2006)[3], and the Stern Report (2006) all provide a comprehensive assessment of existing and potential future climate change impacts. Other reports, for example, the IPCC special report on climate change and biodiversity (2002) and the Arctic Impact Assessment Report (2005), have assessed specific impacts in more detail. There are also thousands of reports and studies that these studies have drawn on to formulate their assessment of climate change and its impacts.

The type and extent of different impacts will vary across regions and over different timescales. For example, the higher latitudes will experience much more rapid warming than lower latitudes with significant effects on permafrost and Arctic and Antarctic ecosystems (see **polar impacts**). The subtropics will tend to experience greater reductions in precipitation than other latitudes (see **water impacts**), face more intense tropical storms, and reach critical temperature thresholds for plant photosynthesis before the higher latitudes (see **land carbon sinks** and **agriculture and food supply impacts**).

Most scientists agree that the impacts of climate change are likely to be overwhelmingly negative and that developing countries are likely to suffer greater impacts than wealthier industrialized countries (see **socioeconomic impacts**). Nevertheless, there are some potential positive impacts for particular regions and communities, at least at low rates of warming. For example, warming temperatures in the mid- and higher latitudes, combined with elevated levels of **carbon dioxide (CO_2)**, could lead to an expansion in agricultural areas, an increase in crop productivity and forest growth, and a reduction in human mortality from extreme cold (particularly in Russia and Canada).

Impacts can be broadly classified into three main categories: geophysical, biophysical, and socioeconomic impacts. The objective of this section is to provide a broad overview of key impacts and to guide readers to other sections of the book that provide a more detailed review of specific impacts.

Geophysical impacts

This category includes the impacts of climate change on the main physical components of earth's climate system. Examples include changes in **ice sheets and glaciers** and the earth's **albedo** (reflectivity), **sea level rise**, changes in the ocean currents and chemistry, including a possible slowing or shut down of the **thermohaline** and ocean acidification (see **marine impacts**).

The loss of sea- and land-based ice will continue this century and result in significant impacts, especially in the polar regions. There is also a significant risk that key temperature thresholds will be breached, such as those leading to deglaciation of the Greenland ice cap, destabilization of the Western Antarctic Ice Shelf (WAIS), and the trigger point for large-scale releases of greenhouse gases from the **methane hydrates.**

Global warming is expected to have significant impacts on the global water cycle and change the quantity, intensity, and seasonal distribution of precipitation events. This will alter the frequency and intensity of droughts and floods (see **water impacts**). Increased heat energy in the oceans and atmosphere is expected to increase the intensity of storms and wind speeds (see **extreme weather events**). These potential geophysical changes will have major implications for humans and other living organisms and also affect the operation of the earth's **carbon cycle**. They could also initiate positive **climate change feedbacks** that could further accelerate **global warming**.

Biophysical impacts

This category includes changes that directly impact living organisms. Climate change is expected to have significant **biodiversity impacts**. Even at a global average temperature increase of only 2°C, up to a quarter of all species face either extinction or significantly reduced habitat ranges. At temperature increases above 4°C, up to two-thirds of species could face extinction (see **biodiversity impacts**).

Increased ocean temperatures and acidification will have major **marine impacts**, particularly for coral reefs and marine system primary productivity. Rising sea levels and more intense extreme weather events will have major **coastal zone impacts** with significant flow on effects for migratory birds, fisheries, and low lying wetlands and delta ecosystems. Changes in temperature and precipitation will have impacts on the primary production of terrestrial plants and forest systems (see **land carbon sinks**) and the prevalence of pests and diseases affecting many different species, including humans.

Much depends on the rate of change in temperature since the faster the temperature changes, the lesser the ability of natural ecosystems to adapt. That said, the ability of natural biological systems to adapt is limited, and many species and ecosystems are highly susceptible to even small changes in climatic conditions. Some species have already been adversely impacted, even resulting in extinctions (see **biodiversity impacts**).

Human and socioeconomic impacts

The final category relates to impacts that directly affect humans. Climate change has far-reaching implications for human livelihoods. Increased temperatures and changing precipitation patterns, particularly during **extreme weather events**, can reduce agricultural crop yields and aggregate food production, especially in lower latitudes. This could potentially result in global food supply deficits, increasing famines (see **agriculture and food supply impacts**). Climate change will also result in human **health impacts**, particularly through more severe heat waves and changes in the incidence and distribution of vector-borne diseases.

Enormous capital is invested in coastal infrastructure, industry, and human settlements that are particularly susceptible to **coastal zone impacts**, with potentially hundreds of millions of people being permanently displaced over the next century. Some nations could mostly or entirely disappear, involving the difficult and contentious politics of nationhood and transnational migration. Global economic production and trade and the future development prospects and climate vulnerability of the majority of the world's population are likely to be adversely affected by the projected level of global warming this century (see **socioeconomic impacts**). The poorest people, and particularly indigenous communities that depend on natural ecosystems for their livelihoods, are the most vulnerable to the impacts of climate change and are expected to bear a disproportionately high-impact burden.

Overall, climate change is expected to have significant negative consequences for humans and result in substantial social and economic costs. Although humans have a much greater ability to adapt and cope with projected climate change impacts than other organisms, poorer countries or communities do not have access to the full array of adaptation options, primarily due to financial resource constraints. Adaptation constraints could have major implications for the world economic order and international relations over the course of the twenty-first century, particularly as those that are likely to face more severe climate impacts (mainly the poor nations) are also those that have contributed least to the problem.

Global warming has already changed the earth's climate with a commensurate range of observed geophysical, biophysical, and human impacts. However, it is the substantial acceleration in global warming projected over this century that will have more far-reaching and substantial consequences on the earth's climate and those organisms that inhabit the planet, including humans.

How do impacts vary with different amounts of global warming?

Most scientists agree that the incremental impact of each 1°C rise in average global temperature is unlikely to be equal. Changes will be nonlinear, such that a global mean temperature rise from 2 to 3°C is likely to have much greater and far-reaching impacts than a rise from 1 to 2°C. There are also likely to be certain thresholds and critical points beyond which an abrupt change may occur, such as the total collapse of an ecosystem or a sudden change in ocean circulation patterns.

Many other existing or emerging environmental problems, such as land degradation, large-scale alterations to hydrological systems (e.g. excessive water diversions from rivers and lakes), and industrial pollution will tend to be amplified by a changing climate.

These factors increase uncertainty and make it difficult to predict the specific impacts of a given increase in global temperature. Nevertheless, sufficient information is now available to enable a reasonably confident assessment of the type and scale of impacts for a given temperature change and to identify potential critical temperature thresholds that will trigger abrupt changes. For example, the current scientific consensus is that the temperature threshold that will trigger the eventual total deglaciation of the Greenland ice cap is believed to lie in the range of 2.5 to 3°C above preindustrial levels (see **ice sheets and glaciers**).

Estimates of global mean temperature change over this century vary from as low as 1.5 to an alarming 10°C. Most studies provide a more limited range of temperature increases, such as follows: 2020 – 1.1 to 1.4°C; 2050 – 1.7 to 2.9°C; and 2080 – 2.4 to 4.3°C.[4] These ranges are typically used to link certain impacts to different levels of global temperature increase.

Most climate impact studies focus on impacts of temperature increases ranging up to 4°C. Few identify likely impacts for higher temperature increases since temperature rise is likely to be less than 5°C by 2100 and because the uncertainty of impacts increases considerably once mean global temperatures increase more than 3–4°C. Most impact assessment studies also usually assume sea levels will increase by 0.5–1.0 m and that ocean pH falls by a further 0.2–0.3 pH points (on top of the 0.1 pH fall already observed) by 2100.

Based on the findings of several major assessment studies,[5] some key impacts for different temperature changes are:

Up to 1°C

Ten percent of global ecosystems will be adversely affected; accelerated melting, retreat, and possible loss of most South American, Himalayan, and African glaciers; increased coral reef bleaching and mortality; reduced crop yields in some regions (especially Africa); accelerated sea ice loss and permafrost thaw; possible further species extinction; several unique biodiversity rich ecosystems will face increased vulnerability to damage or loss, including the Dryandrea forests (Western Australia), the North Queensland rainforests, the Sundabans of Bangladesh and the Fynbos and Karoo in South Africa.

While these impacts are substantial, or even devastating, at the local level, on a global scale the impacts of a 1°C warming will not be significant, and most communities, and many ecosystems, are likely to cope with a temperature change of this magnitude.

Between 1 and 2°C

Decreased agricultural yields become more widespread in the tropics and subtropics with possible yield increases in higher latitudes; 1–3 billion people could experience increased water stress and up to 200 million additional people at risk of hunger; an expected spread and increased incidence of malaria and dengue; increased infrastructure damage and losses due to extreme weather events; over 90% of the world's coral reefs subject to serious damage, and many could be permanently lost; up to half of frogs and many reptiles could face extinction, as well as possible extinction of several Arctic species in the wild and significant losses in global bird populations – overall species loss could reach 20%; 60% of tundra would become unstable; mobilization of the Kalahari sand dunes and significant drying in several regions, particularly North Africa and the Mediterranean; up to 50% loss of Kakadu wetlands (Australia), Dryandrea (Australia), Succulent Karoo (South Africa), and Chinese boreal forests and 40% loss of Queensland rainforests (Australia).

Between 2 and 3°C

Complete loss of the Kakadu wetlands, the Succulent Karoo, Dryandrea forests, and Chinese boreal forests; near total loss of the world's existing coral reefs and widespread collapse of the Queensland rainforest ecosystems; 50% loss of the Fynbos (South Africa); up to half European and all Australian alpine species at risk of extinction; 20–70% loss of migratory bird habitats; and possibly 30–40% of all species at risk of extinction; approaching or exceeding the estimated trigger point for complete deglaciation of Greenland (with eventual 7 m sea level rise); possible destabilization of the WAIS (with eventual additional 6 m of sea level rise); large impacts on global cereal production (with possibly 5–10% losses in crop yields in lower latitudes) and an additional 400 million people at risk of hunger; possibly half the world's population at risk of exposure to dengue fever and an additional 300 million people to malaria; substantial increase in people at risk from heat waves and extreme weather events; increased vulnerability of the Amazon rainforest ecosystem to significant transformation or collapse; widespread thawing of the permafrost resulting in significant increases in methane and CO_2 emissions and total loss of summer Arctic sea ice.

Between 3 and 4°C

Large losses of agricultural production across most regions with the possible exception of the higher latitudes; declines in grain yields in the lower latitudes and increased food prices could place an additional 600 million people at risk of hunger; 60% of the world's population exposed to dengue; potentially more than half of all species at risk of extinction; complete collapse of the Arctic ecosystem and total loss of summer ice over the Arctic Ocean; two-thirds of the Tundra and up to one-quarter of coastal wetland ecosystems lost; possible increased instability of the methane hydrates at higher latitudes with potential large-scale methane releases over the next few centuries; substantial slowing, or even possible shutdown of the thermohaline; noticeable increase in storm intensity and extreme weather events; Amazon rainforest ecosystems may totally collapse and possibly becoming a major source of emissions; significant chance of the terrestrial land carbon pool transitioning from a carbon sink to a carbon source; much higher risk of collapse of the WAIS.

Between 4 and 5°C

Serious and widespread declines in crop yields and some countries would suffer near complete collapse of their agricultural systems (particularly, parts of South Asia, the Mediterranean, and Africa) with major implications for world food security; major socioeconomic impacts associated with collapse or transformation of marine and terrestrial ecosystem changes and health and water impacts, affecting a majority of the world's population; high probability of collapse of the WAIS with a subsequent commitment (combined with Greenland) to long-term sea level rise of at least 10–15 m over the next 300–500 years and also committing many existing coastal marine ecosystems to destruction; complete collapse of the Amazon ecosystem, almost certain transition of the land carbon pool from sink to source; up to two-thirds of all species committed to extinction; much larger risk for significant methane hydrate releases and shutdown of the thermohaline.

Beyond 5°C

Very few estimates of potential impacts of temperature increases above 5°C are contained in the literature due to high uncertainty surrounding the impacts of such high temperature changes. However, based on the extent and magnitude of the impacts that are likely to occur below 5°C, it is expected that the impacts would become much more severe and widespread at temperature changes beyond 5°C. Changes of this magnitude may commit the earth to runaway climate change where emissions from nonhuman-related sources (natural processes) become the self-sustaining driving force for even more global warming (see **climate change feedbacks**). Few climate events of this magnitude have occurred in the earth's past history, particularly at the rate of temperature change that could eventuate over the next two centuries if we are unable to stabilize atmospheric greenhouse gas concentrations at moderate levels.

Climate change has potentially far-reaching consequences for humans and other living organisms that presently inhabit the earth. There is a growing consensus that global temperature increases of greater than 2°C will deliver a range of impacts that could constitute dangerous anthropogenic interference with the planet's climate system and contravene the **United Nations Framework Convention on Climate Change** (see **dangerous climate change**). Temperature increases exceeding 4°C are likely to be catastrophic and must be avoided.

While decision makers have difficulty dealing with issues that may take several centuries to fully manifest themselves, it must be borne in mind that what the global community does in the next 20–50 years will determine the type of temperature increases the planet will be committed to over the next 100–200 years and to impacts that could become progressively worse for many centuries, and possibly persist for several millennia.

See also: agriculture and food supply impacts, albedo, biodiversity impacts, carbon cycle, climate sensitivity, dangerous climate change, extreme weather events, global warming, health impacts, ice sheets and glaciers, land carbon sinks, marine impacts, polar impacts, sea level rise, socioeconomic impacts, thermohaline, water impacts.

Notes

1 IPCC 2007
2 Ibid.
3 Schellnhube *et al.* 2006
4 See Parry *et al.* 2004, Hare 2006, and Warren 2006
5 Ibid.

Further reading

IPCC 2007, 2002; Stern 2006; Hare 2006; Parry *et al.*

CLIMATE SENSITIVITY

Climate sensitivity is a term used to describe the relationship between the rise in atmospheric greenhouse gas concentrations and changes in global mean surface temperature. Although the precise **Intergovernmental Panel on Climate Change (IPCC)** definition of climate sensitivity refers to the global temperature change that would result from a doubling of atmospheric **carbon dioxide (CO_2)** concentrations above preindustrial levels (namely from 280 to 560 parts per million [ppm]), it is more often applied to a doubling of **carbon dioxide equivalent CO_2e** concentration of all **anthropogenic greenhouse gas emissions** (from around 300 to 600 ppm CO_2e). The concept of climate sensitivity can also be more generally applied to yield the change in global mean temperatures resulting from any change in (CO_2e) concentrations, for example, from current levels to 500 ppm, 750 ppm, or any other concentration level. The use of CO_2e provides a more accurate representation of climate sensitivity as it takes into account the radiative forcing of all **greenhouse gases**.

Climate sensitivity (for double preindustrial concentrations) is thought to lie somewhere in the range of 1.6–11°C, but the precise value is not known.[1] In its Fourth Assessment Report, released in 2007, the IPCC concluded that climate sensitivity is estimated to be between 2 and 4.5°C, with the most likely value of 3°C.[2]

By 2007, atmospheric CO_2 concentrations had risen to 382 ppm[3] and CO_2e concentrations had reached 455 ppm. Atmospheric CO_2e concentrations are currently increasing at an average rate of around 2.5 ppm/year and, based on current projections of anthropogenic greenhouse gas emissions to 2050, annual concentration growth rates are expected to increase further: this is likely to bring forward the date concentrations reach double preindustrial levels to as early as 2040 (see **future emissions trends**).

Due to the ability of the oceans to absorb vast quantities of heat, there are significant time lags between the emission of greenhouse gases and when global temperatures reach their equilibrium levels. This complicates climate sensitivity modeling and estimates of the timing of temperature impacts. Over the past 50 years, the oceans are estimated to have absorbed around 80% of the warming caused by greenhouse gases.[4] Even if we were to reduce emissions to a level that stabilized concentrations at today's level, it would be at least 30–50 years, and possibly longer, before temperatures would approach equilibrium levels.

The observed rise in global mean temperature since 1900 has been around 0.76°C. However, when the full temperature impact of greenhouse gas concentrations of 455 ppm CO_2e are realized, scientists estimate that temperatures will increase by at least another 0.5°C,[5] and possibly by as much as 1°C.[6] This is termed the residual *commitment to global warming*. If the cooling effect of **aerosols**, which have, so far, masked the full radiative forcing of present greenhouse gas concentrations (see **aerosols** and **global dimming**), or the potential impact of positive and negative **climate change feedback** effects, such as the reduced ability of the land and oceans to absorb CO_2 (see **ocean carbon sinks** and **land carbon sinks**) are taken into account, the warming effect of present (2007) concentrations is likely to be somewhere in the range of 1.5–2.0°C at equilibrium levels.

Given the significant uncertainties surrounding the warming impact of specific greenhouse gas concentrations, it is not possible to define a precise figure for climate sensitivity at double preindustrial CO_2e concentrations. This will remain an area of ongoing research and considerable uncertainty for some time to come. The greenhouse gas concentration–temperature relationship is expected to be nonlinear. In other words, each ppm increase in greenhouse gas concentration will not yield the same temperature change. The change in temperature will be affected by both the absolute level, and the rate of change, of concentrations. Global temperature change depends on a complex set of interactions in the climate system that we do not yet fully understand – in particular, the interaction of the land and ocean with the atmosphere and the temperature thresholds that could trigger positive **climate change feedbacks**. These include decreases in the earth's **albedo** resulting from reduced ice cover, accelerated releases of **methane** from wetlands, permafrost and **methane hydrates**, or changes in the ability of the oceans and land to absorb carbon. It is only recently that the impacts of these potential climate change feedbacks have begun to be quantified.

To simulate these complex interactions, scientists have developed sophisticated models that attempt to mimic the operation of the earth's climate system and the role of different feedback effects. While the ability of these models to predict future temperature changes continues to improve, they still produce a wide range of possible climate sensitivity outcomes, from as low as 1.5°C to as high as 11°C due to different model structures and weights attached to different feedback mechanisms. Scientists are also using combinations of different models (ensembles) to refine estimates of climate sensitivity under different assumptions.

A number of recent studies have derived climate sensitivity estimates and probability distributions. For example, Stainforth et al.[7] and Meinshausen[8] derive probability distributions of specific temperature outcomes associated with a doubling of preindustrial concentrations. In a majority of modeling exercises undertaken to date, the most frequently occurring climate sensitivity result has generally been in the range of 2.5–3.5°C, broadly consistent with the IPCC estimate. However, the probability distribution of different climate sensitivity outcomes is not even, and there is a greater likelihood that the temperature increase will be above this range rather than below it, what scientists call the *long tail effect*. For example, there is a 5% chance that climate sensitivity will be only 1.5°C (or 1–2°C below the mean value) and an equal chance that it will be as high as 11°C (7–8°C above the mean value).

Ongoing research, requiring more complex and sophisticated models with significantly greater computing power, should help narrow the estimated climate sensitivity range. To overcome computational constraints, one recent experiment combined the computing power of around 100,000 home computers around the world (the climateprediction.net experiment) to run a model. The results of this experiment revealed that the climate sensitivity of a doubling of atmospheric CO_2e concentration would range from 1.9 to 11.5°C, with 3.4°C being the most likely climate sensitivity outcome, slightly above the IPCC's most likely estimate.[9] Another recent study, based on actual observed values from satellite data corrected for solar irradiance variations (independent of any model approaches), estimates that a doubling of CO_2e concentrations would yield a global mean temperature increase between 2.3 and 4.1°C, a narrower range than many modeling estimates, but consistent with other findings.[10]

The present scientific consensus suggests that climate sensitivity is most likely in the range of 3.0–3.5°C. Many scientists consider that a temperature change of this magnitude would result in **dangerous climate change** and significant damage costs. If temperatures increase by this much, it will mean that global temperatures will be the highest they have been in at least 3 million years, and well above the levels experienced by human civilization to date.[11]

Climate sensitivity is an important concept in understanding the impact of **anthropogenic greenhouse gas emissions** on global temperatures. The degree of sensitivity has an important bearing on the costs and benefits of stabilizing greenhouse gas concentrations at a given level and hence the timing and magnitude of the response by the global community to reduce emissions. It will also have an important bearing on the direction of future climate change negotiations and commitments. If climate sensitivity is low, say 2°C, then climate change is likely to be more manageable and incur relatively lower damage costs, although they may still be significant. This may result in a less aggressive policy response by the international community. However, if climate sensitivity is high, say 4–5°C, then impacts would be very significant indeed, and the damage costs much greater. This would warrant much more urgent action to keep greenhouse gas concentrations below double preindustrial levels.

It is important to put the magnitude of the greenhouse gas concentration stabilization task in context. If no major action to limit emissions is undertaken, then, based on IPCC "business as usual" greenhouse gas emission scenarios, concentrations could be double preindustrial levels CO_2e by as early as 2040 and up to 1,000 ppm by 2100. Our present understanding of climate sensitivity at such high concentrations is too limited to know with any certainty what the likely global temperature change would be, but some models forecast temperature increases of 10–15°C during the next century if concentrations reach triple preindustrial levels. The consequences of temperature change of this magnitude would likely be catastrophic and, based on paleoclimatic records, well beyond maximum global temperature levels of the past 100 million years.

See also: albedo, carbon dioxide (CO_2), carbon dioxide equivalence (CO_2e), climate change feedbacks, dangerous climate change, methane (CH_4), methane hydrates, nitrous oxide (N_2O), radiative forcing.

Notes

1 Stainforth *et al.* 2005
2 IPCC 2007
3 NOAA 2007
4 Levitus *et al.* 2005
5 Meehl *et al.* 2005
6 Wigley 2005
7 Stainforth *et al.* 2005
8 Meinshausen 2006
9 Stainforth *et al.* 2005

10 Tung and Camp 2007
11 Hansen *et al.* 2006

Further reading

IPCC 2007; Stainforth *et al.* 2005.

COASTAL ZONE IMPACTS

Coastal zones are particularly susceptible to climate change as they are subject to several different, and often reinforcing, physical and biological climate-related impacts. These include sea level rise, higher risks of exposure to extreme events (such as intense storms, storm surges, and flooding), salt contamination of fresh-water resources, and climate-related changes to marine and terrestrial ecosystems. Humans are also vulnerable to coastal zone changes as one-quarter of the world's population resides within 100 km distance and 100 m elevation of the coastline, with significant increases in the number of people living in coastal regions expected over the next half century.[1] Many of the world's major cities and industrial facilities are located on the coast.

Human activities have already had a significant impact on coastal zones, and will continue to do so with or without climate change. Urbanization, pollution, changes to nutrient loads, conversion of marshlands and mangroves to other uses, overfishing, coral reef degradation, changes in natural sediment loads and erosion patterns, and alterations to hydrology and drainage patterns are just some of the impacts humans have had on coastal zones. Global warming will, however, introduce an additional set of physical and ecosystem impacts that are likely to affect the livelihoods of millions of people.

The three principal areas of climate change concern are **sea level rise**, marine and terrestrial ecosystem and **biodiversity impacts**, and increased exposure to **extreme weather events**.

Sea level rise

Estimates of how much sea levels will rise over this century vary considerably – they range from 0.5 to 1.5 m (see **sea level rise**). While uncertainty remains, it would be prudent to plan for sea level rises of at least 1 m, even if this does not in fact eventuate until sometime next century. As we are already committed to ongoing, and potentially significant, sea level rises over the coming centuries, the adaptation measures humans adopt to accommodate a 1 m rise will be needed next century in any case. It is likely that we are already committed to a global mean sea level rise of at least 2–3 m over the next few centuries, even if atmospheric greenhouse gas concentrations are stabilized at today's levels.

There are two important considerations on coastal zone impacts from sea level rise: (1) the change in average sea levels and (2) the level of variation in sea levels during extreme weather events (for instance, in the form of storm surges). Increases in average sea levels will result in a permanent loss of land, while extreme weather events can cause temporary rises in sea level of several meters during a storm.

The most significant impacts of average sea level rise will be experienced in those countries that have a large proportion of their land surface less than a few meters above current sea level. Many low-lying atoll nations, such as Kiribati, Tuvalu, the Maldives, and the Marshall Islands, are particularly vulnerable. For example, Majuro Atoll, the Marshall Islands' most populous island, would suffer the permanent loss of 80% of its land area with a 1 m rise in sea level.[2] Even at the lower end of IPCC global temperature projections for the twenty-first century, it is likely that these countries will disappear entirely within the next 150 years. Larger and more populous countries such as Bangladesh and the Netherlands are also highly vulnerable to sea level rise. Bangladesh would lose approximately 20% of its surface area with a 1 m sea level rise,[3] while 70% of the population of the Netherlands would be threatened by a similar rise.[4] In North America, a 1 m sea level rise would result in the loss of 640,000 km² of land, much of it densely populated Gulf Coast and Florida, while Europe has 140,000 km² within 1 m of sea level and 20 million inhabitants in these areas.[5] While these represent significant areas of land and include several major cities, a 1 m rise would not cause large-scale loss of land on a global scale – most countries could be expected to accommodate a rise in sea level of this magnitude. But over the next few centuries, as the sea rise inexorably continues, an ever-increasing area will be subject to permanent loss.

Although increases in average sea levels result in permanent loss of land, it is the short-term changes in sea level during extreme weather events that are of more immediate concern to coastal zone communities. Coastal zones are exposed to much more intense storm events than inland areas (principally due to the fact that intense tropical storms originate over the oceans and lose intensity rapidly once they cross the coast). During intense storms, and particularly in the tropics and subtropics, the large drop in atmospheric surface pressures can raise sea levels by up to 1 m or more in a very short period of time. This can cause the complete submergence of low-lying coastal floodplains, deltas, and island atolls.

Following these events, sea levels return to normal relatively quickly but often leave behind a trail of destruction. Storm surges can result in large-scale infrastructure damage and loss of life. In addition to rises attributable to low atmospheric pressure, wind-driven storm surges, and flooding from the intense precipitation that typically accompanies these storms, can raise sea levels by several meters and extend well inland to areas that would normally not be subject to inundation under average conditions. For example, during Hurricane Katrina in 2005, New Orleans, USA, experienced a storm surge that lifted water levels 10 m above normal levels.[6]

Changes in sea levels can significantly modify erosion and deposition patterns. Those areas that are already subject to high-erosion regimes are expected to erode further. In other areas, enhanced deposition of eroded sediments may reduce future vulnerability to some degree. Erosion and sedimentation patterns are also affected by changes in hydrological regimes. The expected rise in intense precipitation events will tend to increase the risk of coastal zone flooding, particularly in delta areas at the mouth of major rivers (e.g. the Mekong). In some areas, the increased frequency and intensity of droughts and reductions in long-term average precipitation, combined with human-engineered changes to river hydrology (such as dam construction), are expected to reduce sediment flow to the ocean and thereby result in increased subsidence of river deltas (e.g. the Nile and Indus deltas).[7] Such subsidence could increase the risk of exposure to sea level rise and storm surges.

Overall, the impact of climate-induced changes to sea levels, and the interactions of other climate impacts (particularly extreme weather events), will vary according to local specific circumstances and characteristics. Nonetheless, most coastal zones are likely to experience adverse impacts, and these impacts will increase in line with future changes in sea levels.

Ecosystem impacts

Climate change will affect terrestrial and marine ecosystems in several different ways. Marine ecosystems, and coral reefs in particular, will be adversely affected by changes in water temperatures and ocean chemistry; there will also be significant loss of important fish breeding grounds in coastal estuaries (see **marine impacts**). Terrestrial coastal zone ecosystems will be vulnerable to loss of habitat, increased exposure to extreme weather events and changes to hydrological regimes (see **biodiversity impacts**).

Coral reefs and marine species

Changes in ocean temperatures and ocean acidification are expected to have significant impacts on marine ecosystems (see **marine impacts**). Coral reefs are particularly vulnerable and are expected to be subject to significant damage at even low levels of warming (less than 2°C). These reef systems support a wide variety of marine biodiversity and also act as important natural wave protection barriers for coastlines, particularly during storm events.

Changes in sea levels and water temperatures will also affect the productivity and survival of seagrasses and mangroves, and the fish and other marine organisms that depend on them. Changes in water temperature will, in general, adversely impact on the productivity of plankton and the health, location, and abundance of fish populations (see **marine impacts**). Ocean acidification will affect both corals and other calcifying organisms, and this will, in turn, have flow-on effects for other marine species. All these impacts will affect the health and structure of marine ecosystems and the livelihoods of those that depend on marine resources, particularly fisheries, aquaculture, and tourism. For example, significant damage to coral reefs is expected to be detrimental to the tourism industries of not only many island nations, particularly the Pacific and Caribbean islands, but also Australia (home to the Great Barrier Reef and one of the world's 34 biodiversity hotspots).

Coastal wetlands

Wetlands and tidal estuaries play an important role in coastal ecology and support important marine species' breeding grounds and habitats for migratory birds. These systems are highly susceptible to sea level rise, changes in erosion and sediment deposition patterns, and changes in hydrology and drainage. Estimating the impact of global warming on coastal wetlands depends on assumptions about the magnitude and rate of increase of both sea level and water temperature. It also requires assumptions about changes in sedimentation patterns and the rate and ability of coastal wetland ecosystems to migrate.

Some coastal ecosystems may be able to adapt if sea level rise is slower than expected, or if high sedimentation rates enable wetlands to migrate inland. Other areas – such as the Port Royal wetlands of Jamaica – are bound by topography and have little migration potential. Over time, new wetlands will develop as new areas are inundated, but this process can take many centuries. If sea level rise is rapid, progressing at more than 6–8 mm/year (60–80 cm per century), the ability of many wetland ecosystems to adapt and migrate will be limited.[8]

Coastal wetlands are generally more susceptible to sudden changes in sea levels, as occurs during extreme weather events, than to slow changes in average sea levels. However, most impact assessments are based on changes in average sea levels and, furthermore, assume a linear relationship between atmospheric temperature and sea level rise when a nonlinear coupling might be more appropriate. Consequently, some scientists believe that existing assessments may underestimate the impact on coastal wetlands.[9]

The IPCC estimates that 20% of the world's coastal wetlands could be lost by 2080 at mid-range warming of 2–3°C.[10] Other studies estimate that up to 25% of the world's existing coastal wetlands could be lost by 2100 if temperatures exceed 3.5°C.[11] A recent comprehensive study of the literature on impacts to 2100 found that there was significant regional variation in expected impacts and concluded that some areas are clearly more susceptible than others.[12] Key findings of this study were that sea level rises below 20 cm and temperature increases below 1°C are not expected to have a significant impact on coastal wetland systems. At temperature increases of between 1 and 2°C, vulnerable wetland systems, such as the Sundarbans in Bangladesh and the Kakadu wetlands in Australia, will be significantly affected and could lose 50% of their current range. At temperature increases of 2–3°C, many wetlands areas in Europe (particularly the Baltic and Mediterranean regions) and North America could suffer significant impacts this century and might lose 50% or more of their current extent. Above 3°C, wetland impacts become more global in nature and would result in 95–100% loss of the Kakadu and Sundarban wetlands and extensive losses in the Mediterranean, Middle East, eastern United States, and the Baltic.

Of particular concern is the impact of climate change on coastal mangroves as they perform an important biological function and provide important fish breeding grounds, erosion control and protection from storms. Human activity, including conversion of mangrove systems to shrimp farms and clearance for ports and urban development, has already had a significant impact on mangrove forests. As mangroves are very susceptible to changing sea levels, water temperatures, and sedimentation patterns, climate change is expected to adversely affect them. The ability of mangroves to adapt will vary regionally, and some will not be able to migrate and will completely die off.[13]

What will be the impacts on humans?

Coastal zones are home to large numbers of people (200 million of whom live within 1 m of current sea levels). Two-thirds of the world's cities with populations over 5 million are within 10 m of current sea levels, and in China, 144 million people (11% of the population) live in this zone.[14] Over the next few centuries, the global population exposed to sea level rise is expected to increase to between 300–1,000 million.[15]

Sea level rise and storm surges will have major repercussions for coastal infrastructure and the livelihoods of millions of people. The amount of coastal infrastructure at risk from even small changes in sea levels and storm surges is significant. More than 30 of the world's major cities are very susceptible to sea level rise of 1 m, including the major financial centers of London, Tokyo, and New York. Other vulnerable cities include Shanghai, Calcutta, Mumbai, Lagos, Bangkok, and Amsterdam.

Coastal areas are also home to large concentrations of industry (such as oil refineries and steel mills) and port infrastructure that is essential to international trade. London has $ 220 billion of assets built on its low-lying floodplain, and Miami has $ 900 billion worth of infrastructure at risk of damage by hurricanes, sea level rise, and storm surges.[16] Coastal infrastructure can also be adversely affected by salt intrusion, which can undermine building foundations and contaminate water supplies. Coastal damage from storms is expected to increase with the greater intensity of **extreme weather events**. The destructive power of wind is the cube of wind speed, and a 5–10% increase in hurricane intensity could result in a doubling of GDP loss in the United States alone – equivalent to 0.13% of GDP.[17]

Climate change-induced changes to sea levels, storms, ecosystems, and hydrology will affect those that live on coastal floodplains, particularly in developing countries where communities are more vulnerable and have fewer **adaptation** options. In general, coastal floodplains house a disproportionately high concentration of poor people – the wealthier sections of the community are likely to live on higher ground. For example, approximately 35 million of Bangladesh's poorest people live in areas that are likely to be lost with a 1 m rise in sea levels and 247 million people in Least Developed Countries (see **UNFCCC**) live within 10 m of current sea level.[18]

Coastal communities are also expected to suffer from several climate-sensitive health outcomes, such as an increased incidence of malaria, dengue, and shell fish diseases,[19] and the effects of toxic algal and phytoplankton blooms.[20] Projections of health impacts are complex because of the substantial differences in the sensitivity of different diseases to increasing temperatures (see **health impacts**).

Adaptation options

As we are already committed to ongoing sea level rise of possibly several meters over the next few centuries, humans will need to implement a range of **adaptation** measures to cope with these changes. Response measures might include relocating people, infrastructure, and industry; coastal protection measures; improved early warning systems; and faster emergency responses. For some communities, particularly in the poorer developing countries, the options may be very limited – they will need either to retreat from affected areas completely or simply try to cope with the new conditions. If there is a need to relocate significant numbers of people (as might be the case for Bangladesh and several other delta areas), this could lead to potential conflicts and stresses in areas that absorb the migrating people.[21] For other communities, particularly in the developed world, the range of adaptation options is likely to be greater.

The adaptation response will depend to a large extent on the resources available to communities to invest in coastal protection and the value of the infrastructure at stake. For example, The Netherlands is highly susceptible to sea level rise as considerable portions of the country are close to or below sea level. Over the centuries, the Dutch have constructed massive dikes to hold out the sea and to reclaim land. Given that The Netherlands is a densely populated and wealthy country, increasing the height of the dikes and building additional protection systems is likely to be a viable adaptation response as it would be for other vulnerable cities (for instance, London, New York, or Tokyo), at least this century. Where valuable assets exist, investing in sea walls, storm surge barriers, and other protection systems may prove an economically viable option, at least for small levels of sea level rise (perhaps a few meters). However, the costs of constructing sea barriers and other coastal defences are significant, and the further the seas rise, the more expensive it will be to preserve existing assets in low-lying areas.

Overall, coastal zones are highly susceptible to climate change and face a complex array of impacts. The type and urgency of adaptation responses will vary between coastal zones. Much depends on how much and how quickly the seas rise and other climate impacts manifest themselves. To date, most impact assessments have focused only on the expected sea level rise for this century. However, it is also important to consider the long-term implications for coastal zones as the seas will continue to rise for many centuries. Even if global temperature increase can be kept below 2–2.5°C, the seas are still likely to rise by several meters over the next few centuries. If temperatures rise by 3°C, our commitment to sea level rise would most likely be at least 8 m over the next 500 years (see **sea level rise**), and possibly more. In terms of human civilization, 500–1,000 years is relatively brief. London has been a city for more than a thousand years, but much of present-day London is likely to be under water in another 500 years. It is important to factor in long-term impacts (those beyond 2100) when developing climate change mitigation and adaptation policies.

See also: biodiversity impacts, climate sensitivity, extreme weather events, global warming, health impacts, marine impacts, sea level rise.

Notes

1 Small and Nicholls 2003
2 IPCC 2002
3 Stern 2006
4 Ibid.
5 Ibid.
6 Ibid.
7 IPCC 2002
8 Hare 2006

9 Ibid.
10 IPCC 2002
11 Nicholls *et al.* 1999
12 Hare 2006
13 IPCC 2002
14 Holden 2007
15 Lowe *et al.* 2006
16 Stern 2006
17 Nordhaus 2007
18 Holden 2007
19 IPCC 2007
20 Peperzak 2005
21 Stern 2006

Further reading

Hare 2006; Stern 2006; IPCC 2002, 2007.

DANGEROUS CLIMATE CHANGE

Dangerous climate change is a term that refers to the degree of climate change that would contravene Article 2 of the 1992 **United Nations Framework Convention on Climate Change (UNFCCC)** – hereafter referred to as the Convention. Article 2 states that

> the ultimate objective of the Convention … is to stabilize greenhouse gas concentrations at a level that would prevent *dangerous anthropogenic inter-ference* with the climate system. Such a level should be achieved within a time frame sufficient to:
>
> - allow ecosystems to adapt naturally to climate change;
> - ensure that food production is not threatened; and
> - enable economic development to proceed in a sustainable manner.[1]

Although Article 2 of the Convention does not actually use the term *dangerous climate change*, it is commonly used in the context of "dangerous anthropogenic interference" and refers to the "man-made" emissions forcing climate change beyond natural variation. The difficulty is that the UNFCCC does not actually define what constitutes "dangerous anthropogenic interference" – there are no clearly defined limits or thresholds that would mark a breach of any of the three principles.

Indeed, it may not be possible to objectively define such thresholds. Consideration of what constitutes "dangerous" change is ultimately a value judgment: it depends on who and what is being adversely affected, who is making the judgment, and what is considered to be "acceptable" or "unacceptable."

Some would assert that we are already committed to a level of climate change that will have dangerous consequences, while others suggest that we still have sufficient time to reduce emissions to a point where "dangerous" impacts can be avoided. Some commentators believe the contribution of humans to climate change has been exaggerated and/or does not pose a serious threat to society. However, as the body of scientific evidence on climate change and its likely impacts have grown, the number of these "climate skeptics" appears to be dwindling.

More clearly defining what would, in the eyes of most of the global community, be considered *dangerous climate change* is fundamental to determining how the global community should respond and how urgent this response should be. In effect, it is the principal determinant of future climate change policy and action.

Defining dangerous climate change

To be of relevance, the concept of "dangerous climate change" needs to be assessed against a wide range of potential impacts and social, economic, and ethical criteria. Clearly the three guiding principles contained in Article 2 of the Convention should hold considerable weight in this assessment since the UNFCCC has been signed by 192 countries, demonstrating their agreement with these principles. However, for the global community to decide whether or not any of these principles have, or will be, clearly breached requires specific knowledge and agreement on key thresholds or impact levels that are considered dangerous. Impacts need to be categorized and valued to identify the potential risks, costs, and trade-offs. They also need to include not only the consequences for humans – both for this generation and future generations – but also for other species, which have no say in the decision-making process at all.

In essence, there is a continuum of points at which climate change could be considered dangerous by different stakeholders – there is no single, definitive point and, arguably, there is no "safe" level of climate change. Potential impacts also need be viewed in the context of the level of scientific uncertainty about how much global temperatures will increase and what impacts will result from different temperature changes. This depends on how sensitive the climate and dependent ecosystems are to different levels of greenhouse gas concentrations and on how much these atmospheric concentrations rise (see **climate sensitivity** and **future emissions trends**).

Any assessment of "dangerous climate change" must also be scale dependent. Climate change impacts can be viewed at global, regional, national, and local levels. Understandably, much of the international debate on defining dangerous climate change has tended to focus on the global-scale impacts. These include the commitment to sea level rise, climate change feedback mechanisms (such as the shutdown of the **thermohaline**, a transition of the **land carbon sink** to a source of emissions, and large-scale greenhouse gas releases from the permafrost and **methane hydrates**), trigger points for the total deglaciation of Greenland or collapse of the West Antarctic Ice Sheet, total loss of coral reefs, and potentially the loss of complete ecosystems. The **IPCC** "Burning Embers" approach presented in its third assessment report was an attempt to indicate possible global risks and impacts at different levels of warming. While this provides a useful benchmark to guide decision makers, it remains a very broad framework that does not illustrate in detail who will be affected and when. Many would agree that these global-scale impacts are unacceptable and would constitute dangerous climate change: consequently, it may be easier to achieve global consensus on these impacts than on those viewed at a lower scale.

Many of the most serious consequences of climate change will actually be felt at the national or local scale and will not be equally distributed within nations or across the globe (see **socioeconomic impacts**). Some people and some regions are more vulnerable than others and could face a disproportionate burden, while some communities may even experience net benefits at low levels of warming. For example, the disappearance of small island nations as sea levels rise or an increase in the incidence of drought in sub-Saharan Africa would obviously be viewed as dangerous for those that suffer these impacts. But viewed from a global perspective, such localized impacts may be considered tolerable and given little weight in the decision-making process.

The further one moves down the spatial scale, the more likely it is that the threshold of dangerous climate change will be breached. For example, for the Golden Toad of Costa Rica, the first documented species to have become extinct due to climate change, the dangerous climate change point was reached in 1989.[2] But few would value or even worry about its demise, and it is likely to be considered irrelevant in the global decision-making process.

Defining dangerous climate change requires adequate and reliable information on the physical processes at work, the potential impacts at different levels of global warming, who will be affected and how, to what extent we can accommodate these impacts, and what the costs and benefits are of different response options. Scientists, economists, sociologists, ecologists, and a range of other disciplines all have a role to play in providing information to decision makers. The scientific community can provide an objective analysis of possible impacts across different temperature changes and greenhouse gas concentration levels, but the importance of these impacts is largely a political decision, based on ethics, equity, and perceived costs and benefits. As IPCC chairman Rajendra Pachauri so aptly states, "What is dangerous is essentially a matter of what society decides. It is not something that science alone can decide."[3]

Some suggest conventional costs–benefit analysis can be employed.[4] This has many attractions, not least its extensive use in a wide range of other policy making areas, and many studies have adopted this framework. It is clear that cost–benefit analysis can be a very useful tool for decision makers, provided sufficiently accurate information on costs and benefits can be derived. However, it also has many weaknesses as it assumes that all things can be given an economic value and that a dollar is worth the same to each individual. These issues have major equity implications, which are discussed in more detail below.

In addition to cost–benefit analysis, decision makers might use other quantitative indicators, such as the number of species lost, the increase in morbidity and mortality rates, increases in regional vulnerabilities (for instance, the number of people subject to serious climatic events and coastal zone flooding), and the amount of land lost. All have advantages and disadvantages, and decisions would need to be based on a wide range of variables and assessment criteria.

At what point would the guiding principles of Article 2 be breached?

As a starting point in determining what constitutes dangerous climate change, it is useful to review at what point, based on what we presently know, any or all of the three principles of Article 2 would be breached, and thereby contravene the Convention.

Allowing ecosystems to adapt naturally

A considerable amount of scientific evidence regarding the impacts of climate change on natural ecosystems has been amassed, since the Convention was signed. Climate change has already resulted in ecosystem impacts, including species extinction, and these impacts are expected to become more pronounced and severe as global temperatures increase further. The general scientific consensus is that a temperature increase of 1°C will result in negative ecosystem impacts and biodiversity loss but is unlikely to be catastrophic or lead to widespread ecosystem collapse. At temperatures of 1–2°C, the impacts on ecosystems and biodiversity become more pronounced and at 2–3°C become significant and widespread, with up to one-third of all species vulnerable to extinction. Above 3°C, the impacts are considered to be very significant indeed, and many ecosystems could face total collapse (see **biodiversity impacts** and **marine impacts**). For natural ecosystems, the IPCC concludes that a global mean temperature increase of 2°C or more would most likely result in unacceptable losses in biodiversity and ecosystem functions.[5] Overall, it appears that for most ecosystems the point at which climate change is considered dangerous would occur at or before a 2°C increase in global mean temperature.

In terms of rates of change that ecosystems may be able to cope with, which is the actual guiding principle enunciated in Article 2, most scientific studies suggest that many ecosystems, whose ability to migrate or adapt is not restricted, may be able to accommodate temperature increases of up to 0.1°C, and possibly as much as 0.2°C per decade (see **biodiversity impacts**). Given that the current decadal rate of change is around 0.2°C and that global temperatures this century are likely to increase by at least 2°C, and possibly more if emissions continue on their current trend, it would appear that the threshold for dangerous climate change for ecosystems has already been breached, or soon will be.

Ensuring food supplies are not threatened

Climate change is likely to significantly change the productivity and location of agricultural production over the course of this century. Impacts on agriculture and food supplies will stem from changes in temperature, precipitation, and carbon dioxide concentrations (see **agriculture and food supply impacts**). Climate change has already resulted in changes to precipitation patterns (see **water impacts**) and **extreme weather events** (especially heat waves) and has affected agricultural production in several regions – notably sub-Saharan Africa and the Mediterranean region. Although it is difficult to isolate the impacts of climate change on global food production from a range of other factors (urbanization, population growth, land degradation, and economic change), there appears to be, as yet, little discernible impact at the global scale.

As temperature rise and precipitation changes become more pronounced, the effects on agriculture will become more apparent. These impacts will vary considerably across regions – some regions will experience declines in yields and production, while other regions may experience increases. The extent of these changes will be influenced by the strength of the carbon dioxide (CO_2) fertilization effect and changes in precipitation patterns (see **agriculture and food supply impacts**). The higher latitudes (above 30°N and below 30°S) are likely to benefit at low levels of warming, but the tropical and subtropical regions are expected to experience lower yields. For temperature increases below 3°C, the production increases in the higher latitudes may offset expected declines elsewhere, depending on the strength of the CO_2 fertilization effect. However, beyond 3°C, negative impacts are expected to become widespread across most regions (see **agriculture and food supply impacts**), and food supplies could be, in the terminology of the Convention, "threatened."

Food supply problems are most likely to arise as a result of a redistribution of production between regions, from the poorer countries in the tropics and subtropics to the wealthier countries at higher latitudes, rather than from an inability to produce enough food to feed the world's population. The poorest countries are much more dependent on rain-fed agriculture, and their economies and livelihoods are much more vulnerable to climate change than the wealthier nations. As a result, the point at which many low-latitude developing countries consider climate change to be dangerous is likely to arise much sooner than for countries in the higher latitudes. For tropical countries then, this principle could be breached somewhere between 1 and 3°C. The fact that the perception of "dangerous" varies considerably with latitude on an issue as fundamentally important as food supply underlines the difficulty of reaching a global agreement on this principle.

Enabling economic development to proceed in sustainable manner

This is the most difficult of the three guiding principles to assess as the economic costs of reducing emissions to limit climate change has generally been the main argument put forward against introducing aggressive emission reduction measures. The key point of contention is whether the costs of reducing emissions to stabilize greenhouse gas concentrations at a particular level of global warming (e.g. 2°C) exceed or are less than the impact costs associated with that level of warming. In short, is it cheaper to reduce emissions or adapt to their consequences?

While it is clear that the costs of reducing emissions increases as greenhouse gas concentration stabilization targets become more stringent (see **stabilization targets**), it is also clear that the costs of impacts rise significantly for each additional degree of warming. For example, a stabilization target of 475 ppm[6] **carbon dioxide equivalent (CO_2e)** would be more difficult and costly to achieve than a target of 500 or 550 ppm CO_2e. The actual costs of reducing emissions relative to impact costs incurred at different levels of warming remain highly controversial – different studies offer very different results (see **mitigation** and **socioeconomic impacts**). Estimated costs depend on how costs and benefits are measured and valued, how costs borne today are compared with lower climate impact costs at some stage in the future, who is required to reduce emissions and by how much, and the interpretation of sustainable economic development.

The concept of sustainable development is central to the climate change debate. The 2000 Johannesburg World Summit on Sustainable Development (WSSD) identified five areas where progress is essential if a basic quality of life is to be attained for the majority of the world's population. These are water, energy, health, agriculture, and biodiversity. These are also the areas where many of the negative impacts of climate change will manifest themselves most strongly. Each of these areas is dealt with separately in other sections of this book. Nonetheless, it is clear that water, health, agriculture, and biodiversity will all be significantly affected, even at relatively low levels of warming.

There is no single, universally agreed definition of development, though the most commonly used measure is Gross Domestic Product (GDP) per capita – often adjusted to an index of purchasing power in each country. Whether increases in GDP are sustainable or not depends on whether the economic activities used to generate this GDP is achieved in a manner that does not reduce the ability of present and future generations to meet their needs. GDP is often criticized as a measure of welfare as it often fails to reflect the true costs and benefits of economic activity (such as environmental degradation or loss of biodiversity), nor does it indicate how GDP is distributed across different segments of society – it is not equity based. In relation to climate change, many of the impacts that are likely to eventuate (such as biodiversity loss) are not given a cost or value. Furthermore, using only GDP impacts as an assessment criterion assumes that each dollar is worth the same to each individual. This is clearly not the case as the marginal benefit of an extra dollar to a Bangladeshi farmer is much greater than the same dollar to a rich New York financier. Nonetheless, while GDP is a relatively crude measure of development, most studies that review the economics of climate change use changes in GDP as the main benchmark against which costs and benefits are assessed.

Even if we accept that GDP is an adequate criterion to use in the assessment process, it does tend to mask underlying distributional impacts and important equity issues. Absolute damage impacts – as measured by changes in per capita GDP – in Burkina Faso and the USA, for example, may be of similar magnitude but would be of considerably different relative importance when compared with each country's total GDP. For example, a US$ 100 decline in per capita GDP for Burkina Faso would have a much more significant impact than the same drop in per capita GDP for the USA. So when using changes in global GDP as a basis for determining relative costs of emission reductions versus climate impacts, it is important to recognize the equity issues that arise. Aggregate changes in global GDP do not reflect the full human welfare impacts of climate change. For decision makers to make rational judgments on whether climate change is dangerous in a sustainable development sense, they need to have access to a broader range of indicators beyond just GDP.

With moderate climate change (up to 2°C), the impact on GDP for developed countries is likely to be proportionately less than for many of the poorer developing nations in the tropics and subtropics. Some countries (e.g. Canada) may actually receive a net economic benefit at low levels of warming. If the aggregate GDP gains for the wealthy nations (which account for a large proportion of global GDP) were to equal the losses experienced by the poorer developed countries (which account for only a very small proportion of global GDP), then it would appear that global net economic welfare has remained unchanged, when, in fact, this would not be the case in terms of the number of people suffering a net economic loss. These issues will be important in future international climate change negotiations.

There have been a number of recent reports and studies that have attempted to quantify the costs of climate change impacts and have compared them with the costs of reducing emissions to avoid these impacts (see **socioeconomic impacts**). These estimates vary according to the studies, but the emerging consensus is that once global mean temperatures exceed 2°C, the socioeconomic impacts of climate change are likely to be negative on a global basis, particularly for the poorer developing countries. Furthermore, climate change impact costs, in terms of reductions in global GDP, are likely to exceed the GDP costs incurred in reducing emissions to a level that would avoid these impacts in the first place (see **socioeconomic impacts** and **mitigation**). Overall, it appears that the principle of enabling economic development to proceed in a sustainable manner may also be breached at global mean temperature increases of more than 2°C.

Is there consensus on what constitutes dangerous climate change?

It is also apparent that to some degree these principles could be counteractive as adhering to one principle (for instance, enabling ecosystems to adapt naturally) might compromise another (notably, economic development). Overall, however, it appears that the three guiding principles contained in Article 2 of the Convention are likely to be breached at relatively low levels of global warming and possibly at or less than 2°C.

Attempts at defining a specific temperature threshold that would constitute dangerous climate change have been made. For example, one recent scientific paper identifies 2.85°C as the median threshold point for dangerous climate change.[7] In reality, given the continuing uncertainties that prevail, it is not really possible to define a specific threshold. Nonetheless, reaching global consensus on a temperature range that is considered dangerous will be important for guiding international action to address climate change. The European Union has called for a maximum temperature rise target of 2°C to avoid dangerous climate change, and many others have called for a similar target range and some for even lower thresholds.[8] Given that the global community could be committed to a rise in global mean temperatures of 2°C by or before 2050 (see **future emissions trends** and **stabilization targets**), it is clear that avoiding dangerous climate change is a daunting challenge that will entail difficult and contentious trade-offs over the coming decades. Considerations such as how much climate change can be tolerated, who bears the climate change impacts, and who should reduce emissions and by how much are all important considerations that should underpin future international climate change agreements.

See also: climate change impacts, future emissions trends, mitigation, socioeconomic impacts; stabilization targets; UNFCCC.

Notes

1 UNFCCC 2005, p. 33
2 Flannery 2005
3 Pachauri 2005
4 See Tol 2003
5 IPCC 2002

6 ppm is parts per million by mass, the standard measure of atmospheric concentration.
7 Mastrandrea and Schneider 2004
8 Izrael and Semenov 2006

Further reading

Tol 2003; Mastrandrea and Schneider 2004; Izrael and Semenov 2006; Yamin *et al.* 2006;
Stern 2006.

EL NINO SOUTHERN OSCILLATION (ENSO)

The El Nino Southern Oscillation (ENSO) cycle of alternating warm (El Nino) and
cold (La Nina) events in the tropical Pacific is the dominant interannual climate variation on earth. It is a naturally occurring ocean–atmosphere interaction which results
in a 2–7 year cyclical fluctuation of atmospheric and oceanic temperature and circulation patterns in the Pacific Ocean.

El Nino refers to sea surface temperature anomalies in the eastern tropical Pacific.
The name originates from the warm ocean current which periodically flows along
the coast of Ecuador and Peru, commencing around Christmas, disrupting the local
fishery and lasting several months, prompting local fishermen to call it El Nino (the
Christ Child).[1] Southern Oscillation refers to the atmospheric circulation response in
the Pacific–Indian Ocean region, measured by the difference in temperature anomalies between Tahiti and Darwin. ENSO is the linked interaction between El Nino
(oceanic cycles) and the Southern Oscillation (atmospheric cycles) – referred to as
an ocean–atmosphere coupled cycle.

ENSO and climate

Normally, the difference between high atmospheric pressure centered over the
Eastern Pacific and low pressure centered over Indonesia and Australia causes trade
winds from East to West along the Equator. During an ENSO, these prevailing trade
winds weaken and the equatorial countercurrent strengthens, causing warm surface
waters around Indonesia to flow eastward. Since warm water is less dense than cold
water, the easterly flowing waters overlie the cold waters normally found on the
eastern Pacific Rim. This shift in warm water disrupts and prevents the upwelling of
cold, nutrient-rich waters that are essential to biological production and thus reduced
fish catches in Peru and Ecuador.

ENSO events change precipitation patterns, typically bringing drought to Australia,
Indonesia, and neighboring countries but heavy rains and flooding in the island
states of the central Pacific and west coast of South America. These changes in
precipitation patterns also lead to changes in atmospheric circulation and weather
patterns in all ocean basins and across all continents, through what are termed teleconnections. Strong ENSO events are causal factors in flooding in Northwest United
States, increased frequency of tropical cyclones making landfall in Japan and the
Korean Peninsula,[2] monsoonal failures in India[3,4] and East Africa[5], with associated

increases in the risk of forest fires in regions that become drier (see **extreme weather events** and **marine impacts**). La Nina events have roughly the opposite effects, though less pronounced. The most significant impact of strong La Nina events is the tripling of likelihood of major Atlantic hurricanes striking the United States. It also brings above average rainfall to Australia and Indonesia.

ENSO has a major influence on climate and weather and hence an important influence on global environmental conditions and economic production (especially agriculture and fisheries). During ENSO events, primary production in the tropical Pacific decreases significantly in response to weakened upwelling of nutrient rich water. It also reduces oceanic uptake of **carbon dioxide (CO_2)** (see **carbon cycle**).

Past ENSO events

Historical climate reconstructions derived from studies of coral accretion show that ENSO has existed for at least the past 130,000 years,[6] operating even during glacial times when regional and global temperatures were substantially lower than today.[7] By studying boreal tree ring data, scientists have been able to compile a reasonably accurate record of ENSO patterns for the past 1,000 years. These data correlate well with instrument recordings over the past two centuries.

Since the 1950s, there have been eight well documented major ENSO events. Although ENSO has exhibited considerable interdecadal variability over the past century, a systematic change has been observed since the 1976–1977 ENSO – events have become more frequent, intense, or persistent.[8] The 1997–1998 ENSO was the strongest and the 1982–1983 the second strongest event in the well-documented data dating back to 1800.[9]

ENSO events and climate change

During the warm Pliocene 3 to 5 million years ago, atmospheric CO_2 levels were well above normal, comparable to current levels. During this period, near permanent ENSO conditions are believed to have prevailed. This resulted in very dry conditions in what is now Southeast Asia and Australia and warmer and wetter conditions in the Eastern Pacific. Evidence from the Eocene "hothouse" conditions that prevailed 35 to 55 million years ago, when atmospheric CO_2 levels were near double pre-1750 levels, also suggests a strong ENSO cycle prevailed.[10] Scientific evidence suggests that the twentieth-century ENSO events are stronger and more frequent compared with ENSO events of previous cool (glacial) and warm (interglacial) times, at least over the past 130,000 years.[11]

Although past history suggests that stronger El Nino cycles generally coincide with warmer global conditions, it is too early to conclude that recent **global warming** is the cause of the more frequent and intense events observed over recent decades. Given the absence of reliable long-term data (only a small number of data points are currently available), it will require several more cycles (and hence decades) to draw reliable statistical conclusions about the link between ENSO and global warming. Recent computer modeling for the period to 2100 shows only a slight increase in the frequency and intensity, and some studies conclude that **global warming** will have very little influence on ENSO in the medium term.[12]

However, if past history repeats itself and a warmer planet does coincide with stronger and more frequent ENSO events, this would have a significant, and possibly devastating, impact on agriculture and ecosystems in Australia, Southeast Asia, East Africa, and India. It will also result in widespread shifts in regional climate patterns across the globe through teleconnections. Overall, there could be greater global uncertainty for rain-fed crops, fisheries, drinking water supplies, and commensurate economic activities (see **water impacts** and **agriculture and food supply impacts**).

What is not in dispute is that ENSO events superimposed on temperatures already elevated through **global warming** will result in more **extreme weather events**. Where ecosystems are already climate stressed, ENSO events can push them beyond sustainable threshold levels and result in the collapse of populations and catastrophic losses of biodiversity or keystone species (see **biodiversity impacts**). Entire ecosystem regions may take decades to recover from such losses, or they may be permanently altered. Such events appear to have already occurred, with the loss of 16% of the world's reef-building coral with the 1997–1998 ENSO event exacerbated by elevated background sea surface temperatures.[13]

See also: biodiversity impacts, carbon cycle, climate change impacts, extreme weather events, marine impacts, thermohaline.

Notes

1 Glantz 2001
2 Fudeyasu *et al.* 2006
3 Kumar *et al.* 2006
4 Kane 2006
5 Weldeab *et al.* 2007
6 Ourbak *et al.* 2006
7 Tudhope *et al.* 2001
8 IPCC 2007
9 Jones *et al.* 2001
10 McPhaden *et al.* 2006
11 Tudhope *et al.* 2001
12 Oldenborgh *et al.* 2005
13 McPhaden *et al.* 2006

Further reading

IPCC 2007; McPhaden *et al.* 2006; Glantz 2001.

EMISSIONS INTENSITY

Emissions intensity is a measure of the quantity of greenhouse gas emissions per unit of economic output. A country's Gross Domestic Product (GDP) is the most commonly used measure of economic output. However, other measures, such as the quantity of emissions per physical unit of production (e.g. emissions per tonne of steel, cement, or aluminium), can also be used at the industry or sector level, both within and between countries.

As GDP is the principal measure used in deriving emission intensity statistics, it is important to be aware of some of the shortcomings of GDP as a metric. GDP is only an approximate measure of the economic value of productive activities and only a broad indicator of wealth. The accounting and valuation procedures that are used to derive GDP statistics result in only some activities being valued and accounted for, and not others. For example, GDP only measures the values of goods and services exchanged in the market and excludes activities such as subsistence agricultural production and barter trade (which can be significant in some countries). GDP statistics also do not account for many externalities generated by economic systems, such as pollution, loss of biodiversity, and climate change. Moreover, some clearly unwelcome environmental impacts, such as the costs of cleaning up an oil spill, are registered as positive additions to GDP, rather than negatives.

Despite these shortcomings, GDP is a widely accepted measure of economic output. Accounting procedures are consistently applied across most countries, and data can be presented in common currency units after being converted to purchasing power parity values (purchasing power parity adjusts GDP to reflect what a standard currency unit, such as one US dollar, can buy in different countries).

Emissions intensity figures are usually presented in terms of total greenhouse gas emissions per unit of GDP (converted to **carbon dioxide equivalent** units – CO_2e) but are also sometimes presented for just **carbon dioxide (CO_2)** emissions per unit of GDP.[1] Emission intensity figures that are based on all greenhouse gas emissions more accurately reflect a country's actual emissions intensity and should be used where data are available.

Factors that determine emissions intensity

The emissions intensity of a country is primarily determined by the structure of its economy, its primary energy supply, the energy intensity of economic activities, and, to a lesser extent, its land-use and agricultural practices.

Economic structure

A broad range of economic activities can contribute to GDP, and these are usually categorized according to sectors such as mining, energy (coal, oil, and gas production; electricity generation; petroleum refining), heavy industry (such as iron and steel, nonferrous metals, cement, glass, and chemicals), food processing and manufacturing, finance and services, tourism, transport, and agriculture and forestry, and so on. The relative contribution of each of these sectors to national GDP defines a country's economic structure.

Economic structure has a major bearing on emissions intensity. Some sectors generate more greenhouse gas per unit of GDP than others. Sectors such as energy, heavy industry, transport, agriculture, and manufacturing are generally more emissions-intensive than tourism, finance, and other service-sector activities. Those countries that generate a significant portion of their national income from energy-intensive activities will generally have higher emission intensities than countries that generate a large proportion of their national income from finance, services, and tourism.

In general, as economies expand and mature, the dependence on agriculture and heavy industry tends to fall and the contribution from the services sector tends to rise. As a consequence, high-income countries (such as Japan, the United States, and most European economies) are often less emissions-intensive than low- and middle-income developing countries, although there is considerable variation between countries due to the effects of other variables that influence emissions intensity.

Primary energy supply mix

The types of energy and technologies that a country uses to meet its demand for energy services has a major bearing on emissions intensity, particularly CO_2 emissions. Energy can be sourced from **fossil fuels** (coal, oil, and gas), **renewable energy** sources (hydropower, wind, solar, geothermal, biomass), and **nuclear power**. The contribution of each of these different sources defines a country's primary energy supply mix. Countries usually source their energy from a variety of sources, and the relative contributions of each source varies considerably between countries, primarily influenced by a country's endowment of natural resources (e.g. the size and type of fossil fuel resources or the existence of significant hydropower resources).

Countries that source a large proportion of their primary energy supplies from fossil fuels, particularly coal (the most carbon-intensive **fossil fuel**), will generally have higher emission intensities than countries that source a large proportion of their energy from low- or zero-emission sources such as nuclear, natural gas, and renewable energy. China, South Africa, Ukraine, and Australia, for example, source a high proportion of their energy from coal and, as a consequence, have CO_2 emission intensities that are higher than countries such as France (which has a large nuclear contribution) or Brazil (which has a large contribution from renewable energy sources).

For most countries, CO_2 emissions from **fossil fuel** use normally account for the largest share of national greenhouse gas emissions and, for some countries (e.g. Japan and Singapore), can account for up to 90% of total emissions. Furthermore, energy-related CO_2 emissions usually account for a greater proportion of the emissions in high-income countries than in low-income countries (see **emissions per capita**). There are, however, several notable exceptions to these general rules. For instance, nonenergy emissions account for more than half of the emissions of Brazil, Argentina, New Zealand, and several other countries.

Energy intensity

All economic activities use energy, but some use more than others. Energy intensity is the amount of energy required to generate a unit of GDP. The energy intensity of an economy is determined by its economic structure (especially the contribution from the energy and heavy industry sectors); the efficiency with which primary energy is converted to energy services (heating, refrigeration, transport, etc.); and, to some extent, the country's prevailing climate (countries that have long, cold winters generally use more energy than countries with less extreme climates).

Energy efficiency has an important bearing on energy intensity and has been a major contributor to declining energy intensities in many countries in recent decades. Major energy efficiency improvements have been achieved in Japan, China, and many European Union countries, while in others the gains have been more modest. The United States, for example, uses approximately 25% more energy to generate each unit of GDP than most other high-income countries. Americans tend to use more energy to heat and light their homes, drive larger cars, and operate industrial processes that are generally less efficient than those employed in Japan and Europe.

Land use and agricultural practices

Emission intensity is also influenced by a country's level of dependence on agriculture and forestry, the type of agricultural activities undertaken, and how land resources are managed. Agriculture is a major source of **methane** (mainly from livestock, manure, and rice cultivation) and **nitrous oxide** (mainly from fertilizers). Forestry practices, deforestation, and land-use management (e.g. the extent of overgrazing and soil tillage) are also sources of CO_2. Nonenergy greenhouse gas emissions generally account for a higher share of total emissions in low- and middle-income countries, relative to high-income countries, due to the higher dependence on agriculture.

As methane and nitrous oxide emissions have higher **global warming potentials** (GWP) than CO_2, those countries that have large agricultural sectors also tend to have high emissions intensities. For both Argentina and New Zealand, for instance, non-CO_2 emissions account for more than half of their total greenhouse gas emissions. Both have relatively high emission intensities, even though their energy-related CO_2 emission intensities are low compared with most high-income countries.

CO_2 emissions from land-use change, mainly deforestation, can be a major source of emissions for some countries. If such emissions are excluded for Indonesia and Brazil, which together account for nearly half of the world's CO_2 emissions from deforestation, their national emission intensities fall significantly.

Variations in emission intensities across countries

There is considerable diversity among countries. Unlike emissions per capita or aggregate emissions, emission intensity is not closely correlated with income. Table 7 contains emission intensity data for selected high-, middle- and low-income countries (with greater than US$ 15,000; US$ 5,000–15,000; and less than US$ 5,000 per capita incomes, respectively).

As is evident from Table 7, emission intensity levels can vary by a factor of three or more between countries. As expected, those countries that have carbon-intensive primary energy supply mixes, or high energy intensities, are also generally those that have the highest emission intensities. Low- and middle-income developing countries generally have higher greenhouse gas intensities than high-income developed countries,

Table 7 Emission intensities for selected countries

Country	GHG intensity tCO$_2$e/$mGDP	CO$_2$ intensity/ $mGDP	Energy intensity (tonnes oil equivalent/$mGDP)	Energy supply mix (CO$_2$/tonne oil equivalent)
Ukraine	2,369	1,368	569	2.4
Pakistan	1,074	382	257	1.6
China	1,023	675	219	3.1
South Africa	1,006	823	258	3.2
Australia	977	630	210	3.0
Indonesia	799	513	241	2.1
India	768	410	200	2.0
South Korea	729	633	258	2.5
United States	720	579	230	2.5
Brazil	679	263	146	1.8
Argentina	659	319	145	2.2
Germany	471	400	161	2.5
United Kingdom	450	363	152	2.4
Japan	400	369	157	2.3
France	344	244	171	1.4
Average developed	633	511	212	2.4
Average developing	888	549	224	2.5

Source: WRI (2005).

Note: Income data are based on purchasing power parity (PPP).

reflecting differences in economic structures. Most countries that have low emission intensities also generally have low energy intensities. This is partly due to the structure of their economies and also partly due to their more efficient use of energy (particularly in the case of Japan and many European countries).

The Ukraine has by far the most emission-intensive economy (four times that of Japan), primarily due to Ukraine's high dependence on coal and heavy industry and also due to low energy efficiency. Other emission-intensive economies include China, Australia, and South Africa, largely due to their very carbon-intensive energy mixes.

With the notable exception of the United States, Australia, Canada, South Korea, and, to a lesser extent, New Zealand, most high-income countries have the least emission-intensive economies. France has the lowest emission intensity of the developed world, primarily due to a large contribution from nuclear power and also its relatively low energy intensity.

Amongst the low- and medium-income countries, there is considerable diversity. Most have moderate to high emission intensities relative to high-income countries, primarily due to high agricultural shares in GDP. On average, the emission intensity of low-income developing countries is approximately one-third higher than the average for high-income countries.[2]

Trends in emission intensity

Overall emission intensity has fallen in most countries over the past few decades. Over the period 1990–2002, emission intensities declined, on average, by 23% in high-income countries and 12% in low- and medium-income countries.[3] These declines were largely the result of falls in energy intensity (achieved mainly through improved energy efficiency) and structural changes in economic structure and also partly the result of changes in primary energy supply. For example, several European countries have increased the use of natural gas (the least carbon-intensive fossil fuel) as a substitute for coal and oil. The most significant falls in emission intensity took place in China (51%) and Poland (43%), followed by Germany (29%) and the United Kingdom (29%).[4] Emission intensities have, however, increased in a few countries, notably in Saudi Arabia (by 45%, largely due to increased energy intensity) and Indonesia (by 22%, due to increased use of coal and deforestation). Overall, global emissions intensity fell by approximately 15% over the period 1990–2002.

While the overall decline in emission intensities has helped to reduce emissions growth below what it could otherwise have been, these gains have been more than offset by increases in GDP and population. As a result, absolute emissions have continued to increase at an average rate of 1.4% per year since the early 1990s.[5]

Significant untapped energy efficiency potential remains in all countries. If the uptake of energy efficiency measures can be accelerated, declines in energy intensity could make a significant contribution to offsetting the emission increases driven by future growth in GDP and population (see **energy efficiency**). The cost-competitiveness of alternative low-emission technologies, such as **renewable energy**, **nuclear power,** and **carbon capture and storage**, are also expected to improve. While the increased uptake of these technologies will help to reduce the greenhouse gas intensity of primary energy supply, the impact on emission intensities will be relatively slow to take effect as the energy supply mix typically changes only very slowly, over periods of 20–30 years or more.

Although considerable potential exists to reduce emission intensities, recent projections for the short to medium term (up to 2030) indicate that the reliance of fossil fuels, and coal in particular, is likely to actually increase marginally without major changes in existing policies. The share of fossil fuels in global primary energy supply is projected to increase from 80% in 2004 to 81% in 2030 (see **future emissions trends**), and energy-related CO_2 emissions will grow faster than total energy consumption, rising by 55% by 2030.[6] The increased use of coal has already altered intensity trends in some countries. In China, for example, the long-term decline in carbon intensity has been reversed since 2002.[7]

Emission intensity changes arising from nonfossil fuel sources are more difficult to project. Such changes are driven by a range of factors often unrelated to trends in GDP (such as demographic changes, agricultural trade policies, the intensification of agriculture, and the creation of national parks and forest reserves), and obtaining reliable long-term data can be problematic. Deforestation rates, while still significant, have stabilized in recent years and in some countries have begun to fall. The share of agriculture has also declined in some countries, including China and India (both large emitters). Changes in population and agricultural commodity prices will also be important determinants of the direction and magnitude of emission intensity trends from nonfossil fuel sources.

In the coming decades, the extent to which the rate of decline in emission intensities can be maintained will depend largely on continued improvements in energy efficiency and on efforts to move to less carbon-intensive primary energy supplies. These factors will, in turn, be influenced by trends in the costs of alternative energy supplies, end-use technologies, and by the greenhouse gas **mitigation** policies adopted by governments.

Emission intensity targets

Reliable data on emission intensities are not only essential for predicting future emissions, but they also have important implications for international climate change negotiations. Currently, emission intensity targets have no formal place in either the **UNFCCC** or the **Kyoto Protocol**. However, several countries, and the United States in particular, have suggested that emission intensity targets could form part of the commitments made by countries under future international climate change agreements. The adoption of sectoral targets – for example, emissions per tonne of output of a given product or activity – has also been suggested as a possible option that could be included in a post-Kyoto framework (which is currently being negotiated).

If sector-based targets do form part of a post-Kyoto framework, then emission intensity data would form the principal basis by which progress in achieving the targets would be assessed. However, sectoral targets are best suited to industries that produce a relatively uniform product (steel, aluminium, cement, and several other products) and are not particularly well suited to industries or sectors that produce a diverse range of products. If extended across all sectors, intensity targets would be much more difficult to negotiate and track due to the significant associated data and monitoring requirements: they would add considerable complexity to any international agreement.[8]

Of the numerous factors that contribute to emissions intensity, not all are within the ability of national governments to control. Furthermore, emission intensities are not particularly transparent indicators of greenhouse gas production as they can be distorted by changes in economic structures, international trade, and carbon leakage issues (where production facilities move to other countries – see **emissions per capita**). For example, in 2002 the US Government announced a target to reduce emission intensity by 18% by 2012. Although this seems like an ambitious target, in fact it largely reflects the ongoing intensity trend the United States had experienced over the previous 10 years (when emission intensity declined by 18.4%): as a result, the target may simply reflect business-as-usual conditions. When other factors such as population and GDP growth are factored in, absolute US emissions would still increase by a projected 14% above 2002 levels by 2012.[9]

Emission intensity targets as a possible basis for an international agreement to reduce emissions have major drawbacks. First, intensity targets do not guarantee that emissions would be reduced as GDP and population growth could outweigh the gains from emissions intensity. This drawback could possibly be overcome if it were to be agreed to reduce emission intensity by a certain percentage below the combined growth in GDP and population. Second, emission intensity targets would need to be comprehensive in terms of international coverage to ensure that the targets are not achieved through carbon leakage to countries that do not have intensity targets (see **emissions per capita**). Finally, intensity targets face a range of administrative, measurement, and monitoring complexities relative to other emission targets. In general, emission intensity targets are not considered a workable basis for any international agreement to stabilize atmospheric greenhouse gas concentrations, but they may have a role at the national level.

In summary, continuing declines in global emission intensity levels will be essential if emissions are to be reduced sufficiently to stabilize greenhouse gas concentrations at a level that avoids **dangerous climate change**. However, stabilization of concentration levels will require emission intensities to fall at a much more rapid rate – possibly twice as fast – than they have over the past two decades. In order to stabilize emissions, let alone atmospheric concentrations (which are determined by past as well as ongoing emissions), emission intensity declines would need to equal the combined effects of GDP and population growth on global emission levels. That is, if GDP grows at 2% per year and population by 1% per year, then emission intensity would need to fall by 3% per year just to stabilize emissions. Obviously, this will require policies and measures additional to those currently in place – see **stabilization targets**).

See also: anthropogenic greenhouse gas emissions, emissions per capita, fossil fuels, future emissions trends, stabilization targets.

Notes

1 (1) CO_2 data typically include emissions from fossil fuel use and cement production but exclude CO_2 emissions from land-use change and forestry; (2) Official greenhouse gas emission data only account for those gases covered under the **Kyoto Protocol** and do not account for non-Kyoto greenhouse gases such as chlorofluorocarbons, halons, or ozone (see **anthropogenic greenhouse gas emissions**).
2 WRI 2005
3 Ibid.
4 Ibid.
5 Stern 2006
6 IEA 2006a
7 WRI 2005
8 Ibid.
9 Ibid.

Further reading

Stern 2006; IEA 2006a; WRI 2005.

EMISSIONS PER CAPITA

Each person on the planet emits, or is responsible for emitting, greenhouse gas emissions into the atmosphere each year. The quantity of emissions varies considerably from individual to individual, according to where they live, what they produce, and what they consume. It can vary from less than one tonne of **carbon dioxide equivalent (CO_2e)** emissions per year (e.g. as might be the case for a poor rural villager in Bangladesh) to possibly thousands of tonnes of CO_2e/year for very wealthy individuals who consume large quantities of goods and services and who travel extensively. The quantity of emissions each individual is responsible for emitting to the atmosphere (either directly or indirectly) is often referred to as their "greenhouse gas footprint" or, sometimes, just the "carbon footprint."

Individuals can calculate their approximate greenhouse footprint using Web-based emission calculators, but few people actually know the size of their greenhouse gas footprint. Individual emissions data are not collected or published since the data collection and processing needed to produce emissions information at this level of detail would be formidable and well beyond current data collection systems. Instead, individuals' emissions data are usually presented as a simple national per capita average (usually expressed as CO_2e/capita), derived by dividing a country's total annual greenhouse gas emissions (its national **greenhouse gas inventory**) by its population.

The two principal factors governing emissions per capita are income per capita (derived by dividing a country's Gross Domestic Product (GDP) by its population) and the **emissions intensity** of the economy. Emissions intensity is determined by the mix of different economic activities that generate a country's GDP (the reliance on agriculture, heavy industry, services, and other economic sectors) and the greenhouse gas intensity of the country's primary energy supplies (a function of the relative shares of coal, oil, gas, nuclear and renewable energy sources).

Emissions per capita are closely correlated with income per capita. In general, as incomes increase, so do greenhouse gas emissions, primarily due to higher per capita fossil fuel energy consumption. As incomes rise, so, too, do the level of car ownership, the propensity to travel, the number of household appliances owned, and the size of houses – these all consume energy and generate greenhouse gas emissions. For people living in high-income industrialized countries, fossil fuel-related CO_2 emissions usually account for more than three-quarters of their greenhouse gas footprints: in the case of the average Japanese citizen, the figure is over 90% (see Table 8). By comparison, energy-related CO_2 emissions typically account for less than two-thirds of the greenhouse gas footprint of people living in developing countries, and often less than half.[1]

For many developing countries, **methane** and **nitrous oxide** emissions generally account for a greater proportion of the greenhouse gas footprint than in high-income countries. This is primarily due to the importance of agriculture to developing economies: the agricultural sector typically accounts for between 15% and 30% of GDP and employment in most developing countries, compared with 1–3% for most high-income countries.[2]

Table 8 contains data on emissions per capita for selected countries and clearly demonstrates the relationship between income per capita, emissions per capita, and energy-related CO_2 emissions. The selected countries provide a representative mix of high-income (greater than $15,000/capita), middle-income ($5,000–15,000/capita), and low-income countries (below $5,000/capita).[3]

Table 8 Emissions per capita for selected countries (2000 data)

Country	GHG emissions (CO_2e)/capita	Energy sector CO_2/capita	% Emissions from energy	US$ GDP/capita (2002 PPP)
Australia	25.6	17.3	67	27,256
United States	24.5	20.4	83	34,557
Canada	22.1	17.1	77	28,728
Russia	13.2	10.6	80	7,993
Germany	12.3	10.4	84	26,141
United Kingdom	11.1	9.4	85	25,139
South Korea	11.1	9.9	80	16,570
Japan	10.4	9.5	91	25,788
South Africa	9.5	7.9	75	9,750
France	8.7	6.2	71	26,090
Argentina	8.1	3.9	48	10,664
Mexico	5.2	3.9	75	8,662
Brazil	5.1	2.0	39	7,480
China	3.9	2.7	69	4,379
Indonesia	2.4	1.4	58	3,057
Pakistan	2.1	0.8	38	1,941
India	1.9	1.0	53	2,572
Average developed	14.1	11.4	81	22,224
Average developing	3.3	2.1	63	3,806
Average world	5.6	4.0	71	6,980

Source: WRI (2005).

Note: Income data is based on purchasing power parity (PPP).

It is evident from Table 8 that there is considerable variation in per capita emissions between countries. Per capita emissions in most high-income countries range from 10 to 25 tCO_2e/year, with Australia, the United States and Canada possessing by far the highest emissions per capita. Per capita emissions for most other high-income European and Asian countries are approximately half those prevailing in North America and Australia, generally in the 10–15 tCO_2e/capita range. This is, in part, due to lower energy consumption per capita and also due to less emissions-intensive economies (see **emissions intensity**). France, for example, has a similar average income per capita to Australia, but only around one-third the level of greenhouse gas emissions per capita, mainly due to a large contribution from **nuclear power** (which generates more than three-quarters of France's electricity) in contrast to Australia's reliance on coal-fired electricity. New Zealand is unique amongst the high-income countries as more than half of its per capita emissions (19 tCO_2e) emanate from non-CO_2 sources, principally methane from livestock.[4] Other notable exceptions include Qatar (67.9 tCO_2e/capita) and the United Arab Emirates (36.1 tCO_2e/capita): both have very high emissions per capita, due to the importance of oil and gas production in their national economies, but account for only a very small proportion of global emissions.

Amongst middle-income countries, emissions per capita are usually between 5 and 8 tCO_2e/capita, with Russia having the highest per capita emissions (13 tCO_2e). However, there is considerable diversity amongst such countries, particularly in terms of the relative contribution of different greenhouse gases to their national emission profiles. For example, South Africa and Argentina have similar incomes and emissions per capita, but more than half of Argentina's emissions emanate from nonenergy sources (principally methane emissions from livestock), while for South Africa, which is heavily dependent on coal, nonenergy-related emissions account for only one-quarter of the national total (similar to high-income countries). In the case of Brazil, only 39% of national emissions come from fossil fuel-related sources, principally due to the large contribution of hydropower to Brazilian electricity supply, the relatively high share of **biofuels** in transport energy supply, and the large contribution of emissions from land-use change (primarily as a result of large-scale deforestation).

Per capita emissions for the low-income developing countries are generally very small relative to high-income countries. Most average less than 3 t CO_2e/capita, and several only 1 tCO_2e/capita. China has the highest emission per capita of the low-income countries (3.9 t CO_2e/capita in 2000), due to its relatively high dependence on coal. China's rapid economic and energy demand growth over recent decades has resulted in an emissions per capita level similar to that of middle-income countries such as Brazil and Turkey.

The most common characteristic of low-income countries (excluding China) is a very low level of fossil fuel-related CO_2 emissions/capita, often less than one-tenth of that found in countries such as Australia, Canada, and the United States. Low-income countries simply consume very little fossil fuel energy. One-third of the developing world's population (1.5 billion people) still do not have access to electricity, half (2.3 billion) still depend on biomass for their primary cooking fuel needs, and car ownership is negligible.[5] The number of motor vehicles per 1,000 people is usually fewer than 20 in most low-income countries, compared with 500–600 per thousand in most high-income countries, and over 750 per thousand in the United States.[6]

Reducing the number of people who lack access to basic modern energy services (such as electricity and/or modern gas cooking fuels) is essential to improving the livelihoods of the world's poorest people. It will also be crucial to achieving the Millennium Development Goals (MDGs) agreed by world leaders at the 2000 Johannesburg World Summit on Sustainable Development. As growing numbers of people gain access to modern energy services, and their per capita incomes grow, the contribution of fossil fuel-related CO_2 to their greenhouse gas footprint is likely to increase, unless they are able to source this energy from low- or zero-emission technologies.

Data reliability

By and large, per capita emissions statistics do tend to reflect the relative contributions of the citizens of different countries and regions to the buildup of greenhouse gases in the atmosphere, but data reliability issues exist. Most emissions per capita statistics provide only a relatively crude measure of who is contributing most to **global warming** as they mask some important underlying factors. Consequently, care must be taken when using official emissions per capita data for cross-country comparisons.

Population

In countries with relatively fast population changes (such as India), census data that is five years old may underestimate total number of people by 5–10% or more, with consequent impact on per capita data. However, given the need of governments to know where, and how many people they govern, and the fact that populations usually trend in predictable ways, reasonable estimates of population can generally be established. As such, the denominator of population in per capita emissions estimates generally does not present a major data distortion.

National emissions inventory data

The coverage and quality of emissions data has improved considerably since the early 1990s, but some data are still based on estimates rather than direct measurement. Data for energy-related **carbon dioxide (CO_2)** emissions are generally the most accurate and up-to-date, but data for **methane** and **nitrous oxide** tend to be subject to some uncertainty, particularly for many developing countries. Furthermore, the inventory data from which emissions per capita statistics are compiled are rarely up-to-date and sometimes 5–10 years old (see **greenhouse gas inventories**). They only provide a static snapshot of a dynamic situation that prevailed several years or more in the past.[7] Nonetheless, relative intercountry emissions rankings change only relatively slowly and available data (even if several years old) can still provide a reasonable reflection of current greenhouse gas footprints.

Accounting for international trade flows

A potentially greater distortion of people's true greenhouse gas footprints arises from the exclusion of international flows of "embodied emissions" (the emissions associated with the production of specific products). At present, emissions data are based on official national greenhouse gas inventory data, which only account for emissions physically occurring within national boundaries. National inventories do not include emissions embodied in imported products and services, even though the citizen of the importing country, by creating demand for the product, is at least in some part responsible for the emissions associated with the product's manufacture and transport, even though manufacture and transport emissions occur in a different country. National emission inventories also do not include most of the fuel used by international aviation and shipping (termed "international bunker fuels") – they do not appear in anyone's per capita emission statistics.

In the case of some internationally traded goods and services, the level of embodied emissions can be substantial: depending on the quantities involved, they can significantly distort emissions per capita data for some countries. For example, the production of refined metals (particularly aluminium), cement, and liquefied natural gas (LNG) consumes large quantities of energy and, depending on the primary energy source utilized, can result in significant greenhouse gas emissions. Producing meat, dairy products, and other animal products can also be relatively emissions intensive: livestock accounts for approximately one-third of anthropogenic **methane** emissions. In countries where these products account for a significant share of economic output, national emissions per capita will tend to be higher than in countries where they account for only a small proportion.

The extent to which these products are consumed domestically or exported will influence the greenhouse gas footprint statistics of people living in those countries. For example, if all the production of a good (e.g. steel) is consumed in the producer country, the associated greenhouse gas emissions would be included in the country's national greenhouse gas inventory and, therefore, appropriately reflected in the country's per capita emissions data. However, if all the steel produced in a country is exported, then the emissions associated with the steel production (the embodied emissions) would not be reflected in the consumer country's national inventory and per capita emissions data, but in the inventory and per capita emissions of the producer country. The impact of international trade on the emissions per capita statistics is, in effect, to underrepresent the greenhouse gas footprint of consumer countries and to overrepresent the greenhouse gas footprint of producer countries.

Significant quantities of emissions are embodied in internationally traded goods, and the distortion in emissions per capita statistics can, therefore, be significant. International trade currently accounts for approximately half of global GDP, and, in China, Germany, the Netherlands, and Canada, more than two-thirds of GDP.[8] However, given the complexity and data requirements of tracking emissions embodied in international trade flows, official emission statistics do not reflect these flows. Although it would be technically possible to track the embodied emissions, the complexity and data requirements of doing so means that international greenhouse gas accounting has not yet progressed to this stage.

A number of studies have attempted to adjust national emissions data to allow for trade flows. Those countries that import more embodied emissions than they export register a rise in national emissions when trade adjustments are made. Conversely, countries that export more embodied emissions than they import register a fall in national emissions. One study found that CO_2 emissions increase considerably for some high-income countries, including Japan (17%), France (15%), South Korea (10%), and the USA (5%); for others they fall, notably in Canada (–11%) and Australia (–6%).[9] Emissions for several major middle- and low-income countries also fall significantly, in Russia by –16%, China (–12%), and India (–7%).

International flows of embodied emissions also represent an important aspect of "carbon leakage," the movements in the production of emissions-intensive goods from countries with emissions limitations – for example, Annex I countries under the **Kyoto Protocol** – to those countries that do not have emission reduction targets: the non-Annex I (mainly developing countries). If leakage occurs, then an Annex I country may appear to have met its emissions reduction target by transferring production emissions to another country. Since consumption emissions may be sustained at the old level through increased imports of goods and services, there may be no net benefit to the atmosphere. In fact, transferring production to another country may result in an increase in pollution, since developing countries often have less stringent environmental regulations, and less efficient technologies may be employed.

Of course, moving a major production plant to another country is not something undertaken lightly as the investment in the existing plant would be stranded, and new costs of building a new plant will be high. Many products, such as alumina, aluminium, and cement, include large inputs of natural resources (such as bauxite and limestone, respectively) that are geographically specific and may not be able to be economically transported to a different country. Further, some countries are often less attractive investment sites due to factors such as availability of a skilled workforce, infrastructure availability (ports, roads, railways), and political factors such as civil unrest/conflict and/or corruption. Greenhouse gas emissions would represent only one possible consideration in an investment decision. Unless greenhouse gas emission penalties were very significant they are likely to represent only a minor element in decision making of where to locate emission-intensive industries. For this reason the carbon leakage issue, while it exists, is likely to be generally overstated and other factors influencing locational decisions are likely to be more important.

Although emissions per capita data may not fully reflect the extent to which people in different countries contribute to the build-up of greenhouse gases in the atmosphere, it is clear that the greenhouse gas footprints of people in high-income countries are significantly larger than those of people in low-income countries. This has important implications for the ongoing international climate change deliberations and underpins the negotiating positions of many non-Annex I developing countries (see **Kyoto Protocol**). It also underlies the concept of "differentiated responsibilities" contained in the **United Nations Framework Convention on Climate Change (UNFCCC)**. Many developing countries argue that emissions reduction commitments should be based on the "Polluter Pays Principle" and that responsibility for reducing emissions should therefore be weighted more heavily toward those countries that have, and continue to, contribute most to the buildup of greenhouse gases in the atmosphere.

It is also often argued that, based on fundamental equity principles, each individual on the planet has an equal right to the atmosphere and, if there is to be an allocation of entitlements to emit greenhouse gases into the atmosphere, then each person should receive an equal allocation of these emission rights. By way of illustration, if there was unanimous international agreement that greenhouse gas emissions should be reduced to a level equal to the natural rate of uptake by the earth system (10–20 billion tonnes CO_2e/year[10]) and that this allowable emission level be distributed equally amongst today's global population (approximately, 6.5 billion people), then each individual would receive an allocation of around 2–3 tCO_2e/year. While some countries have average per capita emissions below this level, most are presently above it and some countries have per capita emissions ten times this amount.

Equal per capita emissions allocations underlie the "contraction and convergence" framework put forward by organizations such as the Global Commons Institute.[11] Under this approach, annual emissions per capita in different countries would be allowed to converge toward similar levels over time and possibly roughly equate to the rate at which the natural systems can absorb the excess greenhouse gases in the atmosphere (thus stabilizing concentrations). This would require contractions in emissions by some countries and allow increases in emissions in others. Some also suggest that this should form a basic principle underlying the allocation of emissions caps in a global **emissions trading** system, should one eventually be established.

Elements of the principle of contraction and convergence have merit and should, on equity grounds, hold some sway in the international negotiation process. However, a single equal allocation of emissions rights across the globe is somewhat simplistic and may not necessarily lead to an efficient outcome. Different countries have different resource endowments, different population growth rates, and different opportunities for cost-effective emissions reductions. Countries are also likely to face different transitional constraints and adjustment burdens. These differences would, at least to some extent, need to be reflected in any negotiated agreement to ensure that they did not present perverse incentives or excessive burdens to particular countries. The allocation of emissions rights would also need to take into account international flows of embodied emissions. Understandably, there is considerable resistance to the contraction and convergence principle among countries that have high per capita emission levels.

Future trends in emissions per capita

Emissions per capita for most countries are expected to increase. Global average emissions per capita are also expected to rise, at least for the next few decades (see **future emissions trends**). Given that providing the poor with access to modern energy services will remain an important development objective, it is unlikely that low-income countries will be required to restrict their emissions to current levels, particularly if countries with higher per capita emissions do not significantly reduce their own emission levels.

Overall, the growth in emissions per capita is expected to be higher in low- and middle-income countries than in high-income countries, due to the higher expected per capita income, and hence energy, growth rates. Based on current trends, average emissions per capita in many developing countries are expected to nearly double over the period up to 2030. However, even if such a doubling occurs, it will result in levels that are still only one-third to a half of the average levels expected in the high-income industrialized countries.[12] The extent to which emissions per capita in developing countries can be constrained to levels consistent with **stabilization targets** (in the order of 2–3 tCO_2e/capita per year) will be largely determined by the ability of these countries to access cost-effective low-emission technologies and, thereby, avoid the emissions-intensive development pathway that have characterized most high-income countries.

It is clear that to avoid **dangerous climate change** average emissions per capita, on a global basis, must be reduced to around half of present levels. In some countries (such as the United Kingdom and several other European countries), emissions per capita levels have already fallen to some extent over the past decade, but for most countries they continue to rise. It is also clear that to have any chance of stabilizing at a relatively safe level basically all countries will need to contribute to emissions reduction efforts over the coming decades. However, the responsibility for these reductions will, to a large extent, need to be proportional to the level of emissions per capita. This will present a formidable political challenge and require a high level of international cooperation and commitment.

See also: anthropogenic greenhouse gas emissions, dangerous climate change, emissions intensity, future emissions trends, greenhouse gas inventories, stabilization targets.

Notes

1 WRI 2005
2 Stern 2006
3 In relation to greenhouse gas emissions, countries are often grouped into different categories: developing and developed, Annex I and non-Annex I (for the Kyoto Protocol), economies in transition, or according to per capita income. For the purpose of comparing the relationship between emissions and income, countries are classified into income classes. Although these income classes generally coincide with other classification approaches, particularly in terms of emissions per capita, some non-Annex I countries (such as South Korea, Singapore, and Kuwait) fall in the high-income category, while some Annex I countries are not considered high-income countries (e.g. Russia).
4 WRI 2005
5 IEA 2006a
6 World Bank 2006b
7 For example, the most recent comprehensive data set available for analysis in this book is more than 7 years old (year 2000 data).
8 World Bank 2006b
9 Ahmad 2003
10 Through the operation of the carbon cycle, the oceans and the land combined are currently absorbing an estimated 4–5 Gt of carbon/year – see ocean carbon sink and land carbon sinks.
11 See Global Commons Institute
12 Holtsmark 2006

Further reading

Stern 2006; IEA 2006a; WRI 2005.

EMISSIONS TRADING

Emissions trading is one of a range of policy instruments available to mitigate greenhouse gas emissions. It involves putting a limit on the quantity of greenhouse gas emissions (usually converted to their **carbon dioxide equivalent [CO_2e]**) that can be emitted to the atmosphere over a set period of time – the "emissions cap." Trading scheme participants (typically firms or production facilities) are then permitted to buy and sell emission allowances in order to meet their emissions cap. Emissions trading operates like a commodity market where the price of the commodity is determined by the market demand relative to its supply: in this instance, it is greenhouse gas emission allowances that are the commodities traded. It is the price of the emission allowance (often referred to as the *carbon price*) that provides the financial incentive to limit emissions.[1]

Emissions trading and **carbon taxes** (the other principal market-based mitigation policy instrument) have attracted considerable attention from climate change policy-makers due to their potential to deliver emission reductions at lower cost than other policy alternatives, such as traditional regulatory approaches. Trading in greenhouse gas emission allowances has been considered a potentially viable policy instrument following the success of the US Sulphur Dioxide (SO_2) Allowance Trading System introduced in the 1990s. This trading scheme enabled participants (mainly power stations) to meet SO_2 emissions reduction targets between 1995 and 2000 at half the estimated cost of achieving the same reductions through traditional regulatory means.[2]

Emissions trading alone will not be sufficient to achieve the required emission reductions to stabilize atmospheric greenhouse gas concentrations, but it is rapidly emerging as one of the principal instruments governments will employ. Trading in greenhouse gas emission allowances and credits (the *carbon market*) is already underway, and there are multiple systems, of varying design and coverage, which have been operating since 2004, and several more are planned to commence operation before 2012.

The global carbon market has grown rapidly in recent years, primarily driven by the European Union Emissions Trading System (EU-ETS) and the **Kyoto Protocol** flexibility mechanisms (International Emissions Trading, the **Clean Development Mechanism [CDM]**, and **Joint Implementation [JI]**). In 2004, the value of trades were less than US$1 billion, but this grew rapidly to reach nearly US$10 billion in 2005, $22 billion in 2006, and exceeding $30billion in 2007.[3] The size of the carbon market is expected to continue to grow strongly in coming years.

Market-based policy instruments

An important characteristic of market-based instruments, and emissions trading in particular, is that they provide flexibility in terms of where, how, and when emission reductions are achieved – they let the market decide on the most cost-effective means of reducing emissions. While both emissions trading and **carbon taxes** establish a cost penalty on greenhouse gas emissions, they differ in how this is achieved. With emissions trading, the quantity of emissions is fixed and the market determines the carbon price. With a carbon tax, the carbon price is determined exogenously (set externally by government) and the market decides the quantity of emission reductions.

In theory, both emissions trading and carbon taxes are capable of delivering efficient economic outcomes. Since the early 1990s, there has been ongoing debate on the relative merits of each. Both have advantages and disadvantages depending on how, and to which emission sources, they are applied.[4] In general, emissions trading is the preferred instrument for relatively large point sources of emissions, while carbon taxes are potentially better suited to large numbers of small emission sources.[5]

From an economic perspective, emissions trading offers significant dynamic economic efficiency benefits relative to many other policy instruments, particularly for large stationary emission sources (such as fossil fuel power stations).[6] Potential advantages of emissions trading systems are that they:

- guarantee a specific emission abatement outcome (set by the emissions cap);
- stimulate innovative responses from the market and potentially deliver emissions abatement at least cost; and
- can provide a new source of government revenue (if permits are sold) without increasing the overall tax burden if revenues are used to offset other taxes and charges.

Potential disadvantages of emissions trading are that:

- the carbon price is not known in advance, and permit prices can vary significantly over a relatively short time period (which complicates investment planning and risk assessments);
- can be administratively costly to establish;
- they are not well suited to reducing emissions from small-scale (diffuse) sources (as administrative overheads and transaction costs are often too high);
- although well suited to some sectors, extending coverage to include all sectors and gases faces a range of administrative and technical constraints; and
- if permits are allocated free, economic distortions and dynamic efficiency losses can arise.

Types of trading systems

There are two main types of emissions trading systems: "cap and trade" and "baseline and credit."

Cap and trade

This involves setting a specific limit (a cap) on the quantity of emissions that can be released to the atmosphere from a specified group of emission sources over a given time period. These allowable emissions (emission permits) are allocated to, or purchased by, facilities covered by the trading scheme and represent an exclusive property right to emit a given quantity of **greenhouse gases**. Normally one permit is equivalent to one tonne CO_2e. To comply with the scheme rules, each facility must have sufficient permits to match the greenhouse gases emitted during the trading period. For example, if a facility emitted 1,000 tCO_2e over the trading period, it would need to surrender 1,000 permits to the body administering the scheme (usually governments) in order to be in compliance. Facilities are free to trade (buy and sell) allowances to meet their needs in the most cost-effective manner. Entities that have insufficient permits to cover their emissions are deemed to be "noncompliant" and subject to penalties.

Emissions reductions should, in theory, be delivered at least cost as facilities that are able to reduce emissions at a cost lower than the prevailing emission permit price tend to do so, while those with emission reduction costs greater than the prevailing market permit price would find it cheaper to purchase permits from others rather than reduce their own emissions. In cases where an entity is allocated excess permits for free, or had purchased more permits than required to cover their emissions over the compliance period, surplus permits can be sold to other facilities facing permit shortfalls – thus making a profit. The higher the permit price, the stronger the financial incentive to reduce emissions.

Several cap and trade systems are presently operating, with the EU-ETS being the largest. Under the **Kyoto Protocol**, the International Emissions Trading flexibility mechanism is a cap and trade system that could account for a significant share of the carbon market if countries with emission targets choose to sell their surplus allowances (see **Kyoto Protocol**). International Emissions Trading is unique in that it only involves trade between Annex I country governments that have ratified the Protocol – the emission allowances cover a nation's entire emissions, rather than the emissions of an individual facility. It is not a true market-based trading mechanism as it does not access the full allocative forces of the market since only Annex I countries are involved. Nonetheless, it could provide a potentially cost-effective option for countries to meet their Kyoto targets.

Baseline and credit

This system has no specific emissions cap but instead uses an emissions performance profile (the baseline) that facilities must equal, or perform better than, in order to comply with the scheme. Those that perform better (produce fewer emissions) than their baseline generate emission credits (the difference between actual and baseline emissions). These credits can then be sold to other entities that are unable to meet their emission compliance commitments.

While baseline and credit systems can be applied at the firm level (as is the case with the New South Wales Greenhouse Gas Abatement Scheme in Australia) they are typically project based. Specific activities are implemented to reduce greenhouse gas emissions (or remove CO_2 from the atmosphere, see **biosequestration**) relative to a baseline emissions scenario without the project. The Kyoto Protocol's **Clean Development Mechanism** and **Joint Implementation** are project-based baseline and credit systems.

A crucial determinant of the effectiveness and environmental integrity of baseline and credit systems is the determination of the baseline. A baseline typically reflects a "business-as-usual" emissions scenario, where there are no specific actions adopted to reduce emissions. The baseline could, however, also embody an increasingly stringent emissions performance target (such as an **emissions intensity** target).

Baseline and credit systems face a number of weaknesses and disadvantages relative to cap and trade systems. The primary difficulty is that baselines are inherently uncertain since they require predictions of future emissions that would occur in the absence of the project or performance target. Inaccurate determination of the baseline may result in more credits being issued than is warranted, thereby diminishing the environmental integrity of the system. Furthermore, baseline and credit systems neither guarantee a specific emissions outcome, since improvements in relative emissions performance per unit could be more than outweighed if the number of units produced is increased, nor do they generate government revenue flows. In general, baseline and credit systems are much better suited to project-based approaches than economy wide systems involving private firms.

Rigorous procedures for validating baselines and verifying credits generated are needed to ensure the trading system's environmental integrity. Both the CDM and JI have implemented such procedures, although some critics have questioned whether the emissions reductions from some projects are "additional" (beyond "business as usual"). Other baseline and credit systems that generate emission credits for the "voluntary" credit market have also been often criticized for similar reasons.[7] These shortcomings have, to some extent, undermined the credibility of baseline and credit approaches and the confidence of credit buyers.[8] Several emission quantification standards have been proposed to address these credibility issues, but their effectiveness remains contingent on appropriate implementation and independent oversight.[9]

Cap and trade and baseline and credit systems differ in design, but they can be linked under the same scheme. For example, International Emissions Trading under the **Kyoto Protocol** is a cap and trade system that allows credits (permits) to be purchased from both JI and the CDM baseline and credit systems.

Cap and trade system design

To be effective and administratively efficient as a **mitigation** policy instrument, the trading system should embody several important principles:

- *Maximum emissions coverage:* To ensure a common price signal across the economy and to access the full range of cost-effective abatement opportunities, as many emission sources as practicable should be covered by the cap.
- *Provide a clear market incentive to reduce emissions:* The cap must be sufficiently stringent so as to enable permit prices to reach a level that drives emission reduction actions and technological innovation.
- *Be grounded within a clear, long-term policy framework:* To enable participants to plan and implement long-term emission reduction strategies, a credible and predictable policy environment is essential, while maintaining sufficient flexibility to respond to changed circumstances and knowledge over time.
- *Maximize administrative efficiency:* The system should be designed so that it minimizes administrative costs to market participants and the system administrator.
- *Accurate emissions accounting:* Emissions accounting, monitoring, and acquittal procedures must be consistent across all entities and sufficiently rigorous to ensure the environmental integrity of emission reductions.
- *Economic efficiency and equity:* System design should endeavor to minimize economic loss or dislocation across different sectors of the economy (or economies in a multi-country scheme) and ensure that particular entities or sectors do not shoulder a disproportionate burden relative to other entities, both within and outside the scheme's jurisdiction.

In practice it is difficult to satisfy all these design principles. Even if an emissions trading system is considered the optimal instrument, the political implications of imposing an emissions constraint on industry often tends to result in design compromises, such as exemptions or free permit allocations (reducing economic efficiency benefits). Trading schemes are likely to impact on the relative cost competitiveness of some firms more than others and could also reduce the value of a firm's assets by attaching future emission cost liabilities – creating what is termed "stranded assets." Furthermore, the recognition that not all countries and industries will be covered by the emissions cap means there is a possibility that some greenhouse gas emitting production processes (particularly emissions-intensive activities like primary metal production) may relocate to jurisdictions not covered by the system. This may result in a reduction in emissions in one country but an increase in emissions in another – with little net benefit to the atmosphere (often termed "carbon leakage"). Policy makers are acutely aware of these issues and, consequently, the design of emissions trading systems often reflects a mix of economic efficiency and political pragmatism.

Key issues to consider in implementing a scheme include the following: determining the system boundaries (which gases and sectors to include); setting the emissions cap; the mechanism for permit allocation/distribution; how emissions will be monitored, verified, and acquitted; and the system's compliance and enforcement provisions.

Determining the boundaries

Cap and trade schemes can be applied at various scales. Schemes can cover multiple countries (such as the International Emissions Trading), a specific region or group of countries (such as the EU-ETS), the national level (such as the UK Emissions Trading System or that proposed for Australia), or at a subnational level (such as the emissions trading scheme proposed for several states in North Eastern United States). A cap and trade system can also be applied within an individual organization such as that adopted by British Petroleum.

For maximum efficiency, a system should encompass as many greenhouse gas sources and sinks as possible and cover all **greenhouse gases**, but in practice this is difficult to achieve for two main reasons. First, system administrative costs per permit are generally inversely proportional to the size of the emission source – administrative costs per tonne are low for large emissions sources (such as a coal-fired power station) and high for small sources (such as an individual household or small business) and could even exceed the permit price by a considerable margin. Thus, most cap and trade systems encompass facilities that have emissions that exceed a specified minimum threshold (generally in the order of at least 10,000 tCO_2e/year). For example, in the European Union there are millions of different emission sources but only around 60,000 large emitters (accounting for around 50% of EU emissions) are covered by the EU-ETS up to 2012.[10]

Second, there are a range of technical difficulties in accurately accounting for emissions from some sources, particularly non-CO_2 emissions. For example, measuring the **methane** emissions from livestock, or **nitrous oxide** emissions from agriculture, faces many technical constraints, and uncertainty levels are generally high (often ±50%). By contrast, measuring the CO_2 emissions from fossil fuel consumption is much more accurate (generally, ±5% or better). It is essential that accounting integrity is maintained across the system (namely, that a tonne of emissions from one source equals a tonne of emissions from another), so trading systems tend to only cover gases and emission sources that can be reliably quantified and monitored.

The inclusion of sources that are currently subject to considerable uncertainty (such as livestock emissions) can compromise the systems accounting integrity.

Setting the emissions cap

Determining an appropriate emissions cap for a trading system is a critical decision for policy makers. Typically a cap will be set at the national or subnational level (although the European Union has established a multi-country cap) and will be influenced by the specific emissions reduction objective. For example, a government may wish to establish an emissions trading system to meet an emissions target agreed under the **Kyoto Protocol** or to achieve a specific national objective (e.g. achieving a 20% reduction in emissions by 2025, or some other target).

A stringent cap, where the number of permits is considerably less than "business as usual" emissions would result in higher permit prices (as firms bid against each other for the limited supply of permits) and potentially at higher economic cost. A "soft" cap, where the number of permits available is at or near "business as usual" emissions, is likely to result in low permit prices but few emission reductions. Inevitably governments would aim to strike a balance between setting a cap that is sufficiently stringent to provide an adequate abatement incentive without incurring excessive economic cost and dislocation.

To be effective, policy makers need to provide a predictable long-term policy environment enabling firms to confidently invest in new technology, some of which may have investment pay-back periods spanning several decades. For example, the long-term goal could be a 50% cut in emissions by 2050, but this may be broken up into several trading periods, each with a progressively lower cap.

Emission caps should, however, remain flexible over the short-to-medium term (5–10 years) to accommodate changed circumstances and incorporate lessons learnt in previous trading periods. Higher than expected abatement costs in the initial trading periods may warrant a temporary adjustment to a less stringent cap for a limited period to minimize structural adjustment costs and to provide more time for the development and deployment of new technologies. Alternatively, the need to respond to new scientific information that suggests emissions need to be reduced more rapidly than initially envisaged may warrant a more stringent cap. Mechanisms such as banking unused permits for future periods, or borrowing permits from the next trading period, are possible options for smoothing out and/or avoiding sudden or extreme fluctuations in permit prices, but both have disadvantages and need to be structured with care.[11] Allowing entities to access emission credits generated outside the system boundary (such as from the **CDM**) can help moderate permit price fluctuations and reduce the economic costs of meeting the cap. Nonetheless, any short-term adjustments in caps, banking, borrowing, or use of external credits should remain consistent with the long-term emissions reduction objective.

Permit allocation

Essentially there are two methods of allocating permits: selling permits (usually by auction) or allocating them for free (or some combination of the two). There are advantages and disadvantages with each, but auctioning is likely to deliver a more economically efficient outcome.

Free permit allocation, based on historical emission levels of the entities (grandfathering), has been the most common approach adopted to date. Allocations can also be based on efficiency benchmarking, where permits are provided only up to what is viewed as industry best practice efficiency.[12]

Free allocation has several disadvantages, including:

- providing a possible windfall profit for some entities as they receive an asset of value (emissions allowances) free of charge, which they can then sell – in effect providing a windfall gain to the entity receiving the allocation;
- weakened market incentives to reduce emissions relative to a situation where entities are required to buy permits;
- a potential market entry barrier for new market entrants if they need to acquire permits (normally through purchasing from existing emission permit holders) to enable them to operate, therefore facing a relative cost disadvantage;
- the risk that too many permits are issued relative to actual emission levels; and
- perverse incentives where high emitting facilities are maintained due to the potential emission rights associated with their continued operation.

While free allocation has many disadvantages relative to auctioning, it offers some benefits, at least in the initial phases of a trading scheme. The main attraction is that it minimizes the sudden cost burden existing industries face if all permits are auctioned, which could have adverse impacts on their market competitiveness (particularly if their competitors in other countries do not face an emissions constraint) and result in significant structural adjustment costs for some firms.

By comparison, auctioning permits can overcome many of the dynamic market efficiency problems associated with free allocation and also provide a source of new revenue for governments. The revenue generated from the sale of permits can be used to offset other forms of taxation (with associated economic efficiency benefits), help fund the development of low emission technologies, or even be used to help compensate those entities or individuals that are most adversely affected by the introduction of the trading system. Understandably, most major emitters prefer free allocation over auctioning. Policy makers are likely to face intense political lobbying from emission-intensive industries seeking free permit allocations.

Emissions accounting and acquittal systems

The allocation of emission permits effectively creates a commodity where previously none existed. As with any other commodity, an agreed system of measurement to ensure accuracy, quality, and comparability is required. An essential accounting requirement for all cap and trade systems is that all participants must have in place accurate and verifiable **greenhouse gas inventories** that meet the standards and guidelines set by the scheme administrator. System-wide accounting and acquittal procedures are important to ensure double counting and abuse of the system does not occur. In recent years, internationally consistent inventory accounting standards for entities (such as ISO14064[13]) have emerged that can be used as the basis for system emissions accounting rules. Widespread adoption of such standards facilitates system-wide accounting consistency and greater linkages between systems.

Establishing accurate and consistent emission inventories for all participants can be administratively costly and time consuming and is one reason why coverage is often limited to facilities that exceed a minimum emissions threshold. Once inventories and reporting systems have been established, the ongoing scheme administrative costs tend to fall.

To ensure environmental and market integrity, rigorous inventory auditing and verification systems are also required. Independent verifiers are usually employed to conduct audits to ensure that facilities correctly account for, and report on, their emissions. Periodic verification entails a cost to scheme participants.

Enforcing compliance

To ensure that emission reduction targets are achieved, the penalty to facilities exceeding their cap must be sufficiently punitive to act as a deterrent. Noncompliance penalties are normally set at a much higher level than the expected permit price over the trading period.

The compliance penalty effectively sets a ceiling on the price of abatement action since if the cost of abatement is higher than the penalty, rational firms would prefer to pay the penalty. Some trading systems have used a relatively low penalty price (such as the NSW GHG Abatement Scheme penalty of AU\$ 12/tCO_2) to act as a "safety valve" against excessive compliance costs. The drawbacks of setting low penalties include the following: increased risk that firms will continue to emit and simply pay the penalty; mitigation options are limited to those activities that cost less to implement than the compliance penalty; and reduced incentive for investment in longer-term abatement technologies. The EU-ETS Phase 2 adopted a much higher penalty of €100/tCO_2e, providing a relatively strong compliance incentive.[14]

Future prospects for emissions trading

The carbon market is destined to grow substantially over the period to 2012 and beyond, but its size will be highly dependent on the type of post-**Kyoto Protocol** agreement that is negotiated by the international community. The growth of the carbon market will be influenced by the number of countries that adopt binding emission caps, the stringency of these caps, the proportion of global emissions covered by caps, the extent to which credits generated by project-based "baseline and credit" systems such as the **CDM** are included, and possibly marketable credits resulting from avoided deforestation (see **biosequestration**). The voluntary credit market is also projected to grow as greater numbers of consumers choose to voluntarily offset their emissions.

Overall, emissions trading offers considerable promise as a mitigation instrument and, in theory, should deliver least cost emissions abatement in many circumstances. However, it is still too early to judge whether trading will deliver a superior economic outcome relative to other policy instruments. No cap and trade system with fully auctioned permits has been operating for a sufficient period of time to ascertain whether it has actually driven cost-effective abatement.

For emissions trading to provide a common global carbon price signal, it would require the establishment of a truly global cap encompassing all **greenhouse gases** and all major emission sources. In reality, such a comprehensive global system is unlikely to eventuate in the short to medium term since emission caps for most developing countries are unlikely to be adopted for some time (see **UNFCCC**). Furthermore, it is difficult to foresee agreement on a global emissions cap (possibly based on a specific **stabilization target**) with an allocation of emission allowances across all countries before 2020. Such a system would also require universal alignment of emissions inventory accounting, verification standards, and noncompliance penalties. Establishment of the legal and administrative infrastructure to manage such a system would take many years and possibly decades.

However, it is plausible that a more comprehensive global system could emerge before 2030 if large emitting countries that currently have no agreed emission caps (such as the United States, India, China, and Brazil) were to agree to caps on their emissions (thus providing sufficient global emissions coverage). In the meantime, emissions trading is likely to be confined to the national or regional level schemes in developed countries (with the notable exception of International Emissions Trading under the **Kyoto Protocol**). The participation of developing countries in emissions trading is expected to be largely confined to project-based systems. Careful monitoring of "carbon leakage" issues and ensuring the "additionality" of credits generated through such baseline and credit systems (see **CDM**) will be required to deliver meaningful reductions in global emissions.

See also: carbon taxes, Clean Development Mechanism, Joint Implementation, Kyoto Protocol, mitigation.

Notes

1 See Ekins and Barker 2001 for an overview of these instruments.
2 Burtraw 1996
3 Roine and Tvinnereim 2007
4 See Hepburn 2006
5 See Stern 2006 and Hepburn 2006 for further discussion.
6 Tietenberg 2006
7 Elgin 2007
8 GLOBE-Net 2007 and Davies 2007
9 ISO 2006
10 EC Directive 87, 2003
11 See Tietenberg 2006
12 See Grubb and Nuehoff 2006 and Neuhoff et al. 2006
13 ISO 2006
14 World Bank 2007b

Further reading

Stern 2006; Hepburn 2006; World Bank 2008; Grubb and Nuehoff 2006; Ekins and Barker 2001.

ENERGY EFFICIENCY

Energy efficiency relates the output of an energy service (heating, lighting, cooling, transport) or a transformation system (electricity) to its energy input. For example, a compact fluorescent light (CFL) is more energy efficient than a standard incandescent bulb as it produces the same amount of light (the energy service) but consumes one-quarter of the energy: more *useful* energy output is obtained per unit of energy input.

It is useful to distinguish between *energy efficiency*, which refers to the amount of energy required to meet a specific demand for energy services (such as lighting or heating), and *energy conservation*, which refers to a reduction in energy service demand (such as turning off a light when not needed or turning down the thermostat in winter). Energy efficiency measures aim to obtain the same energy services from less primary energy input. In contrast, energy conservation aims to reduce the demand for energy services. Both have a major role to play in reducing energy consumption and, in turn, greenhouse gas emissions.

The global economy is grossly inefficient in the way it uses energy. Large amounts of energy are wasted in energy transformation, powering inefficient appliances (even when in the off-stand-by mode), poorly designed buildings and production processes, maintaining comfortable indoor temperatures in buildings, or transporting people and goods from one location to another. This comes at a huge economic cost to businesses and households. Total preventable energy waste (that which can be cost-effectively avoided) costs the global economy more than US$ 1 trillion per year.[1] Energy efficiency is worth doing regardless of the greenhouse gas emission reduction benefits.

Energy efficiency is the most socially beneficial and least cost means of achieving large-scale **carbon dioxide (CO_2)** emission reductions in the medium term (at least out to 2030). The opportunities for saving energy and money are enormous. However, not all gains in energy efficiency necessarily result in equivalent reductions in greenhouse gas emissions as it depends on the greenhouse gas **emissions intensity** of the energy source displaced. For example, a 10% reduction in electricity use would certainly lower energy bills, but if the electricity was sourced predominantly from renewable energy facilities then emission reductions would be much lower than if sourced from coal-fired electricity.

Although energy efficiency offers significant economic and emission reduction benefits, it has, over the past two decades, received much less attention than supply-side mitigation options such as **nuclear power**, **carbon capture and storage** and **renewable energy**. Governments are often aware of the potential benefits, and many have implemented at least some policies and measures to achieve energy efficiency gains. For example, in OECD[2] countries over the past two decades, technology improvements, often driven by energy efficiency standards, have delivered a 70% efficiency increase in domestic refrigerators, a 35% efficiency increase for washing machines, and a 40% increase for domestic air conditioners.[3] The energy consumption of many appliances (particularly domestic refrigerators) could potentially be halved again by 2025. Nonetheless, the savings captured so far represent only a tiny fraction of what could be achieved cost-effectively. Energy efficiency programmes usually account for only a tiny fraction of government energy sector expenditures – most such expenditures go to support supply-side options and some to directly subsidize the fossil fuel industry.

The main advantages of energy efficiency are as follows:

- *Cost-effectiveness:* On a least-cost life cycle basis (the cost of delivering energy services over the lifetime of equipment or buildings), efficiency measures can reduce energy use by 50–70% and much of this at negative cost (i.e. such measures actually save money).
- *Scale:* Of the suite of available greenhouse gas mitigation options, around half of the reductions required to stabilize global emissions by 2050 can be delivered by energy efficiency, compared with 5–10% from nuclear power and 1–2% from solar power.[4]
- *Time frame:* Emission reductions can be delivered far more quickly through energy efficiency than through changing the existing configuration of the energy supply system (such as building new low-emission power plants). The turnover (the time between equipment purchase and replacement date) for appliances and equipment is much shorter than for most energy production facilities.
- *Ancillary benefits:* Energy efficiency improves economic efficiency and competitiveness, provides energy security and system reliability benefits, and can deliver a wide range of comfort and health benefits (worker productivity is usually 10–15% higher in an energy-efficient office building than in an inefficient one[5]).

If energy efficiency is so economic, why have we not captured this potential already?

While energy efficiency offers huge potential economic and other benefits, the uptake of energy efficient equipment and practices has been constrained by a number of factors. These include the following:

- *Lack of awareness:* When purchasing equipment and appliances, or designing and constructing production facilities and buildings, consumers and firms are often not aware of the cost-effective energy efficiency gains and technologies that are available.
- *Energy is cheap* relative to the energy services it provides: it is rarely a consideration in purchasing decisions. For most manufacturing and service industries, often less than 1% of the delivered cost of a product or service is embodied energy cost (except for a few notable exceptions, such as aviation and energy-intensive metals production).
- *Split incentives:* appliance manufacturers, or those who construct or own buildings, are not the ones that have to pay the energy bills. Consequently, they are more interested in delivering the cheapest possible product rather than lowering the lifetime running costs.
- *Initial purchase and construction costs* can be higher (though not always), even though the life cycle costs (cost of purchase, operation, and disposal) can be significantly lower – this can be a deterrent to consumer uptake.

- *Diffuse savings:* Energy savings usually come from a large number of small actions rather than big chunks – organizations are often unwilling to invest the time and effort to realize such individually small savings.
- *Organizational inertia:* Those that make design, budgeting, and purchasing decisions in firms tend to be removed from the operational level. Engineers on the factory floor often know where energy can be saved, but this information does not reach the finance officer and, even if it does, is usually covered under maintenance budgets rather than capital budgets and are subject to different financial thresholds.
- *Perverse incentives and administrative constraints:* Organization budgets are often based on previous years' expenditure (particularly in the public sector), and reducing energy bills may mean next year's budget is cut.
- *Political inertia:* Vested industry interests and lobbying are often successful in blocking or watering-down energy efficiency policy initiatives, even though this reduces net socioeconomic welfare for the general public.

This represents a long list of constraints, most of which are not economic or technical in nature.

Energy efficiency opportunities

Opportunities to save energy exist in all economic sectors and activities. These can be subdivided into five principal categories: energy supply and distribution (mainly electricity generation and petroleum product production), residential, commercial, industrial, and transportation.

Supply-side efficiency

The production of electricity and heat for supply to other sectors accounts for one-quarter of global greenhouse gas emissions.[6] Although significant efficiency gains have been achieved since the 1970s, the current stock of power plants converts energy to electricity and heat much less efficiently than what is technically and economically possible. For example, current state-of-the-art coal-fired plants have a conversion efficiency of 43–45% (and potentially 50% by 2015), compared with the 35% efficiency of the current global stock.[7] New combined-cycle gas turbines can achieve 55–60% efficiency, compared with the present global average of 42%.[8] If the global average conversion efficiency of coal-fired power plants were to increase from 35% to 45%, CO_2 emissions from coal generation would be reduced by around 20%.[9] The potential contribution to emission reductions from supply-side efficiency improvements is significant, but due to the long operating lives of generating plants (up to 50 years or more), it may take several decades to achieve these gains (see **fossil fuels**).

Residential sector

The residential sector includes the consumption of energy for running appliances, communication equipment, heating/cooling, refrigeration, lighting, water heating, and cooking. It accounts for approximately one-quarter of global electricity consumption and 10% of global greenhouse gas emissions – even more in wealthier countries, where household disposable income is higher and appliance usage and space heating/cooling more prevalent.[10]

Appliances

Residential appliance and equipment energy use is the fastest-growing area of electricity demand in most countries. Many new appliances that have entered the market over the past decade or so (such as MP3 players, mobile phones, portable computers – all with battery chargers that are often left plugged in and consuming power even when not actually charging the battery) have significantly increased appliance electricity demand. New technologies often emerge to replace old technologies, but these are not always more energy efficient. For example, the two new types of flat screen TV technology, LCD and Plasma, consume substantially different amounts of energy. LCD TVs are much more efficient than the old cathode ray tube technology, while plasma TVs are much worse – they consume 2–5 times the amount of energy as LCDs for providing essentially the same service.

Somewhat paradoxically, the single fastest-growing area of appliance energy use is standby power – the energy that appliances use when they are in "stand-by" or "off" mode, accounting for 3–10% of total residential electricity consumption in developed countries (mainly for computers, audio systems, and communication equipment).[11] For example, the average VCR uses 19 times more energy over its lifetime when it is in the off mode than when actually playing videos. In OECD countries, televisions consume 4.5% of residential appliance energy consumption in the on mode and 7.5% in the standby mode.[12] Standby power consumption can be reduced by up to 75%, to as little as 1 or 2 watts – compared with present average consumption of 7.5 watts, and 30 watts for some brands – at little or no cost.[13]

While there have been impressive recent gains in appliance energy efficiency further potential remains. The International Energy Agency estimates that, for OECD countries alone, if all presently available cost-effective appliance technologies were introduced by 2020, then carbon dioxide emissions would be more than 500 million tonnes/year lower than current "business as usual" projections (equivalent to taking 200 million cars off the roads). This could be achieved while making consumers financially better off. In terms of cost per tonne of carbon dioxide avoided, these reductions could be delivered at negative €169 in Europe and negative $65 in the United States – in other words, at a positive benefit to society.[14] Similar cost savings are available across all countries. As developing country appliance demand is expected to grow rapidly over the next two decades, particularly in China and India, maximizing appliance efficiency is essential to reducing long-term greenhouse emissions trajectories.

Lighting

Significant energy efficiency gains are also possible in residential lighting. If current lighting consumption and efficiency trends continue, total lighting energy demand is expected to grow by 80% by 2030 and result in an additional 3 billion tonnes of CO_2 emissions per year.[15] Lighting demand varies considerably between countries, with the average American citizen consuming 100 megalumens of light a year, compared with 50 megalumens for the average European citizen and just 3 megalumens for the average Indian.[16] Most lighting is provided very inefficiently. Converting coal to electricity to power incandescent lights in the home is only 3% efficient – most is lost as waste heat at the power station, during transmission to the home and heat from the light bulb.[17] Candles and kerosene lamps are the most inefficient lighting sources of all, providing less than 1% of global lighting supply but accounting for 20% of lighting-related CO_2 emissions.[18]

Replacing incandescent lamps (which provide 10–15 lumens/watt) with CFL lamps (60–80 lumens/watt) can quadruple lighting efficiency. CFLs also last up to ten times longer (paying for themselves through savings in lamp replacement costs alone) and provide significant financial benefits to consumers. Some governments have already announced plans to phase-out incandescent light bulbs altogether. In Australia, for example, incandescent bulbs will be banned from sale from 2010 onward. The benefits of such schemes vary between countries according to the lighting technologies they use. In the United States, the world's largest residential sector energy consumer, the gains would be significant (twice what solar power is likely to contribute to global electricity supplies by 2030, and at a small fraction of the cost) as 90% of all household lighting hours are still provided by incandescent bulbs.[19]

Increased use of daylight (through skylights and building design) can also deliver significant gains. Many homes are also often excessively lit (for instance, in unoccupied rooms), and simple, cost-effective lighting controls – such as motion detectors and ambient light sensors – can substantially reduce energy waste. Overall, existing cost-effective lighting and control technologies could reduce lighting energy consumption by 30–60% compared with present levels.

Residential buildings

Improvements in residential building design and construction materials also offer significant mitigation potential. Technologies such as passive solar designs (see **solar power**), energy-efficient windows, and improved thermal insulation can provide substantial cost-effective energy savings. Passive solar designs can reduce heating and cooling loads by up to 50%, at no additional construction cost.[20] The choice of windows can also have a major effect on energy consumption since 30% of heating or cooling energy can be lost through windows. In cold climates, double- and triple-glazed windows can substantially reduce heat loss. In warmer climates, low-emittance coatings and external shading can be used to reduce cooling loads. High-efficiency windows can reduce 25–30% of heat loss and 6–30% of heat gain, but few consumers are aware of these savings when choosing windows and often have no choice at all. Builders usually choose the windows and, as they do not live in the house or pay the energy bills, there is little incentive for them to choose energy-efficient designs (unless they are regulated to do so). Insulation is widely used in residential housing but generally well below the level that would give maximum cost-effective savings – again builders often install the minimum rather than the maximum amount of insulation. New, high-efficiency houses (using current existing cost-effective measures) can consume 75% less energy than standard homes.[21]

Often the most efficient and cost-effective policy option for delivering residential sector energy efficiency gains is through the adoption and enforcement of mandatory minimum energy performance standards and efficiency labeling since this is where consumer awareness, information deficiencies, and split-incentive barriers are the greatest. Residential consumers are also much less responsive to energy price changes than consumers in the industrial and power sectors: a 10% power price increase typically has little effect on residential energy demand.

While many countries have adopted appliance energy efficiency standards and labeling, these standards have been relatively soft (usually well below what is technically and cost-effectively possible) and, once introduced, are often not regularly updated to reflect technological advances. In some cases, they are also poorly enforced. The wide variation in coverage, stringency, and implementation across countries has resulted in lower savings than would have been achieved if there had been international consistency and harmony of standards[22] since appliance manufacturers are presented with conflicting signals and incentives. Furthermore, countries with no standards are often dumping grounds for inefficient products that can no longer be sold in countries with mandatory standards.

Minimum energy efficiency standards for buildings are less common. While some countries have introduced mandatory residential building efficiency standards, many others have yet to do so or have only weak and poorly enforced standards. Policy makers often face intense lobbying from the building industry which argues that energy efficiency standards increase housing costs and rents and is an excessive burden on the housing industry and consumers. Initial construction costs can indeed be higher (but not always) with energy efficiency standards in place, but they are generally modest and are recouped many times over during the life of the building.

Commercial sector

Many of the cost-effective energy efficiency measures and technologies that can be applied in the residential sector are also applicable to the commercial sector, and similar efficiency gains can be achieved. The commercial building sector (including appliances and equipment) consumes approximately 15% of global electricity production and is responsible for nearly 6% of global greenhouse gas emissions.[23]

The thermal energy load from appliances, computers, and lighting can have a significant influence on building cooling and heating energy consumption. Inefficient appliances and low-efficiency lighting can generate significant amounts of heat and require more energy for cooling, as well as larger and more expensive air-conditioning systems. The choice of windows also has a greater energy impact for commercial buildings relative to residential buildings, as a larger proportion of the external surface area of a commercial building is windows – installing high efficiency windows can reduce commercial building heating/cooling energy consumption by up to 30%.[24]

Commercial buildings are generally long-lived assets (50 years or more) and, once built, are more often renovated than replaced: on average, major energy consuming equipment is replaced every 20–30 years. As a result, the choices made at the time of construction have a major bearing on the building's energy consumption over its lifetime. The best way of achieving maximum cost-effective energy efficiency is to implement energy efficiency measures during construction – retrofitting buildings is more expensive and offers less cost-effective energy efficiency gains. As a majority of the commercial buildings that will exist in 2050 have yet to be built, the potential for energy efficiency savings is very large indeed. Cost-effective technologies can reduce commercial building energy consumption by 70% compared with today's average building.[25]

As with the residential sector, the adoption of minimum energy efficiency performance standards, combined with building energy efficiency labeling and promotional programs, is the most effective means of achieving energy efficiency gains. However, the adoption and/or enforcement of mandatory commercial building energy performance standards is not widespread, particularly in developing countries (where most of the new commercial buildings will be constructed over the next 30 years). Adopting building efficiency standards, at least up to the point where the costs and benefits are equalized, offers huge potential energy efficiency gains.

Industry sector

Industry accounts for approximately one-fifth of global greenhouse gas emissions.[26] There are significant opportunities for cost-effective energy efficiency improvements through better equipment, production process design, and combined heat and power (where energy is used to generate electricity and the waste heat used to meet the demand for process heat).

The energy efficiency of most industrial processes can be improved by 15–50% through the application of existing cost-effective technologies. Often 15–30% can be saved on actions with payback periods of 2–3 years, while in some cases reductions of 50% or more can be achieved.[27] In cement manufacture, converting from wet to dry kilns can reduce energy consumption by 60–70%.[28] The options for efficiency gains vary considerably between industries, but industrial motors and process heat are the two areas that offer the most potential. The design of the production process also has a major bearing on industrial energy consumption and poorly designed processes (such as the layout of machinery, the length and number of bends in pipes, or the distribution of compressed air) can significantly increase energy wastage, even when the most efficient motors and furnaces are used.

Unlike the commercial and residential sectors, where minimum energy performance standards are the most effective policy option, they are often less effective in the industrial sector (with the exception of electrical motors and boilers) due to the large variation in products and manufacturing processes. Companies often invest in new manufacturing processes and equipment based on 6–8 year payback periods (or sometimes more), but few are willing to invest in energy efficiency measures that have payback periods of greater than 2–3 years and sometimes as low as one year. For this reason, some OECD governments have, or are planning to introduce, mandatory energy audits and energy efficiency reporting requirements. At present, most government initiatives in this area are voluntary and the implementation of efficiency measures is well below what is cost-effective. The introduction of mandatory measures – for example, the requirement to implement all energy efficiency measures that have a payback period of four years or less, which still represents an attractive rate of return on investment, would deliver very large energy savings and emission reductions. However, these measures are usually strongly resisted by industry.

Some companies have adopted progressive energy efficiency strategies. For example, between 1995 and 2005, Du Pont, a major chemical manufacturer, cut energy consumption by 7%, reduced greenhouse gas emissions by 70% (mainly through reductions in non-CO_2 industrial gas emissions), and saved $ 2 billion dollars while at the same time increasing production by 30%.[29] There are numerous other examples of the benefits of progressive corporate initiatives (see **technology, structural change and organizations**) but, by and large, only a very small proportion of industrial enterprises have taken energy efficiency seriously.

Transport sector

The transport sector accounts for around 14% of global greenhouse gas emissions (nearly all transport emissions are CO_2), three-quarters from road transport.[30] In developed countries, the contribution of transport to greenhouse gas emissions is higher: in the United States, for example, transport accounts for one-third of carbon dioxide emissions and consumes 70% of imported oil.[31] Transport energy demand is the second fastest-growing area of energy use after residential consumption. Transport is also almost completely reliant of **fossil fuels**, so each unit of energy saved delivers emission reductions.

Despite more than a century of technological improvements, the automobile is astonishingly inefficient in terms of energy use. On average, only 13% of primary energy input actually reaches the wheels – 87% is lost in the engine, in the drive train, through idling, or through running accessories – and, of the energy delivered to the wheels, more than half is lost as heat from tyres to the road and air: only 6% actually propels the driver to the destination.[32] Since the 1970s, there have been significant improvements in engine and drive train efficiency (mainly due to improved combustion technologies such as fuel injection, turbo chargers, and improved metals and alloys), but average vehicle fuel consumption has remained largely unchanged. Most of the efficiency gains have been consumed by making vehicles larger, more powerful, and adding more accessories and luxury fittings.

Significant opportunities exist to increase transport efficiency. By far the most important measure is reducing vehicle weight and drag (through improved aerodynamics). For every unit of energy saved at the wheel, through reduced weight and drag, up to six units of energy are saved in the engine and drive train.[33] Advanced polymer composites (such as carbon fiber) and new metal alloys can substantially reduce weight without sacrificing strength, durability, and safety. Ultra-light body technology can double the energy efficiency of today's hybrid electric vehicles without raising their costs significantly.

Hybrid and alternative-fuel vehicles can also help to reduce emissions. Sales of hybrid vehicles have grown rapidly since they were first introduced in the late 1990s (in 2007 Toyota sold their millionth hybrid Prius model). Such vehicles can provide 20–50% reductions in fuel use per kilometer, depending on the model and the split between city and country driving. However, they are still relatively expensive and without subsidies are only marginally economic. Technology improvements, and higher oil prices, will undoubtedly improve hybrid economics over the next decade. In the longer term, fuel cells offer substantial energy efficiency improvements, but these are unlikely to be commercially available or cost-competitive before 2025.[34]

Due to political sensitivities, governments are very often reluctant to introduce mandatory efficiency requirements or limits on vehicle size and weight. While this is slowly changing (particularly in the European Union), most governments have to date relied on voluntary agreements with auto manufacturers – which have rarely been met. Fiscal measures such as variations in taxes and registration charges according to vehicle weight and fuel efficiency, as well as petroleum product taxes, can also help influence consumer choice. Moving people from private vehicles to public transport systems (termed "modal shift") offers much potential but also faces political hurdles and consumer resistance.

In the air transport sector (the fastest growing area of transport fuel demand and emissions), fuel used per passenger kilometer has fallen considerably in recent years and new commercial passenger jets use less than half the fuel/passenger kilometer of those flying in 1970. However, future potential advances in engine fuel efficiency are limited, and most of the savings will come from further reductions in aircraft weight, better flight scheduling, and improved air traffic control.

What is the potential contribution of energy efficiency to greenhouse gas emission reductions?

Due to the sheer magnitude of the untapped economic potential and the shorter time frame in which emission reductions can be achieved, energy efficiency is clearly the most important greenhouse gas mitigation option in the short to medium term (out to 2030). Some studies estimate that in addition to direct energy cost savings by consumers, every dollar invested in end-use efficiency through appliance, lighting, and building measures translates to 2 dollars of expenditure avoided in electricity generation and distribution capacity: in short, fewer power plants need to be built.[35] Even a 10% increase in energy efficiency would release a significant amount of capital for other productive purposes.

Nonetheless, achieving major emission reductions through energy efficiency will require proactive efforts on the part of governments and consumers to overcome existing barriers and constraints. As outlined above, the main constraints, unlike those associated with most other energy sector **mitigation** options, are unrelated to energy prices or cost-competitiveness. Instead, energy efficiency suffers from what is termed "market failure," the inability of markets to deliver an efficient economic outcome. For this reason, there is justification, and some would say a responsibility, for governments to implement policies and measures to help overcome this market failure. This is particularly true in the case of information barriers, where information campaigns and energy efficiency labeling schemes can help overcome the lack of consumer awareness. Mandatory appliance and building efficiency standards, and possibly mandatory energy audits and efficiency measures for industry, are also needed.

In a purely economic sense, it would be justified to make standards increasingly stringent over time, at least to the point where the costs equal the gains in social economic benefits: as yet, no country is anywhere near this point. Arguably, mandatory standards could be increased up to the point where the costs equal the next cheapest mitigation option. If taken to this level, energy efficiency gains (and subsequent emission reductions) would be very large indeed.

The introduction of a cost penalty on CO_2 emissions would provide a boost to energy efficiency implementation, but more so in the industrial and power sectors, which are much more sensitive to energy prices. **Carbon taxes** and **emissions trading** are potentially effective policy instruments in the industrial sector, while in the residential, transport, and commercial sectors, they are much blunter policy instruments and are unlikely to have a major impact on energy demand, unless set at very high levels.

Estimates of projected emission reductions from energy efficiency out to 2050 vary according to assumptions about take-up rates, energy prices, and future government policies. Nonetheless, all major mitigation cost modeling studies conclude that in the short to medium term energy efficiency is likely to deliver the largest emission reductions at the lowest economic cost. The latest IEA projection studies suggest that given appropriate policy settings by governments, 80% of the cost-effective CO_2 reductions out to 2030 (over 5 billion tonnes CO_2/year) could be delivered through energy efficiency measures.[36] Long-term IEA projections out to 2050, based on the introduction of more aggressive government climate change and energy security policies, suggest that energy efficiency would provide between 45% and 53% of cost-effective reductions. This dwarfs what any other mitigation options could feasibly deliver.

See also: emissions intensity, fossil fuels, future emissions trends, mitigation.

Notes

1 Lovins 2005b
2 The Organization of Economic Cooperation and Development is a grouping of 24 industrialized countries.
3 IEA 2003
4 IEA 2006b
5 Lovins 2005b
6 WRI 2005
7 IEA 2006b
8 Ibid.
9 Derived from data in IEA 2006b
10 WRI 2005
11 IEA 2003
12 IEA 2001
13 IEA 2006b
14 IEA 2003
15 IEA 2006c
16 Ibid.

17 Lovins 2005b
18 IEA 2006c
19 IEA 2006b
20 Ibid.
21 Ibid.
22 IEA 2003
23 WRI 2005
24 IEA 2006b
25 Ibid.
26 WRI 2005
27 IEA 2006b
28 Ibid.
29 Lovins 2005b
30 WRI 2005
31 Lovins 2005b
32 Ibid.
33 Ibid.
34 IEA 2006b
35 IEA 2006a
36 Ibid.

Further reading

IEA 2003, 2006b, 2006c; Lovins 2005; WRI 2005.

EXTREME WEATHER EVENTS

Extreme weather events are climatic episodes that exhibit significant variation from average climatic conditions. These events can take many different forms: heat waves, extreme cold snaps, powerful storms and cyclones, intense precipitation events, flooding, and drought. While such phenomena form part of the natural climate system, many scientists believe that **global warming** will increase their future intensity and frequency.

In the past few decades, there has been a series of record-breaking extreme weather events. These have included the 2003 heat wave in Europe (the most severe so far recorded), which killed 35,000 people; the two hottest years on record (1998 and 2005); an intense El Nino event in 1997–1998; and a series of floods, extended droughts, and powerful tropical storms. Climate change is believed to have contributed to the severity of these events.

Extreme weather events can have serious impacts on humans and natural ecosystems. Attention is often focused on the impacts of extreme events on humans, but changes to prevailing ambient climate conditions can have a major impact on the survival of animal and plants species (see **biodiversity impacts**). Sudden heat waves, floods, and cold snaps can cause sharp rises in mortality rates of animals, particularly livestock, while droughts can lead to prolonged impacts on ecosystems, particularly aquatic ecosystems, and large-scale animal migrations. While this section will largely discuss the impacts of extreme events on humans, other sections of this book highlight climate change impacts on other living organisms.

During the 1990s, more than 2 billion people were affected by weather-related disasters – three times more than in the 1970s – and economic losses increased five-fold to US$ 629 billion.[1] For the period 1980–2004, the economic costs of weather-related natural disasters have been estimated at US$ 1.4 trillion.[2] As global population grows, and more infrastructure is built, the damage costs of extreme weather events will continue to increase, with or without further climate change. However, damage costs will inevitably be higher in a world with global warming. The socioeconomic implications, including the impact on global **finance and insurance** markets, are expected to be significant (see **socioeconomic impacts**). Major insurance companies are already actively lobbying for strong international action to curb global warming.

Extreme weather events impact most heavily on the poorest countries of the world and the poorest communities within those countries. The poor are more likely to live in substandard housing and in areas more susceptible to weather-related disasters (e.g. floodplains). They are also less likely to have insurance to cover crop and property losses or the financial resources to fund **adaptation** responses (e.g. constructing storm-proof housing or stockpiling food and medical provisions). Poor communities can find themselves caught in "vulnerability traps," where an inability to recover from weather events can progressively increase their exposure to subsequent weather extremes.

Has climate change already increased the intensity and frequency of extreme weather events?

This question has been the subject of lively debate among both scientists and politicians. Events such as the 2003 European heat wave and a series of intense tropical storms, like hurricanes Katrina, Rita, and Wilma (all in 2005), have attracted considerable media attention. While there appears to have been a general upward trend in the severity and frequency of extreme events in recent years, it is not yet possible to determine conclusively whether a sustained trend exists and, if so, whether it is due to climate change. Extreme weather events are so infrequent that consistent data must be collected over periods of at least 30–50 years to determine underlying trends, and it is likely to be another 10–20 years before enough conclusive evidence has accumulated. Nonetheless, a growing body of scientific modeling and observations suggests that climate change has increased – and will continue to increase – the intensity and frequency of extreme weather events.

Types of extreme weather events

Extreme weather events can be subdivided into three main categories: destructive storms, floods and droughts, and temperature extremes.

Destructive storms

These events are characterized by high winds and intense precipitation and can occur at all latitudes across the globe. However, it is in the tropics and subtropics (generally ±30° of the equator) that the most destructive storms usually occur. These intense tropical storms are interchangeably referred to as cyclones, hurricanes, or typhoons.

Tropical cyclones are characterized by destructive high-speed winds (sometimes exceeding 250 km/hour) and intense precipitation (sometimes more than 1,000 mm of rain can fall in less than 24 hours). Storms receive names once wind speeds exceed 62 km/hour and become cyclones (or hurricanes/typhoons) once they exceed average wind speeds of 119 km/hour. Above this wind speed (measured at the edge of the storm's eye), they receive a category storm scale ranking from 1 to 5 (known as the *Saffir-Simpson Hurricane Scale*). A Category 1 hurricane has wind speeds of 119–153 km/hour, while the most severe Category 5 ranking is given to storms with wind speeds exceeding 249 km/hour. The most intense Category 5 storm so far recorded was Hurricane Wilma, which hit the coast of Cuba and Mexico in 2005.

The intensity of tropical storms is related to the amount of energy transferred between the oceans and the atmosphere. There is a high statistical correlation between ocean surface temperatures and tropical storm intensity.[3] Warm ocean temperatures make heat energy available to fuel tropical storms. Surface water temperatures of 26.5°C or more are required before storms can form. So much heat energy is drawn from the surface oceans during their formation that they often leave water temperatures several degrees cooler after they pass. As climate change warms the atmosphere and oceans, the conditions for tropical storm formation will occur more often, and over a larger geographic range, which could lead to an increase in both storm intensity and frequency.

While it is not yet possible to conclude that global warming has increased storm intensity, there is mounting scientific evidence to suggest that it has. Studies of Atlantic storm data reveal hurricane intensity has increased by as much as 50% over the past 30 years.[4] There does not appear to be discernible change in the total number of storms in the Atlantic over the past few decades, although there does appear to be an increase in the number of Category 4 and 5 storms.[5] Severe tropical storms have occurred on numerous occasions in the past in this region, and their frequency has tended to vary over cycles spanning several decades. Some scientists consider that the lack of consistent and reliable historic data means that it is too early to judge whether storm frequency has increased.[6] The number of cyclonic storms over the north Indian Ocean has, however, shown an increasing trend in the last three decades.[7]

The spatial distribution of tropical storms has also changed. In some regions, storm activity has increased (e.g. in the South China Sea and North America), while in others it has decreased (e.g. in the Bay of Bengal and the East China Sea). Tropical cyclones are also extending over a wider range. In 2004, a hurricane (Hurricane Catarina) made landfall in southern Brazil for the first time ever: it left 33,000 people homeless and caused US$ 350 million of damage.[8] Previously, sea surface temperatures in this region had been too low to enable storms to form.

The 2005 Atlantic hurricane season achieved a record for the number of named tropical storms in a single year (26 – four above the previous record of 22) and also the highest number of cyclones (14 compared with the previous record of 12). The most heavily reported event of the season occurred in August 2005, when Hurricane Katrina devastated the city of New Orleans. Katrina is the most costly natural disaster experienced by the United States to date, with an estimated damage cost of up to

$150 billion and the loss of more than 1,000 lives. At one stage, Katrina reached Category 5 status but was downgraded to a Category 3 storm by the time it made landfall. While Katrina may have formed with or without global warming, the scientific consensus is that Katrina's intensity was higher than it otherwise would have been. When Katrina was forming, water temperatures in the Gulf of Mexico were around 30°C (which is 2–3°C above normal for that time of year).[9]

The **El Nino Southern Oscillation (ENSO)** is known to affect the frequency, distribution, and intensity of tropical storms, so any perceived changes in storm frequency and intensity must be viewed against a background of regular climatic cycles. In strong El Nino events, the frequency and intensity of tropical storms in the Atlantic decrease, while during La Nina events they tend to increase.[10] In the Northwest Pacific, the number of cyclones that hit the Indochinese Peninsula, and their duration and intensity, are usually much higher during El Nino events than normal years.[11] Some scientists believe that climate change will change the pattern of El Nino and La Nina cycles, which could also affect storm patterns (see **ENSO**).

Floods and drought

While tropical cyclones tend to attract the most media attention, it is floods and droughts that have the greatest impacts in terms of loss of life and livelihoods, particularly in the developing world. The IPCC predicts that the frequency and intensity of both floods and droughts will increase this century.[12]

Flooding is the single most frequent natural disaster, accounting for 40% of all extreme weather events. A warming climate will intensify the water cycle and increase the occurrence of floods. The number of major floods has increased over the past two decades, particularly in Europe and South Asia.[13] North and central Europe are likely to experience wetter winters and drier summers in the future, and winter flooding is expected to become a much more frequent event. Latin America, Eastern Africa, and South Asia have also seen a significant rise in intense precipitation events and flooding since the 1980s. In early 2006 Bolivia experienced one of the most intense periods of precipitation so far recorded that triggered flooding and mudslides and displaced 10,000 people.[14] Over the period 1950–2000, the average monsoon rainfall has remained relatively constant, but the frequency of intense rainfall events has risen (averaging a 10% increase per decade) and the number of moderate events has declined.[15]

Floods result in major losses of infrastructure (roads, railways, and bridges), crops and livestock, and human life. The health impacts of floods can often have the largest effect on humans. Floods can lead to a major increase in water-borne diseases (such as cholera and diarrhea) and increase the prevalence of vector-borne diseases (malaria, dengue, and yellow fever). For example, following the Mozambique floods in 2000 that devastated parts of the country, there was a five-fold increase in the number of malaria cases and similar rise in diarrhea and other water-borne diseases.[16] They often cause more death and economic dislocation than the actual direct impact of the event itself.

Droughts differ from floods in that their impacts are slow to emerge but can prevail for a long period of time. They also tend to affect much larger spatial areas than floods and storms. Droughts are the single most important cause of famine. Over the past century, millions have perished from drought-induced famines and disease, and many economies have been severely impacted, particularly in the poorer developing countries. Some regions of the world have experienced extended periods of drought in recent decades, most notably the Sahel in sub-Saharan Africa, where there has been widespread famine and loss of life since the 1970s. Climate change is thought to have contributed to the consistent failure of the Sahel monsoons, though it is unclear whether the Sahel will become wetter or drier over the coming decades (see **water impacts** and **global dimming**). Other regions, particularly southern Australia, have also experienced a significant fall in average annual rainfall, and the incidence of drought has increased.

El Nino stages of the **ENSO** cycle are linked to severe droughts and monsoon failures in Australia, India, East Africa, and flooding in Northwest United States and the west coast of Latin America (see **ENSO**). Drought also leads to a greater risk of large-scale forest fires. The 1997–1998 El Nino event coincided with large-scale forest fires in Southeast Asia, which was the main contributing factor behind the highest single annual rise in **carbon dioxide** concentrations so far recorded.[17]

Extreme heat and cold

More frequent and prolonged heat waves will be a noticeable feature of climate change during this century. There is also likely to be a reduction in the number of extreme cold periods, which will benefit some countries in the higher latitudes. Although average temperatures will change relatively slowly, the frequency of extreme temperature events is expected to become more pronounced. It is the temperature peaks that cause the most damage and stress.

Increases in the number of hot days and the occurrence of abnormal temperature extremes have already been recorded for the twentieth century. The upward trend is expected to continue during the twenty-first century, in line with projected climate change. Global warming has already increased the probability of extreme heat events in Europe between two- and four-fold and could increase the likelihood of these events 100-fold during the twenty-first century, with a substantial shortening of the return time for extreme heat waves.[18] Statistically, the 2003 European heat wave was a very extreme event: temperatures were up to 6°C above long-term averages, well outside what would be considered the normal range of variability.

Human mortality and morbidity increase significantly during heat waves (see **health impacts**). Heat waves caused more fatalities in Australia during the twentieth century than any other natural weather event.[19] Extremes of heat can also adversely affect crop yields, overload energy systems, reduce hydro- and wind power generation, and lead to more extensive wildfires. Warmer temperature episodes can also have major impacts on forest ecosystems, resulting in accelerated forest dieback from heat stress and the spread of tree pests and diseases. Portugal lost 14% of its forest cover during the 2003 European heat wave, with losses estimated at more than 1 billion euros.[20] There were also significant losses of crops, livestock, and forest resources, and the soils in the region experienced a net loss of carbon.

Global warming-induced changes in regional weather patterns can produce extremes at both ends of the temperature spectrum. While warming has led to significant glacial retreat in Peru, it has also prompted the arrival of a new weather extreme that the locals call *friaje*. This recent phenomenon occurs when icy winds from the South Pole travel up the central Andes. In 2004, a sudden cold snap caused temperatures to plummet in mountain regions to –35°C, killing 50 people and leaving tens of thousands suffering from bronchitis and pneumonia. This event resulted in the loss of more than half of Peru's Alpaca herd and dealt a severe economic blow to many Andean communities.[21]

Adaptation options

Global warming will increase the frequency and intensity of extreme events, and people will need to accommodate and cope with these changes. To minimize the future impacts of extreme storms, there are a range of potential strategies and options available. The most important is to reduce anthropogenic greenhouse gas emissions into the atmosphere as rapidly as possible so that the rate and magnitude of warming is constrained. The more the planet warms, the greater the risk of more intense and frequent weather extremes.

Nonetheless, due to past emissions, we are already committed to further global warming over the next few decades, even if we completely cease emissions today. This will mean a continuation of the upward trend in extreme events. It is therefore vital to implement measures that reduce people's risk and vulnerability, through improved disaster preparedness, emergency response, and relief capabilities. Measures should include, among others, establishing comprehensive early warning systems; adopting and enforcing appropriate building codes and land-use zoning/planning; improving emergency response capabilities and evacuation plans; stockpiling food, medicines, and emergency supplies; public education; and strengthening health and preventative medicine capabilities. Research shows that for every dollar spent on disaster reduction measures, the costs of disaster recovery are reduced by between $ 4–10, depending on the region.[22]

Cuba is a good example of the benefits of forward investments in disaster prevention and preparedness. Although a poor country, Cuba's investment over the past two decades has paid worthwhile dividends. During Hurricane Wilma in 2005 (the most intense tropical storm yet recorded), authorities evacuated 640,000 people from high-risk areas and put into action a range of emergency response measures. The result was that only one person died and losses were lower than what would normally be expected from a hurricane of this intensity. Of the six major hurricanes to hit Cuba over the period 1996–2002, only 16 people have lost their lives. This can largely be attributed to efficient early warning systems, clear lines of decision making, well-equipped rescue teams and emergency stockpiles, clear evacuation plans and procedures, and effective land-use planning and construction standards.

In summary, it appears that climate change has led to an increase in the number and severity of extreme climate events over the past two decades, although it is still not possible to discern a general increase in tropical storm frequency. Projected climate change over this century is expected to accentuate this trend. This will have significant effects on humans and natural ecosystems. Although it is anticipated that the loss of life and property, and increased economic damage, will continue to rise, measures can be implemented that can reduce risk and vulnerability.

See also: anthropogenic greenhouse gas emissions, ENSO, finance and insurance, health impacts, global warming, water impacts.

Notes

1 International Red Cross
2 Epstein and Mills 2005
3 Hoyos *et al.* 2006
4 Emanuel 2005
5 Webster *et al.* 2005
6 Landsea *et al.* 2006
7 Goswami *et al.* 2006
8 Simms and Reid 2006
9 Schiermeier 2005
10 McPhaden *et al.* 2006
11 Fudeyasu *et al.* 2006
12 IPCC 2007
13 Epstein and Mills 2005
14 Simms and Reid 2006
15 Goswami *et al.* 2006
16 Epstein and Mills 2005
17 Page *et al.* 2002
18 Stott *et al.* 2004
19 Epstein and Mills 2005
20 Ibid.
21 Simms and Reid 2006
22 Ibid.

Further reading

Epstein and Mills 2005; IPCC 2007; Landsea *et al.* 2006; Simms and Reid 2006.

FINANCE AND INSURANCE

The global financial system, and its institutions, will play a central role in managing climate risks and financing climate change **mitigation** and **adaptation** measures. Financial markets influence investment decisions, create markets that price and trade economic externalities associated with climate change (such as greenhouse gas emissions and climate change impact risks), influence consumer and corporate behavior, and drive innovation. The efficiency and effectiveness of climate change policies will, in part, depend on the response of the financial system.

Climate change is expected to present both threats and opportunities for the global finance and insurance markets. The major threats to the financial system will largely stem from economic and insurance losses incurred by **climate change impacts** (particularly **extreme weather events**), government policy/regulatory changes that change key market variables; and the economic losses from the effect of climate change on human mortality and morbidity (see **health impacts**). Potential opportunities will arise from the growing market for risk management services and products (e.g. increased demand for insurance), the growth in new markets and investment opportunities (such as low emission technologies), and opportunities for new financial products and services (such as those emanating from the rapid growth in carbon trading).

The financial system consists of all financial institutions, markets, and instruments. Its principal role is to facilitate the distribution and usage of economic resources.[1] The financial market provides a mechanism through which buyers and sellers exchange items of value, such as financial securities (e.g. bonds), commodities (such as wheat and steel), currencies, and other financial assets.[2] Financial markets raise capital via stocks and bonds in the capital markets, underpin international trade and finance through the currency markets, and risk transfer in the derivatives and insurance markets – they facilitate capital flows from participants who have excess funds to those who require funds. The annual value of global financial market transactions is currently around US$ 120 trillion.[3]

In relation to climate change, the financial markets and institutions will perform several important functions, which include:

- providing mechanisms for managing climate risk, principally through the derivative and insurance markets;
- financing (through traditional bank financing, equity, and venture capital) the development and deployment of low emission technologies necessary to reduce global greenhouse gas emissions;
- raising capital to fund adaptation measures;
- influencing corporate governance and investment strategies of firms through their investment/lending guidelines (especially mutual and pension funds) and public climate risk disclosure requirements; and
- developing and supporting financial instruments (such as carbon trading mechanisms) to facilitate cost-effective mitigation responses.

Managing climate change risks

The financial system plays an important role in managing risk, particularly through the insurance market, but increasingly through the derivative markets.[4] Risk is the combination of the probability of an event and the consequences of that event – the higher the probability of an event occurring, and the larger the expected negative impact of that event, the higher the risk exposure.[5] Climate change is becoming an increasingly important risk that governments, firms, individuals, and the global financial system will need to accommodate. How finance and insurance markets accommodate these risks, and especially how they price risk, will have an important bearing on the impact of climate change on the global economic system. Climate change risk can be subdivided into two broad categories: *physical* risk and *policy* risk.

Physical risks stem from **climate change impacts**, such as more frequent and intense **extreme weather events**; changes in the supply of food commodities due to crop failures from droughts, floods, heat waves, and insect infestations (see **agriculture and food supply impacts**); changes in fish stocks (see **marine impacts**);

changes in human mortality and morbidity (see **health impacts**); disruptions to energy and transportation systems; potential conflict from human migration; and a range of other climate-induced macroeconomic changes (such as the flow on effects of impacts on financial markets and the cost of capital).

Policy risks mainly arise through changes in government policies and regulatory frameworks to address climate change, particularly those designed to limit greenhouse gas emissions but also adaptation measures (e.g. changes to building codes and land-use zoning); investor attitudes toward certain firms and industries; changes in consumer behavior and purchasing patterns; and exposure to liability claims.

The extent and timing of these risks remain uncertain but are likely to increase over time as the impacts of climate change begin to manifest themselves. Organizations and individuals will have to identify their climate risk exposure and adopt measures to manage these risks.[6] Options include direct measures (such as constructing more storm-resistant buildings and infrastructure) or indirect measures (such as accessing insurance). In some cases they may simply bear the risk (costs) of climate change impacts (willingly or unwillingly) or rely on governments or international agencies to accept risk on their behalf (e.g. taxpayer-funded safety nets such as government flood insurance schemes in the United States or disaster relief facilities funded by foreign aid).

Role of insurance markets

Insurance is the world's largest industry with annual revenues in excess of US$ 3 trillion dollars and accounts for around 8% of global GDP.[7] Around half of global premiums are nonlife, such as property and liability insurance, with the remainder principally from life and health insurance.[8] The insurance industry plays an important role in maintaining the stability of the global economy by helping to spread risks over time, between sectors and geographic regions, and across greater numbers of people – pooling risk through insurance reduces the impact of a particular event on specific individuals, communities, and organizations. Climate change will impact either directly or indirectly on the entire industry and will transform the way the industry operates in future.[9]

Private insurance is the single most important means by which individuals and organizations manage residual risk (risk which is difficult to independently self-manage). They can insulate themselves (hedge) against the risk of losses from a particular event by transferring risk (by means of a legal contract) to an insurer in exchange for payments (premiums). These payments are determined by the price insurers attach to different categories of risk (the risk price). The insurer holds the payments in a fund that is used to compensate those entities that experience financial loss as a result of the insured event.[10] For instance, a home owner might pay monthly premiums for storm damage insurance, and the level of those premiums will be determined by the insurance company's assessment of the likelihood of a damaging storm and the expected costs of this damage. The magnitude of expected future losses determines how much money needs to be held in reserve to meet such losses.

In comparison to most other markets, which set product prices based on known costs, the insurance industry sets product prices in advance of costs incurred.

Insurers employ models to help determine future costs (potential claims) they are likely to encounter. Premiums are generally based on historical data of the magnitude and frequency of claims incurred for specific events. However, insurers occasionally incur large unexpected losses through catastrophic natural events (such as floods and earthquakes) – they occur so infrequently that insurers have great difficulty in determining such losses in advance. To cover these events, private insurers also insure themselves through either private or public reinsurance mechanisms that are willing to accept a proportion of the risk of catastrophic losses or through other financial instruments such as catastrophe bonds and weather derivatives. This provides insurers with access to additional capital to cover events that they could not afford to raise in the financial markets by themselves.

Derivative markets are the second most important financial instrument for managing risk. Derivatives (such as futures and options markets) trade risks associated with the price of an asset changing, such as wheat futures, carbon credits, oil futures, and so on. Different actors are willing to accept different levels of risk that the price of a commodity, security, or some other asset will change in value (either up or down) at some point in time in the future – they make profits or losses according to which direction price changes. As climate change is expected to increase the volatility of commodity and other securities in coming decades, particularly as a result of extreme weather events, the derivatives markets will be an important means of managing these risks.

Trends in insurance losses

Owing to the types of events that tend to be insured against, the insurance industry is particularly susceptible to the impacts of climate change: weather events represent the largest proportion of losses sustained by insurers.[11] As the frequency and intensity of extreme events will increase over the coming decades (see **extreme weather events**), the potential magnitude of losses sustained by insurers will increase, as will climate-related health and life insurance claims, essential input supply interruptability insurance losses (transport, commodities, water, and energy), and liability claims.

Over the past half century, the frequency of "great natural disasters" has increased significantly, from a global mean of two per year in the 1950s to an average of seven per year over the decade to 2005 (normalized to take into account population increase and the increase in the amount and value of assets at risk). This increase has all been weather related as there has been no discernible change in the frequency of geophysical events.[12]

Over the period 1960–2005, weather-related economic losses have increased seven-fold and insured losses by 25-fold, with 2004 and 2005 the two highest years for weather-related insurance losses.[13] Since 1995, insurance losses have exceeded US$ 100 billion in several years.[14] There is mounting evidence that insurers may have underpriced climate risk since the mid-1990s as industry insurance payouts have exceeded premium income in several years and the number of insurance companies that have gone into receivership has also increased.[15]

Estimates of potential future losses vary widely, but all major studies indicate that they are likely to increase substantially over the coming decades. Some studies suggest that the scale of losses could exceed US$ 150 billion/year by 2015[16] and as high as $US 1 trillion in some years by 2040.[17] Estimating future losses many decades in the future is, of course, subject to considerable uncertainty and much depends on how quickly climatic conditions change and the extent of adaptation and risk reduction measures put in place. For example, one recent study estimates that potential future losses from some weather-related events could be halved by implementing cost-effective risk mitigation measures (such as improved building standards and flood control measures).[18] Nonetheless, the expected increase in insurance losses will have important implications for global financial markets as more capital will need to be held in reserve to cover large-scale losses (thus unavailable for productive investments elsewhere).[19]

Climate change will also most likely change the types of insurance cover the industry is willing to provide. Areas that are most prone to weather-related disasters, such as the Gulf of Mexico, are likely to find it increasingly difficult to obtain insurer cover in future or be subject to stringent liability clauses. In these cases governments may intervene to provide cover, effectively transferring the risks from private insurance markets to tax payers. For example, the increasing risks of flood damage in some areas of the United States has led many insurers to withdraw from offering insurance cover and resulted in the establishment of publicly subsidized insurance schemes (such the National Flood Insurance Program, which provides coverage in excess of US$ 500 billion).[20]

Having access to affordable insurance cover is important to maintaining economic stability as it smooths out the economic impact of sudden losses and spreads the risk. However, the ability to access or afford insurance is unevenly distributed across countries. Developed high-income countries have much better insurance coverage than most developing countries and, as a result, are much better insulated against the risk of adverse climate change impacts. On average, insurance covers 40% of economic losses in high-income developed countries and the proportion of economic losses covered by insurance in developing countries is very low, averaging less than 3%.[21] Furthermore, although the average annual weather-related economic loss is much higher (nearly four times higher) for developed countries than developing countries (as the value of their assets is less), the economic impact of these events is much more significant for developing countries. For example, between 1985 and 2000, these economic losses amounted to 13.4% of the combined GDP of developing countries, but only 2.5% for developed countries.[22] The lack of insurance cover means that developing countries must absorb the losses themselves, which can have devastating economic consequences and exacerbates poverty.

The potential impacts of climate change-induced economic losses, particularly extreme weather events, has led to calls for new and innovative measures to provide affordable insurance cover to developing countries. The importance of insurance in assisting developing countries to manage climate risk is recognized in international climate change agreements and negotiations. The need to expand the role of insurance in managing climate change impacts is specifically mentioned in the **United Nations Framework Convention on Climate Change (UNFCCC)**[23] and the **Kyoto Protocol**,[24] and a range of initiatives have been proposed over the past decade. These include proposals for an "international insurance pool" under the

UNFCCC, the Climate Impact Relief Fund, and the Climate Change Funding Mechanism proposed by Germanwatch.[25]

Insurance industry role in financial markets

In addition to the insurance industry's climate change risk management role, it is also a major player in global capital markets. The industry is the single largest investor in the global economy, controlling more than $ 16 trillion worth of capital assets, many of which are vulnerable to climate change.[26] In future, insurers will need to source new sources of capital to cover the potential of rising losses and will increasingly tap into global financial markets, which can more readily absorb large-scale catastrophe losses without creating major economic disruptions. This could divert capital from productive investments elsewhere and could have major flow on effects for the global economy, particularly through changes to the price of capital (interest rates).

The insurance industry is acutely aware of the problems climate change could present for their industry and the global financial markets. Major industry players, such as Swiss Re and Munich Re, have been actively engaged in raising public awareness of looming climate threats and contributing to the international climate debate. The insurance industry is also an active participant in a range of initiatives to increase the level of public disclosure of the greenhouse gas footprints and climate change policies of major companies.

The pooling of industry risk and financial reserves will greatly assist the industry to manage climate change-related losses. For example, the insurance industry's unwillingness to provide insurance cover in certain regions could result in a reduction in construction of new building in some areas and even a retreat of habitation from some zones – in effect, a market-induced adaptation response.

Emerging business opportunities

If the international community is to stabilize atmospheric greenhouse gas concentrations at a level that avoids **dangerous climate change**, then global emissions will have to be reduced by at least 50% below present levels by mid-century (see **stabilization targets**). This will require substantial increased investment in low emission technologies and **energy efficiency** (see **mitigation**). Establishing appropriate market signals and financial incentives, such as introducing a cost penalty on greenhouse gas emissions, will be important to delivering these outcomes. The global financial system will play a critical role in this process – including raising the required investment capital to fund mitigation and adaptation measures, developing new financial services and products to support these activities, and especially through facilitating the expansion of the rapidly growing international carbon market (see **emissions trading**).

Accessing sufficient capital to finance the start-up of new low emission technology industries (e.g. commercializing renewable energy technologies) has been a constraint in the past, particularly in the absence of certainty over future climate policy settings. New companies have often had to rely on nontraditional sources of financing, such as venture capital, to get started. However, since the mid-1990s, an increasing number of alternative financing sources have become available to the **renewable energy** and **energy efficiency** industries. Venture capital funds (which assist new products in industries to enter the market) have played a significant role in stimulating investments in low emission technologies and energy efficiency. In the United States alone during 2005 venture capital funds invested $739 million in renewable energy technology development, an increase of 36% over 2004.[27] These investments make an important contribution to funding higher risk ventures where other sources of funding may not be available – they are essential to driving technological innovation and deployment (see **technology, structural change and organizations**).

Other financial institutions, which have previously had little activity in the clean energy market (such as hedge, mutual, and pension funds) have also become actively involved. This has, in part, been driven by the rapid increase in the size of the renewable energy market, but it is also evident that the increased engagement of a broader range of financial institutions has also contributed to the industry's rate of growth. In 2007, the market for renewable energy technologies had reached US$150 billion/year.[28] The market for renewable energy is projected to increase ten-fold before mid-century.[29] In total, the International Energy Agency projects that $13 trillion will need to be invested in clean energy technologies and energy efficiency by 2050 to ensure global emissions are kept at or below current levels.[30] This is a significant amount of finance that capital markets will need to raise and manage.

Following the ratification of the **Kyoto Protocol** in 2005 and the establishment of the EU Emissions Trading Scheme, a significant international carbon trading market has emerged (see **emissions trading**). This includes both the compliance market (to achieve mandatory emission limits) and the voluntary markets (such as the Chicago Climate Exchange and a range of private carbon offset facilities). The explosive growth in the carbon market, which tripled from around $10 billion in 2005 to over $30 billion in 2006, has required a substantial increase in the range of financing services offered by financial institutions to support the market (project due diligence processes, insurance, risk management services, among others).[31] Carbon trading is likely to develop into a major new commodity market over the coming decades and involve very large international financial flows.

In addition to the clean energy and carbon markets, insurance companies and banks are also beginning to offer a range of new financial services and products. While these are mainly directed at managing their clients' risk exposure, they also often provide a market incentive to reduce emissions and also promote adaptation responses. In recent years, new products that link insurance coverage and premiums to climate risk reduction measures, such as constructing cyclone-proof housing, linking premiums to building **energy efficiency**, and new car insurance products based on 'pay-as-you-drive' pricing structures encouraging consumers to consume less energy and drive less, are now being offered.[32] Some banks have also introduced adjustable mortgage rates according to home energy efficiency ratings.

Investor response to climate change

Private and institutional investors are becoming progressively more active in guiding the investment and asset management approaches adopted by firms in relation to climate change. Analysts, regulators, investors, and rating agencies are increasingly scrutinizing the climate risk management practices of firms and particularly their exposure to carbon regulatory risk (especially new government policies to limit emissions). Shareholders and investors are demanding greater public disclosure of the greenhouse gas emission footprint of firms and are increasingly structuring their investment portfolios based on sustainability considerations.

Examples of new initiatives by the financial markets to drive more sustainable investment patterns by firms include the following:

- Investor Network on Climate Risk (INCR): a network of institutional investors and financial institutions dedicated to promoting better understanding of the financial risks and investment opportunities posed by climate change, now including more than 50 institutional investors that collectively manage over $ 3 trillion in assets.[33]
- CERES: a coalition of US institutional investors, pension funds, environmental and religious organizations, and other public interest groups, with over $ 200 million in assets that is working with companies and investors to address sustainability issues – their activities include filing shareholder motions demanding action on climate change and publishing a blacklist of companies that are lagging on environmental issues.[34]
- Carbon Disclosure Project (CDP): launched in 2000 and involving 125 institutional investors (with combined assets of US$ 21 trillion) working with major corporations to publicly disclose details on their greenhouse gas emissions – by 2007, more than 1,000 large corporations were reporting their emissions through the project.[35]
- Enhanced Analytics Initiative (EAI): whose members agree to use part of their budget to reward brokers that publish research on climate change financial issues or brand management – its members have US$ 4 trillion of assets under management.

Financial markets are, and will increasingly, play an important role in driving the transition to a less emissions-intensive economic system as well as the response of companies to climate change. To a large extent this will be largely independent of government policy decision making and, in fact, is likely to partly drive government climate policy decisions. While insurance itself will not be the solution to climate change, it does provide a means of better managing the risks, spreading the costs of climate change across a greater number of people, and promoting and providing incentives for the adoption of adaptation measures to reduce climate vulnerability.

In conclusion, climate change will have major implications for the global financial system, particularly the insurance industry. Climate change presents a new set of risks, constraints, and opportunities. Finance markets will play a critical role in facilitating, and driving, the structural adjustments required to deliver a low emission economy.

See also: adaptation, agriculture and food supply impacts, climate change impacts, emissions trading, extreme weather events, health impacts, mitigation, technology, structural change and organizations.

Notes

1 Merton 1992
2 *Oxford Dictionary of Finance* 1993
3 McKinsey and Company 2005
4 Derivative markets provide a means by which different entities can trade risks associated with the price of a good or asset changing, up or down, at some stage in the future – they provide a mechanism for managing risks.
5 Australian Federal Department of Environment and Heritage 2005
6 See Rogers 2002 for a technical discussion of risks and risk management.
7 Epstein and Mills 2005
8 ABI 2005
9 Reo research Report 2007
10 ABI 2005
11 Epstein and Mills 2005
12 Hoeppe and Gurenko 2006
13 Ibid.
14 Epstein and Mills 2005
15 See Reo research Report 2007
16 Epstein and Mills 2005
17 UNEP 2007
18 ABI 2005
19 Reo research Report 2007
20 Bowers 2001
21 Hoeppe and Gurenko 2006
22 Freeman and Scott 2005
23 UNFCCC, Article 8
24 Article 3.14 KP
25 See Linnerooth-Bayer and Mechler 2006 for a discussion of these issues.
26 UNEP 2007
27 Cleantech 2006
28 UNEP 2008a
29 Stern 2006
30 IEA 2006b
31 World Bank 2007b
32 Reo research Report 2007
33 See http://www.incr.com
34 See CERES and *Financial Times*, February 14, 2007
35 See Carbon Disclosure Project

Further reading

Association of British Insurers 2005; UNEPFI 2007; Epstein and Mills 2005; Hoeppe and Gurenko 2006.

FOSSIL FUELS

Fossil fuels are carboniferous materials that were formed over millions of years through the decomposition and chemical change of organic matter in oxygen-free environments. They are, in effect, ancient stored sunshine as the energy they contain originated from the conversion of sunlight to carbohydrates through plant photosynthesis, some as long ago as 350 million years. It takes approximately 20–25 tonnes of ancient buried plant material to provide the fossil fuel energy contained in one liter of oil equivalent.[1]

The hydrocarbons in fossil fuels are flammable and when combusted in the presence of oxygen release heat energy. The main by-products of combustion are not only **carbon dioxide (CO_2)** but also **nitrous oxide (N_2O)**, air-borne particulates, small quantities of several other gases, and, in the case of coal, ash residues. In aggregate, the earth's crust is estimated to contain around 5,000 billion tonnes of fossil fuels, although less than half could feasibly be extracted. There are also estimated to be approximately 10,000 billion tonnes of **methane (CH_4)** (the main constituent of natural gas) in **methane hydrates**, but it is not yet known whether humans can extract commercial quantities of hydrocarbons from this source (see **methane hydrates**).

The ability of humans to harness the energy contained in fossil fuels (initially coal) to provide mechanical power (via the steam engine) was one of the basic building blocks of the Industrial Revolution. Until 1900, the quantity of fossil fuels consumed was relatively small, but consumption has since grown enormously. The commencement of mass production of the automobile and the large-scale expansion of electricity production and distribution systems were the main contributing factors underpinning this growth. Consumption continues to increase unabated: more fossil fuels have been consumed in the past 30 years than in the preceding two centuries, and consumption could double again by 2050.[2] CO_2 emissions from fossil fuel consumption increased by 145% between 1970 and 2004.[3]

Fossil fuels are the world's main source of primary energy, providing 80% of commercial energy supply.[4] This energy is vital to the global economy: if fossil fuel production were to suddenly cease, economic activity would come to a near standstill in most countries.

Although fossil fuels have made a major contribution to human development and prosperity, and will continue to do so for the foreseeable future, their use has also resulted in a range of adverse environmental impacts. Fossil fuel combustion is the single largest contributor to atmospheric air pollution and urban smog and is responsible for a significant number of human deaths each year. Respiratory illness related to particulate emissions from fossil fuel consumption is estimated to be responsible for the deaths of more than half a million people annually.[5] Many thousands of coal miners lose their lives through mining accidents and coal dust-related diseases. Fossil fuels cause acid rain (through the emission of sulphur and nitrogen compounds), which has adversely affected large tracts of forest and aquatic ecosystems over the past half century. They also emit significant quantities of heavy metals, contribute to water pollution, and raise several other fuel cycle waste issues. However, it is their contribution to increases in atmospheric greenhouse gases concentrations that is the primary concern.

Fossil fuels are the single largest source of **anthropogenic greenhouse gas emissions**, accounting for 60% of global emissions (on a **carbon dioxide equivalent [CO_2e]** basis).[6] In terms of the contribution to the three principal greenhouse gases in 2004, they accounted for 75% of global CO_2 emissions, 30% of methane emissions, and 9% of N_2O emissions. In 2004, fossil fuel combustion released 28 billion tonnes of CO_2 to the atmosphere. Based on the most recent International Energy Agency projections, fossil fuels could contribute 34–40 billion tonnes CO_2/year by 2030 and as much as 50–60 billion tonnes CO_2/year by 2050, if current trends and policies are maintained.[7] Increases in emissions of this magnitude would accelerate the buildup of greenhouse gases in the atmosphere and will almost certainly result in **dangerous climate change**.

Fossil fuels can occur in solid (peat and coal), liquid (crude oil, gas liquids, tar sands, and shale oil), or gaseous (natural gas) forms, each of which can contain varying quantities of hydrocarbons.

Coal

Coal was formed over millions of years through the slow accumulation and decomposition of organic residues. Over time, as these residues were buried even deeper, the pressure and heat caused biochemical changes. First, organic residues were transformed into peat (the lowest-grade fossil fuel), then lignite (brown coal), bituminous coal, and finally to energy-dense, shiny black, anthracite coal (the oldest and highest-valued coal).

Coal is the most abundant fossil fuel, with proven economic reserves of approximately 900 billion tonnes (gigatonnes – Gt). Three-quarters of this tonnage is located in just five countries (United States, Russia, China, India, and Australia), although significant deposits are also found in more than 20 other countries.[8] Coal is the second-largest source of primary energy and, in 2004, accounted for 25% of global energy supplies, 40% of global electricity production, and 41% of energy-related CO_2 emissions.[9] At current extraction rates, known economic coal resources would last approximately 150 years, but significant increases in consumption, as are projected to occur over the next few decades, could result in earlier depletion.

There are three principal categories of coal: brown coal (including peat), black bituminous coal, and black anthracitic coal. Only very small quantities of peat are used for commercial energy production. Brown coal is used almost exclusively for electricity generation at or near the mine site, due to its low heat value per tonne. Global consumption was around 0.6 Gt in 2004, and only a tiny fraction is traded internationally. Brown coal is also the most greenhouse gas-intensive form of coal in terms of CO_2 emissions per unit of energy produced – 20–30% more than black coal. Europe and the Economies in Transition (former Soviet Bloc countries) account for most brown coal consumption.

Black bituminous and sub-bituminous coal (often referred to as steaming coal) is the most common type of coal traded and accounts for approximately 70% of global coal consumption (4 Gt in 2004). It is primarily used for electricity generation and to provide heat in industry. Each kilogram of black coal burned releases 0.95–1.0 kg of CO_2.

Black anthracitic coal (often referred to as coking coal) is the highest-value form of coal (it has the highest heat content) and is mainly used for metallurgical purposes, principally for making iron and steel. It accounts for approximately 15% of global coal consumption (1.2 Gt in 2004) and produces similar greenhouse gas emissions to black bituminous coal.

In 2004, global coal consumption of all forms was 5.6 billion tonnes, with China (1.9 Gt) and the United States (1 Gt) being the two largest consumers and producers, accounting for half of global consumption.[10] Coal is the least internationally traded of the main fossil fuels (87% is consumed domestically). Australia and Indonesia are the two largest coal exporters.

Oil

Oil deposits were mainly formed through the burial and decomposition of marine organisms on the ocean floor over millions of years. Some of the carbon-rich waxes and oils that were contained in these organisms were squeezed out under pressure and migrated through porous sands and rock. While most escaped back into the environment, some encountered nonporous rocks and strata and became trapped under pressure and formed pools and reservoirs of oil and gas. When oil drilling penetrates these pressurized structures, the pressure forces the oil and gas to the surface.

Oil can be found in several forms and of varying quality and can be subdivided into two principal categories: conventional and nonconventional oil. Conventional oil includes crude oil (the most common form), natural gas liquids, gas condensates, and heavy bituminous crude and accounts for nearly all global oil production. Nonconventional oil (often called syncrude) can be extracted from tar sands and oil shale. These deposits need further processing (usually requiring steam or other sources of heat) to yield liquid fuels. Liquid fossil fuels can also be manufactured through reformulating natural gas or synthesic gas from coal.

Total known economic reserves of oil are estimated to be around 1,300 billion barrels. At current consumption rates, known economic oil reserves will last 40–50 years but possibly less if consumption increases in line with projections. It is difficult to provide accurate estimates of global oil reserves, and how long they could last, as new deposits are constantly being found and technology improvements (such as enhanced oil recovery) are enabling more oil to be economically extracted from existing reservoirs. Economic reserves are also a function of oil prices: as oil prices increase, economic reserves increase. Almost two-thirds of known economic reserves are located in the Middle East, with Saudi Arabia alone accounting for 20%. Other major deposits, in order of importance, are located in Canada (mainly non-conventional oil in tar sands), Venezuela (mainly in the form of heavy bituminous oil), Russia, and Nigeria. Based on known and inferred (yet to be discovered) resources, approximately 2,300 billion barrels could ultimately be recoverable.[11]

Oil is the single most important source of primary energy, accounting for 35% of global supply in 2004.[12] In 2005, world oil consumption was 84 million barrels/day (mb/d).[13] The top five producers in 2005, in order of importance, were Saudi Arabia (9 mb/d), Russia (9 mb/d), United States (5 mb/d), Iran (3.9 mb/d), and China (3.6 mb/d). Conventional oil is less greenhouse intensive than coal, producing 75–80% of the CO_2 per unit of energy produced. In 2004, oil accounted for 39% of energy-related CO_2 emissions.

Production of nonconventional crude currently accounts for less than 2% of global production but is expected to quadruple its share to 9 mb/d by 2030.[14] Just over half of this production will come from Alberta, Canada, which has vast tar sand deposits (ultimately recoverable resources are estimated at 315 billion barrels of oil).[15] The remainder is expected to come from gas-to-liquids and coal-to-liquids production facilities. World oil shale resources are very large and estimated at around 500 billion barrels, mainly located in the United States and, to a lesser extent, Australia. Currently, no oil shale production facilities are expected to enter commercial production over the period to 2030, but this could change if oil prices continue to rise.[16]

While nonconventional oil provides some energy security benefits (by lessening dependence on OPEC producers), it is a very greenhouse-intensive source of oil. A liter of petroleum product derived from tar sands produces twice the greenhouse gas emissions of a liter of conventional oil, while shale oil is up to five times more intensive.[17] Petroleum products derived from gas-to-liquid plants are three times more greenhouse intensive than conventional crude, while coal-to-liquids is 10 times more greenhouse intensive.[18] A significant expansion in global nonconventional oil production would lead to a large increase in carbon dioxide emissions.

Natural gas

Natural gas is formed in a similar manner to oil, and the two are usually found in conjunction with each other (most oil deposits have associated gas lying above the oil layer). Traditionally, natural gas was viewed as a low-value by-product of petroleum production: most was just vented into the atmosphere or flared, and some still is. However, since the 1960s, the demand for natural gas has expanded considerably, driven by energy security concerns (following the OPEC oil embargoes of the 1970s), the advent of high-efficiency gas turbines that can utilize cheap gas for electricity generation, and environmental considerations in response to acid rain and particulate pollution.

More recently, natural gas has attracted interest due to its greenhouse gas benefits. Natural gas is a clean-burning fuel and is the least greenhouse-intensive fossil fuel, producing only 65–75% of the CO_2/unit of energy produced compared with coal, depending on the source. However, some natural gas deposits have relatively high concentrations of CO_2. It is common practice to separate and vent the CO_2 to the atmosphere (which increases the full fuel cycle greenhouse intensity of natural gas) or to reinject it into the gas reservoir to maintain reservoir pressures (see **carbon capture and storage**).

Natural gas is the second-largest fossil fuel energy resource after coal, currently accounting for 21% of global primary energy supplies. Proven economic resources are estimated to be 180 trillion cubic meters (tcm) and at current rates of consumption would last approximately 60 years – or 40 years if future demand projections prove correct.[19] Over half (56%) of known resources are located in just three countries (Russia, Iran, and Qatar). Ultimate recoverable gas resources are estimated to be around 300 tcm.[20] In 2004, global gas consumption was 2.8 tcm, and consumption is expected to grow by 2% per year over the period to 2030 – the fastest growth rate of all fossil fuels – to reach 4.7 tcm.[21] Russia and the United States are the two largest gas producers and the United States and European Union the two largest consumers. Natural gas is primarily used for electricity generation (approximately

40%), and the rest for process heat in industry and space/water heating and cooking in the commercial and residential sectors. It is usually delivered to the point of use by pipeline or by ship in the form of liquefied natural gas (LNG).

Due to its lower greenhouse gas intensity than other fossil fuels, natural gas is expected to play an important role in limiting greenhouse gas emissions over the coming decades. This will mainly be achieved by replacing coal with gas in electricity generation and industrial process heat, and also to some extent in replacing electricity use for space heating and cooking in the residential sector. Natural gas is also the least sensitive to a carbon cost penalty, imposed via **carbon taxes** or through **emissions trading**.

Future consumption

To date, humans have consumed less than one-fifth of known recoverable fossil fuel resources.[22] It is difficult to predict how much of the remaining fossil fuel resources will actually be consumed as this will depend on future growth in energy demand, the cost and availability of nonfossil fuel energy sources, and the success of the international community's efforts to decarbonize the global economy. Governments have already begun to introduce measures to try to limit fossil fuel consumption (e.g. the European Union's Emissions Trading Scheme), but it is not yet clear the extent to which stringent policy measures will be introduced, nor when or how successful they will be in reducing fossil fuel use. However, what is clear is that without a radical policy shift by major fossil fuel-consuming nations, fossil fuel use will continue to rise for several more decades and possibly much longer. So, too, will fossil fuel-related greenhouse gas emissions.

Based on 2006 International Energy Agency (IEA) projections, fossil fuel consumption will increase 1–2% per year over the period to 2030. The IEA usually produces two sets of projections, based on different assumptions about likely future government policy settings. The first is termed the "Reference Scenario," based on existing policies, energy demand trends, and continued technology and energy efficiency improvements (sometimes also referred to as the "business as usual" scenario). The second is termed the "Alternative Scenario," in which governments are assumed to adopt more aggressive climate and energy security policies, aimed at reducing greenhouse gas emissions and their reliance on fossil fuels (see Table 9).

No matter which scenario is chosen the dependence on fossil fuels is not expected to change much at all. In terms of energy-related CO_2 emissions, there is an increase from 28 Gt in 2004 to 40 Gt in 2030 (Reference Scenario) or 34 Gt (Alternative Scenario). The 2007 IPCC projections are broadly consistent with the IEA projections but deliver a wider range of potential emission outcomes (depending on the IPCC emission scenario adopted).[23]

Beyond 2030, projections are much less certain, but under business as usual conditions fossil fuel CO_2 emissions could reach 58 Gt by 2050 (more than double 2004 emission levels).[24] Under alternative long-term IEA scenarios (which assume rapid technology advance and much higher uptake of **energy efficiency** measures), fossil fuel CO_2 emissions in 2050 could be reduced to 25–30 Gt, broadly similar to 2004 levels (see **mitigation**).[25]

Table 9 Projected fossil fuel consumption to 2030

Fossil fuel	2004 Actual consumption	2030 Reference scenario	Average annual growth rate (%)	2030 Alternative scenario	Annual average growth rate (%)
Coal (Gt/year)	5.6	8.9	1.8	7.1	0.9
Oil (mb/day)	84	116	1.3	103	0.9
Gas (tcm)	2.8	4.7	2.0	4.1	1.5
Global primary energy demand (million tonnes oil equivalent)	11,200	17,100	1.6	15,400	1.2
Fossil fuel share (%)	80	81	—	77	—

Source: IEA 2006a.

What are the prospects for reducing emissions from fossil fuel use?

The main fossil fuel mitigation options include implementing end-use **energy efficiency** measures; obtaining primary energy supplies from **renewable energy** sources (e.g. solar, hydro, wind, and geothermal) or **nuclear power**; increasing the efficiency with which fossil fuels are converted to useful energy; moving to less greenhouse-intensive fossil fuels (interfuel substitution); or by preventing the release of CO_2 to the atmosphere through **carbon capture and storage**.

Coal conversion efficiency

At present, existing energy conversion technologies transform energy to electricity, heat, and motive power (transport) very inefficiently. The global average conversion efficiency of coal-fired electricity generating plants is around 35% (most of the energy is lost to the atmosphere as waste heat). Technologies are already available on the market that can significantly increase conversion efficiencies and offer considerable greenhouse gas mitigation potential over the next few decades.

A majority of existing coal-fired power stations (85%) utilize subcritical pulverized coal technologies, and some have conversion efficiencies as low as 30%.[26] Most new coal-fired power stations are using supercritical steam technologies (currently 11% of operating plants), and these have conversion efficiencies of around 42% (a 15% improvement over subcritical steam technologies). New ultrasupercritical technologies (2% of the existing stock) can achieve 45% conversion efficiencies, while advanced supercritical and integrated gasification and combine cycle (IGCC) technologies are expected to achieve conversion efficiencies of up to 50% by 2015 (often referred to as "clean coal" technologies).[27]

Conversion efficiency improvements could significantly reduce fossil fuel use. For example, if China's coal-fired power plants were to achieve the same average conversion efficiencies as those currently operating in Japan, China's consumption of coal would be reduced by 20%.[28]

Although raising global average conversion efficiencies would significantly reduce CO_2 emissions, it may take several decades. On average, a coal-fired power plant has an operational life of 40–50 years, and as a result, the choice of technology has a major bearing on future emission levels. China is currently commissioning a new coal-fired power plant each week, and large capacity additions are also planned in the USA, India, Japan, and Europe. Significant opportunities exist for increasing the efficiency of coal-fired plants in industrialized countries, as much of the existing European and US stock is ageing and will need to be replaced in the next 10–20 years. Unfortunately, a majority of the new plants still opt for supercritical, or even subcritical, technologies rather than ultrasupercritical and IGCC technologies (which are 15–20% more expensive). These investment decisions mean that the world will be locked into a higher base level of emissions for many decades than would have been the case if the most efficient technology had been deployed. To increase the uptake of high-efficiency coal technologies, either the cost of ultrasupercritical and IGCC plants will need to be reduced and/or a carbon cost penalty will need to be introduced to improve their economics.

Gas conversion efficiency

Rapid advances in the conversion efficiencies of natural gas-fired power stations have been achieved in recent years, primarily through the development of the combined cycle gas turbine (CCGT). The newest state-of-the-art CCGT plant can achieve conversion efficiencies of up to 60%, compared with the global average gas plant efficiency of 42%, and only 33% in Russia.[29] Upgrading the existing stock of gas-fired power plants to new CCGT plant efficiency standards would also deliver significant emission reductions, as would substituting coal-fired power plants with gas-fired plants since gas is a much less greenhouse-intensive fossil fuel: new CCGT plants produce around half the amount of CO_2/kwh associated with new state-of-the-art coal-fired power plants.[30] Russia, for example, is currently the largest user of natural gas, but it could reduce gas consumption by 30% if it had plants as efficient as the average plant in Europe.[31]

The principal barrier to increased use of CCGT plants is cost. While such plants are cheaper and quicker to build, they are much more sensitive to fuel costs than coal-fired plants: fuel costs account for 70% of gas electricity costs compared with 40% for coal. CCGT plants are competitive with coal in some locations, but in general, coal is still the cheapest means of generating base-load power. The rapid increase in natural gas demand in recent years, partly due to its superior environmental credentials, has increased gas prices. If the cost of gas relative to coal widens over the next few decades (as is expected), the amount of coal-fired capacity substituted by gas is likely to be lower (though this could change if a CO_2 cost penalty is introduced).

Substituting gas for electricity in end-use applications such as space and water heating, cooking, and process heat in industry can achieve significant reductions in emissions, depending on the carbon intensity of the electricity displaced. Electricity-to-gas substitution has been underway for many years and is expected to continue over the medium term.[32]

Carbon capture and storage is a technology that offers considerable potential for reducing CO_2 emissions from the power sector and also in the oil and gas industry.

However, this is currently an expensive **mitigation** option and is unlikely to be deployed on a large commercial scale without the introduction a significant carbon cost penalty, possibly in the order of US$ 30–50/tonne CO_2 (see **carbon capture and storage**), and advances in technology.

Overall, increases in fossil fuel power plant energy conversion efficiencies, inter-fuel substitution, and carbon capture and storage can make a major contribution to greenhouse gas emission reductions over the period to 2050. The IEA estimates that 20–30% of the reductions in energy sector greenhouse gas emissions to 2050 can be achieved through these measures.[33]

In summary, based on current projections, the quantity of fossil fuels consumed and the world's dependency on fossil fuels are both set to increase over the next 20–30 years unless significant policy changes are made. As the IEA aptly puts it, "the world is not on course for a sustainable energy future."[34] The IPCC suggests that humans will need to first stabilize fossil fuel greenhouse gas emissions and then reduce them significantly – probably to less than one-quarter of present levels – by the latter half of this century if we are to avoid dangerous climate change (see **stabilization targets**).[35] This represents a challenging task indeed.

The principal limiting factor is the cheapness of fossil fuels, particularly coal. Current fossil fuel market prices are a poor indicator of the true cost of fossil fuels to society: prices do not include health costs, the costs of acid rain or surface ozone, or the contribution that fossil fuels make to global warming impact costs (see **socio-economic impacts**). In economic terminology, fossil fuels do not incorporate the externality costs of their production and consumption.

One way to address this is to introduce these costs artificially, through the introduction of a cost penalty on fossil fuels (for example, through **carbon taxes**, an **emissions trading** system or some other fee). The alternative is simply to regulate or restrict fossil fuel use: for example, through requiring all fossil fuel power plants to capture and store their greenhouse gas emissions. To date, governments have been reluctant to introduce cost penalties or direct use restrictions on fossil fuel usage, particularly for power generation and automobiles. This will need to change if significant cuts in greenhouse gas emissions are to be achieved by 2050 and beyond.

See also: carbon capture and storage, carbon dioxide, emissions trading, energy efficiency, future emissions trends, mitigation, stabilization targets.

Notes

1 Dukes 2003
2 WRI 2005
3 IPCC 2007
4 IEA 2006a
5 WRI 2007b
6 WRI 2005
7 IEA 2006a
8 Ibid.
9 Ibid.

10 US Geological Survey 2000
11 IEA 2006a
12 Ibid.
13 Ibid.
14 Ibid.
15 Ibid.
16 IEA 2006b
17 Ibid.
18 Ibid.
19 IEA 2006a
20 US Geological Survey 2000
21 IEA 2006a
22 Olah *et al.* 2006
23 IPCC 2007
24 IEA 2006b
25 Ibid.
26 Ibid.
27 Ibid.
28 Ibid.
29 Ibid.
30 Ibid.
31 Ibid.
32 IEA 2006a
33 IEA 2006b
34 Ibid.
35 IPCC 2007

Further reading

IPCC 2007; IEA 2006a, 2006b; WRI 2005.

FUTURE EMISSIONS TRENDS

Anthropogenic greenhouse gas emission levels will, over the coming decades, be very much dependent on how the global economic system evolves and how seriously the global community responds to climate change. Population expansion and economic growth, political and cultural trends, technological change, and the impact of, and response to, climate change will all influence future greenhouse gas emissions levels. Considerable uncertainty surrounds the future trajectories of these variables, and the further one looks into the future the more uncertain things become. Nonetheless, it is important to understand what drives greenhouse gas emissions and what the future may look like under different circumstances.

Emission scenarios

Given the wide range of possible future outcomes, analysts often develop scenarios of future developments, with each scenario incorporating a different set of assumptions about critical variables. The **Intergovernmental Panel on Climate Change (IPCC)** scenarios contained in the Special Report on Emission Scenarios (SRES) are often employed as the reference framework of how human society could evolve, although other research groups have also formulated their own scenarios.[1]

The IPCC SRES outlines four different families of scenarios, each of which depicts a different global development pathway. The four families are based on different assumptions about population growth, rates of technology development and deployment, and economic development trajectories. The scenarios are labeled A1, A2, B1, and B2. While none of these scenarios explicitly considers the impact of climate policy-induced mitigation measures, they do implicitly integrate environmental and sustainable development considerations that would affect emissions levels. A brief description of these scenarios is given below.

A1

This depicts a world where there is continued rapid economic growth but declining rates of growth in global population. Under this scenario, global population peaks around mid-century and then declines slowly for the remainder of the century. It is also assumed that there is a large degree of convergence between countries and regions in terms of incomes per capita, fertility, cultural interactions, and communications and that rapid technological change underpins these trends. This scenario includes three subscenarios based on different assumptions about the types of technologies deployed. One focuses on continued reliance on fossil fuel-intensive economic structures (the A1F1 scenario); the second considers the rapid development and deployment of climate-friendly technologies, such as renewable energy and energy efficiency (the A1T scenario); and the third scenario envisages a balanced mix of both fossil fuel reliance and climate-friendly technologies (the A1B scenario). Under this family of scenarios, global emissions (and atmospheric greenhouse gas concentrations) grow substantially in the case of A1F1 and moderately in the A1T scenario. The A1B scenario falls roughly between these two scenarios. The growth in the global economy is the major underlying driver of emissions.

A2

This scenario assumes a greater diversity of development pathways, characterized by much slower convergence in fertility rates, more fragmented technology development across regions, and greater self-reliance and preservation of cultural identities (less globalization). Population continues to grow, is higher than in the A1 scenario, and is a key driver of future emission levels.

B1

This is similar to the A1 scenario, but there is a much greater drive toward sustainability (based on a service and information economy), more emphasis on global solutions to environmental problems and a greater focus on international equity issues. This entails significantly reduced materials intensity and the use of more resource-efficient technologies. It results in lower greenhouse gas emissions and atmospheric greenhouse gas concentrations than most other scenarios.

B2

This scenario shares many of the characteristics of the A2 scenario (such as locally and regionally driven development and sustainability actions), but with lower population growth and diverse, but less rapid, technological change.

These scenarios each deliver very different emission outcomes over the period up to 2100. While the SRES projections prepared in 2000 were recently revised by the IPCC in its 2007 Fourth Assessment Report to reflect changes in some of the underlying assumptions, the new projection results are, nonetheless, similar to the original SRES projections. The 2007 "post-SRES" projections indicate that emissions in 2100 could be anywhere between 25 and 135 $GtCO_2e$/year, from half to nearly triple present emissions levels, with atmospheric CO_2e concentrations of anywhere between 445 and 1,130 ppm.[2] The A1T and B1 scenarios generally deliver emissions at the lower end of the range, while the A1F1 (fossil fuel-intensive scenario) and A2 (high population scenario) tend to deliver emissions and concentration levels at the higher end of the range. The different scenarios also exhibit considerable variation in the emissions of different gases. For example, the A1F1 scenario delivers much higher energy-related CO_2 emissions than the other scenarios, while the A2 scenario delivers the highest **methane (CH_4)** and **nitrous oxide (N_2O)** emissions (largely driven by population growth and the expansion of agriculture).

Given the large range of possible outcomes, one could reasonably conclude that these projections are only of marginal benefit to decision makers trying to identify, and agree on, appropriate policy responses to climate change. However, the scenarios do demonstrate to decision makers how different development pathways can influence future emissions levels. Based on what has actually been observed since the SRES scenarios were originally published in 2000, it is evident that the fossil fuel-intensive A1F1 scenario appears to be the scenario that best depicts the present global development pathway. Unless measures are adopted to shift away from this development pathway, greenhouse gas concentrations are likely to exceed 550 ppm CO_2e by 2030–2040, which will result in **dangerous climate change** (see **stabilization targets**).

How reliable are future emissions projections?

Projecting future greenhouse gas emissions is, understandably, subject to considerable uncertainty. Factors such as population growth rates, the quantity and type of energy and natural resources people consume, the type of food they eat, and how they manage their wastes are all important determinants of future emissions levels. Furthermore, each of the greenhouse gases is influenced by different, but often interrelated, causal factors, many of which humans can directly influence. Nearly every human activity has some direct or indirect effect on greenhouse gas emissions.

Emission projection results are heavily dependent on the assumptions made and models employed. Recent improvements in modeling capabilities have led to more robust projections, particularly for the short to medium term (to 2030). These improvements are attributable to the increased coverage and accuracy of global emissions data, improved understanding of the factors that drive greenhouse gas emissions, and a better understanding of the potential rate of development and deployment of low greenhouse gas emissions technologies.

Emissions projection models come in two principal forms: "bottom-up" technology-based models and "top-down" macroeconomic models. Initially these two modeling approaches produced widely diverging results, but as they have become better integrated (through combining the attributes of both approaches) the projection results have become more closely aligned. Combinations of different models are often used to prepare multigas projections (see **stabilization targets**).

The reliability of projections declines as the projection period increases. Emissions projections over the short to medium term (10–20 years into the future) are considerably more robust than projections of emissions over the longer term (30–50 years into the future). Emissions projections beyond 50 years are much more speculative as periods greater than 50 years provide sufficient time for major changes in the size and structure of the global economy to occur, particularly in response to technological change and new government policies.

The reliability of projections also varies between the different greenhouse gases. Projections of energy sector-related emissions, especially CO_2, are generally much more robust than projections of emissions from other sources such as agriculture, forestry, and wastes (see **greenhouse gas inventories**). Our understanding of the underlying factors driving energy sector emissions, particularly the strong correlation between energy-related CO_2 emissions and growth in Gross Domestic Product (GDP), is also much more developed than for other emission sources.

As end-use technology mixes (the major determinants of energy sector emissions) change only relatively slowly, and in a predictable manner, short- to medium-term projections of energy sector emissions are reasonably robust (see **fossil fuels**, **nuclear power**, and **renewable energy**). This is not the case for agricultural and land-use change emissions, where the driving factors are much more complex and are subject to a wide range of social, political, trade, and poverty-related factors that can fluctuate significantly over time.

Major drivers of future emission

Analysis of past trends indicates that **anthropogenic greenhouse gas emissions** are largely determined by two principal variables: population and the level of **emissions per capita**. Emissions per capita are, in turn, determined by the level of economic output (incomes) and the **emissions intensity** of the national economy.

There is significant diversity between countries, but as a general rule, the larger a country's population, and the higher the level of GDP/capita, the greater the level of greenhouse gas emissions. In total, just 25 countries account for over 80% of global emissions, and just five (United States, China, Russia, India, and Japan) account for half of all emissions.[3] What happens in these 25 countries over the next 30–50 years, particularly in the United States, China, and India, will largely determine future global emissions.

Population

Population growth will be a major factor driving greenhouse gas emissions over the next 30–50 years. By 2005, global population had reached approximately 6.5 billion and is expected to grow to anywhere between 8 and 11 billion by 2050, with 9 billion being considered the most probable outcome (corresponding to an average growth rate of 0.75% per year).[4] The extent to which each additional person on the planet contributes to global emissions depends very much on where they are born and live. For example, each additional person born today in the United States has a much greater impact on global emissions than a person born in China (more than five times greater) or Bangladesh (more than 25 times greater). The reason for this is that the average US citizen has a much higher level of emissions per capita (25 tCO_2e/capita) than the average Chinese citizen (less than 5 tCO_2e/capita) or Bangladeshi citizen (around 1 tCO_2e/capita).[5]

In absolute terms, most of the population increase over the next few decades is expected to occur in the developing world, particularly in Africa and South Asia, and this will increase the contribution to global emissions of these regions. The developing world accounts for approximately three-quarters of the world's population, and they also have higher fertility rates (number of children born per adult female) than high-income industrialized countries. The United Nations projects that fertility rates in most developing countries will tend to fall over the next 30–50 years as per capita incomes increase (fertility rates tend to fall as income per capita rises): fertility rates could halve by 2050, to levels approaching those that prevail in high-income industrialized countries today.[6] These fertility changes are expected to moderate the impact of population growth on global emissions over the next 30–50 years, particularly beyond 2050. Nonetheless, it is clear that by 2050 there will be considerably more people on the planet, and this will be the major factor driving greenhouse gas emissions levels for at least the next 40–50 years. Measures that can reduce the rate of growth in global population will be important to greenhouse mitigation efforts.

Trends in emissions per capita

The growth in per capita emissions will be the primary underlying driver of global emissions levels over the period up to 2050. There tends to be a close correlation between GDP per capita and emissions per capita – as incomes grow so do emissions (see **emissions per capita**). This correlation is particularly strong for energy-related CO_2 emissions: as per capita incomes rise, so does the number of appliances people own, the size of houses people live in, and the more people use private cars and air transport – all of which result in higher energy consumption. The correlation between GDP per capita and emissions from nonenergy-related sources (land-use change, agriculture, and wastes) is still positive but less pronounced.

In projecting future emissions levels driven by changes in per capita incomes, it is important to consider both the absolute change in aggregate GDP (the size of the economy) and the emissions intensity per unit of GDP. Over the period 1950–2000, global GDP expanded at an annual average rate of 3.9%, while CO_2 emissions grew at an annual average rate of 3% and non-CO_2 emissions (in CO_2 equivalent terms) by 1.5%.[7] However, since 1990, there have been declines in **emissions intensity** in many countries, which has reduced the annual growth rate in global emissions to 1.2% and for CO_2 emissions to 1.7% per annum.[8]

In the medium term, out to 2030, the global economy is expected to continue to expand strongly at an estimated annual rate of 3.4%.[9] This will be the most important underlying driver of global emissions. Emissions per capita are expected to grow most rapidly in developing countries. At present, average per capita emissions in the developed world are approximately 14 tCO_2e and have been relatively stable for the past decade and have actually declined marginally in some countries. Emissions per capita in the European Union have fallen 5% since 1990, while in the United States, they have increased by only 2%.[10] It is expected that over the medium term, emissions per capita in the industrialized countries could fall another 5–10% and possibly further if more stringent climate policies are introduced.[11]

Emissions per capita in the developing countries are less than one-quarter of those prevailing in industrialized countries (averaging only 3.3 tCO_2e/capita). However, based on current trends, emissions per capita are expected to nearly double over the period up to 2030, in line with the growth in per capita incomes.[12] Much of the growth in developing country emissions will come from China and India, the world's two most populous countries, both of which are expected to maintain rapid GDP growth rates over the next decade. As a result, the developing country share of global emissions is expected to rise from 48% of global emissions (2000) to 55% by 2025.[13]

Emissions trends by greenhouse gas

The projected growth rates for each of the six major greenhouse gases covered by the **Kyoto Protocol** vary considerably, depending on the underlying pattern of economic growth and development. The expected trends in emissions are as follows:

Carbon Dioxide (CO₂)

This is the most significant anthropogenic greenhouse gas and accounted for 77% of global emissions in 2004 (on a **CO_2 equivalent [CO_2e]** basis).[14] Of these emissions, 70% originated from fossil fuel use and most of the remainder derived from land-use change activities (mainly deforestation and the decay of peat and forest industry residues). Over the period up to 2030, CO_2 is expected to be the fastest-growing source of emissions, with a projected annual growth rate of 1.7% per year (compared with 0.5% per year for non-CO_2 emissions).[15]

Under business as usual conditions, energy sector-related CO_2 emissions are expected to increase to approximately 40 $GtCO_2$ by 2030, although projections range from 34 to 43 $GtCO_2$,[16] and 58 $GtCO_2$ by 2050.[17] Future CO_2 emissions from land-use change and forestry are less certain but are expected to remain around present levels (7.5 $GtCO_2$/year) for the next decade or so before declining to around 5 $GtCO_2$/year by 2050,[18] but emissions from fires and peat decay may increase.[19] Beyond 2050, projections are much less certain, but emissions from deforestation are expected to be much lower than present levels as most of the remaining forest cover will already have been cleared.

In the absence of concerted mitigation efforts, total CO_2 emissions from all sources are expected to be approximately 45–50 $GtCO_2$ by 2030 and 60–65 $GtCO_2$ by 2050.[20] As CO_2 emissions are expected to grow at a faster rate than other gases, carbon dioxide's share of total greenhouse gas emissions is expected to rise over the coming decades.

Methane (CH_4)

This is the second most important anthropogenic greenhouse gas, accounting for 14% of global emissions.[21] The major anthropogenic CH_4 sources are livestock and manure management (33%), the energy sector (29%), waste management (23%), rice (11%), and the remainder (4%) largely accounted for by biomass combustion.[22] The coverage and quality of data on CH_4 emissions and sources and future projections are less reliable than for CO_2 (see **methane**).

Although methane emissions have grown, only very slowly since the early 1990s they are expected to grow at a slightly accelerated rate over the next few decades, driven largely by the expansion of livestock numbers (due to increased meat demand) and by significantly increased fossil fuel production and consumption. By 2020, CH_4 emissions are projected to reach 8 $GtCO_2e$, 39% above 2000 levels.[23] Most of the increase in emissions is expected to come from low- and middle-income developing countries, where growth in agricultural production and energy consumption is projected to be higher than in the high-income industrialized countries. Projections data for the medium to longer term are limited, but the **IPCC** expects that methane emissions could stabilize or even decline beyond 2050, in line with global population changes.[24] Overall, due to the lower growth rate of methane emissions relative to CO_2, methane's share of global emissions (on a CO_2e basis) is expected to fall marginally over the medium term.

Nitrous Oxide (N_2O)

This is the third most important anthropogenic greenhouse gas covered under the Kyoto Protocol, accounting for 8% of global greenhouse gas emissions in 2000. The agricultural sector is the dominant source of N_2O (notably from soils, manure, and agricultural residues), accounting for approximately 80% of N_2O emissions (see **nitrous oxide**). Fossil fuel combustion, industrial processes, and sewage account for the remaining 20%.[25] As with methane emissions, data and projections are much less reliable than those for CO_2.

N_2O emissions are expected to follow a similar path to methane but will grow at a slightly higher rate. By 2020, N_2O emissions are expected to be approximately 1.3 $GtCO_2e$, 45% above 2000 levels.[26] The agricultural sector is expected to remain the major source of N_2O emissions over the coming decades, largely driven by population growth in the developing world. In the longer term, as with methane emissions, growth is expected to moderate substantially as global population growth rates fall and best-practice agricultural practices take hold, particularly with regard to fertilizer application regimes.

Synthetic gases

The present contribution of **synthetic gases** covered under the Kyoto Protocol is small (approximately 1% of global emissions), although some are very potent long-lived greenhouse gases. While emissions of HFCs are expected to fall, over the medium term emissions of PFCs and SF_6 are expected to rise (see **synthetic gases**).

Emissions of synthetic gases controlled by the Montreal Protocol, Chlorofluorocarbons (CFCs) and halons, have made a significant contribution to the rise in atmospheric greenhouse gas concentrations between 1950 and 2000, but aggregate emissions have fallen considerably in recent years. In 2004, emissions of these gases were only 20% of 1990 levels.[27] Their production is expected to be completely phased out before 2030. Most CFCs have a relatively short atmospheric residence time compared with many other man-made greenhouse gases, and their atmospheric concentrations are expected to decline over the next 30–50 years. Precursor emissions that result in tropospheric **ozone** are likely to increase over the coming decades as fossil fuel consumption rises (see **ozone**). Ozone emissions are usually not included in global emission projection data.

Overall, the ongoing expansion in the global economy, population increase, and a continued high dependence on fossil fuels are all likely to lead to a significant increase in greenhouse gas emissions over the next 30–50 years. The strong link between GDP growth and greenhouse gas emissions is expected to persist, at least in the short to medium term. As a result, global greenhouse gas emissions are expected to continue to rise by between 1% and 1.5% per year, at least up to 2030.[28] In the longer term, it is possible that the link between GDP and emissions could be broken if the global economy can be "decarbonized," but this will require significant changes in existing policy settings and rapid technological change and innovation.

Unless policies and measures can be introduced to curtail the rate of growth of emissions, the **IPCC** expects global emissions to be between 59 and 94 $GtCO_2e$ by 2030 (depending on the scenario), with the most likely mid-range estimate being around 70–75 $GtCO_2e$/year (compared with 49 $GtCO_2e$ in 2004), and by 2050 could reach as much as 85 $GtCO_2e$/year.[29] Increases in emission levels of this magnitude will most likely result in atmospheric greenhouse gas concentrations reaching 550 ppm CO_2e by 2035–2040 and 630 ppm CO_2e by 2050.[30] This would commit the world to a temperature increase of 3–4°C above preindustrial levels, or possibly higher, and would almost certainly result in **dangerous climate change**. It may also trigger a positive **climate change feedback** from the release of additional large quantities of CO_2 and methane from the **land carbon sink** and, in the longer term, from the marine **methane hydrates**.

For the global community to avoid this level of warming, it will need to introduce much more aggressive actions to reduce greenhouse gas emissions over the next few decades than those adopted so far. These measures would need to be sufficient to ensure that atmospheric concentrations stabilize at below 550 ppm CO_2e and possibly below 500 ppm CO_2e. To stabilize concentrations at or below 550 ppm CO_2e

would require global emissions levels to be reduced by at least 30% below present levels by 2050, and over 80% by 2100 (see **stabilization targets**). However, as the underlying economic and social drivers of greenhouse gas emissions are expected to remain strong, at least for the next 20–30 years, policy makers face a major challenge in reducing emissions.

See also: anthropogenic greenhouse gas emissions, carbon dioxide (CO_2), emissions intensity, emissions per capita, methane (CH_4), nitrous oxide (N_2O), stabilization targets, synthetic gases.

Notes

1 IPCC 2000
2 IPCC 2007
3 WRI 2005
4 UN Population Division 2005
5 WRI CAIT 2007a
6 UN Population Division 2005
7 Stern 2006
8 Ibid.
9 IEA 2006a
10 WRI 2005
11 Holtsmark 2006
12 Ibid.
13 WRI 2005
14 IPCC 2007
15 Stern 2006
16 Estimates vary – see IEA 2006a, EIA 2005, and IPCC 2007
17 IEA 2006b
18 Houghton 2005
19 IPCC 2007
20 Based on data from IPCC 2007, Houghton 2005, and Stern 2006
21 WRI 2005
22 Scheehle and Kruger 2007
23 Ibid.
24 IPCC 2000
25 Scheehle and Kruger 2007
26 Ibid.
27 IPCC 2007
28 Stern 2006
29 Ibid.
30 Ibid.

Further reading

IPCC 2007; Stern 2006; WRI 2005.

GLOBAL DIMMING

Global dimming is a term describing the reduction in incoming sunlight (or insolation) reaching the earth's surface in the last 50 years. Studies of sunlight data show that insolation reaching the earth's surface decreased by around 5% during the second half of the twentieth century. The main cause has been determined to be higher levels of **aerosols** in the atmosphere and increased cloud cover, both of which reflect solar energy back to space, cooling the planet. Scientists believe that the cooling effect of **aerosols** has masked the real warming potential of **greenhouse gases** in the atmosphere and kept global temperature increases below what would otherwise have been the case. These findings mean that projections of future global temperature change could underestimate the real warming impact of **anthropogenic greenhouse gas emissions**.

It has been known for some time that **aerosols** have a cooling effect on the planet. Airborne particulates from fossil fuel emissions, industrial processes, and fires reflect light that would otherwise reach the earth's surface, as well as changing cloud formation patterns that result in more reflective clouds that are more widespread and more persistent. Global dimming is, in effect, a combination of the direct negative **radiative forcing** resulting from aerosols and the indirect effect of these **aerosols** on cloud formation, both of which increase **albedo**.

The global dimming effect was first brought to the attention of the scientific community in the 1980s when Gerry Stanhill (the person to coin the term "global dimming") was updating sunlight data he had collected in Israel in the 1950s. When comparing the original data with that of the 1980s, he found that the amount of sunlight reaching Israel had fallen by an astounding 0.22% per year.[1] Based on this data, Stanhill published the finding that from 1958 to 1985 the earth had dimmed by 5.3%. In 1997, a similar effect was detected from research conducted on solar irradiance in Germany,[2] and subsequent research using US data over the period 1961–1990 indicated that the globe had dimmed by 4% over the period.[3]

Further research was undertaken in the Maldives starting in the late 1990s. Over four years air samples were taken over the northern Maldives (which is shrouded by an air stream heavy in particulates emanating from India) and from the southern Maldives (where the air mass originates from the clear skies of the southern ocean region). Comparing the two sets of results it was found that the northern islands were receiving about 10% less sunlight than those in the south.[4] In an unrelated research effort undertaken by Australian scientists who were studying pan evaporation rate records dating back 100 years, it was found that the rate of pan evaporation had been falling in the latter half of the twentieth century, which seemed a strange anomaly given that global temperatures had been rising. The primary reason they discovered was that evaporation rates were influenced much more by the amount of sunlight hitting the pan, rather than the other known variables (wind, temperature, and humidity) as previously believed.[5]

The most compelling evidence of global dimming was discovered in the days immediately following the September 11, 2001, terrorist attack on the World Trade Center in New York. The attack led to the cancellation of all domestic air travel in the USA for a three-day period. This meant that the thousands of condensation trails that normally criss-crossed the United States were absent for the first time in decades. Climate scientists had long thought that aircraft condensation trails may contribute to reduced sunlight penetration but had previously been unable to test their theories. After analyzing temperature records from thousands of weather stations across the United States from the three-day condensation-trail free period (September 11–13, 2001), it was found that the average temperature across the United States increased by around 1°C.[6] From a climate perspective, this is a very large change indeed, and much larger than expected. However, condensation trails also have a warming effect through the injection of water vapor at high altitudes, which is expected, on a global basis, to counterbalance the cooling effect of contrails (see **greenhouse gases**).

The significance of global dimming is that it has masked, by around 40%, the full **radiative forcing** effect of the **greenhouse gases** in the atmosphere (see **aerosols**). This has a major impact on estimates of future global temperature changes. The future strength of global dimming remains uncertain. It is anticipated that **mitigation** action, over the course of this century, will eventually result in a reduction in total **fossil fuel** consumption and that urban air quality concerns will lead to more widespread use of cleaner burning fossil fuel technologies, both of which will reduce atmospheric aerosol concentrations and thus global dimming (see **albedo** and **aerosols**). Studies have indicated that the earth did in fact brighten during the 1990s, but it is not clear if this trend will persist (see **aerosols**).[7]

This eventual brightening may be offset to some extent by expected increase in evaporation due to warmer temperatures, leading to increased atmospheric water vapor concentrations and increased cloudiness (a negative **climate change feedback**), which will contribute to global dimming. Recent scientific research, using recalibrated models that better incorporate global dimming, indicates much higher temperature increases this century than models that do not incorporate global dimming.[8] If the effect of global dimming is as significant as the evidence indicates, it may require an upward revision in **global warming** projections and an increase in the international community's sense of urgency to reduce **anthropogenic greenhouse gas emissions**.

See also: aerosols, albedo, climate change feedbacks, global warming, greenhouse gases, radiative forcing.

Notes

1 Stanhill and Morshet 1992
2 Liepert 1997
3 Liepert 2002
4 NOVA PBS 2006
5 Roderick and Farquhar 2002
6 Travis *et al.* 2002
7 Wild *et al.* 2005
8 Cox 2006

Further reading

Stanhill 2005; Stanhill and Cohen 2001; Roderick and Farquhar 2002; Liepert 2002; Wild *et al.* 2005.

GLOBAL WARMING

Global warming is a term used to describe a persistent increase in the earth's mean surface temperature relative to long-term average conditions (those that prevail over centuries). Global warming results in climate change and the two terms are sometimes used interchangeably, though climate change has a broader meaning and refers to changes in average climatic conditions (temperature, wind, rainfall, and other climatic variables) relative to their long-term average.

Both natural and anthropogenic (human-caused) factors can contribute to global warming, but the overwhelming scientific consensus is that the increase in global temperatures over the past century is primarily due to increased concentrations of **greenhouse gases** in the atmosphere.[1] The concentration of these gases, when converted to their **carbon dioxide equivalent (CO_2e)**, had been relatively stable for 10,000 years but has increased by more than 40% since 1750.[2] They are now higher than they has been for at least 650,000 years, and probably for several million years. **Anthropogenic greenhouse gas emissions**, principally from the burning of **fossil fuels**, deforestation, and agriculture, have been the main contributor to the rise in concentrations.

Between 1850 and 2005, the earth's average surface temperature increased by 0.76°C, most of which occurred in the past 50 years. Temperatures are currently rising at a rate of 0.2°C/decade.[3] The higher latitudes are warming much more rapidly than equatorial regions, and in some subpolar regions, average temperatures have already increased by 2°C or more (see **polar impacts**). These temperature increases have already resulted in a range of impacts, particularly on natural ecosystems (see **biodiversity impacts**). Even if concentrations could be held constant at today's level, the inertia of the climate system means that the earth is committed to an additional warming of 0.5–1°C, a total of as much as 1.5–2°C of warming above preindustrial levels (see **climate sensitivity**).

Since the **Intergovernmental Panel on Climate Change (IPCC)** released its first assessment report in 1990, the body of scientific evidence on global warming has grown enormously.[4] Successive IPCC reports have presented an increasingly compelling case that unless anthropogenic emissions are substantially reduced over the next half century, global warming will accelerate and soon reach dangerous levels.[5] The emerging consensus is that the threshold for **dangerous climate change** is in the order of 2°C above preindustrial levels. Some scientists consider that the threshold could be even lower.

The fact that the earth's mean surface temperature has risen over the past century is not disputed. However, there are some (*climate change skeptics*) that dispute that anthropogenic emissions have been the main cause of the observed warming. The skeptics argue that the observed warming is still within the bounds of natural variation and, furthermore, that even if human's do contribute to global warming, the climate impacts are likely to be small and would not warrant the introduction of costly measures to reduce emissions.[6] It is indeed true that the earth's mean surface temperature can vary considerably over time due to natural cycles, both long term (over millennia) and short term (over several years or decades). Nonetheless, the scientific consensus concludes with very high confidence that these factors alone could not account for the rapid increase in temperatures observed over the past 50 years.[7]

Natural versus anthropogenic warming

A key issue in the global warming debate is determining the anthropogenic contribution to warming, given that natural factors are also known to cause significant variations in global mean temperatures.

Milankovich cycles can cause global mean temperatures to vary as much as 5°C between glacial and interglacial periods. However, these cycles take many centuries to cause perceptible changes to global temperatures. They cannot explain the rapid warming experienced over the twentieth century.

On occasions in the past, major cataclysmic events (such as periods of intense volcanic activity or meteorite strikes) have resulted in rapid changes to the earth's mean temperature (both cooling and warming). For example, during the Late Pleistocene Thermal Maxima (LPTM) 55 million years ago, temperatures soared more than 6°C above present levels due to the sudden release of large quantities of greenhouse gases to the atmosphere (see **methane hydrates**). Scientists believe that temperatures rose rapidly during the LPTM, probably over just a few thousand years. It is possible that humans could deliver a similar amount of warming over just a few centuries if emission trends persist.

Mean global temperatures can also vary up and down over periods ranging from several years to several decades as a result of regular climatic cycles (such as the **El Nino Southern Oscillation**), and variations in the amount of solar radiation reaching the earth's surface. The resulting changes to global mean temperatures are generally small and do not have a lasting impact on longer term global average temperatures. For example, variations in the amount of solar energy reaching the earth over the past century have added an estimated 0.12 Wm^{-2} to atmospheric **radiative forcing** compared with the estimated 2.64 Wm^{-2} increase attributable to human activities.[8] Overall, natural factors can only account for a very small proportion of the recent increases in global temperatures.

It is also evident that human activities (such as emissions of **aerosols** and human-induced changes to earth's **albedo**) have to some extent offset the full warming effect of increased atmospheric greenhouse gas concentrations, possibly by as much as 40% (see also **global dimming**).

In addition to the large absolute increases in global mean temperatures over the past century, the rapid *rate* of current global temperature change is unprecedented in at least the last 650,000 years, and earth's history has only seen similar rates of change with rare cataclysmic events that have occurred in the past.[9] The current rate of global temperature change is important, as it is most likely beyond the ability of ecosystems to adapt, and will also test the ability of humans to adapt (see **biodiversity impacts** and **adaptation**).

Future warming

Though nearly all scientists agree that global mean temperatures will rise over the coming decades, some uncertainty surrounds how quickly, and by how much, mean temperatures will change.

The principal factors that will determine temperature increases this century are as follows: the rate at which atmospheric greenhouse gas concentrations rise and the level at which they are stabilized (assuming they stabilize this century), the sensitivity of global temperatures to changes in greenhouse gas concentrations (see **climate sensitivity**), and the strength of positive and negative **climate change feedbacks**.

Greenhouse gas concentrations

Stabilizing concentrations at the lowest possible level is the single most important means of limiting future temperature increases. Concentrations will only stabilize when emissions fall to a level equivalent to the natural rate at which the earth can remove them from the atmosphere. To equal the natural removal rate and stabilize concentrations, anthropogenic emissions need to fall 80–90% below current levels (see **stabilization targets**).

If current emission trends persist, atmospheric greenhouse gas concentrations (in **CO_2e**) are expected to exceed double preindustrial levels before mid-century and possibly before 2040 (see **future emissions trends**). This will have far-reaching implications for both natural and human systems (see **climate change impacts**) and almost certainly represents **dangerous climate change**, contravening key elements of the 1992 **United Nations Framework Convention on Climate Change (UNFCCC)**.

To limit global warming to 2°C would require concentrations to stabilize below 500 ppm CO_2e. Although achieving this stabilization target is technically possible, given current emission trends, the structural inertia of the global economic system (see **technology, structural change and organizations**), and the lack of any international agreement to reduce emissions to the extent necessary to achieve such a target (see **Kyoto Protocol**), it appears very unlikely that concentrations will be stabilized below 500 ppm CO_2e (see **stabilization targets**).

To have a 50:50 chance of keeping temperature increases between 2 and 3°C above preindustrial levels, concentrations should not exceed 550 ppm CO_2e. Emissions need to fall to at least 60–70% below current levels by mid-century to meet this target.

If concentrations exceed 600 ppm CO_2e, which is possible before 2050, the risk of exceeding 3°C increases significantly. Temperature increases greater than 3°C will result in far-reaching **climate change impacts** and significantly increase the risk that strong positive **climate change feedbacks** will come into play.

Climate sensitivity

Projections of future warming are heavily dependent on the relationship between changes in atmospheric greenhouse gas concentrations and global mean temperatures. The IPCC projects that a doubling of preindustrial concentrations will most probably raise average global temperatures by 3°C, and other studies estimate 3.5°C. However, the full range of **climate sensitivity** estimates is 1.5 to 11°C for a doubling of concentrations.[10] If global temperatures are more sensitive to increased greenhouse gas concentrations than currently assumed, then estimates of future warming will need to be revised upward, and more stringent **stabilization targets** would need to be adopted to achieve an agreed temperature outcome.

Climate change feedbacks

The possibility that global warming could trigger strong positive **climate change feedbacks** (those that amplify global warming) remains the most worrying aspect of continued increases in atmospheric greenhouse gas concentrations, particularly as they are subject to so much uncertainty about when major positive feedbacks might be triggered. What is clear is that the risks rise significantly as temperatures rise.

The principal positive feedbacks include reductions in the ability of the land to absorb CO_2 from the atmosphere, and even the transition of the land system from a carbon sink to a carbon source (see **land carbon sinks**); a long-term decline in the uptake of CO_2 by the oceans (see **ocean carbon sinks**); and, most worryingly, the possibility that warming could trigger substantial releases of greenhouse gases from the vast **methane hydrate** deposits contained in the permafrost and ocean floor sediments. It is not known, with any certainty, what the critical temperature thresholds are for a land carbon sink-to-source transition or for major releases from the methane hydrates. However, once global temperatures exceed 2°C, the risks of initiating major feedback effects increases, and beyond 3–4°C above preindustrial levels, it is possible that a runaway climate change threshold could be breached, where self-reinforcing feedbacks (rather than anthropogenic emissions) become the major driver of atmospheric greenhouse gas concentrations.

While uncertainty remains in regard to how much global mean temperatures will rise this century, the balance of evidence suggests that the world is on track to experience temperature increases of at least 3°C and possibly 5°C (or higher). The consequences of a rise in global temperatures above 4°C are potentially so severe that these should be avoided at all costs (see **climate change impacts**).[11]

See also: anthropogenic greenhouse gas emissions, carbon sinks, climate change feedbacks, climate change impacts, climate sensitivity, dangerous climate change, greenhouse effect, stabilization targets.

Notes

1 IPCC 2007
2 Ibid.
3 Ibid.
4 IPCC 1990
5 IPCC 1996, 2001, and 2007
6 See Lindzen 1992, 2006; Singer and Seitz 1998
7 IPCC 2007
8 Ibid.
9 Ibid.
10 Ibid.
11 Lynas 2007

Further reading

IPCC 2007; Stern 2006; Lynas 2007.

GLOBAL WARMING POTENTIALS

A global warming potential (GWP) is an index that compares the climate impact of an emission of a greenhouse gas relative to emitting the same amount of **carbon dioxide (CO_2)**. Since there are numerous anthropogenic greenhouse gases emitted to the atmosphere, each with different atmospheric residence times and physical properties, it is important to be able to quantify and assess each gas's relative contribution to **global warming**.

GWPs are determined as a ratio of the climate impact of an instantaneous (i.e. "pulse") emission of a greenhouse gas relative to that of emitting the same amount of CO_2. For instance, if 1 kg of **methane (CH_4)** were released to the atmosphere, the GWP provides the number of kilograms of CO_2 that must be released to result in the same **radiative forcing**, that is, the global warming impact.

Since different greenhouse gases have different atmospheric residence times, the GWP will change depending on the time horizon. The physical properties of **methane** make it a strong contributor to the **greenhouse effect**, but it is short lived (around a decade), whereas the physical properties of CO_2 result in it being a weaker **greenhouse gas**, but it has a longer atmospheric residence time (averaging 100 years). Thus, the 20 year GWP of **methane**, as the amount of CO_2 needed to provide the same **radiative forcing** as the release of 1 kg of **methane** averaged over 20 years (56 kg CO_2), is much higher than over 500 years (6.5 kg CO_2).

The **International Panel on Climate Change (IPCC)**[1] provides GWPs for selected greenhouse gases at 20, 100, and 500 year time horizons, shown in Table 10. For **UNFCCC greenhouse gas inventories**, including national inventories under **Kyoto Protocol** reporting, a 100-year time horizon is used as it is near the average of CO_2 (the base unit) atmospheric residence time and it is appropriate to the time frames that must be considered in discussions of climate change.

GWPs are used to provide a relevant comparison of **radiative forcing** between emissions of different **greenhouse gases** by converting relevant non-CO_2 gases to a **carbon dioxide equivalent (CO_2e)** by multiplying the emitted mass of the greenhouse gas by its GWP. For instance, using the 100-year GWP time horizon, a comparison of the climate impact of emitting 10 tonnes of CO_2, 10 tonnes of **methane,** and 10 tonnes of SF_6 can be made by multiplying each gas by its GWP from Table 11. Thus, there would be 10 tonnes of CO_2e from the emissions of CO_2, 210 tonnes of CO_2e from **methane**, and a massive 239,000 tonnes of CO_2e from the emissions of SF_6. Having such an equal basis for comparison enables policy makers to target emission reduction measures that will have the biggest impact on mitigating climate change.

Table 10 Greenhouse gas global warming potentials

	Lifetime (years)	GWPs (various time horizons)		
		20 years	100 years	500 years
Carbon dioxide	5–200	1	1	1
Methane	12.0 ± 3	56	21	6.5
Nitrous oxide	120	280	310	170
Sulphur hexafluoride	3,200	16,300	23,900	34,900

Table 11 100-year global warming potentials of Kyoto Protocol gases[3]

Gas	Chemical formula	Estimated atmospheric lifetime (years)	GWP
Carbon dioxide	CO_2	5–200	1
Methane	CH_4	12	21
Nitrous oxide	N_2O	120	310
Sulphur hexafluoride	SF_6	3,200	23,900
Hydrofluorocarbons (HFCs)			
HFC-23	CHF_3	270	11,700
HFC-32	CH_2F_2	4.9	650
HFC-41	CH_3F	Not available	150
HFC-43-10mee	$C_5H_2F_{10}$	15.9	1,300
HFC-125	C_2HF_5	29	2,800
HFC-134	$C_2H_2F_4$ (CHF_2CHF_2)		1,000
HFC-134a	$C_2H_2F_4$ (CH_2FCF_3)	14	1,300
HFC-143	$C_2H_3F_3$ (CHF_2CH_2F)	Not available	300
HFC-143a	$C_2H_3F_3$ (CF_3CH_3)	52	3,800
HFC-152a	$C_2H_4F_2$ (CH_3CHF_2)	1.4	140
HFC-227ea	C_3HF_7	34.2	2,900
HFC-236fa	$C_3H_2F_6$	Not available	6,300
HFC-245ca	$C_3H_3F_5$	Not available	560
Perfluorocarbons			
Perfluoromethane (tetrafluoromethane)	CF_4	50,000	6,500
Perfluoroethane (hexafluoroethane)	C_2F_6	10,000	9,200
Perfluoropropane	C_3F_8	2,600	7,000
Perfluorobutane	C_4F_{10}		7,000
Perfluorocyclobutane	c-C_4F_8	3,200	8,700
Perfluoropentane	C_5F_{12}		7,500
Perfluorohexane	C_6F_{14}	3,200	7,400

Do GWPs change?

Since the behavior of greenhouse gases in the atmosphere is not completely under-stood, the **IPCC** regularly updates the scientific basis for the assessment of climate change, including GWPs.[4] For **synthetic gases** in particular, the uncertainty around their decay rates in the atmosphere are high since most have only been introduced into the environment in the last 50 years. Generally the GWPs do not change from year to year, though the average lifetime of **methane** has been revised upward from 21 to 23, and nitrous oxide's GWP has been revised downward from 310 to 296.[5] However, since critical international agreements such as the **Kyoto Protocol** have brokered quantitative emissions commitments on the basis of earlier GWPs, the GWPs reported by the **IPCC** in 1996[6] will continue to be used until at least 2012.

See also: carbon dioxide equivalent (CO_2e), global warming, greenhouse effect, greenhouse gas inventories, greenhouse gases, Kyoto Protocol, Intergovernmental Panel on Climate Change (IPCC), radiative forcing.

Notes

1 IPCC 1997
2 Ibid.
3 Ibid.
4 IPCC 2006b
5 Ibid
6 IPCC 1997

Further reading

IPCC 2007, 2006b.

GREENHOUSE EFFECT

The greenhouse effect is a natural process that maintains the mean surface temperature of the globe approximately 33°C warmer than it would otherwise be, at 18°C. Without the greenhouse effect, the earth's surface temperature would be –15°C and inhospitable to life. The principle that gases in the atmosphere trap the sun's heat is well understood and was first elucidated by Svante Arrhenius in 1896.[1]

All objects emit electromagnetic radiation, with the wavelength of the radiation dependent on the temperature of the object. Hot objects (such as the filament in an incandescent light) emit short-wave radiation, and cooler objects (such as the earth) emit longer-wave radiation. The sun's electromagnetic radiation covers a broad spectrum, from the very short wavelengths of gamma and x-rays (less than one nanometer – 10^{-9} m, or one millionth of a millimeter) to the very long wavelengths of microwaves (0.1–100 mm) and radio waves (100 mm to tens of meters).[2] Visible light comprises a small fraction of the total radiation spectrum and covers the range of wavelengths from approximately 400–800 nanometers. The longest visible wavelength is red and the shortest is violet, and the spectra on either side of visible light are hence known as ultraviolet (UV – shorter wavelengths from 1 to 400 nanometers) and infrared (IR – longer wavelengths ranging from 800 nanometers to 0.1 mm).

Solar radiation peaks strongly at approximately 600 nanometers: much of this energy we see as visible light and feel as warmth from infrared radiation. Greenhouse gases absorb approximately 20% (67 Wm^{-2}) of the total incoming energy (342 Wm^{-2}) and allow the remaining 80% (275 Wm^{-2}) to pass through the atmosphere. A further 22.5% (77 Wm^{-2}) of the incoming energy is reflected back to space by clouds, leaving approximately 58% (198 Wm^{-2}) of the total incoming energy to reach the surface of the earth. Of the energy that reaches the surface, some is reflected back to space, according to the surface **albedo**. Globally averaged, approximately 15% (30 Wm^{-2}) of the energy reaching the surface is reflected.[3]

Energy reaching the earth's surface that is not reflected warms the earth, which then also emits radiation: *terrestrial radiation*. Being cooler, terrestrial radiation occurs at longer wavelengths (approximately 4,000–40,000 nanometers) in the infrared portion of the spectrum. The greenhouse gases in the atmosphere are much

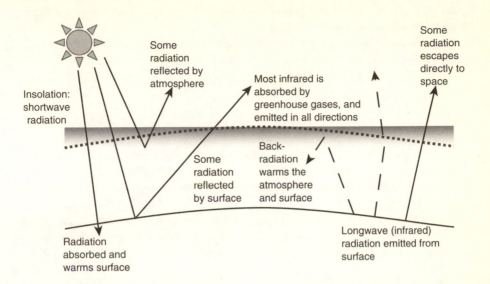

Figure 4 The natural greenhouse effect

more effective at absorbing infrared than visible light, absorbing almost 90% of terrestrial radiation. This absorbed energy is reemitted by the **greenhouse gases** in all directions, including back to the earth's surface where the cycle repeats itself. This reemission back to the earth constitutes the natural greenhouse effect, which results in warming of the earth's surface and lower atmosphere (see Figure 4).

When a gas molecule is struck by electromagnetic radiation, it absorbs the incoming photon (packet of energy): electrons surrounding the nucleus of the molecule jump to a higher excitation state. This can be thought of as electrons moving to a new orbit at a greater distance from the nucleus. When returning to the original, lower excitation state (closer electron orbit), energy is released from the molecule, sometimes at a different wavelength.

Different molecules absorb different wavelengths of energy, depending on their molecular size and characteristics. For example, **methane (CH$_4$)** absorbs most energy at peaks of 3,000 and 8,000 nanometers (in the infrared, IR, radiation range), **nitrous oxide (N$_2$O)** peaks at 4,000 and 8,000 nanometers (IR range), and **carbon dioxide (CO$_2$)** absorbs energy across multiple wavelength peaks in the IR range.[4] Since these gases absorb terrestrial radiation and reemit it, they are called **greenhouse gases**. The other major naturally occurring greenhouse gas is water vapor, which has multiple peaks from the edge of the visible light spectrum (800 nanometers) throughout the IR spectrum (to 0.1 mm).

The term "greenhouse effect" is somewhat inappropriate since actual greenhouses primarily maintain their warmth by suppressing convection and the mixing of warm and cold air. The atmospheric greenhouse effect, in contrast, is a radiative effect, the result of downward electromagnetic reemission by the atmosphere. Nonetheless, the term is widely used, and apt in the sense that transparent coverings – such as the glass of a greenhouse or the **greenhouse gases** of the atmosphere – result in an increase in temperature at the surface.

The amount of terrestrial radiation absorbed by the atmosphere is partly dependent on the type and concentrations of **greenhouse gases** in the atmosphere.

Type of greenhouse gas

Each greenhouse gas can only absorb terrestrial radiation at particular wavelengths. This leaves open some "windows," in which radiation can escape directly to space. Across the entire IR spectrum, the **greenhouse gases** water vapor and **carbon dioxide** can absorb some radiation, but between approximately 8,000 and 9,000 nanometers and 10,000 and 12,000 nanometers, their ability to absorb radiation is much weaker, leaving these "windows" largely open. Some of the **synthetic gases** can absorb in these otherwise open windows, notably sulphur hexafluoride, which peaks at approximately 11,000 nanometers. These gases thus have a very high **radiative forcing** and hence high **global warming potential**.

Amount of greenhouse gases

While the wavelength of radiation absorption is determined by gas type, how much of the radiation is actually absorbed is determined by greenhouse gas concentrations in the atmosphere. The natural (nonanthropogenic) background concentration of **greenhouse gases** in the atmosphere (approximately 260 ppm **carbon dioxide equivalents CO_2e**) is sufficient to absorb 90% of outgoing terrestrial radiation. Increasing the concentration of **greenhouse gases** above the background level, as humans have been doing on a large scale over the past century, acts to increase the fraction of terrestrial radiation that is absorbed and reradiated.

Anthropogenic greenhouse gas emissions enhance the natural greenhouse effect in two distinct ways: by adding synthetic **greenhouse gases** to the atmosphere that absorb energy at wavelengths in otherwise open terrestrial radiation windows and by increasing the concentration of naturally occurring **greenhouse gases** in windows that are naturally partially closed. These two contributions change the radiation budget of the earth and alter the natural equilibrium of the climate system, which has been relatively stable, with natural variations (see **Milankovich cycles**), for at least the last 650,000 years, and probably much longer. Anthropogenic greenhouse gas emissions have disrupted this equilibrium, and the **global warming** now being observed is a physical response of the earth system to restore equilibrium.

Capturing the 10% of terrestrial radiation that escapes the natural greenhouse effect does not seem to be a dramatic change. However, capture of this full 10% would result in surface temperatures more than 10°C higher than the current global average – an increase beyond anything seen in human history or, indeed, in the last several million years.[5]

See also: albedo, carbon dioxide equivalents (CO_2e), global warming, global warming potentials, greenhouse gases, Milankovich cycles, radiative forcing.

Notes

1 Arrhenius 1896
2 Reusch 2004
3 Kiehl and Trenberth 1997
4 Keen 2006
5 Pidwirny 2006

Further reading

IPCC 2007.

GREENHOUSE GAS INVENTORIES

A greenhouse gas inventory is a compilation of estimates of **anthropogenic greenhouse gas emissions**, using a stated method, a specified boundary, and a particular time period. The objective of greenhouse gas inventories (hereafter referred to as "inventories") is to provide a transparent, relevant, accurate, and complete picture of emissions that is consistent over time and comparable with other inventories compiled for the same purpose.

Inventories can be prepared at a national, subnational, company, project, or individual level depending on their intended purpose. National inventory data is essential to the process of coordinating an international response to climate change and underpinning international climate change negotiations. National inventories provide a means of assessing the relative greenhouse gas emissions and trends of different countries and when aggregated provide a means of estimating whether absolute emissions are increasing or decreasing, and thus whether progress is being made toward stabilizing atmospheric greenhouse gas concentrations and avoiding **dangerous climate change**.

The responsibility for compiling an inventory depends on the purpose of the inventory. In the corporate sector, inventories can be compiled for an individual facility (such as a power station) for reporting under an **emissions trading** scheme or some other reporting requirement, or for a company (such as an electricity company) or for a group of companies. These inventories can be used for corporate reporting, formulating greenhouse gas reduction strategies, and assessing the company's risk exposure to government mitigation policies. Specific projects that aim to reduce greenhouse gas emissions (see **Clean Development Mechanism** and **Joint Implementation**) must also establish emission inventories to demonstrate that emission reductions have actually occurred. Inventories may also be compiled at a smaller scale for sporting or political events,[1] travel,[2] or for calculating the "carbon footprint" of an individual or household.[3] These inventories may be calculated using publicly available standards[4] and tools[5] or with the assistance of specialist greenhouse gas accounting consultants.

National inventories are compiled by governments, usually with inputs from several ministries, for reporting under the **United Nations Framework Convention on Climate Change (UNFCCC)** and **Kyoto Protocol**.

Inventory accounting principles

There are five basic principles[6,7] for the compilation of an inventory:

- *Transparency:* Information should be presented regularly in a clear, factual, and coherent manner cognizant of the needs of users of the information. Data and information should be obtained, recorded, compiled, analyzed, and documented in a manner that enables verification. Assumptions, references, and calculation methodologies should be appropriately documented, and enable reported data to be replicated by another party.
- *Completeness:* All greenhouse gas emissions and removals within the inventory boundaries should be identified and quantified according to the requirements of the relevant program or system in which the inventory will be used. Any exclusions of emissions should be reported and justified, and all information material to users of the inventory should be reported in a manner consistent with the declared boundaries, scope, time period, and objectives of inventory reporting.
- *Consistency:* An inventory needs to be comparable over time, and any changes in the basis of reporting, and the consequences of these changes, must be clearly stated and justified to the users of the inventory.
- *Accuracy:* Quantification of greenhouse gas emissions should be systemically neither over nor under the true value of emissions or removals. Uncertainties should be quantified and reduced as far as practicable to ensure the inventory is sufficiently accurate to enable users to make decisions with reasonable assurance as to the integrity of the reported information.
- *Relevance:* Inventories should be compiled to ensure they appropriately reflect the greenhouse gas emissions described within the inventory boundary and serve the needs of decision makers using the reported inventory.

Inventory boundaries

Inventories are a summation of all the **anthropogenic greenhouse gas emissions** within the boundary (e.g. national, company, project, or individual level). Inventories are almost always annual – a time boundary of one year. Setting of boundaries for national inventories is relatively straightforward: all emissions occurring within the borders of the country and its territories within one calendar year are to be included. However, there are some emissions that occur in international jurisdictions (such as fuel consumed by international airlines and shipping) where it is not clear which nation is responsible. The process of allocating these emissions across different nations has yet to be resolved, and these emissions are not required to be included in national **UNFCCC** or **Kyoto Protocol** accounting, although it is recommended.

The boundary of company inventories can be more complex since the size, ownership, and structures of companies can vary greatly. Greenhouse gases to be included in a company inventory may be determined by the following: *Control*, where the organization accounts for 100% of the emissions from facilities over which they exert management control; *Equity Share*, where the organization accounts for the percentage of emissions generated from facilities equivalent to its equity share (ownership) of each facility; *Financial Boundaries*, where the organization's emission accounting is based on the parallel financial, tax, or business financial accounting responsibilities determined by the country or jurisdiction in which the entity operates; or *Contractual Arrangements*, in which the responsibility for compiling the inventory is explicitly set out in a contract between relevant organizations involved in a facility or project. Company inventories are mostly annual and often compiled for the financial year (e.g. in the case of North America, the reporting period is April 1 to March 31).

Emissions coverage

In accordance with the principle of completeness, all anthropogenic greenhouse gas sources within the boundary must be included in an inventory. Inventories thus include all *direct* sources: actual greenhouse gas emissions that occur at sources within the organizational boundary. These sources may be related to the combustion of **fossil fuels** for electricity, heat, or steam; physical or chemical processing (such as cement calcination); combustion of **fossil fuels** for transportation of materials, products, waste, or employees; the intentional or unintentional releases of fugitive emissions (such as release of perfluorocarbon cover gases in aluminium smelting); and emissions and removals from land-use or land management practices such as forestry and agriculture.

However, reporting only direct emissions within a boundary may not provide all relevant information to decision makers since this may not provide a complete picture of the greenhouse gas footprint of a particular activity, organization, or individual. For instance, an aluminium smelter reporting all direct emissions from its facility may have a relatively small inventory, but if it is purchasing electricity produced from coal-fired power plants outside the reporting boundary, the greenhouse gas footprint may be many times greater than the direct emissions reported for the smelter itself. Including only direct emissions in the inventory understates the relevant greenhouse footprint of the smelter.

Thus, greenhouse gas accounting conventions normally require that inventories include *indirect* emissions that are caused as a result of activities within the organization boundary, but where the direct emissions occur outside the boundary. Indirect emissions can include electricity used at the facility but generated elsewhere; transportation of materials and waste outside the boundary; outsourced activities (e.g. the emissions associated with the transport of goods to a customer where the transport is provided by another company); emissions from waste generated by the facility that occurs outside the boundary; and emissions from end-use and end-of-life phases of an organization's products. Reporting both direct and indirect emissions provides a complete picture of an organization's greenhouse gas footprint.

Obviously, if all organizations prepared and reported an inventory and included their direct and indirect emissions, there would be double counting since the direct emissions (e.g. from a coal-fired power station) reported by one organization would also be reported as indirect emissions (electricity use) reported by another. The programme authority (such as a government agency that administrates the reporting system) can reconcile any double counting by accounting for direct and indirect emissions separately.

There are two reasons why national inventories do not include indirect emissions. First, since the **UNFCCC** requires national inventory reporting and has near-universal ratification, effectively all global emissions are reported, and reporting only *direct* emissions provides a relatively complete picture of each country's greenhouse gas emissions relevant to international decision making on policy responses. Second, including indirect emissions would greatly complicate national inventory reporting. For example, a country that imports electricity from a neighboring country would need to know the **emissions intensity** of that country's electricity production and where there are multiple countries within an integrated grid (such as Europe), tracking imports, exports, and electricity suppliers would be complex and difficult. Nonetheless, some national governments regard this "direct emissions only" approach as unrepresentative of a nation's true greenhouse gas footprint as it does not adjust national inventories for the embodied emissions in traded goods (see **emissions per capita**).

Quantifying greenhouse gas emissions

Greenhouse gases are colorless and usually emitted at low densities, making direct measurement difficult, though some direct measurements can be done at point sources for calibration or design purposes, or as sampled measurements (e.g. destructive testing of trees in forests to determine carbon content). Since emission sources are not usually directly measured, quantification of emissions is mostly performed using one of two general methods using known relationships between inputs and outputs to a particular process.

When data are readily available from parameters already measured for normal business purposes – such as the carbon and energy content of input fuels – the mass balance approach is typically adopted. A mass balance quantifies the mass of emissions as the difference between inputs (fuel or feedstocks) minus the output products of the process (mass products, waste). For example, in a coal-fired power station, the carbon content and mass of the coal is well known and frequently measured (in order to most efficiently operate the station), and the mass of unburned carbon retained in the fly-ash and remnant boiler dust can be readily measured. Assuming complete combustion, **carbon dioxide** emissions from the station can be calculated from the mass difference of inputs and outputs.

Where input and output masses are not easily ascertained, *emission factors*, which estimate the emissions released to the atmosphere as a known proportion of a process activity or per unit of throughput, are employed. Emission factors may be established for a particular technology, process, or plant using initial measurement and calibration under known conditions at a particular facility, and such specific emission factors generally provide reasonably accurate quantification of emissions. This can be expensive, and more often average or general emission factors are published by governments,[8] industry bodies,[9] or other recognized sources[10] (such as the **Intergovernmental Panel on Climate Change – IPCC**[11]). For example, emissions from diesel combustion in a truck fleet might use a nationally published emission factor for kg CO_2e/litre of diesel use since individual emission factors for each truck in the fleet would not be known and would be impractical to establish.

Emission factors for CO_2 are well developed for most fossil fuel and industrial sources, and **methane** emission factors are generally well established for fossil fuel combustion, but not for ruminants, forests, or peatland emissions. Emission factors are often limited and less certain for **nitrous oxide** and the **synthetic gases**. The uncertainty associated with emissions quantification increases the more generalized the approach used. Direct measurement and mass balance methods typically have low uncertainty. The uncertainty associated with quantification by emission factors varies from relatively low uncertainty for those factors established from empirical evidence for a particular process through to high uncertainty associated with externally supplied average emission factors for international use. The quantification approach selected should be commensurate with the relative importance of the source in the inventory, and the basis and justification for use should be documented in a fully transparent way.

Compiling inventories

While a significant proportion of an inventory may arise from large point sources (such as power stations and oil refineries) that can be included through one of the quantification methods outlined above, there remain a multitude of smaller or diffuse sources that may not be easily identified and quantified. For national inventories, it is difficult to collect all the data required for completeness using "bottom up" approaches (aggregating actual reported emissions from individual entities). National inventories approach diffuse and smaller sources in a "top-down" manner that requires fewer data inputs: using proxies to estimate greenhouse gas emissions (often macro data such as population, household types, vehicle fleet averages, and proportions of activity from different sectors). Known or sampled emissions from activities are extrapolated on the basis of historical behavior patterns and trends to establish current emissions. For example, the **methane (CH_4)** associated with sewage treatment is reasonably well known on a per capita basis, and a reasonable estimate of national methane emissions from human waste can be established by multiplying the per capita emissions by population census data. In contrast, the methane emissions associated with ruminant animals (such as cattle and sheep) using the same approach is much less certain.

Bottom-up approaches provide more accurate inventories since they can account for various technology types, engineering efficiency gains, and impacts of different policies and programs. They are, however, data intensive and are generally only used for company- or project-level inventories. Typically, national inventories will be established using a combination of bottom-up and top-down approaches, depending on the availability and accuracy of data.

In aggregating data for inclusion in an inventory, consistency must be applied, particularly in relation to boundary setting, quantification approaches, and data manipulation and presentation. This does not imply that once an approach is established it cannot be changed. Where more economical data sources, more accurate quantification methodologies, or more effective data analyses are established these should be used, but the use of new approaches must be clearly stated and justified and, where possible, prior inventories recalculated using the new approach. To enhance the usefulness of an inventory, data should be compiled in a manner that allows for disaggregation and analysis according to different relevant parameters, such as by geographical location, sector, process, facility, and gas. This enables performance evaluation, including analysis and identification of trends and opportunities for emission reductions.

Good inventory preparation, both at the national and organizational level, requires rigorous quality management, which may include regular accuracy checks for technical errors, periodic internal audits and technical reviews, periodic management reviews of greenhouse gas information, organization and delivery of appropriate training for inventory development team members, and completion of uncertainty analysis. In addition to quality management, it is good practice and enhances the credibility of inventories to undertake verification. That is, assessment by an independent external party (auditor) who collects evidence to determine whether the inventory is free from material misstatements.

International accounting rules

Under the **UNFCCC**, countries commit (through ratification and becoming a Party to the Convention) to establish and regularly report national greenhouse gas inventories. The **Kyoto Protocol** provides further commitments and obligations (particularly for developed countries) for regular inventory reporting to the **UNFCCC** Secretariat for global compilation and analysis.

There is no single set of internationally accepted rules and procedures for company-level emissions reporting. However, there are two broadly compatible and consistent quantification frameworks that have achieved widespread international acceptance:

- International Organisation for Standardisation, ISO14064.1: Greenhouse gases: Specification with guidance at the organization level of quantification and reporting of greenhouse gas emissions and removals.[12]
- World Business Council for Sustainable Development (WBCSD) and World Resources Institute's (WRI) GHG Protocol Corporate Accounting and Reporting Standards.[13]

The ISO standard provides a clear set of verifiable requirements, while the GHG Protocol provides detailed guidance on how to prepare a company level inventory, supported by a range of calculation tools. The widespread uptake of a consistent quantification framework is essential to ensure a tonne of emission reductions in Company A is equal to a tonne of emission reductions in Company B, irrespective of where they are located. However, the uptake of these inventory quantification frameworks is currently mostly confined to companies in developed countries.

There is also no universally accepted single set of international rules and procedures for establishing project-level inventories, although the quantification methodologies and requirements of the **Clean Development Mechanism (CDM)** are widely recognized and adopted, and both the ISO (14064.2) and the WBCSD/WRI GHG Protocol have developed project-level quantification standards and guidelines. There are, as yet, no generally agreed approaches to establishing inventories for events or individuals.

National greenhouse gas inventories

Reporting national inventories is fundamental to international climate change cooperation. All 192 Parties to the **UNFCCC** are required to provide comprehensive National Communications, which includes national inventories and information on the policies and measures that have been undertaken toward meeting the objectives of the **UNFCCC**. All Parties "develop, periodically update, publish and make available to the Conference of the Parties" national inventories of greenhouse gases "using comparable methodologies."[14]

Compiling an accurate National Communication and national inventory is a challenging and expensive task. Recognizing this, only Annex I Parties were required to submit their first national inventory within six months of the Convention coming into force for the Party, while non-Annex I Parties could take up to 3 years to submit their first inventory and utilize financial resources provided by Annex I Parties. Least Developed Countries make their initial communication at their discretion.

Under the **Kyoto Protocol**, annual inventories are required to be submitted by the 15th of April each year, although the time taken to gather data and compile an inventory means inventories may be up to 2 years in arrears: for instance, an inventory published in March 2007 usually reports emissions occurring in 2005. The **UNFCCC** regularly publishes synthesis reports that summarize the most important information and aggregate greenhouse gas emissions of all Parties.[15]

Parties are given some flexibility as to what analysis and presentation to use, but national inventories are required to follow the guidelines[16] and good practice guidance[17] set out by the **IPCC** in order to ensure transparency, consistency, completeness, accuracy, and comparability between national inventories. Flexibility is provided in quantification methodologies according to the availability of data and national circumstances. Reported inventories must provide overall aggregate emissions as well as breakdowns of emissions by gas, source and sector, and are expected to provide analysis of trends in each of these breakdowns. They typically also provide a justification for any changes in quantification methodologies, a description of quality assurance processes and uncertainty analyses. Most inventories will also include contextual information such as economic, population, and industry trends.

The **Kyoto Protocol** places some further requirements on Annex I Parties, particularly relating to information required to demonstrate compliance with commitments, independent review of inventories (verification), and the accountability for international **emissions trading**, including eligibility for hosting **Joint Implementation** projects.[18]

Nearly all (179) UNFCCC ratifying countries have provided at least one inventory using consistent and comparable approaches based on **IPCC** guidance.[19] There are, however, large differences in the regularity and number of national inventories submitted by Annex I (developed) countries compared with non-Annex I (developing) countries. Since the **Kyoto Protocol** came into force in 2005, all the 41 Annex I Parties have submitted at least one national inventory, and most report every year. In contrast, few non-Annex I Parties have submitted regular inventory reports (see **Kyoto Protocol**). As of early 2008, only Mexico has submitted three inventories, Korea and Uruguay two inventories, and the remaining Parties have submitted only one inventory. This is due in part to lack of financial resources and the technical complexities of compiling national inventories. Developing country inventories often involve a greater level of top-down estimation (due to data scarcity) and are inherently less accurate due to the higher proportion of emissions from agriculture and land-use change sources which are more difficult to quantify.

Non-Annex I inventories can also be more than five years old and, as a result, often only provide a broad approximation of total emissions (especially for non-CO_2 sources). This can constrain effective policy making and also increase the difficulty of projecting future emissions trends across different countries (see also **anthropogenic greenhouse gas emissions** and **future emissions trends**). Much more work is required to improve the accuracy, coverage, and timeliness of inventories on a global basis.

Although national inventories are subject to a number of reliability and coverage limitations, they are, nonetheless, vital to formulating effective climate change policy responses and to underpin international climate change negotiations. Since the original drafting of the **UNFCCC** in 1992, national inventories have provided the cornerstone of knowledge about global greenhouse gas emissions and underpin national policies and emission reduction measures. National inventory data played an essential role in the **Kyoto Protocol** negotiations and will form the basis for determining whether or not countries meet their commitments during the 2008–2012 commitment period. The most recent **UNFCCC** data on Annex I emissions (2005) reveals that, in aggregate, Annex I Party greenhouse gas emissions including land-use emissions are 4.6% below 1990 levels. However, it also indicates that these emission reductions are almost entirely due to substantially reduced economic output and structural adjustments in the Economies In Transition (former Soviet bloc economies) rather than explicit climate policy-related emission reduction measures (see **Kyoto Protocol**).[20]

A review of national inventory and population data also reveals that developing country **emissions per capita** are well below developed country levels and that the contribution of Least Developed Countries (the world's 50 poorest countries) is only 2% of global emissions (see **anthropogenic greenhouse gas emissions** and **emissions per capita**). This inventory data can serve to justify the developing world's limited obligations under the **UNFCCC** and provides an ongoing rationale for the principle of "common but differentiated responsibilities" (see **Kyoto Protocol**). Furthermore UNFCCC inventory data reveals that 31 countries have a negative inventory (only one, Latvia, is an Annex I country): that is, removals of **carbon dioxide** from the atmosphere through afforestation, reforestation, and other land-use changes (see **biosequestration**) are greater than the emissions from **fossil fuels** and other sources. The inventories' trends also clearly show that, while developing country emissions are increasing, the dominance of developed country emissions is unlikely to fundamentally change in the near future (see also **future emissions trends** and **emissions per capita**).

The existence of reliable, consistent, and up-to-date greenhouse gas inventories at both a national and company level is essential to formulating appropriate mitigation response options and policies across different countries, producing more accurate future emission projections and apportioning equitable emission reduction responsibilities for different countries in the international post-2012 climate change negotiations. To improve the reliability and timeliness of inventory data, significant additional financial resources and technical assistance will need to be provided to non-Annex I countries.

See also: anthropogenic greenhouse gas emissions, emissions per capita, future emissions trends, greenhouse gases, Intergovernmental Panel on Climate Change (IPCC), Kyoto Protocol, United Nations Framework Convention on Climate Change (UNFCCC).

Notes

1 Sinha 2006
2 CBC News, May 28, 2007
3 Seaman 2007
4 ISO 2006
5 WBCSD/WRI 2004
6 ISO 2006
7 WBCSD/WRI 2004
8 Australian Greenhouse Office 2006
9 IPIECA 2003
10 WBCSD/WRI 2007
11 IPCC 1997
12 ISO 2006
13 WBCSD/WRI 2004
14 UNFCCC Article 4.1
15 UNFCCC/SBI 2003
16 IPCC 1997
17 IPCC 2006b
18 UN & UNFCCC, Kyoto Protocol Articles 7, 8, and 17
19 UNFCCC 2007b
20 UNFCCC online data, from http://unfccc.int/ghg_data/items/3800.php

Further reading

UNFCCC Handbook 2006, IPCC 2006; WBCSD/WRI 2004; ISO 2006.

GREENHOUSE GASES

Greenhouse gases are those gases in the atmosphere that absorb earth's outgoing long-wave radiation and reemit it back to Earth. The result is the **greenhouse effect** that warms the earth's surface and lower atmosphere. The majority of greenhouse gases are naturally occurring, as is the associated **greenhouse effect**, which ensures that the average surface temperature of the globe is approximately 33°C warmer than it would otherwise be. However, over the past two centuries humans have been emitting large quantities of greenhouse gases into the atmosphere, primarily through burning **fossil fuels** and large-scale deforestation. These **anthropogenic greenhouse gas emissions** have resulted in substantial increases in concentrations of greenhouse gases in the atmosphere, contributing to **global warming**.

Greenhouse gases make up only a very small percentage of the atmosphere by volume. The Earth's atmosphere consists predominantly of molecular nitrogen (N_2)

Table 12 Average dry composition of the atmosphere up to 25 km[1]

Gas	Chemical formula	Percent by volume
Nitrogen	N_2	78.08
Oxygen	O_2	20.95
Water vapor*	H_2O	0–4
Argon	Ar	0.93
Carbon dioxide*	CO_2	0.0382
Neon	Ne	0.0018
Helium	He	0.0005
Methane*	CH_4	0.00017
Hydrogen	H_2	0.00005
Nitrous oxide*	N_2O	0.00003
Ozone*	O_3	0.000004
Synthetic gases*	Various	Trace

Note: Asterisk indicates greenhouse gases.

and oxygen (O_2), which account for approximately 99% of the atmosphere (excluding water vapor, which can vary from 0% to 4%). Nitrogen and oxygen are not greenhouse gases. Most of the remaining 1% is composed of argon, with greenhouse gases (other than water vapor) accounting for less than 0.05% by volume. This naturally occurring background level of atmospheric greenhouse gases is essential to the maintenance of the climate as we know it (see Table 12).

Water vapor, **carbon dioxide**, **methane**, **nitrous oxide,** and **ozone** can all vary spatially and temporally, and all of these gases are created or influenced by human activities.

Anthropogenic versus "natural" greenhouse gases

In the atmosphere, there is no difference in the physical properties or behavior of a naturally occurring molecule of CO_2 compared with a molecule of CO_2 emitted by human activities (anthropogenic), such as burning of **fossil fuels**. The important difference is that the naturally occurring concentrations of greenhouse gases in the atmosphere have been established (with periodic increases and decreases) over millions of years in an interactive relationship with the physical processes of the Earth, establishing an equilibrium between the three active carbon reservoirs: the atmosphere, the oceans, and the terrestrial land system (see **carbon cycle**). **Anthropogenic greenhouse gas emissions** alter this equilibrium since more greenhouse gases are being emitted into the atmosphere than can be taken up by the oceans and land system (see **ocean carbon sinks** and **land carbon sinks**). Furthermore, humans also emit a range of manufactured **synthetic gases**, such as hydrofluorocarbons and sulphur hexafluoride, that do not occur naturally. Emissions of these gases are distinctly different from naturally occurring greenhouse gases. As a result of these emissions, the concentrations of greenhouse gases in the atmosphere have been rising, leading to an enhanced **greenhouse effect**, **global warming**, and associated **climate change impacts**.

Water vapor

The most important greenhouse gas is water vapor (H_2O) and accounts for about three-quarters of the **radiative forcing** in the atmosphere through absorption and reradiation of energy (see **greenhouse effect**).[2] Water has high variability both spatially and temporally but generally has a short residence time (days to weeks) in the atmosphere. It is in a constant state of flux between the vapor, liquid, and solid phases of the hydrological cycle. The importance of water vapor to warming the earth's surface can be felt particularly in desert environments, where, even after a hot day, night-time temperatures can drop rapidly as there is little water vapor in the atmosphere (felt as humidity) to retain the earth's outgoing radiative heat.

Water vapor concentrations in the atmosphere increase with temperature (among other factors), and thus greater global mean concentrations are expected as a result of **global warming**, providing increased **radiative forcing**, a positive feedback. However, higher concentrations of water vapor are more likely to result in the formation of clouds with high **albedo**, reflecting energy away from earth, inducing cooling rather than warming, a negative feedback (see **climate change feedbacks**). There is significant uncertainty regarding the overall balance between negative and positive **radiative forcing** from water vapor. An increase in **radiative forcing** of 0.07 Wm^{-2} above preindustrial levels is estimated from stratospheric water vapor primarily from the breakdown of methane and a much lesser contribution of 0.002 Wm^{-2} from the direct injection of water vapor from aircraft at high altitude. High altitude aircraft also have a global cooling (negative feedback) effect as they leave persistent condensation trails (contrails) that reflect energy back into space (increasing **albedo**) and absorb the energy of outgoing long-wave radiation from earth (see **greenhouse effect**), a positive feedback. The positive feedback is expected to dominate this balance, with an overall effect from contrails being a 0.01 Wm^{-2} contribution to **radiative forcing**. These factors together represent slightly more than 5% of total anthropogenic **radiative forcing**.

The changes in water vapor concentrations in the troposphere arise primarily from irrigation and evaporation, contributing 0.03 Wm^{-2}, or a further 2% of positive **radiative forcing**; however, this is more than offset by changes in **albedo**, for example, through the conversion of forest to irrigated rice paddies.[3]

Anthropogenic greenhouse gases

Humans emit over 50 different greenhouse gases to the atmosphere, although some are emitted in only tiny quantities. International agreements have been made to control most, but not all emissions. The **Kyoto Protocol** explicitly lists four greenhouse gases and two groups of gases that must be included in a national **greenhouse gas inventory** – these are termed "Kyoto gases" and include **carbon dioxide (CO_2)**, **methane (CH_4)**, **nitrous oxide (N_2O)**, sulfur hexafluoride (SF_6), hydrofluorocarbons (HFCs), and perfluorocarbons (PFCs) (see **anthropogenic greenhouse gas emissions**). The latter three of these gases occur only anthropogenically, while the first three also occur naturally. These are listed in Table 13.

Table 13 Kyoto gases and their chemical formulae

Gas	Chemical formula
Carbon dioxide	CO_2
Methane	CH_4
Nitrous oxide	N_2O
Hydrofluorocarbons (HFCs)	
HFC-23	CHF_3
HFC-32	CH_2F_2
HFC-41	CH_3F
HFC-43-10mee	$C_5H_2F_{10}$
HFC-125	C_2HF_5
HFC-134	$C_2H_2F_4$ (CHF_2CHF_2)
HFC-134a	$C_2H_2F_4$ (CH_2FCF_3)
HFC-143	$C_2H_3F_3$ (CHF_2CH_2F)
HFC-143a	$C_2H_3F_3$ (CF_3CH_3)
HFC-152a	$C_2H_4F_2$ (CH_3CHF_2)
HFC-227ea	C_3HF_7
HFC-236fa	$C_3H_2F_6$
HFC-245ca	$C_3H_3F_5$
Perfluorocarbons	
Perfluoromethane (tetrafluoromethane)	CF_4
Perfluoroethane (hexafluoroethane)	C_2F_6
Perfluoropropane	C_3F_8
Perfluorobutane	C_4F_{10}
Perfluorocyclobutane	$c\text{-}C_4F_8$
Perfluoropentane	C_5F_{12}
Perfluorohexane	C_6F_{14}
Sulphur hexafluoride	SF_6

A range of other **synthetic gases** are covered under the Montreal Protocol, which limits the emissions of Ozone Depleting Substances (ODSs), predominantly chlorofluorocarbons, which are also powerful greenhouse gases (see **synthetic gases**). Several other greenhouse gases, such as tropospheric ozone or precursors to greenhouse gases are not covered under any international agreement that restrict their use (see **ozone**).

Carbon dioxide accounts for approximately 20% of the natural **greenhouse effect** and 63% of the anthropogenic enhancement to the **greenhouse effect** to date, with a **radiative forcing** of 1.66 Wm^{-2}. CO_2 is emitted primarily from the combustion of **fossil fuels** and deforestation and is the most important anthropogenic greenhouse gas. Atmospheric CO_2 concentrations have increased by 37% since 1750 and had reached 382 ppm by 2007 (see **carbon dioxide**).[4]

Methane is a potent greenhouse gas arising from waste, livestock, and fossil fuel extraction and is the second most important of the Kyoto gases. It reacts in the atmosphere and consequently has a relatively short atmospheric residence time of around a decade. It has a **global warming potential** (GWP) of 21.[5] Though absolute quantities in the atmosphere are relatively small at 1,774 parts per billion (ppb), they have risen by more than 150% since 1750. Methane accounts for the majority of non-CO_2 (18% of total) contribution to anthropogenic **global warming** (see **methane**).[6]

Nitrous oxide is emitted primarily from agricultural fertilizer application and fossil fuel combustion (see **nitrous oxide**). It is the third most important of the Kyoto gases, a relatively stable gas, with an average atmospheric lifetime of 120 years, and has a **GWP** of 310.[7] While its atmospheric concentration is low at 319 ppb in 2005, it is 18% higher than preindustrial levels and accounts for 0.16 Wm^{-2} **radiative forcing** or 6% of the total anthropogenic contribution to **global warming**.[8]

The **synthetic gases** (SF_6, HFCs, and PFCs) covered by the **Kyoto Protocol** (excluding ODSs) are characterized by low reactivity, high molecular masses, and high GWPs. To date, their contribution to **global warming** has been relatively modest (approximately 1%), but their stability and long atmospheric lifetimes mean their impact will persist for hundreds, and in some cases, thousands, of years. The synthetic gases covered under the Montreal Protocol have contributed almost 17% of the total anthropogenic **radiative forcing** to date.

Non-controlled greenhouse gases

In addition to the gases covered by the Kyoto and Montreal protocols, there are several other gases that are not covered by any international agreement and include **ozone**, several additional manufactured compounds like ethers, halogenated ethers, and non-methyl volatile organic compounds (NMVOCs).

Ethers are carbon-based molecules used as catalysts and chemical reagents in specialized manufacturing industries, as fuel additives (such as dimethyl ether [DME]) and as paint thinners and solvents.[9] Ethers generally oxidize rapidly when exposed to air and consequently have very short atmospheric lifetimes: they make only a negligible contribution to **global warming**. Ethers may be combined with fluorine to make more stable gases, which are widely used in medical gas delivery (such as HFE-7100 ($C_4F_9OCH_3$), with a GWP of 500, and HFE-7200 ($C_4F_9OC_2H_5$), with a GWP of 100). While these gases are potentially potent greenhouse gases, they are produced in small quantities and their contribution to **anthropogenic greenhouse gas emissions** is also very small.

NMVOC emissions (such as those of ethane, butane, pentane, and octane) generally result from the volatilization of hydrocarbon liquids and gases exposed to the air – for example, refueling an automobile. NMVOCs are greenhouse gases that react rapidly with oxygen in the atmosphere to form CO_2: as a result, they have short atmospheric lifetimes and make a very small contribution to anthropogenic greenhouse gases. They are also very difficult to quantify and track and are excluded from **greenhouse gas inventories** – though some countries voluntarily report on estimated quantities of NMVOCs. NMVOCs also act as catalysts to form ozone in the lower troposphere, which is an important greenhouse gas.

Precursor emissions

There are other anthropogenic emissions that contribute to **global warming** but which are not themselves greenhouse gases; they are known as "precursors." Precursor gases react in the atmosphere to form one or more greenhouse gases. The most prevalent of these are carbon monoxide (CO), which is formed from the incomplete combustion of carbon (for instance, in automobile engines) and which oxidizes in the atmosphere to form CO_2, and oxides of nitrogen (NO and NO_3), which can oxidize in the atmosphere to form N_2O. There are several others, but, like NMVOCs, their volumes are relatively small, and accurately quantifying their greenhouse gas impact is difficult. Since CO damages human and animal respiratory systems, and oxides of nitrogen can form nitric acid (and acid rain), there are other policy incentives to reduce their emissions, in addition to their contribution to climate change.

Natural and anthropogenic combustion processes produce **aerosols**, which can reduce **global warming** due to changes in **albedo**. However, such processes are complicated by **feedbacks** related to the formation of clouds, and their overall contribution is not well understood. **Aerosols** are not included in greenhouse gas inventories, except occasionally as supplementary information.

See also: aerosols, anthropogenic greenhouse gas emissions, carbon cycle, carbon dioxide (CO_2), carbon dioxide equivalent (CO_2e), climate change feedbacks, fossil fuels, global warming, global warming potential, greenhouse effect, greenhouse gas inventories, methane (CH_4), nitrous oxide (N_2O), ozone, radiative forcing, synthetic gases.

Notes

1 Aguado and Burt 2004
2 Holper 2001
3 Ibid.
4 IPCC 2007
5 IPCC 1997
6 IPCC 2007
7 IPCC 1997
8 IPCC 2007
9 Air Liquide 2007

Further reading

IPCC 2007, 2006b.

HEALTH IMPACTS

Climate has always been a major determinant of human health. The prevailing climate determines the range of temperatures that individuals experience, their exposure to different types of disease, and the incidence of extreme weather events, such as storms, floods, and droughts. These factors determine the extent and type of health risk people face and the coping measures they adopt to minimize these risks.

Direct climate-related health impacts include illness or injury (morbidity) or death (mortality), arising from weather events. *Indirect* climate impacts are mediated through vector-borne diseases (such as malaria and dengue fever), allergies and asthma, food-related diseases (bacteria and pathogens), and through malnutrition (food shortages caused by floods, droughts, and insect infestations). Changes in the prevailing climate will change the magnitude and mix of health risks that individuals are exposed to. The magnitude of impacts is very dependent on the degree of variability in weather. Although humans may cope with small changes in underlying averages, it is the sudden shifts in extremes that usually have the largest health impacts. With climate change, the degree of climate variability is expected to increase and with it the probability of more frequent **extreme weather events**.

To assess the health impacts of climate change, we must first distinguish between those impacts that arise from climate change and those that arise from the complex array of other variables that affect human health, including natural climate variability. Not every flood, drought, or heat wave or outbreak of disease can be attributed to human-induced climate change.

There is growing evidence that human health has already been affected by human-induced climate change. There have been changes in the frequency and intensity of heat waves (which are associated with heat-related mortality), changes in the seasonal and geographical distribution of precipitation (resulting in more floods and droughts and increased storm intensity), and changes in the incidence and geographical range of diseases and allergies. The World Health Organization has calculated that climate change resulted in an additional 150,000 human deaths in 2000.[1]

Over the course of the twenty-first century, humans are expected to be subject to even greater and more rapid change in temperatures, with more heat waves, greater year-on-year climate variability, and increased frequency and intensity of **extreme weather events**. These changes could have significant repercussions for human health, both positive and negative. For example, there may be more heat-related deaths but fewer cold-related deaths or the introduction of new health problems into some areas (such as increased exposure to malaria or Lyme disease) with corresponding reductions in existing health problems in others. Climate change is also expected to affect the health of many other species and ecosystems (see **biodiversity impacts** and **marine impacts**), which can exert feedback effects on humans through the supply of clean water, food, medicines, and other ecosystem services.

Temperature effects

Temperature can have both direct and indirect impacts on human health. The direct impacts arise mainly from extremes of heat and cold, which can cause illness and sometimes death. The indirect effects can arise from changes in the incidence and distribution of vector-borne diseases; the occurrence of photochemical smog and forest fires; changes in the reproduction patterns of rodents and pests; crop failures; and a greater incidence of food contamination.

Changes in average temperature of a few degrees over several decades, as are expected with global warming, are unlikely to result in significant direct health impacts. Humans currently inhabit a wide range of climatic zones, stretching from the hot equatorial regions to the extreme cold of the higher latitudes. Although some areas are already approaching the limits of human tolerance, humans can generally acclimatize and adapt to changing conditions, given time. However, human health is much more susceptible to sudden changes in temperature extremes, such as those that occur during heat waves or sudden cold snaps. For example, in 2003 Europe was struck with an extreme heat wave that claimed the lives of an estimated 35,000 people.[2]

Heat waves can be a major cause of morbidity and mortality. When air temperatures equal or exceed body temperature, external heat can increase core body temperature and result in severe dehydration and heat stroke, sometimes resulting in death. Hospital admission rates generally rise significantly during heat waves. During the 2004 heat wave in Brisbane, Australia, for example, when temperatures soared to 5–6°C above average maxima for more than a week, ambulance call-outs increased by more than 50%.[3] Heat waves tend to particularly affect young children, the sick, and the elderly but can also have a disproportionate impact on people with outdoor occupations. The full health impact of heat waves is often underestimated as recorded data mainly relate to mortality, and statistics related to heat-induced morbidity are limited.

Heat waves can also indirectly affect human health. Extended periods of hot weather are often accompanied by elevated levels of photochemical smog in urban areas (elevated particulate and surface ozone levels) and forest fires (resulting in high particulate levels). Surface ozone and particulates can both cause severe respiratory problems. Urban areas are also more susceptible to heat-related health risks due to the *urban heat island effect*, where the mass of buildings, concrete, and asphalt tends to amplify ambient heat levels. On hot days, urban areas are generally several degrees higher than that of surrounding rural areas. The continued rapid growth of cities, particularly in the developing world, is expected to accentuate the heat-related impacts of climate change on health.

Elevated temperatures can also increase the risk of bacterial contamination of food – and hence food poisoning, such as that caused by salmonella.[4] The incidence of diarrheal diseases and the number of bacteria-carrying flies is also known to increase as temperatures rise.[5] People living in low-income areas and urban slums are generally more susceptible to heat waves as they often live in poor-quality housing that offers little protection, have limited access to good quality water services and sanitation, and already suffer from a range of other health problems.

As the planet warms, the number of heat waves is expected to increase and so will heat-related morbidity and mortality, the incidence of cardiorespiratory diseases and diarrhea.[6] There has already been an observed rise in the number of heat wave episodes in many regions over the past two decades, and this trend is expected to accelerate over the coming century.[7] Heat wave mortality in California's major urban centers, for example, is projected to increase by between 100% and 1,000% over this century, depending on the city.[8]

Elevated average temperatures are likely to have some benefit as they are expected to reduce the incidence of cold-related mortality, though these benefits will be largely confined to the higher latitudes.[9] For most developing countries in the tropics and subtropics, there are not expected to be any temperature-related mortality benefits. Overall, the scientific consensus is that climate change-induced increases in heat-related mortality will far outweigh any reductions in cold-related mortality.

Floods and droughts

Climate change is expected to increase the intensity and frequency of floods and droughts (see **extreme weather events**). Storms and high winds already result in significant health impacts through injury and loss of life and are likely to increase over this century. However, the most significant health impacts are expected to arise from the increased intensity and frequency of droughts and floods (see **water impacts**).

Floods

In addition to crop and infrastructure damage, floods can cause significant injury and loss of life through drowning, electrocution, and landslides. Human alterations to the landscape, such as deforestation and urbanization, can increase flood risks and health problems.[10] Floods can also result in contamination of waterways, through sewage overflow and the flushing of toxic chemicals and other wastes, and through providing more breeding sites for insect vectors such as mosquitoes. These factors can lead to outbreaks of disease and respiratory problems. In the 2001–2002 Mozambique floods, 1,800 lives were lost as a direct impact of the floods, but in the months that followed, up to 8,000 additional people died from infectious diseases such as malaria and diarrhea.[11] In the US city of New Orleans, following Hurricane Katrina (2005), there was a significant increase in respiratory diseases and skin infections associated with toxic chemical contamination of the flood waters and the spread of molds.[12]

The loss of crops, livestock, and food stores due to floods can lead to food shortages and, subsequently, malnutrition. This can, in turn, increase people's susceptibility to disease. Floods can also result in the migration of affected populations to other areas, which not only places these areas under increased stress, and potentially introduce conflict, but can also result in the spread of disease to other areas. The trauma of floods can result in increased social stress, suicides, and mental disorders, which can have long-lasting health and social impacts.[13] The greater incidence of flooding in Europe since the late 1980s has been accompanied by increased depression and mental illness associated with these events.[14]

Overall, the developing world generally suffers higher flood-related morbidity and mortality than developed countries as more people tend to live in high-risk flood zones and fewer people have access to early warning and emergency services. As climate change is expected to increase the frequency and severity of flooding over the coming decades, there is expected to be an associated increase in morbidity and mortality.

Drought

The health impacts of drought are largely indirect and take much longer to manifest themselves than those associated with storms and floods. The main health impact arises from the increased incidence of starvation and malnutrition resulting from reduced food production (see **agriculture and food supply impacts**). Malnutrition increases vulnerability to sickness and disease and can result in high mortality rates. As with floods, droughts can also result in large-scale population movements, with similar social stresses and conflicts in areas that are recipients of drought refugees.

Droughts can also lead to outbreaks of infectious diseases through degraded drinking water quality and sanitary conditions. In the early stages of drought, vector-borne diseases can be a problem as stagnant, nutrient-rich pools of water form and provide perfect breeding grounds for vectors such as mosquitoes (see below). These risks tend to fall as the drought progresses.

As climate change is expected to significantly increase the risk of drought in the mid- and lower latitudes, there could be significant health repercussions as this is where the bulk of the poor (the most susceptible) in the developing world live, especially if food production falls to the extent predicted.

Vector-borne diseases and allergies

The range, reproduction rate, and length of breeding season of many disease-carrying insects (*vectors*) are largely determined by temperature and moisture. Insects tend to thrive in warm, moist conditions. As temperatures increase with global warming, and precipitation patterns change, so will the reproduction rates and geographical ranges of many disease-carrying vectors. Higher temperatures can also reduce the incubation time of certain vector-borne parasites. The malaria parasite (*Plasmodium falciparum*) incubation time, for example, is halved if average temperatures increase from 20 to 25°C.[15] Outbreaks of malaria and dengue often coincide with periods of above-average temperature and rainfall. The incidence and seasonal variation of allergies, asthma, and other respiratory diseases are also known to be sensitive to climate.

Vector-borne diseases

A wide range of serious diseases are transmitted by vectors such as mosquitoes, ticks, sand flies, and fleas. These include malaria, yellow fever, dengue, West Nile virus, Lyme disease, and encephalitis. Climate change has already led to observed changes in the incidence and distribution of many vector diseases. There has been a migration of vector-borne diseases to higher altitudes (which were previously vector disease free) in Africa, Papua New Guinea, Latin America, and Asia. For example, in the past, the mosquito that carries dengue and yellow fever was typically found only at altitudes of less than 1,000 m; however, in recent years it has been found at altitudes of 1,700 m in Mexico and 2,200 m in Columbia.[16]

Malaria is the most serious vector-borne disease in terms of mortality. Each year 300–500 million cases of malaria are recorded, and it is estimated that malaria results in the deaths of 1–3 million people each year: on average, 3,000 children die from malaria each day.[17] Around 40% of the world's population live in areas subject to malaria. Malaria also has significant morbidity impacts, which can result in large economic losses through reduced productivity and increased strain on the health system (see **socioeconomic impacts**). Although there has been some progress with treating the disease over the past half century, there is growing evidence that new drug-resistant strains have developed.

The impact of climate change on malaria is an area of some controversy. While it is generally agreed that there is likely to be an increase in the number of people at risk of exposure in areas adjacent to current malarial zones, there is less agreement about the numbers at risk on a global scale. A wide range of variables can influence malaria transmission, and the interactions of these variables are complex and not well understood. Furthermore, many estimates are based on current exposure dynamics and do not adequately take into account the full range of potential adaptation responses, which may lead to overestimates of malarial impact. Some areas, particularly southern and eastern Africa, are expected to experience an increase in numbers at risk, while in other areas, for example, parts of West Africa and Latin America, may experience a decline. Nonetheless, as a whole, Africa (which accounts for 85% of all malarial deaths), is expected to see its malarial range expand over the coming decades, and an additional 20–70 million people could be at risk by the second half of this century.[18] Malaria may also extend into some areas of Europe and Central Asia. While uncertainty remains, few studies conclude that there will be a decline in numbers at risk of exposure.

Dengue is another vector-borne disease that affects large numbers of people each year. It thrives in warm, moist conditions, and outbreaks in many countries are usually most frequent during the rainy season. Outbreaks can also occur during drought conditions, when water tanks and other stationary pools of water can provide suitable breeding areas. While it has a lower mortality incidence than malaria, it is associated with high levels of morbidity. The current consensus is that the incidence of dengue is likely to increase with climate change over this century. Its range is expected to extend into areas of Australia and New Zealand, and some estimates put the potential number at risk as high as 3.5 billion in the latter part of this century.[19]

Since the early 1990s, the prevalence of several other vector-borne diseases has increased and expanded in geographical range.[20] For example, West Nile virus, an encephalitis (brain inflammation) disease, is known to spread during drought periods, when warm weather and nutrient-rich pools of water form and provide suitable breeding sites for mosquitoes. The disease originated in Uganda, and was, until recently, largely confined to Africa. However, it has since spread to many other countries including the United States, Canada, Romania, and the Czech Republic. The meningitis area of West and Central Africa may also expand. There have been numerous outbreaks reported since the mid-1990s, and climate change is expected to lead to an increase in this disease's range.[21]

Lyme disease has emerged as a serious health issue in the northeast and northern central United States and is also common in Europe. In both cases, the deer tick is the main vector. Untreated Lyme disease can affect the nervous system, cause joint swelling, and lead to heart problems and chronic disability. The general warming that will accompany climate change, particularly in the form of warmer winters, is expected to increase the spread and prevalence of Lyme disease, including a likely expansion into Canada.[22]

Allergies and respiratory diseases

Climate change is increasing the prevalence of asthma and other allergic diseases. Climate-induced changes to the life cycles and numbers of many plants have altered patterns of pollen production and dispersal, a major contributor to allergies. Many plants are flowering earlier and producing more pollen. Elevated concentrations of **carbon dioxide (CO$_2$)** are also believed to contribute to asthma and pollen production. One study has concluded that CO$_2$ concentrations of 560 ppm could increase the amount of pollen produced by ragweed, a known contributor to asthma and allergies, by 40–60%.[23] Molds and fungi are also expected to thrive in warmer and wetter conditions: increased spore production could contribute to an increase in allergies and respiratory ailments. Asthma and allergies are one of the leading causes of chronic illness in North America and Europe, costing the United States economy alone $ 18 billion per year.[24]

What are the expected health impacts of climate change over the twenty-first century?

While there remains much we do not know about the interaction between climate and health, the conclusion of IPCC Fourth Assessment Report (2007) is that the expected impacts are overwhelmingly negative. Impacts are expected to be relatively modest up to 2030 but will thereafter become progressively more apparent as the planet warms. Malnutrition and diseases such as diarrhea and malaria are expected to be the greatest contributors to increased climate-related morbidity and mortality.[25]

However, when projecting future health impacts, one must distinguish between *potential* climate impacts and *expected* climate impacts. *Potential* impacts are those that could occur if humans undertake no adaptation or preventative measures to reduce health risks. *Expected* impacts are those that take into account possible adaptation and coping measures. Health impacts must also be assessed against the expected level of *exposure* of the population to changed climatic conditions (how many people will actually be exposed to the changed conditions) and the *sensitivity* of those that are exposed to the changes (this will vary from person to person). There is no single exposure–response relationship that can be used in the impact assessment process. Some individuals and communities are more vulnerable to climate change than others, due to where they live, the access they have to medical and preventative services, and their socioeconomic status. In effect climate change impacts on human health are very much dependent on how humans respond to changing climate conditions.

Adaptation will play a large role in limiting the health impacts of climate change. Measures such as improved health care, better building design, installation of early warning systems, improved emergency preparedness and disaster relief, and a host of other preventative strategies will help alleviate the health risks and impacts of climate change, particularly those associated with extreme weather events. However, there are limits to adaptation, and the ability to adapt is highly dependent on a country's wealth. While many developed countries are likely to cope with, or alleviate, climate change health impacts, this is unlikely to be the case in the developing world. As with the impacts of climate change on agriculture and food production, the poor in the developing world will tend to suffer disproportionately from the health impacts of climate change (see **agriculture and food supply impacts**).

See also: agriculture and food supply impacts, biodiversity impacts, extreme weather events, global warming, water impacts.

Notes

1 McMichael 2004
2 IPCC 2007
3 Steffen *et al.* 2006
4 Fleury *et al.* 2006
5 McMichael *et al.* 2003
6 IPCC 2007
7 Ibid.
8 Hayhoe *et al.* 2006
9 See Kovats and Jendritzky 2006
10 McMichael 2004
11 Cairncross and Alvarinho 2006
12 Manuel 2006
13 Ahern *et al.* 2005
14 Ebi 2006
15 Epstein and Mills 2005
16 Epstein *et al.* 1998
17 WHO 2003
18 Nyong and Niang-Diop 2006
19 Hales *et al.* 2002
20 Epstein 2005
21 Epstein and Mills 2005
22 Ibid.
23 Wayne *et al.* 2002
24 American Lung Association 2005
25 IPCC 2007

Further reading

McMichael 2004; Epstein and Mills 2005, IPCC 2007.

ICE SHEETS AND GLACIERS

Climate change will have significant impacts on the world's ice sheets and glaciers. The amount of ice cover has an important bearing on the earth's climate system through changes to **albedo** and ocean circulation. It also has important implications for marine and terrestrial ecosystems (see **marine impacts**, **biodiversity impacts**, and **polar impacts**) and on humans (see **water impacts** and **socioeconomic impacts**).

Considerable uncertainty surrounds the impact of human-induced global warming on the amount of snow and ice cover as scientists are not sure how sensitive ice sheets and glaciers are to changes in surface temperatures. However, all agree that sea- and land-based ice will continue to contract over this century and beyond and that the rate of contraction is likely to accelerate. This has potentially significant repercussions for global **sea level rise**, particularly over the next several centuries.

Sea ice cover

There has been a marked decline in sea ice coverage and thickness in both the high latitudes of the northern and southern hemispheres over recent decades. Over the past 30 years, the extent of sea ice cover over the northern polar region has contracted by about one million square kilometers: summer ice coverage has declined 15–20%, and the remainder of the ice sheet has thinned considerably over the Arctic Ocean.[1] Some modeling results suggest that current warming trends could see the likelihood of near ice-free summer conditions in the Arctic before 2050.[2] In Antarctica, considerable warming has also been occurring and sea ice coverage has been reduced by nearly 20% over the past 50 years. Antarctic sea ice has migrated more than 1.5° in latitude further to the south.[3]

As sea ice has already displaced its own volume, it does not in itself have a direct impact on the level of the sea, but it does have a significant indirect effect through two mechanisms. The first is the impact on the earth's **albedo**. White ice reflects much more of the sun's energy back into space (up to 90%) than the open ocean (from 10% to 80%, depending on the angle of the sun – see **albedo**). With the reduction in sea ice coverage, the amount of solar energy being absorbed by the earth's surface increases and with it the temperature of the atmosphere and oceans. This contributes to the thermal expansion of the ocean and accelerated melting. These higher temperatures create a positive **climate change feedback** effect, which causes sea ice to recede even further, lowering albedo yet again and accelerating the warming effect. If average polar temperatures increase from their present level of –9°C to say –5°C (which is likely this century), it will result in a strong albedo feedback effect that could trigger a permanent loss in sea ice coverage.[4]

The second means by which declining sea ice contributes to sea level rise is through its impact on the flow rate of land-based glaciers. Glaciers are essentially slow-moving rivers of ice. Most of the natural land drainage points for glaciers in the Arctic and Antarctic are blocked by sea ice. The sea ice acts as a dam, greatly impeding land glacier flow rates in these areas. As sea ice disappears, and ice shelves break away, their effectiveness as dams is reduced. This can result in a significant acceleration in the flow rates of feed glaciers.[5] For example, in 2002 the Larsen B ice shelf collapsed suddenly, due to warming that had thinned and weakened the ice sheet. This collapse has caused the glaciers that feed into the area to accelerate their flow rates. The cause of the Larsen B ice thinning was not just the warmer atmosphere melting it from above but also the warming of the ocean below (in this case, the Weddell Sea which had warmed by around 0.25°C over the previous 30 years). This melting from below led to increased instability of the ice shelf, and some scientists believe that the rest of the Larsen ice shelf is also likely to collapse sometime this century.[6] This is also happening to other ice sheets in West Antarctica, leading to growing concerns about the long-term stability of the entire West Antarctic Ice Sheet (WAIS).

The stability of the WAIS represents the greatest uncertainty surrounding the contribution of Antarctica to sea level rise over this and the next century. If this ice sheet collapsed, it would raise sea levels by 6 m over several centuries.[7] While the IPCC (2007) concluded that this was very unlikely to happen during this century, several recent studies have indicated that the WAIS could be more unstable and that the probability of collapse this century could, in fact, be higher than previously believed.[8] However, much of the marine portion of the Antarctic ice sheet is well grounded on the continental shelf and is believed to be less susceptible to retreat than the northern polar region.[9]

Impact of global warming on glaciers and mountain snow pack

The rapid melting and retreat of glaciers and mountain snowpack is occurring in many places all over the world. Increases in glacier melting and retreat, and even their total disappearance, have been recorded for all equatorial mountain glacier regions, particularly in the Andes. The glaciers in the Himalayas, European Alps, the North American Rocky Mountains, and in the colder regions of Alaska, Patagonia, and Antarctica have also experienced accelerated glacier thinning and retreat. This loss of surface mountain snow and ice cover will also reduce the earth's **albedo** and generate a positive **climate change feedback** effect, particularly in mid-latitude mountain regions.

Ice in the Antarctic Peninsula is responding quickly to regional warming, and satellite data and aerial photographic records show that 87% of the 244 glaciers on the Antarctic Peninsula have retreated over the past 60 years.[10] Glaciers feeding the Amundsen Sea of West Antarctica have also experienced accelerated rates of glacier discharge so far this decade compared with the 1990s, and it is estimated that the current contribution to sea level rise from this area alone has increased to about 0.24 mm/year.[11] The Patagonian ice fields have lost about 10% of their mass since the mid-1990s, and this area is presently contributing about 0.1 mm/year to sea level rise.[12] One significant recent finding is the rate of acceleration in some of the major glaciers in Greenland. Satellite observations conducted over the period 1996–2005 have shown that most of the major glaciers have exhibited significant acceleration, with the speed of the Kangerdlungssaug Glacier increasing from 6 km/year to 13 km/year between 2000 and 2005 alone.[13] Some glaciers, including several in the North of Sweden, are actually increasing in size due to higher precipitation, but these are exceptions to the general trend. At present, the contribution from melting glaciers to sea level is around 1 mm/year.

While most models assume the contribution from glaciers to sea level rise will largely come from surface melting, there are two other mechanisms at work that are not yet fully captured in models. The first such mechanism is melt-water lubrication of the base of the glacier, which reduces friction and increases flow rates. The second is the melting of the buttresses that lock the glaciers to land and the sea bed, mainly due to warmer seas. If the full influence of these two mechanisms is taken into account in modeling, the estimated contribution of glaciers to sea level rise will be larger during the twenty-first century than previously envisaged.

The melting of glaciers and reductions in permanent snow pack outside the polar regions (e.g. those in the Himalayas and Andes) will increase sea levels this century, but overall, their contribution is expected to be relatively small (perhaps a few centimeters). Their contribution will also be of limited duration as most of the glaciers in the mid- and lower latitudes will be gone this century. For example, the Glacier National Park in the United States is likely to have no glaciers left at all by 2050. The more important impact of glacier melting will not be on sea level rise but on those people and ecosystems that depend on glaciers and snow pack for their seasonal water flows. In India, the Andean countries, parts of China, and Southwestern United States, the reduction in glaciers and mountain snow pack will have major implications for water resources (see **water impacts**).

The Antarctic ice sheet

The Antarctic ice sheet first formed about 40 million years ago, but only became permanent around 15 million years ago. Over the last million years, it has expanded and contracted on a regular cycle of about 100,000 years (see **Milankovich cycle**). The Antarctic ice sheet holds a vast amount of water as ice – around 25 million cubic km of ice. If it were to melt completely, it would cause sea levels to increase by 57 meters.[14] The IPCC (2007) concluded that Antarctica is likely to be neutral or experience a net increase in ice mass during this century due to increased precipitation (in the form of snow). Currently, the air over Antarctica is so cold that it holds very little water – so little, in fact, that average annual precipitation for the interior of the continent is only about 50 mm per year, making it one of the driest places on earth. Atmospheric warming is likely to increase the moisture-carrying capacity of the air, and hence precipitation levels. If Antarctica experiences an increase in ice mass, this will reduce the rate of sea level rise during this century.

Some scientific evidence tends to support the IPCC view that the East Antarctic Ice Sheet experienced an annual average increase in mass of around 45 billion tons from 1992 to 2003.[15] However, a 2006 study assessing ice-core data found that there had been no statistically significant increase in ice mass in Antarctica over the past 50 years.[16] Snow accumulation is being offset by the flow of ice into the sea and its subsequent melting. If this situation prevails, the Antarctic ice sheet may actually contract to some degree and, thereby, even make a small positive contribution to **sea level rise** – but considerable uncertainty remains.

The Greenland ice cap

The ice mass covering Greenland is by far the largest potential contributor to sea level rise from the Northern Hemisphere. The Greenland ice cap contains approximately 10% of the world's ice; should it melt completely, average sea level would rise by about 7 m.[17] The large inflow of freshwater from the melting ice could also contribute to a slowing or even a shutdown of the **thermohaline**.

Until recently, the rate of accumulation was thought to be largely in balance with the loss from flows to the sea and iceberg discharge. However, several recent studies have found that the loss of ice mass is much higher than models predicted. The estimates of losses from the Greenland ice sheet vary according to the measuring and modeling techniques applied. At the high end of the scale, one recent study estimated ice mass loss over the 1996–2005 period at 224 cubic km per year,[18] while another found that between 2002 and 2005 the loss was 239 cubic km per year.[19]

Other recent estimates, using a combination of several measurement techniques, yielded a lower figure of around 100–120 cubic km per year over the period 2002–2005.[20] Nonetheless, the evidence suggests that there is presently a net loss of Greenland ice mass. In a warmer climate, snow fall is likely to increase over some parts of Greenland, thus providing a negative **climate change feedback** effect on sea levels. This has already been observed as there has been a gain in mass in the areas above 2,000 m in altitude in recent years, though this has been more than off-set by significant losses in ice mass below this altitude. Current Greenland ice mass loss equates to a contribution to sea level rise of around 0.3 mm/year or 10% of the currently observed increase of 3 mm/year.[21] Future rates of ice mass loss will depend on the amount of warming and changes to the geometry of the ice cap as it recedes.

If local temperatures experience a sustained rise of 2.7°C or more, this would most likely lead to significant melting of the Greenland ice cap: some models predict that this may trigger total deglaciation of Greenland, although this would most likely take more than 1,000 years.[22] Under nearly all emission scenarios, this temperature threshold is likely to be surpassed before the end of this century, and possibly much sooner. The disturbing fact is that several areas of Alaska and Northern Canada have already warmed by over 2°C since the 1950s, and modeling estimates of warming north of the 60° latitude range from between 4°C and 7°C by 2100.[23]

If the Greenland ice cap totally disappears, it is questionable whether it would be able to reestablish itself, even if temperatures eventually restabilized at a level lower than 2.7°C above present levels. It would probably require the onset of a new ice age to reestablish a permanent ice cap. The reversibility of deglaciation of Greenland is still uncertain and remains an area of ongoing research.[24] Given higher degrees of **climate sensitivity** and a possible lower cooling effect from **global dimming** if aerosols are reduced, the atmospheric carbon dioxide concentration at which 2.7°C is reached could be lower than many models currently predict, possibly as low as 500 parts per million **carbon dioxide equivalent** (see **stablization targets**). If this is the case, then the Greenland deglaciation threshold could be exceeded much earlier than anticipated. Thermal expansion of the oceans and Greenland deglaciation will continue long after global temperatures are stabilized. The problem is that once the deglaciation temperature threshold is passed, deglaciation is the inevitable outcome. There is a reasonable chance that this point will be reached sometime this century and possibly by mid-century.

A somewhat disconcerting outcome is that many of the Greenland ice sheet models have tended to underestimate the observed loss of ice from Greenland, and since 2005, several studies have concluded that the future rate of melting of the greenhouse ice sheet could in fact be considerably higher than anticipated. Furthermore the loss of Arctic sea ice has also been more rapid than the models predicted and that Arctic sea ice cover could disappear much earlier than what the most recent IPCC report suggests. A clearer picture of the future of the Arctic sea ice and the Greenland ice cap should emerge over the next decade as more research is undertaken.

See also: albedo, biodiversity impacts, climate sensitivity, marine impacts, polar impacts, sea level rise, water impacts.

Notes

1 IPCC 2007
2 Holland *et al.* 2006
3 Curran *et al.* 2003
4 Winton 2006
5 Rapley 2006
6 Shepherd *et al.* 2003
7 Rapley 2006
8 Ibid.
9 Anderson 2007
10 Cook *et al.* 2005
11 Thomas *et al.* 2004
12 Rignot *et al.* 2003
13 Rignot and Kanagaratnam 2006
14 Rapley 2006
15 Davis *et al.* 2005
16 Monaghan *et al.* 2006
17 Church and Gregory 2001
18 Rignot and Kanagaratnam 2006
19 Chen *et al.* 2006
20 Luthcke *et al.* 2006
21 Ibid.
22 Gregory *et al.* 2004
23 ACIAR 2004
24 Nicholls and Lowe 2006

Further reading

Rapley 2006; Luthcke *et al.* 2006; Nicholls and Lowe 2006; Curran *et al.* 2003.

INTERGOVERNMENTAL PANEL ON CLIMATE CHANGE (IPCC)

The IPCC is the most important source of scientific, technical, and socioeconomic information for the **United Nations Framework Convention on Climate Change (UNFCCC)** and has had a strong impact on the development of the Convention and subsequent international climate change negotiations. The IPCC produces comprehensive Assessment Reports about every five years as well as Special Reports and Technical Papers in response to UNFCCC requests (see **UNFCCC**). The IPCC also maintains the National Greenhouse Gas Inventories Program (see below).

The role of the IPCC is to objectively, transparently, and comprehensively assess the scientific, technical, and socioeconomic information relevant to understanding the risk of climate change, its potential impacts, and options for adaptation and mitigation. The significance of the work was recognized with a Nobel Peace Prize in 2007.[1]

How does the IPCC operate?

The IPCC is a Panel that is open to all member countries of the United Nations Environment Program (UNEP) and the World Meteorological Organization (WMO). The IPCC meets about once a year to establish procedures, direction, and scope of reports being commissioned and to approve reports.

The IPCC does not undertake new research or monitor climate-related data but bases its assessments on published and peer-reviewed scientific and technical literature. A key principle of the IPCC is to be "policy relevant but not policy prescriptive." In a subject as politically charged as climate change, this is essential to bridge the often difficult divide between scientists and policy makers.

Preparing the Assessment Reports is undertaken through three Working Groups:

(i) Working Group I: Scientific aspects of the climate system and of climate change;

(ii) Working Group II: Vulnerability and impacts of climate change on socio-economic and natural systems and options for adaptation;

(iii) Working Group III: Response strategies for limiting greenhouse gas emissions and mitigating climate change.

Reports are written by teams of authors from academia, research centers, businesses, and associations nominated by governments and international organizations and selected according to their expertise. They come from more than 100 countries and are not paid by any central agency. Many undertake this work on a voluntary basis.

Working Group reports must be objective, open, and transparent. This is achieved through a two-stage review: first, an expert "peer-review" and, second, a review by experts and government officials. The outcome of these reviews is then presented at an IPCC Plenary session for an adoption and approval process that is open to all member governments.

The reports stretch to several hundred pages, and in reality, decision makers do not generally read the entire reports. To make the information more accessible, the IPCC also prepares "Summaries for Policymakers." These documents have become much more high profile and are subject to much greater governmental scrutiny because of the political ramifications of their contents. Each individual line of the Summaries is approved by the government delegates to the IPCC in a painstaking process that is part approval and part negotiation. At the end of this process, Assessment Reports are publicly available, and each of them has become recognized as a landmark study.

The IPCC only accepts published and peer-reviewed documents in order to ensure the scientific integrity of Assessment Reports. While this ensures a minimum standard is maintained, climate science and indeed the impacts of climate change are rapidly changing and advancing. As preparing an Assessment Report takes several years, and all contributory papers must be peer reviewed, the most recent cutting-edge research and observations are not reflected in the report. For example, the higher than expected rates of sea ice loss or the accelerating rate of ice loss from Greenland (see **ice sheets and glaciers**) could not be included in the Fourth

Assessment Report (AR4) since they had yet to be published at the time of the Report preparation. By the time the Assessment Reports are published, some of the material is several years out of date.

Another consequence of this policy has been to eliminate government and business reports that are not subject to peer review. The Stern Report (2006) provided a comprehensive study on expected socioeconomic impacts that has been subject to extensive critical review after its publication, yet it cannot be included in IPCC Assessment Reports.[2]

Furthermore, the final approval by government delegates brings with it an element of politicization that cannot be avoided. For example, in the approval process for the AR4, the United States, seeking to downplay the potential costs of climate change, requested to replace "up to 5%" of GDP for 4°C of warming with "1–5% of GDP." Austria, Germany, France, and others, who seek greater action on climate change, supported adding reference to the Stern Report, which estimates GDP losses "at or above 10%." The published AR4 refers to GDP losses between 1% and 5% without mentioning the numbers from the Stern Report.[3]

In order to avoid debate in the Plenary approval process, authors have been inclined only to include information that is well established. As with any science, there is uncertainty in modeling and projections. Considerable time is spent estimating and quantifying uncertainty, where "likely" is a probability of greater than 66%, "high confidence" is greater than 90%, "very high confidence" is greater than 95%, and so forth. While scientists may implicitly understand these descriptors, policy makers and bureaucrats see them as wording to be massaged to give the desired message. This has led to bureaucrats asking for "very high confidence" to be reduced to "high confidence" and/or removed, even though the process of reaching agreement on confidence levels has already been through a rigorous two-stage review, and the wording is deliberate and exact.[4]

Risk is a multiplication of likelihood and impact, so from a risk perspective, a very large impact (say, the collapse of the Amazon rainforest ecosystem [see **land carbon sinks** and **biodiversity impacts**]) that is considered *unlikely* (<33% probability) of occurring could warrant greater attention than a relatively manageable impact (say, 30 cm **sea level rise**) that is *very likely* (>90% probability) to occur.[5] An objective risk assessment would likely suggest greater attention focused on the outcome with the highest overall risk, not the outcome with the highest confidence of occurring. In contrast to the IPCC's conservative approaches, the precautionary principle (see **UNFCCC**) would suggest that even a *very unlikely* (<10% probability) impact should definitely be avoided if the magnitude of the potential impact remains unknown but potentially very large. Unfortunately, medium and lower likelihood impacts have not featured in IPCC Summaries for Policymakers,[6] and as such the overall message of the Summaries tends to understate, rather than overstate, the risks of climate change.

These limitations tend to result in a "lowest common denominator" outcome that does not necessarily convey the gravity of the evidence or the best current scientific understanding. Nonetheless, they do build a minimum consensus on the science of climate change and its potential impacts. Yet even with these limitations and the conservative and staid nature of the format, the findings contained in the IPCC

Assessment Reports have, nevertheless, been cause for considerable concern and have received extensive publicity.

IPCC history

The IPCC was formally established in 1988 and has since produced four major reports and numerous special reports.

First Assessment Report (FAR)[7]

The FAR was finalized in 1990 and was instrumental in motivating action by the United Nations General Assembly to pursue negotiations of a convention on climate change. These negotiations were coordinated with the preparations for the UN Conference on Environment and Development (UNCED) in Rio de Janeiro in 1992.

By the 1992 Rio summit, substantial new information on climate change was available, and the IPCC approved a set of Supplementary Reports that brought the most up-to-date information to the negotiations. The IPCC findings, and heightened public concern over human impacts on the climate, resulted in the adoption of the UNFCCC (the Convention) in 1992 and entry into force in 1994 (see **UNFCCC**). Under the Convention, a Subsidiary Body for Scientific and Technological Advice (SBSTA) was appointed as the main means by which the IPCC findings are channeled to the Convention. In addition to the Assessment Reports, the UNFCCC (via SBSTA) has asked the IPCC to provide scientific technical and socioeconomic information on specific topics. The IPCC responds to such requests through Special Reports or Technical Papers. The IPCC has (as of early 2008) produced 12 Special Reports on subjects varying from Aviation and the Global Atmosphere to Carbon Dioxide Capture and Storage.[8] The IPCC Technical Papers are drawn from material already assessed in Assessment Reports and, as such, are a distillation from the larger reports for relevant groups. Five Technical Papers had been published by IPCC up to September 2007.[9]

Second Assessment Report (SAR)[10]

In 1995, the IPCC approved the SAR, which (among other things) addressed the likely impact of different levels and timescales for stabilizing atmospheric concentrations. It provided scientific, technical, and socioeconomic information that could be used to evaluate what projected impacts constitute "dangerous anthropogenic interference with the climate system," and evaluated adaptation and mitigation options.

The SAR was a landmark document for several reasons. It called for much more substantive measures than had been outlined in the Convention, and it upset many vested interests, who took umbrage that a scientific panel should be dictating policy measures – the so-called climate skeptics began to mobilize. In combination with burgeoning emissions globally, the SAR was influential in the drive to move beyond the Convention and have binding commitments for Annex I countries (see **Kyoto Protocol**).

Third Assessment Report (TAR)[11]

The 2001 TAR included a substantial body of new information. The TAR sought to finally resolve the contention that the observed warming could be natural variation and concluded that "There is new and stronger evidence that most of the warming observed over the last 50 years is attributable to human activities." By providing compelling evidence that the observed warming could not be solely attributable to natural variation, it removed an important political obstacle, namely to discredit the argument that too much uncertainty remained to commit to binding emission reductions. The TAR, and its earlier released Working Group reports, in many ways signaled the beginning of much more bitter politics that now surrounds the **Kyoto Protocol** and its successor agreements.

Fourth Assessment Report (AR4)

The 2007 AR4 provided yet more evidence of observed change and much more detail on the expected impacts, including a greater resolution of regional distribution of impacts and the potential adaptation responses required. More than 2,500 expert reviewers, 800 authors, 450 lead authors, and 130 countries have participated in the elaboration of the AR4, and it represents the most comprehensive assessment of the climate change issue yet produced.

IPCC and National Greenhouse Gas Inventories

Initial work by the IPCC in establishing national and global emissions resulted in a Special Report on "Guidelines for National Greenhouse Gas Inventories."[12] This formed the basis of the UNFCCC process to establish and report national **greenhouse gas inventories**, an essential part of the **UNFCCC** national communications requirement. National inventories are essential to identifying the sources of, and underlying trends in, global greenhouse gas emissions and the benchmark against which progress in reducing emissions is assessed (see **greenhouse gas inventories**).

In 1995, SBSTA requested IPCC to undertake further development of inventory guidelines and spurred the IPCC to establish the National Greenhouse Gas Inventories Program. This program aims to develop and refine an internationally agreed methodology for the calculation and reporting of national greenhouse gas emissions and removals and to encourage the use of this methodology by Parties to the UNFCCC.

The IPCC Guidelines for National Greenhouse Gas Inventories were first released in 1994 and, recognizing that processes are constantly improving and changing, have been updated and revised periodically since. The Revised 1996 Guidelines were important as they formed the basis for calculating the **Kyoto Protocol** differentiated targets. Additional good practice guidance were approved in 2001 and 2003, followed by the fully revised 2006 IPCC Guidelines.[13]

While continual updating and improvement has enabled more accurate national inventories to be compiled, it has also led to some confusion since the **Kyoto Protocol** uses the **global warming potentials** (GWPs) and methodologies of the 1996 Guidelines to establish whether Parties have met or exceeded their targets rather than the updated methodologies. As such, the **Kyoto Protocol** uses a static set of calculation procedures, while the UNFCCC is continually updating. All of the calculations relating to Kyoto targets or flexibility mechanisms of the **Clean Development Mechanism (CDM)** and **Joint Implementation (JI)** use the GWPs of the 1996 Guidelines and will do so until at least 2012 (the end of the Kyoto compliance period).

See also: dangerous climate change, global warming potentials, greenhouse gas inventories, Kyoto Protocol, United Nations Framework Convention on Climate Change (UNFCCC).

Notes

1 The Norwegian Nobel Institute 2007
2 Stern 2006
3 IISD 2007
4 Ibid.
5 IPCC 2007
6 Ibid.
7 See http://www.ipcc.ch/pub/reports.htm
8 Ibid.
9 Ibid.
10 IPCC 1996
11 IPCC 2001
12 IPCC 1996
13 IPCC 2006b

Further reading

UNEP/WMO 2004; IPCC 2007; IISD 2007.

JOINT IMPLEMENTATION (JI)

Joint Implementation (JI) is one of the three flexibility mechanisms defined by the **Kyoto Protocol**.[1] The other two flexible mechanisms are the **Clean Development Mechanism (CDM)** and International Emissions Trading. JI, like the CDM, is a project-based "baseline and credit" emissions trading mechanism and shares many other similarities with the CDM including that the number of credits a JI project generates is calculated as the baseline emissions (those that would have occurred in the absence of the project) minus the project emissions (see **emissions trading**). In contrast to CDM, JI involves trading credits between Annex I Parties (countries with emission targets under the **Kyoto Protocol**), rather than between Annex I and non-Annex I Parties (developing countries without targets under the **Kyoto Protocol**).

JI can be used by Annex I Parties "for the purpose of meeting [their] commitments ..." of emission reduction targets (see **UNFCCC**). As the name suggests, JI is a mechanism by which an Annex I Party can host a project that generates greenhouse gas emission reduction credits that are subsequently acquired by another Annex I Party. Put simply, JI provides flexibility for Annex I Parties to meet their Kyoto commitments without relying solely on reducing emissions domestically.

If a project satisfies the JI project approval criteria, each tonne of **carbon dioxide equivalent (CO$_2$e)** reduced by the project generates a credit, called an Emission Reduction Unit (ERU). ERUs are transferred to another Annex I Party (which is often also the project investor), which can use the ERU as a credit toward its **Kyoto Protocol** target. JI projects can also include carbon sequestration projects (such as afforestation and reforestation) that generate Removal Units (RMUs), which can also be used to meet Kyoto Protocol obligations (see **biosequestration**).

Since an ERU is equivalent to one Assigned Amount Unit (AAU) (see **Kyoto Protocol**), an Annex I Party can exceed its Kyoto target and make up the difference with sufficient emission credits to cover excess emissions. These emission credits can take the form of ERUs or RMUs from JI, Certified Emission Reductions (CERs) from the **CDM**, or AAUs through International Emissions Trading (see **Kyoto Protocol** and **CDM**).

How does JI operate?

JI is administered by the **Kyoto Protocol** Meeting of Parties (MOP) and supervised by the Joint Implementation Supervisory Committee (JISC). The ERUs arising from a JI project may come from any sector of the economy, provided they meet the requirements of:

- ratification of the **Kyoto Protocol** by each Party involved;
- approval of each Party involved; and
- reductions in emissions are additional to any that would otherwise occur.

These requirements were agreed at the third Conference of Parties (COP-3), Kyoto, in 1997, as part of the **Kyoto Protocol** negotiations. However, since JI could only commence after ratification of the Protocol, subsequent elaboration of the JI rules and procedures did not proceed until the first Meeting of Parties (MOP1) in Montreal, Canada, in 2005 (the first meeting after the Protocol had come into force). At MOP1, the JI Steering Committee was formally established and mandated to elaborate the operating procedures for JI.

JI operates in an **emissions trading** system where the total number of Annex I emission allowances remains fixed (capped) by the Kyoto Protocol emissions targets. Since buyer and seller Parties each have emission targets under the **Kyoto Protocol**, any ERUs transferred out of a host country must have the same number of AAUs subtracted from that country's allocation in order to maintain the integrity of the Kyoto targets (see **emissions trading**).[2] By contrast, the **CDM** generates emission allowances that are additional to the fixed quantity Annex I emission allowances.

While it is clearly in the interest of project proponents to maximize the number of ERUs – and hence revenue – generated by a JI project, this does not necessarily align with the host government's interests since any overcrediting of ERUs to a JI project will be subtracted from the host Party's AAUs. The assurance that ERUs do, in fact, represent emission reductions is therefore contingent on the integrity of the host Party's **greenhouse gas inventory**.

The JI project cycle

There are two options available to Annex I countries in terms of how JI projects are approved and monitored, termed Track 1 and Track 2. The steps a JI project proponent must complete prior to being issued ERUs varies according to whether Track 1 or Track 2 procedures are adopted. The choice of procedures also determines which entities are involved in the project cycle.

Track 1

Under this procedure, host Parties can approve projects directly (without the need for independent validation, verification, or JISC involvement) as well as issue ERUs, resulting in faster and simpler approval of projects. However, to utilize the Track 1 procedure, the host Party and the Party who will be the recipient of the ERUs are required to have:

- calculated and recorded their assigned amounts;[3]
- in place a national system for estimating emissions of all **greenhouse gases**;[4]
- in place a national registry and to make any additions to, and subtractions from, the assigned amount as a result of buying or selling ERUs;[5] and
- submitted to the UNFCCC annually the most recent required inventory, including the national **greenhouse gas inventory** report in the common reporting format.

If a host Party fulfills all of these eligibility criteria, Track 1 JI may be applied. That is, "a host Party may verify reductions in anthropogenic emissions ... [from a JI] project as being additional to any that would otherwise occur ... [and] then the host Party may issue the appropriate quantity of ERUs."[6]

Host Parties that do not meet these requirements must follow Track 2 JI. Parties that have met Track 1 requirements may still use Track 2 if they choose.

Track 2

Under this procedure, JI project approval and ERU issuance adopt a similar procedure to the **CDM**: the project must be validated and verified by an independent third party (an Accredited Independent Entity), and ERUs are only issued by the host Party after approval from the JISC.

By the start of 2008, all JI projects utilized Track 2 procedures as there is little incentive for countries to establish the required administrative infrastructure to support Track 1 since independent project verifiers and the JISC are already effectively operating and a clear set of administrative procedures are in place to ensure that emission crediting is accurate and consistent.[7]

For a project to get to a point where it is issued ERUs (or RMUs), a series of specific steps need to be completed:

Preparation of a Project Design Document (PDD)

A PDD is required for both Track 1 and Track 2 procedures and is the responsibility of the project proponent. The project proponent is the entity seeking approval for a JI project and can be from the private or public sector (or a combination of both). As with the **CDM**, project proponents often prepare a Project Idea Note (PIN) that can be used to attract potential investors (generally an Annex I Party that will purchase the ERUs generated by the project and who also often underwrites the project). Formulating a PIN is common but not a formal requirement in the JI project cycle.

The JI PDD adopts a similar format as the CDM PDD and must utilize a baseline methodology approved by the JISC (which include all approved CDM project methodologies).[8] Like the **CDM**, JI projects must demonstrate that they meet the "additionality" criterion: essentially, that the emission reductions would not have occurred without the project (see **CDM** for a discussion of additionality).

Participating Annex I Party Approval

Once a PDD has been completed, the project must be approved by both the host government's Designated Focal Point (DFP) and the DFP of the Annex I Party to which ERUs will be transferred.[9] DFP approval ensures that the project adheres to national priorities. Thus, the host Party can block projects that it deems unsuitable, and this ensures that participation of the Parties is voluntary.

For Track 2, project proponents must complete several additional steps.

Project validation

An Accredited Independent Entity (AIE), which is an independent, third-party organization with relevant expertise in greenhouse gas emission reduction projects, validates the project. Validation involves checking the PDD to ensure that an appropriate methodology is used, that emission reduction estimates are reasonable, and project monitoring plans are sound. The purpose of the AIE validation is to provide a statement on the project's reasonableness regarding future emission reduction estimates. The **CDM** established essentially the same role for Designated Operational Entities (DOEs), and the JISC agreed that DOEs may operate provisionally as AIEs.[10]

PDD submission to the JISC for final determination

When the PDD has been granted DFP approval and has been validated by an AIE, it must be submitted to the JISC for "final determination" (essentially equivalent to project registration under the **CDM**). Once final determination has been granted, the project can be implemented as a JI project.

Project verification

Once the project has been operational for a period of time (usually one year), the project proponent prepares a monitoring report that quantifies the emission reductions delivered by the project. This must be verified (assessment of the project to confirm the validity of the report's assertions) by an AIE who provides a statement (or "opinion") in a verification report. As with **CDM**, to avoid conflict of interest, the AIE that verifies the report must be different from the AIE that validated the project.

Issuance of ERUs

The verification report and monitoring report is submitted to the JISC, which determines whether the host Party is authorized to issue and distribute ERUs. In the case of Track 1, it is the host government that issues the ERUs; in the case of Track 2, they are also issued by the host country, but only after JISC approval.

This cycle repeats, with the project proponent preparing regular monitoring reports, having emission reductions verified, and being issued ERUs.

JI crediting periods

Since ERUs are exchanged for host Party AAUs, JI projects can only generate ERUs for the duration of the five-year commitment period of the **Kyoto Protocol**, 2008–2012. Projects can be recognized with start dates as early as 2000, but ERUs from these projects can only be issued from the start of 2008.[11] This limited crediting period reduces the financial attractiveness of JI projects as they must generate sufficient credits in five years to warrant the investment – this is much shorter than is available to **CDM** projects.

While it is possible that the crediting period of JI projects could be extended beyond 2012, there is no certainty that this will occur. Crediting beyond 2012 will depend on whether any international agreement that is negotiated to follow the **Kyoto Protocol** includes provisions for JI projects.

JI project numbers and locations

Although the **Kyoto Protocol** came into force in February 2005, it took until mid-2006 before the JISC rules and procedures were sufficiently well established to enable projects to proceed. Furthermore, it also took time for Annex I Parties to also establish their DFP procedures (by late 2007, only 27 of the 39 eligible Annex I Parties had formally established DFPs),[12] and it was not until March 2007 before the first JI project had received final determination.[13]

By the start of the Kyoto commitment period, the number of JI projects had grown to nearly 200, which in aggregate could generate as much as 200 million ERUs over the period to end of 2012.[14] Projects have been proposed in 12 of the 15 sectoral categories.[15] Almost half of these projects are **renewable energy**, including biomass/biogas energy, hydropower, and **wind power**. **Methane** capture (from fugitive emissions and landfill) and **energy efficiency** make up almost all of the remainder.[16] The diversity of project types indicates the wide range of cost-effective opportunities for reducing greenhouse gas emissions in Annex I countries. Compared with the CDM project mix, JI projects focus more on fugitive emission reduction projects (e.g. fixing leaks in gas pipelines) and **energy efficiency** in industrial processes and district heating – a reflection of the significant infrastructure upgrades and efficiency opportunities in the major JI project host countries.

While all 39 Annex I Parties that have ratified the **Kyoto Protocol** are eligible to host JI projects (United States is not eligible), it was always expected that the majority of JI projects would be located in the Economies In Transition (EIT) countries of the ex-Soviet Union since their industrial infrastructure was generally less energy efficient and many EIT countries face capital availability constraints and welcome foreign investment and technology inflows.

Initially, the EU accession countries of Eastern Europe led JI initiation of "early mover" projects, but by 2007 Russia and Ukraine (the second- and fifth-largest Annex I emitters in 1990, respectively) had become the two largest JI host countries. Bulgaria, Czech Republic, Poland, Romania, Hungary, and Estonia account for most of the rest. Of the 13 eligible EIT countries, only Croatia and Slovenia (both small emitters) had not accessed JI by early 2008.[17] The only non-EIT JI host countries are New Zealand and Germany, though France is considering JI projects.[18]

As with the **CDM**, there is no penalty under the formal JI process for projects failing to meet the emission reductions projected in their PDDs. If projects exceed their projected ERU production, they will be issued ERUs. These surplus ERUs will be issued provided the monitoring and verification approaches laid out in the PDD are adhered to. The matter of nondelivery is less straightforward. JI is a market mechanism involving contracts between buyers and sellers. In many of these contracts, the buyer requires the seller to guarantee delivery of all, or some proportion, of the ERUs projected in the PDD. If a project fails to deliver these ERUs, the project proponent may be contractually required to supply any shortfall from other sources, such as purchase from the open market, which will most likely be more expensive. This represents a financial risk to project proponents, who, therefore, have a strong incentive to ensure that the project performs as planned.

The ERU market

As with CERs generated through the **CDM**, the demand for ERUs is largely determined by the expected shortfall in Annex I AAUs during the commitment period (see **Kyoto Protocol**), the price, and the buyers' perception of project risks when they are negotiating the Emission Reduction Purchase Agreements (ERPAs) at the outset of the JI project. As ERPAs are usually signed before any ERUs have been generated, buyers face many uncertainties, such as: the project may not achieve final determination with the JISC; future market prices may evolve unfavorably; the project may not produce as many ERUs as expected in the time frame; the project may not be properly managed; environmental and stakeholder opposition to the project may arise; and the host country may not provide a stable investment climate owing to natural disasters, political changes, and conflict. These risks are essentially the same as for **CDM** projects.

However, JI projects face a more limited crediting period (restricted to 2008–2012) than **CDM** projects, which poses a greater risk that any delays in implementation will reduce the potential revenues from ERUs. Despite these different market risks between CERs and ERUs, forward purchase prices (prior to 2008) are similar, and ERUs generally traded in the €5–7 (US$ 7–10) range during 2006–2007.[19]

The JI market is dominated by the governments of just a few countries, mainly the Netherlands, Denmark, and Austria, although several other European countries and Japan have also committed to purchasing ERUs.[20] Unlike the **CDM**, where a small but increasing volume of CERs have been purchased in the voluntary market (e.g. corporations seeking to be "carbon neutral" – see **CDM**), ERUs have been exclusively purchased by, or on behalf of, Annex I governments (e.g. the World Bank has made forward commitments to buy ERUs, which will be later sold to Annex I governments).

What contribution will JI make to meeting Annex I emission commitments?

Given that the likely aggregate shortfall in AAUs for Annex I Parties could be as high as 5 billion AAUs by 2012, there is likely to be reasonably strong demand for ERUs during the commitment period.[21] JI projects could generate as many as 200 million ERUs over the period 2008–2012 (equivalent to 200 $MtCO_2e$ of emission reductions – roughly the emissions of The Netherlands for one year). If these ERUs eventuate, they will make a moderate, but valuable, contribution to Annex I endeavors to comply with their emission targets. More JI projects may commence after 2008, but the JI crediting period of 2008–2012 means that the incentive to develop JI projects diminishes rapidly during the first commitment period.

It appears that JI will be moderately successful as a flexibility mechanism but that it is only likely to contribute less than one-tenth of the emission credits that are expected from the **CDM**. However, the environmental integrity of JI projects is arguably more secure than that of CDM projects since ERUs do not create new credits but transfer credits within a total amount capped by **Kyoto Protocol** commitments.

JI has also facilitated the flow of investment capital and more efficient technologies to many EIT economies, most of which has gone to projects that utilize low emission technologies such as **renewable energy** and **energy efficiency**, which would have been unlikely to proceed in the absence of JI. Thus, it appears that JI will contribute to emission reductions and to the development of more climate-friendly economic systems in many EIT countries.

See also: Clean Development Mechanism (CDM), emissions trading, Kyoto Protocol.

Notes

1 UN & UNFCCC, Kyoto Protocol, Article 6
2 Ibid.
3 Ibid., Article 3

4 Ibid., Article 5

5 Ibid., Article 7

6 UNFCCC 2005, CMP1 Report, http://ji.unfccc.int/Eligibility

7 UNFCCC, JI Home page, http://ji.unfccc.int/index.html

8 UNFCCC CDM Statistics http://cdm.unfccc.html

9 UNFCCC 2005 CMP1 Report

10 Ibid.

11 UNFCCC 2005, CMP1, Decision 9, http://ji.unfccc.int/Eligibility

12 UNFCCC JI Home page, http://ji.unfccc.int/index.html

13 UNFCCC March 27, 2007

14 UNEP Risoe Centre

15 Ibid.

16 Ibid.

17 Ibid.

18 Caisse des Dépôts 2006

19 World Bank 2007b

20 Ibid.

21 UNDP 2006a

Further reading

UNFCCC JI Web site; World Bank 2007b; UNEP Risoe Centre CDM/JI Web site.

KYOTO PROTOCOL

The Kyoto Protocol (the Protocol) is an international agreement linked to the **United Nations Framework Convention on Climate Change (UNFCC)** (the Convention). The Kyoto Protocol's preamble states that "Parties to this Protocol ... in pursuit of the ultimate objective of the Convention ... have agreed ...." Thus, the objective of the Protocol is to contribute to the objectives of the Convention, which include "... to stabilize greenhouse gas concentrations at a level that would prevent dangerous anthropogenic interference with the climate system."[1]

In contrast to the voluntary commitments under the Convention, the Protocol includes quantified greenhouse gas emission limits on Annex I Parties (industrialized, developed countries) that are signatories to the Convention. In aggregate these commitments require Annex I Party emissions to be reduced by at least 5% below 1990 levels by 2010, though there is considerable variation in the emission targets of different countries.

The Protocol does not impose emission limits on non-Annex I Parties (mainly developing countries) in accordance with the Convention's principle of "common but differentiated responsibilities."[2] This principle acknowledges that the largest share of historic emissions (and hence **global warming**) is attributable to Annex I countries (see **emissions per capita** and **anthropogenic greenhouse gas emissions**) and reflects the greater development needs of non-Annex I countries.

Although one of the main objectives of the Protocol is to reduce emissions from Annex I countries, it also serves several other objectives, including:

- creating flexible market mechanisms through which Annex I countries can achieve emission reductions (such as International Emissions Trading, the **Clean Development Mechanism [CDM]**, and **Joint Implementation [JI]**);
- providing an internationally consistent framework through which countries quantify and report their **greenhouse gas inventories**; and
- to introduce a range of instruments and measures to assist non-Annex I countries to reduce emissions and accommodate the impacts of climate change (e.g. the Adaptation Fund, the Special Climate Change Fund, and the Least Developed Country Fund).

The Kyoto Protocol was never intended to completely address climate change, but was, rather, intended as a first step, forming a basis for international cooperation for the longer-term goal of avoiding **dangerous climate change**.

That the Protocol was successfully negotiated just four years after the Convention underscores the importance that the international community attached to climate change and, in comparison to other international conventions and protocols, represented rapid progress. Although negotiated and signed in 1997, it did not come into force until 2005 when sufficient countries had ratified the Protocol. By the commencement of the Protocol's "First Commitment Period" (2008–2012), 175 countries and one regional organization (the European Community, EC) had ratified the Protocol.[3]

How does the Protocol operate?

Countries that wish to be a "Party to the Kyoto Protocol" must first have ratified the Convention (that is, be "Parties" to the Convention). The Protocol operates using the same administrative structures and bodies as the Convention, including the Conference of the Parties (COP); the President and Bureau; Subsidiary Bodies; Secretariat; and other committees, working groups, and expert bodies (see **UNFCCC**).

Parties to the Protocol meet annually in conjunction with the Convention COP. In the wording of the Protocol, these gatherings are termed "the Conference of the Parties serving as the Meeting of the Parties to this Protocol" but generally known as the Meeting of the Parties (or MOP). The COP and MOP have together been called the "COP/MOP," though since 2007 this has been refined to the 'CMP' (Conference and Meeting of the Parties).

There are some differences in operation between the Protocol and the Convention. Parties to the Convention that are not Parties to the Protocol can only participate as observers and not in any decision making. While the Convention must take decisions by consensus, the Protocol has a voting mechanism that enables decisions to be made if a three-fourths majority vote is secured – each Party present at MOP sessions has one equal vote.[4] In practice, the President of the MOP is generally reluctant to take issues to vote but having the option of a vote helps to prevent individual or small groups of countries from blocking changes that the majority of Parties seek.

The role of the Kyoto Parties

The Protocol recognizes the same groups of Parties as the Convention – and all Parties are divided into Annex I (industrialized countries) or non-Annex I (developing countries). The emission targets of Annex I Parties are listed in Annex B of the Protocol, and Annex I Parties are officially termed Annex B Parties. Annex B Parties of the Protocol are identical to Annex I Parties of the Convention, with two exceptions: Belarus, which has yet to establish a 1990 emissions baseline (against which emission commitments can be assessed) and Turkey, which argued that its emissions would need to grow further to meet human development needs, and should not, therefore, have a target. In total, there are 39 Annex B Parties under the Protocol, compared with 41 Annex I Parties under the Convention.

The Kyoto Protocol relates predominately to Annex I Parties, each of which commits to a target expressed as a percentage of its 1990 emissions. While the 15 countries that were members of European Economic Community (EEC) in 1997 undertook to adopt an aggregate collective target of 92% of 1990 emissions (the "EU bubble"), the European Union does not actually have an official target and individual EU member states remains responsible for meeting their own targets.

Annex I emission targets

Targets were originally set for 2010, but it was recognized that judging success against a single year's emissions may not be indicative of an overall trend in emissions. To overcome this, targets were set as an average over the five-year period 2008–2012 (termed the First Commitment Period). Each Annex I Party is assigned a target percentage of the 1990 baseline emissions (in tonnes of **carbon dioxide equivalent, CO_2e**) multiplied by five. This allocation is termed the "assigned amount" and each "assigned amount unit" (AAU) is equivalent to one tonne of CO_2e. To honor its Kyoto commitment, each Annex I party cannot emit more than its assigned amount during the five-year commitment period.

For example, Belgium has committed to reducing its emissions to 92% of its 1990 emission levels (see Table 14). In 1990, Belgium's emissions were 144.3 Mt CO_2e, thus Belgium's average annual emissions target over the period 2008–2012 is $0.92 \times 144.3 = 132.8$ $MtCO_2e$. The aggregate five-year target for Belgium is $132.8 \times 5 = 664$ $MtCO_2e$, or 664 million AAUs. To honor its Kyoto commitment, Belgium cannot emit more than 664 $MtCO_2e$ during 2008–2012. It is the five-year period as a whole that provides the relevant accounting period. Belgium could emit more than the annual average (132.8 $MtCO_2e$) in 2008, and less than 132.8 $MtCO_2e$ in 2010, provided it emits less than 664 $MtCO_2e$ over the entire 2008–2012 period. In reality, Belgium (or any other Annex I party) could exceed its emission target (assigned amount) and still comply with the Protocol by sourcing sufficient additional emission allowances from one of the three Kyoto flexible mechanisms (discussed below) to cover its excess emissions. If it is unable to secure these additional emission allowances, Belgium would not comply with the Protocol.

In order to determine whether or not Parties meet their commitments, the Protocol requires Annex I Parties to prepare national **greenhouse gas inventories** according to predefined methodologies and approaches accepted by the **Intergovernmental Panel on Climate Change (IPCC)**. These inventories are more rigorous than the National Communications required under the Convention and require quantification, reporting, and tracking over time of the **greenhouse gases** listed under Annex A of

Table 14 Kyoto Protocol emissions data including land use, land-use change, and forestry
(LULUCF)[5]

Annex B Parties to the Kyoto Protocol	1990 Base year (MtCO₂e)	Emission target (% of base year)	2005 Emissions including LULUCF (MtCO₂e)	2005 Emissions including LULUCF (% of base year)
Australia	499.9	108	522.2	104.5
Austria	67.1	92	76.3	113.6
Belarus	105.3	—	50.7	48.1
Belgium	144.3	92	143.5	99.4
Bulgaria	127.6	92	52.0	40.7
Canada	473.3	94	729.7	154.2
Croatia	16.7	95	13.1	78.6
Czech Republic	194.5	92	141.0	72.5
Denmark	71.0	92	64.0	90.2
Estonia	33.3	92	12.8	38.6
Finland	49.6	92	38.3	77.2
France	533.3	92	495.4	92.9
Germany	1,199.6	92	965.4	80.5
Greece	105.5	92	132.2	125.3
Hungary	112.6	94	75.7	67.3
Iceland	5.4	110	5.5	100.3
Ireland	55.5	92	69.3	124.9
Italy	437.0	92	469.5	107.4
Japan	1,179.9	94	1,263.9	107.1
Latvia#	5.8	92	−3.6	−161.5
Liechtenstein	0.2	92	0.3	118.4
Lithuania	38.6	92	13.6	35.2
Luxemburg	12.4	92	12.5	100.4
Monaco	0.1	92	0.1	96.8
The Netherlands	215.4	92	214.5	99.6
New Zealand	42.9	100	52.6	122.7
Norway	35.0	101	26.9	76.9
Poland	554.0	94	366.8	66.2
Portugal	63.7	92	89.5	140.3
Romania	249.8	92	116.2	46.5
Russian Federation	3,166.4	100	2,289.2	72.3
Slovakia	69.7	92	47.0	67.5
Slovenia	18.7	92	15.0	79.9
Spain	244.6	92	391.0	159.8
Sweden	68.7	92	63.0	91.8
Switzerland	51.0	92	53.4	104.6
Turkey	126.5	—	222.5	175.9
Ukraine	872.4	100	360.4	41.3
United Kingdom	774.3	92	655.4	84.6
United States Of America*	5,529.2	93	6,431.9	116.3

Notes: Bulgaria, Hungary, Poland, Romania, and Slovenia use earlier years than 1990 for baseline due to lack of data availability in 1990. Data for 2005 emissions inventories (the most comprehensive set currently available) is not available for Croatia, Greece, and Turkey, where 2004 emissions data has been used.
*Has not ratified the Kyoto Protocol.
#Latvia's net sequestration exceeds its emissions.

the Protocol: **carbon dioxide (CO₂)**, **methane (CH₄)**, **nitrous oxide (N₂O)**, hydro-fluorocarbons (HFCs), perflurocarbons (PFCs), and sulphur hexafluoride (SF₆) – the so-called Kyoto gases (see **greenhouse gases**). Parties also include quantification of their emissions and atmospheric removals (sequestration) arising from "direct human-induced land-use change and forestry activities [of] afforestation, reforestation and deforestation since 1990."[6]

Each Annex I Party must have in place before 2007 "a national system for the estimation of anthropogenic emissions by sources and removal by sinks of all greenhouse gases," and the **greenhouse gas inventory** which this system produces must be submitted annually during the commitment period of 2008–2012. In practice, Annex I Parties combine their National Communications requirements under the Convention with their Protocol inventory requirements rather than providing two separate reports. The information submitted is subject to review by expert teams nominated by the Parties and coordinated by the UNFCCC Secretariat. Their role is to "provide a thorough and comprehensive technical assessment of all aspects of the implementation by a Party of [the] Protocol."[7] Any problems identified by the expert review teams are reported to the Secretariat, which circulates this to all Parties.

Non-Annex I Parties to the Kyoto Protocol do not have emission reduction targets or any new commitments beyond those contained in the Convention, based on the principle of common but differentiated responsibilities.[8] Non-Annex I Parties can actively participate through the Protocol's flexibility mechanisms (see below) by hosting emission reduction projects under the **Clean Development Mechanism (CDM)**.

Why do different Parties have different targets?

The Convention and Protocol's principle of "common but differentiated responsibilities" does not apply solely to the distinction between Annex I and non-Annex I countries (industrialized and developing countries) but also seeks to recognize differing circumstances within the group of Annex I Parties.

There are many factors that determine the greenhouse emissions of individual countries, including national economic structure; primary energy supply mix and natural resource availability; energy intensity and energy efficiency of the economy; land-use and agricultural practices; and broader socioeconomic factors such as population and economic growth rates (see **emissions intensity**). A "one-size-fits-all" target for all Annex I Parties would fail to recognize these legitimate differences and could result in large differences between Parties in the cost of meeting emission targets (see **mitigation**).

In initial negotiations, several Parties sought a transparent and rational basis for target setting. For example, Brazil proposed that targets should be set in accordance with their proportional contribution to climate change based on the impact of their cumulative historical emissions since 1840 (see **emissions per capita**).[9] While appealing from an equity standpoint, the proposal apportions liability to specific countries for something that they could not have known the consequences of, and hence should not necessarily be entirely responsible for, their greenhouse gas emissions prior to the 1970s (when the buildup of greenhouse gases in the atmosphere was first recognized as a potential problem), let alone prior to 1900.

A key element of the Protocol negotiators sought was some parity in the cost to each Party of meeting their targets. It was recognized that some Annex I Parties, whose economies and emission inventories are driven by land-use, agriculture, and fisheries (such as New Zealand and Iceland), have fewer cost-effective emission mitigation options than those countries with a higher proportion of emissions emanating from fossil fuel-based (CO_2-intensive) energy, where mitigation technologies are generally more readily available (see **emissions per capita** and **emissions intensity**). Ultimately, no systematic "formula" for target setting was employed to establish Party targets beyond the collective "at least 5% below 1990 levels" of Article 3 of the Protocol – targets were politically negotiated.

Only six of the 39 Annex I signatory countries successively negotiated either an increase in their allowable emissions – Iceland (110%), Australia (108%), and Norway (101%) – or constant emission levels (Russia, Ukraine, and New Zealand, 100%). This was primarily due to specific circumstances they faced, such as a high dependence on agriculture and/or fossil fuel industries – though Australia also argued that allowance should be made for its relatively high population growth rate.

The mainstream media has often portrayed these higher emission targets as overly generous (and many still do), but this is a somewhat simplistic view of the complexity of the emission reduction issue. It is also often asserted that the absence of emission targets for non-Annex I countries undermines the value of the Protocol. However, the diversity of Annex I targets and the absence of targets for Non-Annex I Parties are consistent with the Convention principle of "common but differentiated responsibilities." Any workable international agreement must reflect the differences in national circumstances, such as the ability to pay, the level of economic development, and different costs of emissions mitigation across different countries. In reality there is a strong rationale for differentiated targets set for different Parties under the Protocol.[10]

The ratification process

To provide an incentive for countries to ratify the Protocol and ensure that ratification truly represented global collective action, specific ratification requirements were agreed to in the Protocol. Specifically, the rules for entry into force of the Protocol required:

(i) 55 Parties to the Convention to ratify (or approve, accept, or accede to) the Protocol and
(ii) that the ratifying Annex I Parties should account for at least 55% of total Annex I emissions in 1990.[11]

Due to these two requirements, the Protocol could not enter into force unless at least 16 non-Annex I Parties ratified the Protocol, and a majority of Annex I Parties' emissions were included. These requirements also prevented non-Annex I Parties from bringing the Protocol into force without corresponding ratification by the majority of Annex I Parties and prevented any single Annex I Party from "vetoing" the entry into force of the Protocol since no Party's emissions in 1990 were more than 45% of the Annex I total (see Table 14). Without these requirements a situation may arise where only a small number of countries would shoulder most of the mitigation burden. These provisions provided some security that ratifying countries would not be held accountable unless, and until, there was a critical mass of countries and sufficient coverage of Annex I emissions.

Tuvalu and several other island nations ratified the Protocol in 1998, while Denmark and Finland were the first Annex I Parties to ratify the Protocol, in May 2002. By 2003, the first requirement (55 Parties to ratify) had been met, but the second requirement (at least 55% of Annex I emissions) had not – either Russia or the United States had to ratify to meet the second threshold. In 2001, the United States announced it would not ratify the Protocol, which left entry into force hinging on Russian ratification.[12] The EU in particular put considerable effort into lobbying Russia to ratify, drawing Russia's attention to the potential economic benefits of the flexibility mechanisms of International Emissions Trading and **Joint Implementation**. After several years in which it seemed the Protocol would not enter into force, Russia ratified in November 2004, taking the percentage of Annex I 1990 emissions to 61.6% and exceeding the second threshold. The Kyoto Protocol formally entered into force 90 days later, on February 16, 2005.

Although a majority of countries have ratified the Protocol, there are still 15 countries that have not: 14 non-Annex I countries (four of which have yet to ratify the Convention and therefore cannot ratify the Protocol) and one Annex I Party, the United States.[13] Although Australia had previously announced that it did not intend to ratify, it reversed this decision and ratified in December 2007. None of the non-Annex I parties that have failed to ratify are major greenhouse gas emitters, and they have little bearing on the success or failure of the Protocol. However, the failure of the United States to ratify has serious ramifications for the perceived success or failure of the Protocol. For example, the United States accounted for approximately 40% of Annex I emissions in 2004 – their failure to ratify obviously greatly reduces the impact of the Protocol.

The most important and far-reaching ramification of the United State's nonratification stance is the impact on the post-2012 international climate change negotiations. High- and middle-income non-Annex I countries (such as South Korea and Singapore among others) may see little justification for taking on commitments if one of the world's most affluent countries has not committed to demonstrable action. If **dangerous climate change** is to be avoided, it is essential that Annex I Parties make serious emission reductions and that non-Annex I countries progressively take on commitments to limit their emissions post-2012. However, until such time as all Annex I countries make a serious commitment to reducing their emissions, the willingness of non-Annex I countries to take on commitments is likely to be limited.

Noncompliance penalties

There are no specific financial penalties for Annex I Party noncompliance, even though some parties argued for specific financial penalties during the Protocol negotiation process. It was considered that it would be difficult to enforce these penalties without significantly complicating the Protocol and raising potential conflicts with other international agreements. Nonetheless, the Protocol has one of the most advanced set of compliance rules and procedures of any of the Multilateral Environment Agreements (MEAs) to date.[14] The rules on noncompliance were largely agreed at COP-7 in Marrakech and adopt a consultative and conciliatory approach, rather than a punitive approach, to compliance enforcement. This is a similar approach to the Montreal Protocol (see **ozone**).[15]

There are no provisions for what happens to a Party with identified compliance problems, except that the MOP "shall take decisions on any matter required for the implementation of this Protocol."[16] To address this, the first MOP established a Compliance Committee, consisting of a facilitative branch and an enforcement branch. The facilitative branch aims to help Parties comply by advising on technical, financial, or capacity building measures. The enforcement branch determines whether Parties are complying with their obligations, determines consequences for Parties that are not, and determines whether and how to adjust Parties' inventories based on expert team reviews.

The Protocol's 2008–2012 commitment period has always been intended to be the first of a series of commitments, with ongoing reduction targets. Parties agreed that those with a shortfall of AAUs would carry the shortfall into the next commitment period with a 30% penalty and would provide a detailed compliance plan. For example, if a Party missed its target by 10 MtCO2e (a shortfall of 10 million AAUs), it would start the next commitment period with 13 million fewer AAUs (shortfall plus 30%) than its negotiated target allocation.

There are several incentives and disincentives in place to entice Annex I Parties to honor their commitments. For example, unused emission allowances (surplus AAUs) are allowed to be "banked" (rolled over) for use in any subsequent commitment period – thus providing an incentive for Parties to institute strong mitigation policies for the first commitment period. However, Parties that have a deficit of allowances are not able to "borrow" AAUs from any subsequent commitment period – thus providing a disincentive to postpone action until future commitment periods.

While the commitments of the Kyoto Protocol are theoretically enforceable under international law, any enforcement, and the acceptance of this enforcement by the recipient Party, depends on a political will to do so – the Party must be willing to recognize the international rulings and processes and agree to abide by them. The adoption of any trade or other sanctions against the noncompliant party is not catered for under the Kyoto Protocol and would necessarily overlap with the domain of the World Trade Organization and other multilateral bodies. Overall, implementing noncompliance provisions against noncompliant Parties after 2012 is likely to be problematic. Nonetheless, the political reputation of countries deemed noncompliant could be diminished in the eyes of the international community, which may have both domestic and international political repercussions.

The Kyoto Protocol flexibility mechanisms

To assist Annex I countries to meet their Protocol emission commitments, the Protocol established three "flexibility mechanisms." As the name suggests, these mechanisms offer flexibility to Annex I Parties as to how they meet their emission obligations under the Protocol. Through these mechanisms, Parties with insufficient AAUs are able to purchase AAU equivalent units from other Annex I Parties, through International **Emissions Trading** and **Joint Implementation (JI)**, or from Non-Annex I Parties, through the **Clean Development Mechanism (CDM)**. As the cost of mitigation varies between countries, it may prove cheaper for some Annex I countries to reduce greenhouse gas emissions in other Annex I and non-Annex I countries rather than solely relying on domestic measures – the flexibility mechanisms provide this alternative mitigation option.

The inclusion of the flexibility mechanisms was a contentious issue during Protocol negotiations. The United States actively negotiated for the inclusion of market-based mechanisms, partly based on their experience with delivering cost-effective reductions of sulphur dioxide emissions through an **emissions trading** instrument. Initially most European countries, and many NGOs, were opposed to the inclusion of the flexibility mechanisms as they considered this would enable countries to "buy their way out" of their commitments, rather than undertaking emission reductions. Eventually, negotiators agreed that, provided the overall 5% reduction target of the Protocol was met, it should not matter where reductions are undertaken. The Protocol does, however, state that "trading shall be supplemental to domestic actions for the purpose of ... commitments," though the meaning of "supplemental" is not defined in the Protocol. Ironically, the original champion of emissions trading, the United States, is unable to access the flexibility mechanisms since it has not ratified the Protocol.

Overall, at least 7 billion tradable units could be available through the three flexibility mechanisms by the end of the first commitment period. The eventual supply will depend on the growth in emissions in the EIT countries over the period to 2012 (which will determine the number of surplus AAUs) and the number of CERs generated through the CDM. See **emissions trading**, the **Clean Development Mechanism**, and **Joint Implementation** for more detailed overviews of the flexibility mechanisms.

Will Annex I Parties meet their emission targets?

Although the Kyoto Protocol has other objectives beyond just the "at least 5% reduction" in aggregate Annex I emissions, the Kyoto targets are, nonetheless, perceived as the primary objective and are of considerable symbolic importance. Should Annex I Parties fail to meet their emission reduction commitments, it would seriously undermine the credibility of the Protocol and, more important, potentially weaken the resolve of Annex I and non-Annex I Parties to take on emission limitation commitments beyond 2012, which is essential if **dangerous climate change** is to be avoided.

Based on the most recently available national inventory data (2005) and projected emission levels over the period to 2012, approximately half of the Annex I countries that have ratified the Protocol will face AAU deficits at the end of the first commitment period. As most of these are part of the EU bubble, they can be grouped together. In effect, there are three main AAU deficit entities: the European Union, Canada, and Japan. Several smaller non-EU countries (such as Australia and New

Zealand) may also face AAU deficits, but these represent only a very small fraction of the total expected AAU shortfalls.

Canada faces the most significant compliance problems as, by 2005, its emissions were already 55% above 1990 levels, and more than 60% above their Kyoto emissions target, with their emissions projected to continue to grow strongly. Canada's shortfall could reach 1.5 billion AAUs or more by 2012.[17] Rapid growth in energy consumption and significantly lower than projected carbon sequestration gains from forests are the main contributing factors to this large shortfall. Within the "EU bubble," several countries (especially the United Kingdom and Germany) are likely to easily meet their targets, while others (such as Spain, Ireland, Greece, Italy, and Portugal) are well above their Kyoto targets. The EU surpluses and deficits will be aggregated under the "EU bubble" burden sharing arrangement – on this basis the EU grouping was 4% below 1990 emissions by 2005 (still 4% short of the aggregate EU emissions target). The European Union is expected to have an aggregate deficit in the vicinity of 1 billion AAUs by 2012.[18] Japan was 7% above their emissions target by 2005 and is expected to have a shortfall of up to 1 billion AAUs by 2012. Overall, the aggregate shortfall of Annex I deficit countries is expected to be as much as 3–4 billion AAUs by 2012. If these countries are to meet their emission commitments, they will need to utilize the flexible mechanisms and/or implement much more aggressive domestic mitigation action up to 2012.

In contrast, approximately half of the Annex I Parties are likely to have surplus AAUs that they could trade with Annex I countries with AAU deficits. Russia, with as much as 4 billion surplus AAUs, and Ukraine, with up to 2 billion surplus AAUs, dominate the picture. The large AAU surplus for these countries is the result of significant economic contraction and restructuring that occurred following the breakup of the former Soviet bloc – in some countries, annual emissions in 2000 were less than half of their 1990 base year allocations under the Protocol. In aggregate, over the period to 2012, surplus AAUs from Economies in Transition (EITs) could be as much as 7 billion.[19]

Since these surpluses were not generated by deliberate policies and actions aimed at greenhouse gas mitigation, they are often referred to as "hot air." Some critics have asserted that an excessive reliance on purchasing "hot air" AAUs would significantly diminish the credibility of the Protocol. The "hot air" AAU surpluses are essentially a one-off windfall gain to these countries. In recent years, emissions in most EIT countries have begun to rise again, and this emissions growth will slowly reduce the quantity of surplus AAUs over time, but not until after 2012.

Overall, it appears that the Annex I Parties that have ratified the Protocol will, in aggregate, have more than sufficient AAUs to meet the Protocol's 5% reduction target, and possibly exceed it by a fair margin, given the magnitude of the Russian and Ukrainian AAU surpluses. Whether Russia and/or Ukraine are willing to trade a significant quantity of their surplus AAUs remains unclear: this hinges on the price at which they could sell AAUs to other Annex I deficit countries (possibly combined with other trade or political concessions) and, most important, whether or not a post-2012 international agreement is struck and the types of emission commitments embodied in any such agreement for the surplus AAU countries. They may wish to monetize some of their surplus and/or may choose to carry forward emission allowances to a meet their commitments under a future agreement.

Will the Kyoto Protocol be viewed as a success?

Even though the first commitment period is not yet complete, it appears that Annex I countries will deliver the emission reductions promised in 1997 but not necessarily as a result of explicit emission mitigation actions. Some will view the Annex I efforts at reducing emissions as relatively weak since most Annex I countries have not, as yet, implemented sufficiently aggressive domestic mitigation measures to meet their targets and will need to rely on purchasing AAUs and CERs to ensure compliance with the Protocol. If the EIT economies had not experienced such a significant decline in economic activity since the early 1990s and had the United States, the largest Annex I emitter, ratified the Protocol, it would be highly improbable that the 5% emission reduction commitment of Annex I countries would be honored.

It appears that the Protocol will have some successes and some failures. The Protocol's main successes are likely to include:

- successful creation of the international administrative infrastructure necessary to support international emission reduction agreements;
- establishment of flexible market-based mechanisms that assist in delivering cost-effective mitigation as well as facilitating the transfer of low emissions technologies to non-Annex I countries through the **CDM**;
- a clear and heightened awareness of climate change across the international community, and a clear recognition that something more than Kyoto will be needed if **dangerous climate change** is to be avoided;
- enhanced greenhouse gas inventory development and reporting processes;
- the establishment of several significant financial instruments, such as the Least Developed Country Fund, the Special Climate Change Fund, and the Adaptation Fund, that may assist non-Annex I countries to accommodate the impacts of climate change and potentially operate as a proxy climate change impact compensation mechanism; and
- delivery of the Annex I emission reduction commitment made in 1997, albeit in part through events unrelated to mitigation efforts.

The main failures of the Kyoto Protocol are likely to be the:

- nonratification of the United States, which significantly undermines the political credibility of the Kyoto Protocol in terms of developed country leadership on climate change and reduces the overall emissions coverage, and hence emission reductions resulting from the Protocol;
- expected reliance on "hot air" AAU trades to enable Annex I countries to honor their commitments;
- absence of strong domestic mitigation action in many Annex I countries, and thereby a relatively weak demonstration to developing countries of the political resolve to address climate change;

- relatively weak compliance enforcement penalties that provide only a limited deterrent to Annex I parties not meeting their emission reduction commitments; and
- relatively limited financial contributions (compared with what is needed) to the special financial instruments established to support non-Annex I countries respond to climate change (though this may change before 2012).

While the Kyoto Protocol has clearly made a substantial contribution to establishing the administrative framework and instruments necessary for future global climate agreements, these will not, in themselves, make a significant contribution to reducing global emissions. Nonetheless, the Protocol must be viewed as only the first step in the international community's response to climate change. Without the Protocol, it is unlikely that the international community would be as advanced as it is in terms of establishing a coordinated global response to climate change.

What happens beyond Kyoto?

There is widespread international agreement that substantial cuts in greenhouse gas emissions are urgently required if atmospheric greenhouse gas concentrations are to be stabilized at a level that avoids **dangerous climate change** (see **stabilization targets**). While the Kyoto Protocol is an important first step, it must be followed by much more substantive international action beyond 2012. The critical issue facing the international community is how to move forward from Kyoto and establish an internationally agreed workable approach post-2012. As of early 2008, little clarity has emerged on the form and structure of the post-2012 agreement.

To be effective, a post-2012 agreement must involve more substantial emission reductions than delivered by the Kyoto Protocol. This will require the United States to take on commitments proportionate to its share of global emissions, expanded commitments by the other 40 Annex B countries of the Protocol, and an increase in the number of countries with emission caps to cover a greater proportion of global emissions. It is also evident that any post-2012 agreement will need to devote greater attention to adaptation responses, and the role of the financial mechanisms (such as the Adaptation Fund), particularly for the poorest countries who have contributed least to climate change but will face the brunt of the impacts.

While the Convention's principle of "common but differentiated responsibilities" legitimately requires wealthier countries with the highest emissions per capita to take on a greater responsibility for emission reductions than most developing countries (see **emissions per capita**), it is clear that Annex I countries alone will be unable to resolve the climate change issue. It will require a broad-based approach involving virtually all nations: even if Annex I countries (including the United States) were to halve their current emissions by 2050, if emission growth from non-Annex I countries remained unchecked, aggregate global emissions and atmospheric greenhouse concentrations would continue to rise and, as a consequence, so would global temperatures (see **future emissions trends**).

CMP13 in Bali, Indonesia, heralded the commencement of the formal post-2012 agreement negotiations. It was agreed that these negotiations be largely completed during the two-year period to COP15 (late 2009) – this process is often referred to as the "Bali Roadmap." However, it is clear that considerable, and some may say irreconcilable, differences remain in the negotiating positions of different countries. Some industrialized countries, and the United States in particular, have adopted a position that they will not be party to a post-2012 agreement unless the large developing country emitters also take on emission reduction commitments (notably China, India, and Brazil), though what would constitute a meaningful commitment has yet to be articulated. Most non-Annex I countries maintain the position that they should not be required to accept emission constraints until their basic development aspirations are fulfilled and until the wealthy industrialized countries have clearly demonstrated their resolve to substantially reduce their emissions. The challenge for the international community is to achieve the required global emissions reductions while also enabling developing countries to achieve their legitimate economic development aspirations. This is clearly a difficult and politically complex task.

Another issue that may need to be resolved is the present division of countries into Annex I and non-Annex I categories since some current non-Annex I countries (such as South Korea, Singapore, and several others) are now at comparable levels of economic development as many Annex I countries. Arguably these countries are also now in a position to take on emission reduction commitments. In terms of most other non-Annex I countries, the imposition of stringent emission caps is likely to be problematic, politically unachievable, and unjustifiable given differences in **emissions per capita**. They could, however, possibly take on commitments other than emission caps that would, nonetheless, contribute to reducing global emissions. These could include commitments to reduce rates of deforestation, adoption of mandatory energy efficiency standards for appliances and buildings, improved land-use practices, sustainable transport policies, and a range of other options.

Several important considerations that underpinned Kyoto Protocol negotiations remain relevant to the current negotiating process, including that:

- commitments be equitable and linked to the ability to pay and the relative level of economic development of different Parties to the agreement;
- any agreement should strive to place the international community on track to stabilize atmospheric concentrations at a level that avoids dangerous climate change;
- emission reductions be achieved in a manner that minimizes the socioeconomic cost to the global community;
- mechanisms be established that promote the development and transfer of low emission technologies; and
- that due recognition is given to adaptation, particularly for the poorest nations, which will require a significant increase in the resources to assist countries to cope with the impacts of climate change.

While formulating an effective post-2012 international climate change agreement(s) faces many serious challenges, it is, given the projected pace of climate change over the coming decades, essential that the international community succeeds in reaching agreement. Failure to do so will have substantial adverse consequences for present and future generations.

See also: anthropogenic greenhouse gas emissions, Clean Development Mechanism (CDM), dangerous climate change, emissions intensity, emissions per capita, emissions trading, future emissions trends, Joint Implementation (JI), mitigation, United Nations Framework Convention on Climate Change (UNFCCC).

Notes

1 UN & UNFCCC 2005, Article 2
2 Ibid., Article 3
3 From UNFCCC Web site, http://unfccc.int/kyoto_protocol/status_of_ratification/items/ 26 13.php.
4 UN & UNFCCC 2005, Kyoto Protocol, Article 22
5 UNFCCC, 2007
6 UN & UNFCCC 2005, Article 3
7 Ibid., Article 8
8 Ibid., Article 10
9 UNFCCC 1997
10 See Oberthur and Ott 1999 and Pearce 2006 for further discussion
11 UN & UNFCCC 2005, Article 25
12 Office of the Press Secretary, June 11, 2001
13 Ratification status as at May 2008
14 See Wang and Wiser 2002
15 See Yamin and Depledge 2004
16 COP/MOP 1, decision 27/CMP 1
17 Based on data from World Bank 2005, UNDP 2006a and IEA 2006a
18 Ibid.
19 See World Bank 2005 and UNDP 2006a

Further reading

UNFCCC Handbook 2006; Oberthur and Ott 1999; Yamin and Depledge 2004; Pearce 2006; UNEP/UNFCCC 2002.

LAND CARBON SINKS

The terrestrial land system (the land) both emits and absorbs large quantities of carbon dioxide. The land absorbs (*sequesters*) around 100–120 billion tonnes (gigatonnes, Gt) of carbon (C) – equivalent to 370–440 Gt **carbon dioxide (CO$_2$)** – from the atmosphere each year through photosynthesis by plants. The sum of photosynthesis by all plants is termed Gross Primary Production (GPP). A similar quantity of CO$_2$ is released back to the atmosphere, through respiration of living organisms and microbial decomposition of organic residues in the soil and surface litter of dead plants and through the consumption of plants by other organisms that also emit CO$_2$ back to the atmosphere through respiration. The residual of GPP-less respiration yields Net Primary Productivity (NPP) – this is what determines the amount of carbon retained in the land system (see **carbon cycle**). The sizes of the carbon *fluxes* (exchanges of CO$_2$) between the atmosphere and the land vary significantly from year to year due to changes in climatic conditions. At present, the land absorbs slightly more CO_2 than it emits and is, therefore, a net **carbon sink**.

Estimates of the annual net CO_2 uptake from the atmosphere vary considerably, depending on the methods used to calculate annual changes in land carbon stocks. Due to variability and measurement difficulties, considerable uncertainty surrounds how much carbon *sequestration* exceeds carbon releases to the atmosphere. At present, the net uptake of CO_2 by the land is estimated to be in the range of 6–14 $GtCO_2$/year.[1] Some studies suggest greater uncertainty, and a range as wide as 1–18 $GtCO_2$/year.[2] However, this is offset to a considerable extent by anthropogenic CO_2 emissions from ongoing deforestation and land-use change, which together release 3–9 $GtCO_2$ of carbon each year to the atmosphere.[3] As a result, most estimates of net carbon uptake (including land-use change) generally fall in the range of 4–9 $GtCO_2$/ year. This compares with the global anthropogenic greenhouse gas emissions of around 50 Gt CO_2e in 2005.

While the magnitude of the carbon *fluxes* between the land and the atmosphere are similar to those between the ocean and atmosphere, the land is a much smaller *reservoir* of carbon (containing only around 5% of exchangeable carbon). However, the land system is far more sensitive than the ocean and responds much more rapidly to changes in both atmospheric CO_2 concentrations and global temperatures.

Unlike **ocean carbon sinks**, humans have a much greater ability to influence the amount of carbon stored in the land system. In recognition of this, land-use activities (or more precisely, Land Use, Land Use Change and Forestry – LULUCF) receive special treatment under the **Kyoto Protocol**. Actions such as replanting trees on land that had been previously cleared (reforestation) or planting trees on land that was either not naturally forested or has not been forested for a period of at least 50 years (afforestation) can receive carbon credits from the **Clean Development Mechanism (CDM)** and **Joint Implementation (JI)**.

Climate change and continued deforestation is likely to significantly influence the amount of CO_2 the land system will sequester from the atmosphere over the course of this century. Changes in temperature, water availability, and atmospheric CO_2 concentrations will affect GPP and NPP and could result in the land transitioning from a net sink to a net source in the next hundred years (see **climate change feedbacks**).

The uptake of CO_2 by the terrestrial land system depends on a range of different factors. Key factors include the prevailing climate (e.g. average temperature and rainfall, amount of solar radiation, and seasonal climate changes), the type of ecosystem (tropical forest, grasslands, savanna, temperate and boreal forests), and the extent of human interference across the landscape (agriculture and deforestation). As a result, there is considerable variation in CO_2 uptake across different regions of the globe. It is young, growing forests that are responsible for most of the net CO_2 *sequestration*. Mature forests (sometimes called *old growth forests*) are generally thought to be largely in equilibrium (carbon absorption and release are in balance) and sequester little or no net carbon, although recent research has revealed that net carbon stocks in mature temperate forests have been increasing, mainly in the soil and, to some extent, through vegetation thickening. For example, over the period 1979–2003 old growth forests in southern China exhibited an average increase in soil carbon stocks of 2.2 tonnes of CO_2 per hectare/year.[4]

Temperate and boreal forests

Most of the carbon storage, and net uptake of CO_2, occurs in the vast temperate forests of the northern hemisphere (above 30° north). Boreal forests may also be sequestering carbon but are generally considered to be largely neutral.[5] In temperate and boreal forests, the soil contains about four times as much carbon as the vegetation. Much of the carbon sequestration is associated with the growth of new, regenerating forests on land that was cleared in temperate regions over the past few hundred years. From 1800 to 1950, vast tracts of land were cleared for agriculture and timber production in the United States, Canada, Russia, Australia, and Southern Africa, and, to a lesser extent, Europe. Land-use changes over the past 50–70 years have enabled forest regeneration in some of these areas, particularly North America and Europe (these two regions account for about one quarter of global uptake of CO_2 by the land).[6] Regenerating forests in the United States alone have been absorbing around about 900 $MtCO_2$/year from the atmosphere in recent years.[7]

Most regrowing forests are located in the mid- to high latitudes of the northern hemisphere. Once these forests are reestablished and reach maturity, their ability to absorb CO_2 will diminish. While considerable carbon sequestration potential also exists in the mid-latitudes of the Southern Hemisphere (e.g. Australia and the southern regions of Africa and South America), there is less potential to support forests since the area below 30° south is mostly ocean.

Tropical forests

The tropical regions account for around one-third of the land–atmosphere exchange of CO_2 and are the second largest land carbon reservoir. In a tropical forest, carbon storage is split relatively evenly between the soils and vegetation. Intact tropical forests are estimated to be a net sink of around 4–5 $GtCO_2$ each year at present,[8] but this carbon uptake is largely offset by deforestation. Although tropical forest carbon fluxes are not known with precision, and estimates of the emissions from deforestation vary widely, the tropical forest sink is believed to either be slightly positive or neutral at present.

The most significant tropical forest region is the Amazon Basin. Scientists long believed that the Amazon Basin stores large amounts of carbon in forests and soils for long periods of time (decades or even centuries). They also believed that the Amazon River transported significant quantities of carbon to the Atlantic Ocean for long-term storage. However, recent research has revealed that much of the CO_2 absorbed by the Amazon is recycled to the atmosphere within five years and that the amount transported to the Atlantic is only a fraction of what had been previously thought. Microbes consume much of the dissolved organic and inorganic carbon and emit it back to the atmosphere as CO_2.[9] As a result, tropical forests may play a more limited role as long-term carbon sinks than previously believed.

For tropical forests, most of the future carbon sink potential exists in maintaining the existing forest cover by avoiding further deforestation. Once tropical forests have been cleared, it is very difficult to reestablish them as forests with the same carbon storage ability. Many deforested areas, particularly in the Amazon, end up as degraded lands with low carbon densities. Some turn into tropical deserts with very limited CO_2 uptake potential.

Grasslands, tropical savanna, and scrubby woodlands

In these ecosystems, vegetation is sparse, and upto 90% of the carbon reservoir is usually stored in the soil for long periods of time if undisturbed. Over the past century, many of these areas have been used extensively for agriculture and pastoral activities and considerable land degradation has occurred. This has resulted in large losses of soil carbon, which has mainly been emitted to the atmosphere as CO_2. Most of the grassland and scrubland biomes are also located in relatively moderate to low rainfall regions (Australia, the Sahel and Southern Africa, Central Asia and China, and the southern and central United States). As a result, their primary production capabilities are much lower than that of wetter regions.

Although many grasslands and savannahs have been net sources of CO_2 emissions over the past century, their long-term carbon sequestration potential is significant. Improved land management practices and rehabilitation of degraded land can be important means of sequestering CO_2, but their low uptake rates means that it takes a long time for this to occur. Sequestration per hectare in regenerating grassland areas (on average 1–3 t CO_2/hectare per year) is less than a quarter of what can be achieved by reforestation in wetter regions.[10] However, as rehabilitation activities can be carried out over very large tracts of land, large quantities of CO_2 can be sequestered. Increasing the amount of organic carbon in the soil can also assist in raising the productivity of agricultural systems and the livelihoods of inhabitants. The wider poverty and environmental issues facing many of the dryland regions in the developing world means that their role as future carbon sinks will be dependent on how these issues are addressed.

Tundra

High latitude tundra is generally a sparsely vegetated ecosystem, but significant amounts of carbon can be contained in the soil, mainly as frozen organic matter. Unfortunately, **global warming** is causing permafrost to thaw, which releases significant quantities of organic carbon (in the form of CO_2 and methane) back into the atmosphere – resulting in a positive **climate change feedback** (see **polar impacts** and **methane [CH_4]**).

Future trends

Over the past century, the land system, as a net CO_2 sink, has helped constrain the buildup of CO_2 in the atmosphere. The land will continue to perform this valuable role as long as the quantity of CO_2 absorbed through photosynthesis is greater than the amount released to the atmosphere through respiration. However, the land is not expected to remain a net carbon sink indefinitely as climate change is expected to reduce the net amount of CO_2 uptake by the land over the course of this century and could even become a net source of CO_2 emissions to the atmosphere. Uncertainty remains over the timing and extent of these changes.

Human land-use activities, such as changes to the extent of deforestation, reforestation, afforestation, and agricultural practices, will have a major bearing on the quantity of carbon stored in the land system (see **biosequestration**). There are also several important natural processes that will alter land–atmosphere CO_2 fluxes, primarily, changes in atmospheric gas concentrations; global warming-induced temperature changes; and climate-induced changes to precipitation, fire regimes, permafrost, and pests and diseases that effect CO_2 fluxes.

Increased atmospheric concentrations of CO_2 and other gases

Elevated CO_2 concentrations stimulate primary productivity through the *CO_2 fertilization effect*. CO_2 is an essential ingredient for photosynthesis, and higher levels enhance the efficiency of photosynthesis and improve vegetation growth. Increased CO_2 also enables plants to use water more efficiently (see **agriculture and food supply impacts**). CO_2 fertilization has an upper bound (believed to be around 800–1,000 ppm CO_2 without other nutrient limitations) as plants eventually become CO_2 saturated. The saturation point is influenced by other essential inputs required to sustain photosynthesis, such as nitrogen and phosphorous constraints, which reduce the effective saturation point. So too will any changes in the amount of sunlight reaching the earth's surface through changes in cloud regimes (see **albedo**) or reductions due to atmospheric **aerosols** (see **global dimming**).

Nitrogen availability is an important determinant of photosynthetic production, and insufficient nitrogen limits plant growth no matter how much CO_2 levels rise. Although humans are emitting large quantities of nitrogen to the environment primarily through artificial fertilizers (see **nitrous oxide [N_2O]**), nitrogen levels are not keeping pace with the rate of increase of CO_2. Much of the additional nitrogen falls on agricultural land rather than forests, thus limiting its impact on primary productivity – and therefore potentially constraining the benefits of increased CO_2 on plant growth. Nonetheless, increased nitrogen deposition will tend to increase CO_2 uptake by the land this century, but only marginally (probably less than 1 GtCO_2/year).[11] Even where sufficient nitrogen is available, other nutrient deficits, such as phosphorous, will limit plant growth, particularly in phosphorous-deficient tropical soils. Applying artificial fertilizers may not only stimulate primary production in some forest regions but can also increase soil respiration rates and thus CO_2 emissions, by as much as 20%, potentially offsetting the CO_2 fertilization effect.[12] Increased surface **ozone** also impairs plant production and in some heavily industrialized regions could lead to reduced CO_2 uptake (see **agriculture and food supply impacts**).

Most scientific studies have tended to focus on the impact of elevated CO_2 levels on above-ground vegetation growth, and much less attention has been given to the rate of soil carbon sequestration. However, it is in the soil where 70% of the carbon in the land reservoir is stored. Research has found that elevated CO_2 levels can reduce carbon sequestration in the root zone, thus reducing overall net uptake.[13] As the temperate forests store most of their carbon in the soil, even small reductions in the soil carbon pool may have major feedback effects on atmospheric CO_2 concentrations in the future.

Overall, CO_2 fertilization will tend to increase CO_2 uptake by the land, but once CO_2 concentrations exceed 500–600 ppm, the positive effect is expected to diminish significantly, mainly due to other nutrient limitations.[14]

Higher temperature impacts

Warmer temperatures increase the rate of most chemical and biological processes and also stimulate the growth of trees but only up to a certain point, after which increased temperatures impair the functioning of plant cells and reduce growth. The growth benefits of higher global temperatures will mainly be limited to the mid- and higher-latitude forests (particularly in the cold boreal forest regions) as tropical forests are already close to optimal temperatures for maximum growth. In fact, there is some evidence to suggest that temperature growth limits have already been reached in leaves in the top canopy in tropical forests at the hottest part of the day, where some plants shut down photosynthesis completely.[15] The increased prevalence of extreme heat waves and hotter, drier summers will also constrain the uptake of CO_2. For example, the 2003 European heat wave is estimated to have resulted in a 30% reduction in primary production and that region became a net source of CO_2 during that year (emitting around 2 Gt more CO_2 than sequestered).[16]

As the earth warms, the forest frontier (the tree line) will move northward into tundra regions, increasing CO_2 uptake in these areas and reducing the rate of growth of CO_2 concentrations in the atmosphere. However, this may not reduce **global warming** since land covered by forests reflects a lot less energy back to space than snow-covered tundra. The spread of forests toward the poles will reduce the earth's **albedo**, and the decline in albedo in tundra regions is expected to more than outweigh gains from enhanced CO_2 uptake by northward expanding forests (see **polar impacts**).[17]

Increased global temperatures are also likely to have a major negative impact on the size of the land carbon reservoir, due to its tendency to increase soil respiration rates and through thawing permafrost (see **polar impacts**). As global temperatures rise, the land is likely to emit much more CO_2 from the soil reservoir. Respiration fluxes rise with temperature, particularly at lower levels of warming.[18] There is still considerable uncertainty over the long-term sensitivity of respiration fluxes to temperature as some research has found that as temperature increases, net mineralization of nitrogen may also increase, stimulating plant growth and, therefore, CO_2 uptake.

While uncertainty remains in relation to the interaction of the carbon and nitrogen cycles at higher temperatures, the scientific consensus is that climate change-induced increases in soil respiration will result in a net increase in CO_2 emissions to the atmosphere. This implies that the terrestrial carbon sink has a finite lifetime and that it will eventually transit from being a net sink to being a CO_2 source. It is uncertain when this will occur, but some studies suggest that it could be during this century.[19] Temperature change sensitivity will vary across different regions, and it is generally agreed that the tropics will cease to be a sink before forests in the mid- and higher latitudes.

Other factors

Several other climate change-related factors are also likely to reduce the ability of the land to store carbon in the coming years. These include the availability of water, changes in the frequency and intensity of **El Nino Southern Oscillation (ENSO)** events, the incidence of fire, plant pests and diseases, and changes in the structure of forest ecosystems.

Many of the world's major forest regions are already experiencing high levels of water stress (lack of water). There are certain thresholds of water availability at which the ability of the land to sustain forests is impaired. In the tropical regions, once rainfall falls below 1,200 mm per year, forests can potentially convert to savanna.[20]

The amount of CO_2 uptake by the land can vary significantly from year to year due to annual fluctuations in climatic conditions. The **ENSO** has a major influence on annual climatic variations and rainfall. During strong ENSO events, global average climate conditions are generally warmer and drier than normal, reducing plant growth and the amount of CO_2 absorbed by the land system. Some forests can become net sources of CO_2 during ENSO events.

Drought in tropical forest areas (e.g. in Southeast Asia, the Congo, and Amazon) is known to substantially reduce primary production of the forests and increase dieback. Rainfall has reduced significantly in the Congo and Eastern Amazon over the past few decades, and many scientists believe this could lead to significant positive **climate change feedbacks**. For instance, the drying of the Amazon could result in large-scale releases of CO_2 back to the atmosphere. While droughts have usually been associated with ENSO events, the severe drought that afflicted the Amazon rainforest in 2005 did not coincide with a strong ENSO event and may indicate an underlying drying trend in this region.[21]

The continued draining of temperate and tropical peatlands (primarily for agriculture) and increased incidence of drought and peat fires is likely to lead to large-scale additional oxidation of soil organic matter over this century. One study estimates that up to 350 Gt of additional CO_2 emissions to the atmosphere could emanate from peatlands and wetlands over the next 100 years (equivalent to around seven times anthropogenic CO_2 emissions in 2005).[22] Overall, global warming is expected to increase the incidence of drought and soil carbon oxidation, which will impair the CO_2 sink potential of the land.

FIRE

As forests dry out, their flammability also increases and so does the chance of fire. Growing forests sequester CO_2 slowly over long periods of time, but fire can return large quantities of CO_2 back to the atmosphere very quickly. For example, the massive tropical forest fires in Southeast Asia associated with the strong 1997–1998 ENSO event burned out 20 million hectares of forest and released over 10 $GtCO_2$ to the atmosphere in that year alone.[23] This was one reason why 1998 had the highest increase in atmospheric CO_2 concentrations of any year so far recorded (double the annual average increase of the preceding decade).

Scientists expect the incidence of fire to increase with higher temperatures and more frequent drought. Recent studies have supported this conclusion. A comprehensive study of forest wildfires in the western United States since 1970 has revealed that the occurrence of wildfire showed a four-fold increase over the period and the size of the area burnt a six-fold increase, with the largest change at altitudes over 2,000 m due to reduced snow pack cover and duration.[24] Carbon emissions from forest fires are estimated to have increased to around 10 $GtCO_2$ a year during the first few years of this century, equivalent to around one-third of CO_2 emissions from fossil fuels. While improved fire management techniques may help reduce the risk of fire, the probability of large-scale fire disturbance and sudden releases of CO_2 to the atmosphere will increase with increasing global temperatures. This will most likely accelerate the loss of carbon stored in the land system.

PESTS

The incidence of pests and diseases on existing forests is also considered to be a potential negative effect on the land carbon sink. Increased average temperatures have enabled some pests, which were previously kept in check by seasonal temperature changes, to multiply and spread more widely than previously. For example, spruce bark and pine bark beetle infestations have increased significantly in recent years and have already resulted in the destruction of millions of hectares of forest from Southwest United States right up through British Columbia and Alberta to Alaska.[25] While trees may grow back to replace those that have died, the loss of land carbon stocks due to pests is expected to increase with continued global warming (another positive **climate change feedback** effect).

Finally, climate change is likely to set in train changes in the composition of forest plant species that may reduce carbon density in some forests. In tropical forests, it is apparent that both tree growth and mortality are increasing. When trees fall they create gaps in the forest and faster-growing, light-demanding, species of trees tend to thrive at the expense of shade-tolerant species. Fast-growing species generally have lower carbon densities, and this is likely to reduce carbon density per hectare over time, thus reducing the overall carbon sink quantity of existing forests.

Transition from sink to source

The ability of the land to remain a net carbon sink in the future will be influenced by a wide range of factors. Some factors (CO_2 fertilization and temperature) will tend to enhance photosynthesis and, thereby, CO_2 uptake (negative feedbacks). Other factors such as enhanced respiration rates, permafrost thawing, and reduced water availability will tend to have the opposite effect (positive feedbacks), particularly in tropical forests. The general consensus is that the terrestrial land system will eventually make the transition from being a net sink to a net source of CO_2. When this occurs, the land system will accelerate the rate of increase in atmospheric CO_2 concentrations, rather than suppress them, a positive **climate change feedback**.

There is a critical concentration and temperature threshold beyond which the transition from sink to source occurs, but at present there is considerable uncertainty about when this critical point will be reached. Much depends on how much global temperatures increase for a given increase in CO_2 concentrations (see **climate sensitivity**). To date, many models used to make climate change projections do not fully incorporate these complex **carbon cycle** feedback mechanisms. This may mean that the negative impacts of land feedbacks have been underestimated and the level of future global warming has been understated. Recent research using models that include detailed carbon cycle feedback mechanisms suggests that the land's net uptake of CO_2 is likely to diminish much more rapidly than previously thought. Some model results have shown that the land may transit from sink to source as early as 2050 under a "business as usual" emissions scenario.[26] At present, the general scientific consensus is that the land system is not likely to make the transition from sink to source until later this century or into next century. However, nearly all scientists agree that the role of the land as a CO_2 sink is likely to diminish this century. The sink-to-source transition would be a major turning point. A recent study suggests that the transition from sink to source could add as much as 1–1.5°C by 2100.[27]

See also: agriculture and food supply impacts, albedo, anthropogenic greenhouse gas emissions, biosequestration, carbon cycle, carbon dioxide (CO_2), carbon sinks, climate change feedbacks, climate sensitivity, dangerous climate change, El Nino Southern Oscillation (ENSO), Kyoto Protocol, ozone, polar impacts.

Notes

1 Lewis *et al.* 2006
2 Canadell *et al.* 2007
3 Prentice 2001
4 Zhou *et al.* 2006
5 Gurney *et al.* 2002
6 Heath *et al.* 2005
7 US EPA 2006
8 Lewis *et al.* 2006
9 Magoya *et al.* 2005
10 Conant *et al.* 2001
11 Canadell *et al.* 2007
12 Cleveland and Townsend 2006
13 Heath *et al.* 2005
14 Canadell *et al.* 2007
15 Lewis *et al.* 2006
16 Ciais *et al.* 2005
17 Betts 2000
18 Canadell *et al.* 2007
19 Ibid.
20 Salzmann and Hoelzmann 2005
21 SciDev net June 2006
22 Gruber *et al.* 2004
23 Cochrane 2003
24 Westerling *et al.* 2006
25 Epstein and Mills 2005
26 Cox *et al.* 2000
27 Friedlingstein *et al.* 2007

Further reading

Lewis *et al.* 2006, IPCC 2007, Cox *et al.* 2000, Cochrane 2003.

MARINE IMPACTS

The marine environment encompasses the saline waters of the world's oceans, seas, and coastal estuaries. This is a vast body of water, in places extending to depths of 10 kilometers or more, and is home to a broad array of interrelated biological systems. These systems ultimately rely on primary producers (phytoplankton and other photosynthetic marine organisms that convert sunlight to energy), which form the base of the marine food chain. Primary production supports microscopic zooplankton, crustaceans, fish, marine mammals (such as whales), and reptiles (such as turtles), as well as a host of other organisms. Marine ecosystems also support terrestrial (land-based) organisms, such as sea birds, seals, penguins, polar bears, and humans. Approximately one-quarter of humans' dietary protein is derived from the marine environment. Overfishing and exploitation of marine resources by humans have already had a major impact on the abundance and health of marine ecosystems, but climate change is expected to exacerbate this ongoing decline in marine biodiversity.[1]

Increases in atmospheric concentrations of **greenhouse gases**, and the global warming associated with such increases, can affect marine ecosystems through alterations in ocean chemistry, increases in water temperatures, changes in ocean currents, modifications to nutrient circulation patterns (see **thermohaline**), and reductions in the extent of sea ice coverage. These changes can influence the structure and vitality of marine ecosystems and can, in turn, affect those organisms (including humans) that depend on such ecosystems for their livelihoods.

Changes to ocean chemistry

The oceans play an essential role in the global **carbon cycle**, absorbing and emitting large quantities of **carbon dioxide (CO_2)** each year through a variety of physical and biological processes (see **carbon cycle** and **ocean carbon sinks**). In the process of absorbing atmospheric CO_2, carbonic acid is produced in seawater, though this is neutralized relatively quickly through reactions with dissolved carbonate compounds to form carbonate and bicarbonate ions. Before atmospheric CO_2 concentrations began to rise, the carbonate ions consumed during this process tended to be largely replaced by inputs from the weathering of terrestrial carbonate rocks (such as limestone and chalk) and from the chemical breakdown of skeletons of dead marine organisms. This maintained the pH level of the oceans, which are mildly alkaline, at a relatively constant 8.2 pH (the potential of hydrogen [pH] is determined by the concentration of hydrogen ions and is measured on a scale of 1–14, with 1 being highly acid, 14 highly alkaline, and 7 neutral).

However, the human-induced increase in atmospheric CO_2 concentrations over the past century has led to the oceans absorbing ever-increasing quantities of CO_2. Currently, the oceans are absorbing approximately 7 billion tonnes (gigatonnes, Gt) more CO_2 from the atmosphere each year than they are releasing back to the atmosphere (see **ocean carbon sinks**). As a result, the demand for carbonates to neutralize the additional carbonic acid has outstripped supply, and ocean pH has begun to fall. This trend is often referred to as *ocean acidification*.[2] Ocean acidification is a result of CO_2 enrichment of the oceans, rather than climate change per se.

The additional uptake of CO_2 by the oceans over the past century has already caused average ocean pH levels to fall by around 0.1 pH units, from 8.2 to 8.1.[3] This may not sound significant, but, as pH is measured on a logarithmic scale, this represents a 25% increase in hydrogen ion concentrations. As long as ocean uptake of CO_2 continues to exceed the release of CO_2 to the atmosphere, ocean pH and the oceanic stock of carbonate ions will continue to fall. As the oceans become increasingly saturated with CO_2, their ability to absorb further CO_2 will be reduced. Consequently, the rate of ocean acidification will eventually begin to decelerate. However, the acidification process will only fully come to a halt when carbonate supply and demand are brought back into balance.

Estimates of future ocean pH levels vary according to assumptions about future CO_2 emission trajectories and biophysical processes. The **Intergovernmental Panel on Climate Change (IPCC)** estimates that, based on expected emission trends, ocean pH could fall by another 0.14–0.35 points by 2100 depending on the emissions scenario.[4] This increase would mean that ocean pH could be as low as 7.75 by the end of this century, which would represent an increase in the concentration of hydrogen ions of up to 180% over preindustrial levels. Ocean pH changes of this magnitude have not been experienced for at least 20 million years and possibly longer.[5] Over the next two to three centuries, ocean pH levels could fall to as low as 7 (high emissions scenario) or stabilize at around 7.9 (low emissions scenario).[6]

The most significant impact of ocean acidification is on calcifying marine organisms such as corals, plankton, and other life forms that depend on carbonate ions to form their skeletons and shells. These organisms are vital to marine food chains, and any reduction in their abundance will have flow-on ecosystem effects and could also reduce the amount of carbon exported to the deep ocean for long-term storage (through sinking carbonate skeletons), thereby further reducing the **ocean carbon sink** effect.

The pteropods, which represent an important food source for many marine species (such as whales, salmon, and cod), and corals, which play an essential role in maintaining marine biodiversity, are particularly susceptible to ocean acidification. Both depend on high saturation levels of an unstable form of calcium carbonate (aragonite).[7] As the oceans acidify, aragonite concentrations will fall, and this will impair these organisms' ability to form skeletons.

The greatest impact of acidification will be felt first in higher-latitude cold waters as these waters hold more CO_2 and will therefore more readily become undersaturated in aragonite (see **ocean carbon sinks**). The reduced ability of corals to form carbonate compounds will also inhibit their capacity to grow upward to keep pace with rising sea levels.[8] As sea urchins, starfish, and other echinoderms also construct their skeletons from carbonate minerals, they, too, will be affected by ocean acidification, as will some plankton species (such as coccolithophores).

In addition to inhibiting the formation of bones and shells, ocean acidification may also act to dissolve already-formed body parts. A study of pteropods in Norway, for example, found that their shells started to dissolve after just two days in water at the pH concentrations predicted for 2050.[9]

Impact of water temperature changes

Many marine species are sensitive to even small changes in water temperatures, which can significantly affect mortality, breeding success, and migratory patterns. Temperature changes can also alter ocean currents and nutrient flows, which can serve to reduce primary productivity in some areas and increase it in others. Changes in the location of primary productivity activity can have potentially significant impacts on entire marine ecosystems.

Changes in average global atmospheric temperatures will eventually be reflected in ocean temperatures, though the mass of the oceans is so large that ocean temperature changes only very slowly. There is a considerable time lag between atmospheric warming and oceanic warming. Surface ocean temperatures generally lag several decades behind changes in atmospheric temperatures, and it takes centuries for the deep oceans to reflect surface temperature changes. The overall average ocean temperature change so far experienced has been quite small (approximately 0.06°C), but this has nonetheless been much more rapid than most scientists' expectations given the slow turnover of ocean waters (see **thermohaline** and **methane hydrates**). Surface ocean temperatures have risen most rapidly, by as much as 0.5°C since the 1950s.[10]

However, the rate of warming has not been even across the oceans, and in some regions, particularly the higher latitudes, much greater temperature increases have been recorded. For example, North Sea surface waters have warmed by up to 2°C in the past few decades, which has resulted in some significant changes in North Sea marine ecosystems.[11] The rapid warming has caused the northward movement of some phytoplankton species by up to 1,000 km and has also resulted in a significant increase in phytoplankton mortality. This has, in turn, led to a reduction in the abundance of small crustaceans (copepods) that live off the phytoplankton and to a reduction in the population of North Sea sand eels, which feed on the copepods. The flow-on effects to higher-order organisms in the food chain, such as sea birds, have been quite pronounced. For example, in 2004 several sea bird species in the Shetland and Orkney Islands experienced a massive failure of the summer breeding season. Arctic Skuas and Arctic Terns produced virtually no offspring at all, whereas normally the number of chicks would have been in the order of 100,000 or more.[12] This breeding failure can be directly linked to the decline in sand eel numbers.

Some species in the North Sea have benefited from changes in water temperatures. There has been an explosion in the population of sea cucumbers over the past decade, due to a larger food supply in the form of dead phytoplankton falling from above.[13] This increase in food supply is likely to be only temporary, however. Eventually, as phytoplankton numbers fall off, the sea cucumbers are likely to suffer a decline in food supply.

Corals are very sensitive to water temperature changes. Many reef systems have already been adversely affected by human activities through overexploitation, nutrient pollution from agriculture, and increased sediment runoff from the land. Climate change is, however, likely to represent a much larger threat in the future. Even small changes in water temperature can cause coral bleaching and even death of the coral. Many corals have a unique symbiotic relationship with algae (Zooanthallae). The algae live in the polyps, and their photosynthesis provides the primary source of food for the polyps. If water temperature rises, even by as little as 1°C, the photo-synthetic production of the algae decreases. When the algae no longer provide sufficient food energy to earn their keep (i.e. they become a net energy cost to the coral polyps), the algae are ejected. If the temperature increase persists for long enough, the coral will eventually die.

The frequency and intensity of coral bleaching have become a more common occurrence since the 1970s, and the most severe bleaching events have usually coincided with the El Nino phase of the **ENSO**. For example, in 1998, a particularly warm year dominated by a strong El Nino event, elevated ocean temperatures resulted in severe bleaching and mortality to 18% of the world's living coral cover.[14] Recent studies have estimated that if the trend in atmospheric CO_2 concentrations and surface water warming continues, most of the world's coral reefs will be affected by coral bleaching.[15] Most southern Indian Ocean coral reefs are unlikely to be able to tolerate temperature increases above 1.5°C.[16] Scientists estimate that if surface ocean temperatures increase another 2°C, over three-quarters of the world's coral reefs will be subject to bleaching, and if they rise another 3°C, virtually all coral reefs will suffer irreversible damage.[17] As a wide range of marine species depend on coral reefs for their existence, any reduction in the health and vitality of reef systems will have major impacts on marine biodiversity. Many coastal and island communities depend on coral reefs for fishing, tourism, and coastal protection, and any reduction in reef ecosystem productivity, biodiversity, or beauty could represent a major cost of climate change to these communities.

Changes in water temperatures have also affected the productivity and location of fish populations, and in some areas, there have been significant declines in the abundance of certain species. Cod, for example, have low thermal tolerance, and increased water temperatures can cause significant falls in growth rates and performance. The reduction in cod abundance in the higher latitudes in recent decades has, in part, been linked to temperature-induced physiological changes and also to changes in the quality of food supply.[18] Cod depend on certain types of plankton: as these plankton have migrated north, they have been replaced by less nutritious varieties of plankton moving up from the south.

Overall, there has been a general movement of fish species toward the poles, a finding consistent with the expected impact of climate change. Most of the warm-temperate and temperate fish species have been migrating north at a rate of 250 km/decade, much faster than the migration of land-based species.[19] Warming in the northern Pacific, for example, has caused Sockeye Salmon stocks to migrate north toward the Bering Sea.[20] Such movements do not necessarily mean that the aggregate abundance of fish in a certain region will fall – merely that there will be a change in composition of fish species. Although cod are moving north, for instance, they are being replaced by sardines and anchovies from the south. Regional changes in fish species will require local fishing economies to adapt, as existing fishing vessels may no longer be suitable. Furthermore, while fish can migrate, the geographical boundaries governing the rights to traditional fishing grounds are often fixed by treaties and quota systems. Major changes to the prevalence of certain fish stocks could potentially increase tensions between different fishing nations over access to certain types of fish.

Climate change-induced changes to the ENSO cycle can also affect the abundance and distribution of marine species. During El Nino events, there are typically large falls in the stocks of pilchards, herring, and sardines off the coast of Africa and South America and declines in the fisheries off the west coast of North America.[21] If El Nino events become more pronounced, as some scientists expect, the size and location of the warm pool in the central and eastern Pacific may be affected, with the result that tuna stocks will move. This would have major impacts on many small Pacific island countries, where tuna fishing makes a significant contribution to local economies.

Changes in ocean circulation

Climate change is expected to result in a slowing, or even possibly a shut down, of the **thermohaline**, one of the main drivers of ocean circulation (see **thermohaline**). If this occurs, it will result in thermal partitioning of different layers of the ocean (thermal stratification) and will isolate the surface ocean (where the phytoplankton reside) from the deep oceans (where most of the nutrients are found). Reduced mixing of the ocean waters could reduce the supply and distribution of deep ocean nutrients, which could, in turn, reduce aggregate marine primary production. This would have repercussions for the marine species that directly and indirectly depend on plankton and could potentially reduce the export of carbon to the deep ocean (sea **ocean carbon sinks**).

Reductions in the flow rate of the thermohaline could also reduce oxygen supply to the deep oceans. Ultimately, the deep oceans could become anoxic (depleted in oxygen), killing off deep-ocean aerobic marine species (as happened during a period of rapid global warming 55 million years ago). It could also lead to an increase in sulphate-reducing anaerobic organisms, which produce hydrogen sulphide as a waste product. Hydrogen sulphide depletes soluble iron, an important nutrient source for phytoplankton and the major limiting factor of ocean primary production (see **ocean carbon sinks**). Furthermore, a long-term reduction in ocean primary productivity would reduce the production of oxygen (a by-product of photosynthesis): over time, this might lead to changes in the composition of the atmosphere, with oxygen declining and methane and carbon dioxide increasing in relative importance.

Declining sea ice coverage

Global warming has caused, and will continue to cause, reductions in sea ice coverage. As the productivity of plankton is highest in areas covered by sea ice, particularly at the margins, retreating sea ice is likely to reduce ocean biological productivity in the higher latitudes (see **ice sheets and glaciers**). Sea ice retreat has led to an observed decline in Antarctic phytoplankton productivity with subsequent flow-on impacts: krill populations have nearly halved since the 1950s.[22] This has, in turn, affected whales, which rely on the krill as a primary food source, and penguins, which depend on abundant fish supplies. Declining sea ice coverage has also had impacts on many marine-dependent mammals in the Arctic, such as seals and polar bears, which have lost habitat and suffered a reduction in the length of their traditional hunting period (see **polar impacts**).

Overall, there have already been noticeable climate change impacts on marine ecosystems. While our understanding of how marine systems will respond to climate change in the coming decades is still somewhat limited, it is, nonetheless, evident that marine systems will be adversely affected by climate change. If the findings of recent marine impact studies prove accurate, then even if the low end of the IPCC's global warming projections eventuate, it is likely to have significant and far-reaching effects on the marine environment over the course of this century and beyond. It could also have significant implications for individuals and communities that rely on the marine environment for their livelihoods.

See also: carbon cycle, ice sheets and glaciers, ocean carbon sinks, polar impacts, thermohaline.

Notes

1 IPCC 2002
2 See http://www.iwcoffice.org
3 Orr *et al.* 2005
4 IPCC 2007
5 Caldeira and Wickett 2003
6 Lenton *et al.* 2006
7 Orr *et al.* 2005
8 IPCC 2002
9 Hendersen 2006
10 Barnett *et al.* 2005
11 McCarthy 2004
12 Ibid.
13 Krieger 2004
14 Van Vliet and Leemans 2006
15 Hughes *et al.* 2003
16 Sheppard 2003
17 Woodford 2004
18 Portner and Knust 2007
19 Van Vliet and Leemans 2006
20 IPCC 2002
21 Ibid.
22 Gross 2005

Further reading

IPCC 2007, 2002; Van Vliet and Leemans 2006; Orr *et al.* 2005; Hendersen 2006.

METHANE (CH$_4$)

Methane is a colorless, odorless, combustible gas that consists of one carbon atom and four hydrogen atoms (CH$_4$). It is a hydrocarbon that is solid at temperatures below −183°C and becomes a gas at −163°C. Methane is less dense than air, not very soluble in water, and highly exothermic (heat is released when oxidized). Methane is also a powerful greenhouse gas.

Methane is present in the atmosphere in only very small quantities (0.00018% of atmospheric gases), with current concentration slightly less than 1,800 parts per billion (ppb). Methane is reactive with other gases in the atmosphere and is broken down relatively quickly through a series of complex interacting chemical processes. Methane interacts with hydroxyl radicals (OH) and **ozone** in the lower atmosphere and has important interactions with a range of other gases. The interaction of methane with OH is the primary means by which methane is broken down, eventually ending up as **carbon dioxide (CO$_2$)** and water vapor. Unlike CO$_2$, which is a stable gas, methane's atmospheric residence time is relatively short at around 12 years.

Methane gas can accumulate naturally in geological reservoirs, often in association with oil and coal deposits. However, most naturally occurring methane is stored as frozen **methane hydrates** in the permafrost and ocean floor sediments. Methane can also be synthesized commercially from the distillation of coal (often called "town gas").

Methane is the primary component of natural gas. Natural gas is the third most important **fossil fuel** after oil and coal and accounts for around 21% of global primary energy supplies. It is used mainly as a fuel for electricity generation and heating but has a range of other applications in the chemical industry (e.g. as a feedstock for the production of fertilizers and methanol).

Methane is the second most important of the **greenhouse gases** after CO$_2$. It is a powerful greenhouse gas, and molecule per molecule methane is sixty times more powerful than CO$_2$. However, due to methane's short atmospheric residence time, its warming effect decreases significantly with time. The atmospheric **global warming potential (GWP)** of methane over a 100 year time horizon is 23[1] (under the **Kyoto Protocol**, it is assigned a GWP of 21). In other words, a molecule of methane emitted today will have an atmospheric warming effect 23 times that of a molecule of CO$_2$ over a 100-year period.

The current **radiative forcing** of methane is 0.48 Watts/m^2, or slightly less than one-third that of carbon dioxide.[2] However, methane also has an indirect warming effect through its interactions with other gases in the atmosphere (particularly **ozone** and OH). When the direct and indirect warming effects are combined, it has been estimated that the radiative forcing of methane over the period 1750–1998 may be as high as 0.79 W/m^2, or around half that of CO$_2$.[3] This suggests that its contribution to **global warming** in the past century may have been more significant than previously thought.

Although up to one-third of past warming may be attributable to methane, the rapid rise in CO$_2$ emissions in the past few decades has meant that methane's warming impact relative to CO$_2$ has declined. Methane accounted for about 14% of the warming effect of anthropogenic emissions in 2005 when measured over a 100-year period.[4]

While methane is an important greenhouse gas, it has attracted much less attention than CO$_2$ in international climate change negotiations. Nevertheless, methane emissions could become a much more serious problem at some time in the future as the planet warms. Global warming is likely to increase the extent of permafrost thawing, which will release methane. In the longer term, increases in ocean temperatures could potentially trigger large-scale releases of methane from the vast **methane hydrate** deposits contained in the ocean floor sediments, although significant releases are not expected to occur this century. The more the planet warms, the higher the chances of significant releases from the hydrates, which would lead to further warming and thereby create a significant positive **climate change feedback** effect.

Sources of methane

Methane emissions come from many different sources, around 60% of emissions are anthropogenic and the remainder from natural sources.

Natural sources

Natural methane emissions are largely the result of the decomposition of organic matter in oxygen-free environments. The size of methane emissions from natural sources can be difficult to measure accurately and are subject to considerable uncertainty because the biological processes governing emissions vary significantly with time, space, and ambient environmental conditions.

The **IPCC** estimates that natural sources emit around 260 million tonnes (Mt) of methane each year, mainly from wetlands and peat bogs. Emissions from wetlands are estimated to be around 200–230 Mt/year, accounting for more than three quarters of all natural methane emissions. Other small but important sources include emissions from termite digestion processes (20 Mt), the oceans (10–15 Mt) and **methane hydrates** (10 Mt) mainly from thawing permafrost.

One source that had previously not been fully accounted for are emissions from terrestrial plants. It has generally been thought that methane emissions are due to biological processes in anaerobic (oxygen-free) conditions. However, in 2006 research findings from studies of tropical forests identified significant emissions emanating from plants in aerobic conditions (in the presence of oxygen).[5] The research found that methane emissions from intact plants and detached leaves, when scaled to a global level, could be a source of between 60 and 240 Mt each year. More research is needed to substantiate the results at the global scale, but the current scientific consensus is that the lower end of this range is the most likely magnitude of contribution from plants. Nonetheless, the findings are important to our understanding of the global methane budget.

Anthropogenic sources

Anthropogenic methane emissions are estimated to be around 330–360 Mt per year. The two main anthropogenic sources are emissions from the energy sector and ruminant livestock digestion processes, though waste landfills and rice production are also significant contributors.

Energy sector emissions come mainly from the production of oil and gas, coal mining, natural gas transmission and distribution leakage, and methane released through the combustion of **fossil fuels**. The energy sector is estimated to emit around

100–110 Mt/year. Since 1990 there has been significant progress in reducing emissions from oil and gas production, coal mining, and natural gas distributions systems in many countries. Even though energy sector methane emissions have increased in some countries, particularly in China and India, these have been more than offset by declines in other countries (particularly, Russia and the Ukraine). Nevertheless, coal and petroleum production are projected to nearly double by 2030, which may put upward pressure of methane emissions from the energy sector.

Ruminants (mainly domesticated cattle, sheep, and goats) produce significant quantities of methane through anaerobic digestive processes (enteric fermentation) in the stomachs of the animals. Although methane from enteric fermentation has always been a natural source from wild animals, human activities have led to a large increase in number of domesticated ruminants over the past century. The total number of domesticated cattle, sheep, goats, and buffalo now number more than three billion. Ruminant numbers are determined primarily by human dietary patterns. As incomes increase, so does the consumption of animal products, particularly meat and dairy products (see **anthropogenic greenhouse gas emissions**).

Estimates of methane emissions from ruminants have been derived from tests on animals in controlled environments. Actual emissions in natural conditions vary according to feed quality and seasonal factors. The average cow emits 80–110 kg/year of methane from enteric processes. In total, domesticated ruminants are estimated to emit around 80–100 Mt of methane each year, or about 15% of all methane emissions. In some countries, emissions from ruminants make up a significant percentage of the national **greenhouse gas inventory**. For example, New Zealand has large numbers of sheep and cattle, and ruminant emissions account for nearly half of New Zealand's inventory.

The anaerobic decomposition of organic matter in landfills and organic wastes (e.g. sewage) contributes around 60 Mt/year of methane. Large volumes of organic matter (e.g. food and garden wastes) are buried in thousands of municipal landfills throughout the world each year. Slow decomposition generates methane, which eventually permeates through the layers of refuse and is released to the atmosphere. In recent years, significant reductions in landfill methane emissions have been achieved in many countries due to better sorting of waste streams to remove organic material (much is composted for garden mulch) and through the capture and burning of methane. To capture landfill methane, gas collection pipes are placed through the landfill to drain off excess methane. This is either flared or used to generate electricity. At suitable locations, combining methane capture and electricity generation can prove economically attractive. These projects can generate revenue from electricity sales and, in some cases, also from carbon credit sales (see **Clean Development Mechanism**). Landfill gas capture and flaring can also provide other benefits such as reduced odor and fire risk. The rapid growth in the number of landfill gas capture projects over the past two decades has resulted in a significant fall in emissions from this source.

The cultivation of rice is another important source of anthropogenic emissions. Rice is submerged in water for much of the growing cycle and, in effect, creates an artificial wetland environment. Rice cultivation is estimated to generate around 40 Mt/year of methane. As the area under rice cultivation continues to expand, it is expected that emissions from this source will also increase. Research and development of new rice varieties, which can tolerate drier growing conditions, combined with improved cultivation techniques, are expected to help reduce methane emissions per unit of rice produced.

Another major source of methane is from biomass burning. Human-induced burning of grasslands and forests and wood for cooking and heating generate methane gas as a by-product. At present, it is estimated that human-related biomass burning releases around 40 Mt/year of methane. Changes in land-use practices and improved fuel wood combustion technologies can significantly reduce methane emissions from these activities.

Changes in atmospheric concentrations

As direct measurements of atmospheric methane concentrations only began in 1978 (see Figure 5), continuous time series data is limited. Historical data on methane concentrations are mainly derived from the study of ice cores and carbon isotopes contained in sediments. In the past hundred million years, there have been several large-scale methane emission events that have led to increases in atmospheric concentrations of CO$_2$ (methane oxidizes to CO$_2$ relatively quickly). These events led to significant increases in global temperatures (see **methane hydrates**).

Reasonably reliable historical ice core data date back more than 650,000 years. These records show that atmospheric methane concentrations have varied up and down with the earth's regular glacial and interglacial cycles (see **Milankovich cycles**). As with CO$_2$ concentrations, methane concentrations are normally higher in warmer interglacial periods and lower during glacial periods (ice ages). For example, at the peak of the last glacial period (around 20,000 years ago), methane concentrations were estimated to be as low as 350 ppb but rose rapidly during the early Holocene (around 8,000 years ago) to reach around 700 ppb.[6] In the 8,000 years to 1750, methane concentrations remained relatively stable at around 700 ppb.[7]

Since 1750 methane concentrations have increased by more than 150%, largely attributed to human activities. Concentrations increased relatively slowly up to 1900 but rose rapidly during the twentieth century. By the time atmospheric methane concentration measurements began in 1978, they stood at 1,520 ppb, or more than double preindustrial levels. During the 1980s, concentrations were increasing at an annual average rate of about 15 ppb, but the rate of increase began to slow appreciably in the 1990s, averaging only around 6 ppb/year in the first part of the decade. Since the late 1990s, concentrations have remained relatively constant at around 1,765–1,775 ppb. The 2005 concentration was 1,774 ppb.[8]

The recent leveling off of methane concentrations is welcome news, but the reasons for this stabilization have, until recently, puzzled scientists. Recent research has improved our understanding of methane emission sources and sinks, and this has helped shed some light on this issue. There appear to be both natural and human-related reasons for the recent leveling off in methane concentrations.

One contributing factor has been an increase in the abundance of OH in the atmosphere, which has increased by 1.4% since 1990, driven by the rise in **ozone** abundance attributed to the increased incidence of lightning (which triggers chemical

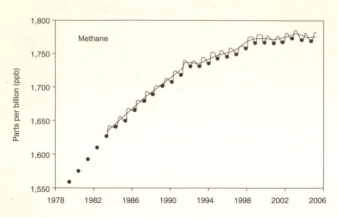

Figure 5 Global average methane concentrations in the atmosphere[9]

Source: US National Oceanic and Atmospheric Administration (NOAA).

reactions that produce OH). Surface warming has resulted in more convective activity (warm moist air ascending to form clouds) and, thereby, more lightning. The existence of more OH means that methane is oxidized more quickly than was the case previously. When combined with increased tropospheric temperatures (global warming), the result has been to reduce the atmospheric lifetime of methane by an estimated 0.17 years.[10] Although there is considerable uncertainty over whether OH concentrations will continue to rise in future, it is evident that changing OH abundance has created a slight negative **climate change feedback** effect by reducing methane's atmospheric residence time.

Changes in the amount of methane emitted from plants may also be part of the reason why methane concentration increases have slowed since the late 1980s. Large-scale deforestation of tropical forests over the past decade has reduced forest cover and, thereby, methane emissions from plants. Tropical forest cover was reduced by 12.3% during the 1990s.[11] Recent research suggests that methane emissions from tropical forests have declined by up to 20 Mt since the early 1990s, which would have slowed the buildup of methane in the atmosphere.[12] When these findings were released, some media commentators asserted, incorrectly, that deforestation reduced greenhouse gas emissions and that reforestation (planting trees) may not be the answer to reducing global warming. Although reforestation of cleared lands may increase methane emissions from plants, the impact on greenhouse gas concentrations is insignificant compared with increased CO_2 sequestration from reforestation activities (see **biosequestration**). Elevated methane emissions from plants would only reduce the net benefit of reforestation by 1–4%. It is also evident that the large increase in CO_2 emissions from deforestation (accounting for around one-quarter of anthropogenic CO_2 emissions) far outweighs any benefits from reduced methane emissions.

Another contributing factor is the fall off in methane emissions from tropical wetlands due to increased drought. Since the late 1990s, many tropical forest regions have experienced drier than normal conditions, which has led to the drying out of some tropical wetlands and bogs.[13] This was most pronounced in Indonesia in association with the strong **ENSO** event of 1997–1998.

Contributing anthropogenic factors are believed to include a substantial reduction in energy sector-related emissions from Russia and other Eastern European countries following the collapse of economic activity during the 1990s, significant progress in reducing anthropogenic emissions from fossil fuel production (reduced venting and leakage from petroleum and coal production), increased capture and flaring of fugitive methane from landfill sites, and better agricultural waste management (mainly improved handling of animal manure).

Projected trends in methane concentrations

Projecting future methane concentrations is subject to considerable uncertainty due to uncertainties over the trends in anthropogenic emissions and the extent of any feedback effects from climate change. Increased lightning activity and OH abundance in the atmosphere may strengthen the negative **climate change feedback**, but this is expected to be outweighed by the expansion in wetland emissions in a warmer and wetter world.[14] Methane emissions from plants, due to enhanced photosynthetic activity (CO$_2$ fertilization effect), are also expected to rise in future.[15] However, by far the greatest uncertainty surrounds the potential contribution from **methane hydrates**. The accelerated thawing of the permafrost will most certainly result in increased methane emissions (see **polar impacts**). Large-scale releases from seabed methane hydrates are considered unlikely this century but remain a significant long-term risk. Overall, natural sources of methane emissions are likely to increase in importance as the planet warms, possibly surpassing anthropogenic sources sometime this century.

Estimating emissions from anthropogenic sources is subject to less uncertainty but remains highly dependent on the extent to which methane emission control measures are introduced. Considerable potential exists to reduce anthropogenic emissions in the energy sector, landfills, waste management, and, to some extent, biomass burning. Opportunities for reducing emissions from ruminants and rice production exist but are more limited. The rapid expansion in energy production and consumption projected for the next 30 years, and the continued expansion in livestock numbers and rice production are expected to result in increased anthropogenic methane emissions over the medium term.

Future projections of methane concentrations have a much lower level of confidence than those of CO$_2$. This is primarily due to uncertainty over how natural methane sources will respond to climate change. It is also compounded by the short atmospheric residence time of methane, which means atmospheric concentration levels can vary significantly over just a few decades if there is a large increase or decrease from the major emission sources. In the short to medium term, anthropogenic emissions are estimated to increase from 6,000 MtCO$_2$e (in 2000) to as much as 8,000 MtCO$_2$e by 2020.[16] Such an increase would result in a rise in atmospheric concentrations. In the longer term, based on the IPCC emission scenarios (see **future emissions trends**), methane concentrations are expected to either increase over this century, by up to 1,000 ppb under the highest emission scenario, or fall by up to 200 ppb under the low emissions scenario.[17] These projections do not include any significant releases from the **methane hydrates**. This presents a very wide range of possible outcomes, from a significant increase in the **radiative forcing** from atmospheric methane to a significant decrease in methane's share of global radiative forcing.

See also: anthropogenic greenhouse gas emissions, carbon dioxide (CO_2), climate change feedbacks, fossil fuels, future emissions trends, global warming potential, greenhouse gases, methane hydrates, polar impacts.

Notes

1 IPCC 2007
2 Ibid.
3 Shindell *et al.* 2005
4 WRI 2005
5 Keppler *et al.* 2006
6 McDonald *et al.* 2006
7 British Antarctic Survey 2006
8 IPCC 2007
9 Data source: U.S. National Oceanic and Atmospheric Administration (NOAA).
10 Fiore *et al.* 2001
11 FAO 2001
12 Keppler *et al.* 2006
13 Fiore *et al.* 2006
14 Shindell *et al.* 2004
15 Keppler *et al.* 2006
16 Scheehle and Kruger 2007
17 IPCC 2007

Further reading

Shindell *et al.* 2005; IPCC 2007; Fiore *et al.* 2006; Scheehle and Kruger 2007.

METHANE HYDRATES

Methane hydrates (also referred to as *methane clathrates*) are deposits of **methane (CH$_4$)** gas trapped within ice in the permafrost and ocean floor sediments. Significant quantities of free methane gas bubbles are also trapped beneath the solid methane hydrate layers in the ocean.[1] Hydrate deposits are generally stable at low temperatures and high pressures, which occur at depths greater than 300 m, but can become unstable as temperatures increases and can dissociate (break down) and release methane gas. One cubic meter of solid methane hydrate can contain as much as 170 cubic meters of methane gas.

Methane hydrates are part of the **carbon cycle**, and their formation and release vary according to different stages of glaciation and deglaciation. Exchanges (*fluxes*) of carbon in the methane hydrates to and from the ocean and atmosphere reservoirs are relatively small compared with ocean–atmosphere and land–atmosphere exchanges. Nevertheless, at certain times in the earth's history, there have been large-scale releases of methane (at times totaling several thousand billion tonnes (gigatonnes, Gt) to the atmosphere. At present it is estimated that around 10–15 million tonnes of methane from the hydrates are being released to the atmosphere each year, mainly from thawing permafrost.

Methane hydrates are vast and are the largest single reservoir of organic carbon. It is estimated that between 5,000[2] and 10,000[3] Gt of carbon are stored in methane hydrate deposits, or up to twice that of all **fossil fuels** contained in the earth's crust. Most of the methane hydrates (more than 95%) are contained in the ocean floor sediments, with the remainder in land-based permafrost. There have been no estimates made of the size of the methane hydrate deposits contained in Antarctic regions.

Unsuccessful attempts at extracting methane from the hydrate deposits were made by Russia in the 1960s and 1970s. More recently Japan has undertaken some preliminary exploration (drilling commenced in 1999) to assess their energy potential. The United States has also allocated funds to assessing the energy supply potential of the hydrates. Extracting methane from hydrate deposits is possible but technically challenging and expensive. Nevertheless, they may provide a source of energy in the future.

How are methane hydrates formed?

Methane hydrates are formed when organic matter is frozen or trapped in an oxygen-free environment. On land, methane hydrates are formed through the freezing of organic carbon rich areas, such as frozen wetlands and peat bogs. Around one-quarter of the earth's land surface is subject to permafrost. Large quantities of frozen organic matter are contained in the permafrost regions in the Arctic regions, particularly Siberia, Alaska, and northern Canada. Since the end of the last ice age (around 14,000 years ago), the permafrost has been thawing and slowly releasing methane and **carbon dioxide (CO_2)** to the atmosphere.

Methane hydrate deposits in the ocean bed are formed by a combination of the sedimentation of organic matter, temperature, and pressure. Most organic carbon deposited in ocean sediments is oxidized relatively quickly to either CO_2 or dissolved inorganic carbon, but when sedimentation rates are high enough (greater than 1 cm/year), organic matter can be trapped in an oxygen-free environment, where methane is produced through anaerobic decomposition. Through a combination of pressure and low temperature, methane hydrate layers are formed. Areas of methane hydrate formation are confined to high latitude land areas (where temperatures are low enough to freeze water and create permafrost) and the continental shelves (where most of the deposition of organic rich material occurs). Sedimentation rates of the continental margins are uneven, and as a result, the size and thickness of methane hydrate deposits vary.

Methane can be found in two forms in the ocean sediments. It is either locked in stable hydrate structures (frozen) or occurs as free methane gas bubbles trapped below the hydrate layers. In high latitudes, ocean temperatures are cold enough for hydrates to form at relatively shallow depths. At lower latitudes, most hydrate layers occur on the continental margins in waters 300–2,000 m deep. At these depths, the temperature and pressure are sufficient to ensure the long-term stability of the methane hydrate structures. The thickness of the hydrate deposits is influenced by the thermal gradient of the underlying earth (temperatures increase as you go deeper into the earth), and some are believed to be more than 500 m thick. At certain sediment depths, the earth is too warm to keep methane hydrates stable and they break down to form water and free methane bubbles underneath the stable hydrate layers.

Past methane hydrates releases

Sudden large-scale releases of methane from the hydrates can cause rapid changes in the earth's climate. By studying paleoclimate records of the abundance of different carbon isotopes in the atmosphere, scientists have been able to identify several major methane hydrate release events in the past. Methane hydrate has a unique carbon isotope signature that is much lighter than carbon isotopes from other sources. Periods which show a high concentration of light carbon isotopes have generally coincided with warm climatic periods. Many scientists believe that the releases of methane from the hydrates have been a major contributor to high planetary temperatures during these periods.

An often cited event occurred around 55 million years ago in the Paleocene called the Late Paleocene Thermal Maximum (LPTM). This was a period of elevated warmth for the planet, and global mean temperatures were around 5–6°C higher than today, and as much as 8°C higher at the poles.[4] The significant rise in global temperature during the LPTM is largely attributed to large-scale releases of methane from hydrates and substantial releases from the **land carbon sink**.[5] It is estimated that around 2,000 Gt of methane were released to the atmosphere in less than 10,000 years, and possibly as little as 1,000 years. Scientists believe that submarine volcanic activity may have been the initial trigger that ignited methane hydrates in the ocean floor sediments. The resulting increase in atmospheric greenhouse gas concentrations raised global temperatures, which eventually triggered additional releases of methane from the hydrates, leading to further global warming, thus creating a self-reinforcing positive **climate change feedback**. It is believed that it took at least 20,000 years for the oceans to absorb most of the excess atmospheric CO_2.[6] This resulted in ocean acidification and oxygen depletion as well as mass extinctions of some marine species (see **marine impacts**).[7] It took more than 100,000 years for the oceans to return to normal alkalinity levels.

In more recent times, scientists also believe that methane releases from the permafrost in the early Holocene period (8,000–12,000 years ago) is one reason why the planet warmed so quickly and there was such rapid deglaciation.[8]

Possibility of future releases

Rises in global surface and ocean temperatures will inevitably lead to increased emissions of methane from the hydrates, although it may take several centuries for large-scale releases to occur. Initially, thawing of Arctic permafrost will be the primary mechanism of methane hydrate dissociation, and this process is already well underway. It is estimated that over a million square kilometers of Siberian permafrost has begun to thaw over the past decade, and it is expected that emissions of methane from this area will increase significantly over this century.[9] Methane emissions from the Siberian permafrost have already increased by 60% since the 1970s and are presently estimated to be releasing an additional 4 Mt/year of methane (or more than 80 $MtCO_2e$) to the atmosphere.[10] In some areas of Alaska where significant warming has occurred, the rate of increase is even higher.

In the oceans, methane can be released through sea floor slumping and/or venting of methane gas bubbles that underlay the hydrates. Sea floor slumping can result in sudden large-scale releases. Slumps can be triggered by earthquakes or through temperature-induced instability of the sediments on the continental margins. As ocean temperatures rise, the sediments expand and become more unstable. When submarine slumps occur, methane hydrates are exposed to warmer ocean temperatures and can lead to rapid dissociation of the hydrates. This allows methane to escape into the ocean, some of which can migrate through the water column to the atmosphere. Marine sediment slumping can also trigger tsunamis with major impacts on coastal regions. As sediments are spread unevenly throughout continental slopes, it is likely that some areas may be more prone to slumping than others.

Venting of methane bubbles occurs through fractures that cause columns (like chimneys) extending through the sediment layers to the ocean floor. As the sediments warm, the pressure of the methane bubbles increases and at a critical point the vents can open and vent methane into the ocean, some of which may reach the atmosphere. When the pressure falls sufficiently, the vents close. In effect, they operate like a steam valve on a pressure cooker. Scientists are uncertain how much methane could be vented through this process or how sensitive these releases are to ocean temperature changes.

Considerable uncertainty surrounds the sensitivity of the methane hydrates to increased temperatures as well as the magnitude of the releases for a given temperature change. Estimates of the warming required to trigger a significant methane hydrate release vary considerably. Some scientists consider that if global temperatures increased by 5°C, it would result in the eventual release of 2,000 Gt of carbon in the form of methane or CO_2.[11] Other scientists consider that methane hydrates are much more temperature sensitive and that up to 85% of the methane contained in the hydrates, and free gas underlying the hydrates (or up to 8,500 Gt of carbon), could be released to the atmosphere if ocean temperatures rise by as little as 3°C.[12] Even the smaller release of 2,000 Gt is equivalent to two and a half times the amount of carbon presently in the atmosphere. Obviously the impact on atmospheric greenhouse gas concentrations, and hence global temperatures, would be dramatic.

Increased surface temperatures will initially impact the land-based methane hydrates deposits or those in the shallower waters of the Arctic Ocean. If global temperatures rise by 3–4°C, as many expect could happen this century, eventually most of the methane contained in the permafrost is expected to be released to the atmosphere, though this may take several centuries. In total, up to 500 Mt of carbon could be released from thawing permafrost. The impact on global temperatures would depend on how quickly the methane was released, but it is possible that releases from permafrost could accelerate global warming during this century.

The hydrate deposits in ocean floor sediments are vast but are relatively well insulated from changes in surface temperatures. Most of the methane is believed to be trapped quite deeply in the marine sediments – it is thought that it would take many centuries for sufficient heat to penetrate the sediment layers and destabilize the hydrates. It is not known with any certainty how temperature sensitive the seabed methane hydrate deposits are or what the release thresholds would be at different locations.

In some areas, ocean surface temperatures have already increased rapidly (e.g. the North Sea[13]), but it will take some time before surface ocean heat penetrates the deeper ocean due to the ability of the oceans to absorb vast amounts of heat and the slow turnover of ocean waters. So far the average temperature increase of the mid-oceans (to depths of 3,000 m) has been relatively small, around 0.06°C, or about one-tenth of the global average atmospheric temperature increase so far. While a 0.06°C temperature rise is small, scientists did not expect ocean temperatures to increase so quickly at this depth, and the deep ocean has warmed by around 0.005°C (at 5,000 m) when it was not expected that there would be any warming at all.

Based on the atmosphere–ocean temperature increases of the past 50 years, a global temperatures rise of 3–4°C this century is likely to result in a rise in mid-ocean temperatures of less than 1°C by 2100. The risk of a temperature change of this magnitude causing significant methane hydrate releases is considered very small. However, the faster surface temperatures rise, the higher the risks of release. As ocean temperatures are likely to eventually catch up to surface temperatures (once they stabilize), significant releases of methane from the hydrates are inevitable, even though this may be several centuries into the future. Nonetheless, some scientists consider that major releases could be sooner than we previously believed and possibly this century.[14]

Any releases will lead to more warming, accelerating the release of even more methane from the hydrates in a positive **climate change feedback**. Once large-scale releases are triggered, they are likely to be self-perpetuating and beyond the ability of humans to control. However, due to the increasing depth of the remaining hydrates, this feedback process is largely self-limiting: temperatures would need to rise to very high levels to enable sufficient heat to penetrate the very deep sediments.[15]

The impact of any methane hydrate releases depends on the size of the release and the time period in which the releases occur. Small releases over longer time periods would be oxidized in the ocean or consumed by *methanotrophs* (methane-consuming organisms in the ocean), and little would be expected to reach the atmosphere as methane. Oxidation of methane and/or consumption by methanotrophs uses up oxygen and can lead to ocean anoxia (oxygen depletion). This would impact negatively on marine species, as has been observed from past events.

Although hydrate methane release is expected to be relatively slow in the initial ocean warming phases, as temperatures rise, a growing proportion of the hydrate deposits are likely to approach their release thresholds. Rather than slow leakage, large-scale slumping could result in sudden and abrupt releases. If large quantities of methane were released quickly from the hydrates, it is likely that much would escape to the atmosphere as methane, rather than CO_2, which could result in a sudden warming jolt. The impact of such a jolt on global temperatures could dwarf the amount of warming associated with the **anthropogenic greenhouse gas emissions** that initially triggered the release.

In summary, it is evident that the global temperature increases projected for this century (2–5°C) will eventually lead to significant releases of methane from the methane hydrates, initially from thawing permafrost, and over the next few centuries, as ocean temperatures catch up with surface temperatures, the likelihood of major releases from the ocean sediments will increase. Although there appears to be a low probability of significant releases from ocean sediments this century, the potential long-term consequences for **global warming** are very significant indeed. The disquieting aspect is that once mid-ocean temperatures increase by 3°C or more, it is inevitable that large quantities of methane will eventually be released.

See also: carbon cycle, carbon dioxide (CO_2), climate change feedback, global warming, global warming potential, marine impacts, methane (CH_4).

Notes

1 Dickens *et al.* 1997
2 Buffet and Archer 2004
3 Kvenvolden 1998
4 Katz *et al.* 1999
5 Pagani *et al.* 2006
6 Malthe-Sorrenssen *et al.* 2004
7 Zachos *et al.* 2003
8 Nisbet 1990
9 *Guardian,* August 11, 2005
10 Walter *et al.* 2006
11 Hornbach *et al.* 2004
12 Buffett and Archer 2004
13 Levitus *et al.* 2000
14 Wood *et al.* 2002
15 Nisbet 1990

Further reading

Buffett and Archer 2004; MacDonald 1990; Walter *et al.* 2006.

MILANKOVICH CYCLES

The Milankovich cycles refer to changes in the earth's orbit around and orientation toward the sun and have an important bearing on long-term changes in the earth's climate. They are named after the Serbian astronomer, Milutan Milankovich, who published the "Canon of Insolation of the Ice Age Problem" in 1941.[1] In this work, he identified three principle cycles in the earth's orbit and orientation toward the sun that influenced climate over time and particularly the cyclical recurrence of cold periods (ice ages).

Orbit variations

About every 100,000 years, the earth's orbit around the sun varies from nearly circular to elliptical (elongated). At present, the earth is in a reasonably circular orbit and the amount of solar energy (sunlight) hitting the earth varies by about 6% between summer and winter. When the earth's orbit is at the maximum elliptical extent, the amount of solar radiation hitting the earth varies by as much as 20–30% between midwinter and midsummer. This has a significant effect on the earth's climate. It is important what time of year the earth is closest to the sun. If it is during the northern summer, it yields hotter than normal summers – it would of course be the opposite in the southern hemisphere.

Axis tilt

About every 42,000 years, the tilt of earth's axis ranges from 21.8–24.4 degrees. The tilt determines the angle that solar radiation (see **greenhouse effect**) actually hits the earth's surface. The greater the angle, the larger the difference between summer and winter temperatures. We are presently near the middle of this range at around 23.5 degrees, but it is slowly decreasing. It will be more than 20,000 years before we are back in the same position, and over the next 10,000 years, the mean temperature variation between winter and summer will be reduced (in the absence of any human-induced changes).

Orientation

About every 22,000 years, the orientation of the earth's axis changes between the Pole star and the Vega star. This determines what time of year summer and winter occur. At present, the northern summer is centered around the July–August period, but in 10,000 years from now, the northern summer will be centered in January–February.

Changes in earth's orbit around the sun and the tilt and orientation of earth's axis have only a very small effect on the amount of solar radiation reaching earth each year (it only varies by about 0.1% during the entire Milankovich cycle). However, this is enough to cause a variation in average global temperatures by as much as 4–5°C. Summer temperatures in the Northern Hemisphere are likely to be low at times when the cycles synchronize and the tilt of earth's axis is small and the orbit is very elongated, and summer occurs when earth is furthest from the sun. Though the timing of each cycle can vary, the synchronization or near synchronization of cycles has been shown from ice-core records to occur about every 120,000 years. In higher latitudes during these periods there is insufficient solar radiation to melt the previous winter's snow and ice, altering earth's **albedo**. This has in the past led to the onset of new ice ages.

How much have the Milancovich cycles contributed to recent observed changes in global mean temperatures?

Greenhouse gas emissions, the Milankovich cycles, and variations in the sun's solar energy output, all have an impact on global climate forcing. Over history these have caused significant shifts in the earth's climate. In the context of the claims of "climate skeptics" that recent variations in climate are a result of natural variation, including Milankovich cycles, it should be noted that Milankovich cycles do lead to variations that affect Earth's climate, but they occur over such long timescales (tens of thousands of years) that these alone cannot explain the magnitude of the recent **global warming** that has been observed. By comparison, changes in atmospheric greenhouse gas concentrations can result in rapid climate changes comparable in magnitude to the Milancovich cycle.

Over the shortest Milankovich cycle of about 20,000 years (5°C warming in 10,000 years, followed by similar cooling over subsequent 10,000 years), global mean temperature increases/decreases by 0.025°C per century. By comparison, the earth's average mean temperature over the past 60 years has increased at a rate of just over 0.1°C/decade, and by 0.2°C/decade since 1980.[2] When converted to a century timescale, this is a rate of warming 80 times faster (nearly 2 orders of magnitude) than from the maximum likely natural variation due to Milankovich cycles. Furthermore, the projected increase in **anthropogenic greenhouse gas emissions** over the coming decades is expected to accelerate the decadal rate of change in global mean temperatures beyond 0.2°C/decade, unless there are substantial reductions on future emission levels (see **future emissions trends**).

Milancovich cycles do influence the earth's climate and cause global mean temperatures to vary by as much as 5°C when the three cycles coincide, but at a much slower rate than current observed warming. The current phase of the Milancovich cycles means that the earth is in a relatively warm period which, with all other factors being equal, could be expected to last for another several thousand years. However, the ongoing progression of the Milancovich cycles is likely to result in a cooling of the planet over the longer term and could initiate another ice age, but this would be many thousands of years into the future.

While the Milancovich cycles do influence climate, they are slow to manifest themselves and unlikely to be observable over periods of less than a century. The sheer length of time involved, and the slow rate of change, implies that the Milancovich cycles are not particularly relevant to the current debate on climate change.

See also: albedo, anthropogenic greenhouse gas emissions, climate change impacts, future emissions trends, global warming, greenhouse effect, greenhouse gases.

Notes

1 Pantic 1998
2 IPCC 2007

Further reading

Hays *et al.* 1976.

MITIGATION

Mitigation refers to human actions that prevent **greenhouse gases** from entering the atmosphere or to actions that remove greenhouse gases from the atmosphere. The primary objective of mitigation activities is to limit the increase in atmospheric greenhouse gas concentrations and, thereby, to reduce the extent of human-induced climate change the world will experience over the coming century and beyond.

Over the past two centuries, and especially since the 1950s, human activities such as **fossil fuel** burning, deforestation, agriculture, and industrial processes have released large quantities of **greenhouse gases** into the atmosphere (see **anthropogenic greenhouse gas emissions**). This has resulted in a 40% increase in atmospheric greenhouse gas concentration since preindustrial times.[1] By 2005, greenhouse gas concentration, converted to its **carbon dioxide equivalence (CO_2e)**, had reached 455 parts per million (ppm). The concentration level is expected to continue to rise by 2–3 ppm CO_2e/year over the next few decades.[2] The concentration of **carbon dioxide (CO_2)**, the most significant greenhouse gas, has risen from a preindustrial level of approximately 280 ppm to reach 382 ppm by 2007 (see **carbon dioxide [CO_2]**).[3]

If current **anthropogenic greenhouse gas emission** growth trends persist, atmospheric greenhouse gas concentration will reach double their preindustrial level before 2050 and triple sometime between 2075 and 2100 (see **future emissions trends**). The current scientific consensus is that a doubling of preindustrial concentrations would ultimately result in a 3°C rise in global mean temperature, and possibly as much as a 5°C increase (see **global warming** and **climate sensitivity**).[4] Thus, a doubling of preindustrial greenhouse gas concentration is likely to exceed the 2°C threshold commonly associated with dangerous levels of climate change and will increase the risk of triggering major **climate change feedback** mechanisms that could further accelerate future warming (see **dangerous climate change**).

The international community, through ratification of the **United Nations Framework Convention on Climate Change (UNFCCC)**, recognizes that atmospheric greenhouse gas concentrations need to be stabilized at a level that avoids **dangerous climate change**. The **Kyoto Protocol** is an important first step by the international community to limit **anthropogenic greenhouse gas emissions**, but it is clear that even if the Kyoto targets are achieved, much greater emission mitigation efforts will be required beyond 2012 if climate change is to be restrained to a manageable level (see **Kyoto Protocol**). While there is a general consensus that future emission levels need to be reduced well below present "business as usual" projections, considerable conjecture still surrounds the extent of emission reductions required, what constitutes a realistic atmospheric concentration stabilization target (see **stabilization targets**), the costs associated with achieving different emission reduction objectives, and the allocation of responsibility across different countries for delivering these emission reductions.

How large is the mitigation requirement?

The extent to which emissions need to be reduced over the period to 2050 and beyond largely depends on what level of climate change the international community considers to be acceptable and manageable. As long as greenhouse gas concentrations continue to rise, the world will be committed to ever-increasing levels of climate change. The future commitment to climate change will depend on when, and at what level, atmospheric concentrations are stabilized.

The present scientific consensus is that to restrain mean global temperature increase to no more than 2°C, atmospheric concentrations would have to stabilize between 450 and 500 ppm CO_2e, depending on how sensitive earth's climate is to elevated greenhouse gas concentrations (see **stabilization targets** and **climate sensitivity**) and the extent of any **climate change feedbacks**. Limiting mean global temperature increase to less than 3°C would require concentrations to be stabilized somewhere in the 500–600 ppm CO_2e range.

Aggregate global emissions now exceed 50 gigatonnes (Gt) CO_2e/year and are increasing by approximately 1 $GtCO_2e$/year (see **anthropogenic greenhouse gas emissions**).[5] Future emissions growth will be driven largely by population increase and by rises in per capita income (see **future emissions trends**). If current emission trends persist, and there are no major changes in existing policies to reduce greenhouse gas emissions (the "business as usual" scenario), global emissions are projected to reach 60–90 $GtCO_2e$/year by 2030,[6] with most projection studies converging on the range of 65–75 $GtCO_2e$/year.[7] By 2050, under business as usual conditions, aggregate global emissions could be 80–100 $GtCO_2e$/year, although emission projections this far into the future are subject to considerable uncertainty. Nonetheless, on current trends atmospheric greenhouse gas concentration is likely to be in the range of 550–650 ppm CO_2e by mid-century and possibly 750–900 ppm by 2100. The world is, therefore, currently on track to exceed the **dangerous climate change** concentration threshold within the next 20–30 years, if it has not done so already.

To stabilize concentrations, aggregate annual global emissions would need to fall to a level that matches the earth's natural capacity to remove greenhouse gases from the atmosphere – estimated to be between 10 and 20 $GtCO_2e$/year (65–75% below present levels). However, in the absence of concerted emission mitigation efforts, global emissions in 2030 are expected to be triple this level and quadruple by 2050. The eventual level at which concentrations are stabilized will depend on how quickly emissions can be brought down to the natural rate of absorption. The longer it takes to achieve this, the higher the concentration level and, consequently, the greater the increase in global temperature (see **stabilization targets**).

To stabilize concentrations in the 450–500 ppm CO_2e range (in order to avoid the 2°C dangerous climate change threshold), most mitigation studies conclude that this would require emissions to be stabilized in the next few years (before 2015) and then reduced rapidly (by at least 3–4% per year) over the coming decades (see **stabilization targets**). Given the projected growth rates of the underlying drivers of emissions, notably population and economic growth (see **future emissions trends**), this represents a formidable mitigation challenge, and one that most analysts consider to be not realistically achievable. As a result, most attention has focused on what emission reductions would be required to stabilize concentrations in the 500–600 ppm range, with 550 ppm CO_2e the most commonly used target. To ensure that the 550 ppm CO_2e concentration level is not exceeded, global emissions would need to fall to approximately 35–40 $GtCO_2e$/year by 2030, 25–30 $GtCO_2e$/year by 2050 and 10–20 $GtCO_2e$/

year by 2100. This translates into emission reductions of at least 30 Gt below business as usual projections by 2030 and at least 50 Gt below by 2050.

What are the main mitigation options?

There are a wide range of options for reducing greenhouse gas emissions. These vary according to the particular greenhouse gas being mitigated. To date, most international attention has focused on measures to reduce CO_2 emissions, especially those related to fossil fuel consumption. Although CO_2 is the most important and fastest-growing source of anthropogenic emissions and is likely to increase in relative importance over the coming decades, emissions of **methane (CH_4)**, **nitrous oxide (N_2O)**, and man-made **synthetic gases** currently account for one-quarter of global emissions, and almost one-third of anthropogenic **radiative forcing** (see **anthropogenic greenhouse gas emissions**).

CO_2 mitigation options

The consumption of **fossil fuels** is the single largest source of CO_2 emissions (approximately 70%), although land-use (mainly agricultural soil tillage) and land-use change (mainly deforestation) are also significant sources (approximately one-quarter). The remainder emanates from industrial processes, such as cement manufacture (see **anthropogenic greenhouse gas emissions**).

Fossil fuels

The options for mitigating fossil fuel CO_2 emissions can be analyzed according to three principal economic activities: electricity generation (the single most important source of fossil fuel CO_2 emissions), transport, and direct use in industrial, commercial, and residential-sector applications.

Fossil fuels currently account for approximately two-thirds of global electricity production, with coal the single largest primary energy source. Mitigation options in electricity production can be divided into three categories: (1) reducing the carbon intensity of fossil fuel-based electricity (through more efficient conversion processes, substitution of coal with less carbon-intensive fuels, primarily gas – see **fossil fuels**), and **carbon capture and storage**; (2) utilizing nonfossil fuel generating technologies such as **renewable energy** (hydro, biomass, geothermal, ocean energy, **solar power**, and **wind power**) and **nuclear power**; (3) reducing the demand for electricity through **energy efficiency** measures.

Transport emissions currently account for 14% of global emissions and approximately one-quarter of energy sector CO_2 emissions and are one of the fastest growing sources of emissions. Unlike electricity generation, transport is nearly completely dependent on fossil fuels (about 2% of transport fuel demand is met by nonfossil fuel sources, mainly ethanol). Mitigation options in the transport sector are mainly limited to **energy efficiency** of motor vehicles and "modal shift" (transitioning from private car to public transport; walking and cycling; and for freight, from road to rail). Fuel substitution, primarily through the use of **biofuels**, natural gas, and electric vehicles (using nonfossil fuel sources of electricity), offers some potential but is expected to play a relatively limited role over the next few decades.

Reducing emissions from direct fossil fuel usage in industry, commercial, and residential sectors can be achieved mainly through **energy efficiency** measures, interfuel substitution (mainly using gas rather than coal, electricity, and oil), and substituting fossil fuels with other energy sources, primarily biomass and solar power (see **fossil fuels**, **renewable energy**, and **solar power**).

Land-use change

Mitigation measures include reducing deforestation (often termed *avoided deforestation*); sequestering CO_2 from the atmosphere through afforestation, reforestation, and land rehabilitation (see **biosequestration**); reducing wetland drainage; and reducing peat burning. Improved agricultural management practices, such as reduced or zero soil tillage also offer some prospects for reducing CO_2 emissions. Some studies suggest that significant reductions in deforestation are available at relatively moderate costs, some as low as $ 5/tonne CO_2 mitigated.[8] However, the potential for reducing emissions from land-use change is constrained by a range of complex social, cultural, and political factors, and cost-effectiveness alone will not necessarily be the greatest consideration.

Methane mitigation options

Methane currently accounts for approximately 14% of global anthropogenic emissions – livestock and manure (mainly from ruminant animals) are the largest anthropogenic sources of methane, accounting for nearly one-third of emissions. Energy-sector emissions (primarily fugitive emissions from coal mines and oil and gas production and distribution) account for approximately 30%; emissions from waste (landfill and waste water treatment) nearly one-quarter; and the remainder stems mainly from rice paddies and biomass burning. Significant progress has been made in recent decades in reducing emissions from waste (mainly from gas capture and flaring) and fugitive emissions from the energy sector (see **methane**). Mitigation options for reducing emissions from livestock and rice production are much more limited and face many economic, technical, and social constraints.

Nitrous oxide mitigation options

Nitrous oxide currently accounts for 8% of global anthropogenic emissions, principally arising from the application of artificial fertilizers in the agricultural sector. Industrial process emissions account for most of the remaining 10% of emissions. Mitigation options in the agriculture sector primarily involve better fertilizer management. In the industrial sector, cost-effective mitigation measures exist and are already being applied (see **nitrous oxide**).

Synthetic gas mitigation options

Although **synthetic gases** (mainly fluorinated gases such as CFCs, HFCs, PFCs, SF_6, and halons) account for only a few percent of global anthropogenic emissions, many of them are potent and long-lived greenhouse gases. Ozone-depleting CFCs and halons are being actively phased out under the Montreal Protocol, through the introduction of a range of new synthetic gases (mainly HFCs). However, many of these substitute gases are also strong **greenhouse gases**, and future mitigation options will involve replacing them, in turn, with more greenhouse-friendly gases and also preventing their release into the atmosphere (see **synthetic gases**). PFCs emanate primarily from the aluminium and semiconductor industries – the principal PFC mitigation option is preventing their release to the atmosphere. Mitigation options for SF_6 (the greenhouse gas with the highest **global warming potential**) mainly focus on better handling and gas recovery techniques (see **synthetic gases**).

The final major greenhouse gas is **ozone**, which is not currently covered under the Kyoto Protocol or any other international agreement. Emissions of surface ozone are mainly associated with **fossil fuel** combustion, so measures that reduce fossil fuel use will also contribute to lowering ozone emissions (see **ozone**).

The potential contribution of the different mitigation options

The magnitude of the mitigation task to stabilize concentrations at or below 550 ppm CO_2e (at least 30 $GtCO_2e$ by 2030 and 50 Gt by 2050) implies that mitigation measures will need to target all **greenhouse gases**, across all countries and regions – in both developed and developing nations. There is no single mitigation option that could deliver emission reductions of the magnitude required over the next few decades to achieve a target of 550 ppm CO_2e or lower.

The contribution from different mitigation options is dependent on a range of variables, including:

- The **stabilization target** that is adopted (e.g. 500 ppm CO_2e, 550 ppm, 600 ppm, or some other target), as the stringency of the target changes so to do the relative contributions different options can deliver, principally as stabilization targets are time constrained. For example, a 500 ppm CO_2e target would require substantial reductions in emissions over the next 20–30 years and, in this time frame, the contribution of some mitigation options (such as **solar power** and **carbon capture and storage)** would be very limited.
- The range of **greenhouse gases** targeted. Modeling suggests that a multigas approach would deliver far greater emission reductions, and at significantly lower cost: estimates range from 20% to 50% cost reduction relative to a CO_2-only strategy.[9]
- Assumptions about the extent of progress in developing low-emission technologies and their rate of deployment.
- The timing and magnitude of any greenhouse gas cost penalty imposed by governments (via emissions trading or a specific emissions tax). Imposing a US$ 100/tonne CO_2e cost penalty will deliver far greater emission reductions over a given period than a US$ 10/tonne penalty.
- The political will to introduce the necessary policies and measures to enable different mitigation options to deliver their full economic potential as well as the response of industry and consumers to these policies.

To put the emission reduction task in perspective, and the potential role of different options, Pascala and Socolow put forward the "Stabilisation Wedges" concept.[10] Although this approach focuses only on CO_2 emissions, it provides a useful means of indicating the extent to which different technologies would need to be deployed to deliver 1 Gt of carbon emission reductions per year (3.7 Gt CO_2) by 2050 (equivalent to one wedge). To keep CO_2 emissions from exceeding current levels over the period to 2050, Pascala and Socolow suggest that seven stabilization wedges would need to be delivered. However, this merely stabilizes CO_2 emissions at current levels and would not stabilize atmospheric concentration (which would still be increasing by more than 2 ppm/year). To stabilize at 550 ppm CO_2e or less by 2050 would probably require the equivalent of nearly 14 CO_2e stabilization wedges.

The stabilization wedge serves as a useful reference unit. Each of the following mitigation actions would deliver one stabilization wedge by 2050: deploying two million 1 MW wind turbines; tripling present nuclear power generating capacity; completely halting deforestation and planting 300 million hectares of trees (see **biosequestration**); increasing the average **energy efficiency** of all the world's buildings by 25%; installing **carbon capture and storage** facilities in 800 large coal-fired power plants; doubling the fuel efficiency of the world's 2 billion cars that will be on the road by 2050; increasing **solar power** photovoltaic cell installed capacity to 700 times current levels; and converting 15–20% of the world's crop lands to **biofuels** production. Obviously, each wedge would require an accelerated rollout of alternative technologies, well beyond what would occur under business as usual policy settings. However, most of these wedges utilize existing technologies, and the deployment rates are considered technically and economically feasible given appropriate market incentives (particularly a cost penalty on emissions) and supporting regulatory frameworks.

Mitigation studies employ different techniques and models to determine the most cost-effective mix of mitigation options. Some use technology-based models (often termed "bottom-up models") and others econometric models ("top-down models"), or combinations of the two. The results, of course, vary according to the assumptions made about the costs of alternative technologies, rates of technological change, government policies (particularly carbon cost penalties and the use of the revenues generated), the level of international participation, and several other important variables. Bottom-up models tend to deliver more optimistic estimates of the contribution of different mitigation options compared with econometric models.[11]

When assessing the potential contribution of different options, it is important to distinguish between what is termed *economic potential* and *market potential*. Economic potential is what could be delivered if the markets operated perfectly and there were no major barriers to implementing the measures: in other words, if all measures were implemented according to the least-cost option. Market potential is what is expected to be delivered when existing barriers (e.g. lack of information, corporate inertia) and policies (energy subsidies and other policy settings that could reduce the uptake of low-emission technologies) are taken into account. Market potential is lower than economic potential.

Estimates of the potential of different options vary significantly. For example, the IPCC estimates that, at a cost of $20/t$CO_2e$ or less, the economic mitigation

Table 15 Estimated emission reduction potential of mitigation options by 2030 and 2050

Mitigation option	2030 Reductions (up to $50/tCO_2e)	2030 (Up to $100/tCO_2e)	2050 Reductions (up to $50/tCO_2e)	2050 (Up to $100/tCO_2e)
Energy efficiency (including transport)	7–10	8–12	12–15	12–18
Renewable energy	0.5–1	1–2	2–4	4–8
Nuclear	1	1–2	1–2	1–3
Fossil fuel switch and biomass cofiring	0.5–1	2–3	1–2	2–4
Carbon capture and storage	0–1	1–2	2–3	3–6
Agriculture (soils/fertilizers/livestock)	1–3	2–4	3–4	3–5
Waste	0.5–1	1	1–2	1–2
Biosequestration/avoided deforestation/biofuels	2–3	3–5	4–5	4–6
Industry (N_2O, CH_4, and synthetic gases)	0.5–1	1	1	1
Total	13–22	20–31	25–35	31–53

Source: Derived from data contained in IPCC 2007, IEA 2006b, Stern 2006, Enkvist et al. 2007, Metz and van Vuuren 2006, and Benitez et al. 2005.

potential is as much as 17 $GtCO_2e$/year by 2030, increasing to 26 Gt/year at US$50/t CO_2e and 30 Gt/year at $100/tCO_2e.[12] Other studies offer slightly lower estimates[13] and some higher estimates (as much as 40–50 $GtCO_2e$ at less than $100/tCO_2e).[14] Table 15 indicates the potential contributions of different mitigation options. The figures are presented as ranges of potential reductions, reflecting the considerable variation in the estimates from different studies.

As is evident from Table 15, the potential contribution from different mitigation options varies considerably according to the time frame considered and the willingness to accept higher mitigation costs. It is also evident that the mitigation goal of 50 Gt by 2050 to stabilize concentrations at or below 550 ppm CO_2e is achievable, but all mitigation options would need to deliver at the upper end of the estimated ranges contained in the studies reviewed.

Out to 2030, **energy efficiency** measures are by far the most important mitigation option, accounting for around half of the mitigation potential. This is primarily due to the length of time required to develop and deploy alternative technologies and the (generally slow) rate of turnover of the existing capital stock (power stations, appliances, cars, etc.). As large-scale electricity-generating plants have lifetimes of 40–50 years, it will take several decades before a majority of existing plant can be replaced. It is estimated that energy efficiency measures could deliver as much as 5–7 $GtCO_2e$ at negative cost (a net economic benefit).[15] Below US$25/tCO_2e, several other mitigation options offer some potential: low-cost reforestation and avoided deforestation; reductions in methane emissions from waste; reduced industrial process emissions; changed agricultural practices; and, to a limited extent, biomass, **biofuels**, and **nuclear power** in countries that face relatively high fossil fuel prices.

The role of **renewable energy** is expected to be quite small over the period to 2030. **Carbon capture and storage** is unlikely to make a meaningful contribution at costs of less than US$30–50/tCO$_2$e. Even at higher abatement costs, the contribution from **renewable energy** and **carbon capture and storage** remains limited, due to the time it would take to rollout these technologies on a large scale.

Out to 2050 and beyond, **energy efficiency** remains the principal mitigation option (potentially accounting for 25–40% of reductions), although the role of other mitigation options becomes more prominent, notably in the form of **renewable energy**, land-use change, **carbon capture and storage**, and, to a limited extent, **nuclear power**. Beyond 2050, technologies may emerge that could change the mitigation and cost mix: **solar power** may take on a more important role, as might nuclear fusion and a renewable-based hydrogen economy, although mitigation estimates remain speculative.

Overall, modeling studies suggest that emission reductions of up to 30 GtCO$_2$e by 2030 and 50 GtCO$_2$e by 2050 could be achievable at costs of less than US$100/tCO$_2$e. This would potentially enable the international community to stabilize greenhouse gas concentrations at 550 ppm CO$_2$e or less. To achieve stabilization at a lower concentration level would require more immediate, concerted action and probably involve higher costs/t CO$_2$e. Any significant delays in the implementation of mitigation strategies will make the task of limiting concentrations at or below 550 ppm CO$_2$e increasingly more difficult and costly.

Mitigation costs

Mitigation costs will vary across regions, between different sectors, and between different gases.[16] Mitigation cost estimates vary according to the assumptions made about which gases are targeted, rates of induced technological and structural change, and the degree of flexibility available in technologies and mitigation options. However, the most important determinant is the concentration target that the international community agrees to adopt and how the responsibility for emission reductions is allocated.

Most model projections indicate that the cost of reducing emissions tends to increase with the stringency of the concentration target adopted. The cost tends to be lower when all emission sources are targeted rather than just energy-sector CO$_2$ emissions. For example, the costs of stabilizing the greenhouse gas concentration in the 500–550 ppm CO$_2$e range is estimated to be around one-third the cost of stabilizing concentrations in the 450–500 ppm CO$_2$e range,[17] while the mitigation costs are generally around one-third lower if all **greenhouse gases** are targeted rather than just CO$_2$.[18] This is largely due to the higher annual rate of emission reduction required to achieve lower stabilization targets: more stringent stabilization targets would require accelerated turnover of the existing capital stock and more rapid structural change.

Emission reduction costs are generally higher in countries and regions that have already achieved low emission intensities through **energy efficiency**, structural change, or other greenhouse gas mitigation measures. Mitigation cost estimates are generally lower when other co-benefits of mitigation (such as reduced health costs from improved air quality, more productive agricultural systems associated with reversed land degradation, and enhanced energy security) are considered. Although co-benefits are often difficult to quantify, mitigation cost reductions of more than one-third have been estimated for some regions.[19]

As most models do not include induced technological change and technology learning curves, ancillary co-benefits, or the full suite of mitigation options, the costs of mitigation are generally overestimated by modeling studies (see **technology, structural change, and organizations**).[20] Furthermore, models do not generally model the potential contribution of voluntary changes in consumer behavior. For example, the economic benefit to individuals of recycling household waste is negligible (and arguably incurs a net cost to individuals when the opportunity cost of their time is included), but citizens in many countries nonetheless actively participate in recycling. Voluntary consumer mitigation behavior might be expected as awareness of the climate change challenge permeates more widely.

Costs are usually expressed as a percentage change in GDP at some future point in time (e.g. 2030 or 2050) relative to a situation without climate change mitigation. Modeling estimates usually do not include the estimated costs of climate change impacts, even though the net cost of mitigation should be assessed in terms of the aggregate cost of reducing emissions to achieve a specific concentration level (e.g. 550 ppm CO_2e) less the estimated cost of the climate change impacts that would occur in the absence of the mitigation measures (the value of avoided impacts). Except for very ambitious concentration targets (e.g. below 475 ppm CO_2e), the net economic benefits of mitigation (which exclude many ancillary and nonmarket benefits) generally exceed the benefits of inaction, and by a considerable margin once concentrations exceed the 500–550 ppm range (see **socioeconomic impacts**).

The more recent mitigation cost assessment studies estimate that stabilization of emissions at or around double preindustrial concentrations range from a small global GDP benefit (of around 1%) by 2050 or possibly a small reduction in 2050 GDP of up to 3%, with a best-guess estimate of between 0.5% and 1% relative decrease in 2050 GDP.[21]

It is important to clearly understand the significance (or in this case the insignificance) of this global cost estimate. A 1% reduction in 2050 GDP is sometimes misinterpreted as a situation where the world (as a whole) would be 1% worse off (in an economic sense) than it is today. However, as global GDP is expected to expand by approximately 3% per year over the coming decades, it is likely to be double current (2008) levels by around 2030 and nearly quadruple current levels by 2050. A 1% reduction in global GDP translates as a one-year delay in quadrupling global economic output. In other words, even if the mitigation cost estimates prove accurate (and many would assert they are overly pessimistic), then the global economy would be quadruple its current size in 2051, rather than 2050 (excluding the costs of **climate change impacts**).[22]

When viewed from this perspective, it is clear that mitigation measures represent a tiny cost imposition compared with the potential costs and risks of climate change. While there will inevitably be costs associated with structural change during the transition period to a less greenhouse gas-intensive economic system, and very large-scale investment will be required, the net costs are small.

In the energy sector alone, the IEA estimates that to maintain energy sector emissions at today's levels until 2050 would require US$ 13 trillion to be invested in low-emission technologies, but this would save nearly the same amount in avoided investment in new power plants and energy production facilities. In other words, such mitigation could be achieved at basically little or no net cost.[23] A number of recent studies also suggest that the induced technology development associated with a concerted global mitigation effort could actually increase global economic activity above what would otherwise be the case (see **technology, structural change, and organizations**).

In summary, it is evident that to maintain the atmospheric greenhouse gas concentration at a level that would constrain global mean temperature increase in the range of 2–3°C above preindustrial levels will require greenhouse gas emissions to be reduced significantly by mid-century. It is also evident that the technologies to achieve these reductions are already available and that these reductions could be achieved at little net cost to the global economy and may even result in an economic benefit. If the cost of **climate change impacts** is included, then mitigation would clearly lead to higher global GDP compared with a situation where no mitigation measures were taken. As such, it appears that the mitigation challenge is not meaningfully constrained by technical or economic factors, but by the political will to instigate the policies and measures to achieve the required emission reductions.

See also: anthropogenic greenhouse gas emissions, biofuels, carbon capture and storage, climate change impacts, dangerous climate change, energy efficiency, fossil fuels, future emissions trends, Kyoto Protocol, nuclear power, renewable energy, socioeconomic impacts, stabilization targets, technology, structural change, and organizations.

Notes

1 IPCC 2007
2 Concentration figures are for Kyoto gases only and exclude CFCs, halons, and ozone.
3 NOAA
4 IPCC 2007
5 The most recent global emission estimate is for 2004 (49 $GtCO_2e$/year), but due to the growth in energy consumption and global GDP since 2004 annual global emission levels are likely to have grown by an additional 3–5 $GtCO_2e$ (to 52–54 $GtCO_2e$/year by 2008).
6 IPCC 2007
7 Estimates of future emission levels vary considerably and depend on assumptions about future growth in global population, global GDP, and the rate and type of technological change (see **future emissions trends**).
8 Grieg-Gran 2006
9 Estimates of the cost reductions of a multigas versus CO_2-only strategy vary considerably, with most indicating a 20–30% reduction and some as much as 50%. For a review of different studies, see Weyant *et al.* 2007.
10 See Pascala and Socolow 2004 and Pascala 2006
11 See Barker *et al.* 2006
12 IPCC 2007
13 See Enkvist *et al.* 2007
14 See IEA 2006b and Stern 2006
15 IPCC 2007

16 See Stern 2006
17 Ibid.
18 See Weyant *et al.* 2007 and Metz and van Vuuren 2006
19 See Syri *et al.* 2001 and van Vuuren *et al.* 2006
20 Stern 2006
21 See Stern 2006 and IPCC 2007
22 See Azar and Schneider 2002
23 IEA 2006b

Further reading

IPCC 2007; Stern 2006; Enkvist *et al.* 2007; Metz and van Vuuren 2006; Azar and Schneider
    2002.

NITROUS OXIDE (N$_2$O)

Nitrous oxide (or dinitrogen oxide) is a colorless gas that consists of two nitrogen atoms and one oxygen atom (N$_2$O). It has a slightly sweet odor and is a mild anesthetic, also known as "laughing gas." It is solid at temperatures below $-91°C$ and becomes a gas at $-88°C$. N$_2$O is about 50% more dense than air and is highly soluble and stable. N$_2$O slowly breaks down through photochemical dissociation in the stratosphere (reacting with **ozone** in the presence of sunlight) to free nitrogen (N$_2$) and oxygen (O$_2$) gas.

N$_2$O is a powerful and long-lived greenhouse gas, though present in the atmosphere in only very small quantities. It has an atmospheric lifetime of 114 years and a **global warming potential (GWP)**[1] of 298.[2] In the 11,000 years prior to 1750 (preindustrial times) atmospheric concentrations of N$_2$O had been relatively stable at around 270 parts per billion (ppb) but have since increased by more than 20% and currently stand at approximately 320 ppb (0.000032% of atmospheric gases).[3]

N$_2$O is the fourth most important contributor to **radiative forcing** of the long-lived **greenhouse gases** after **carbon dioxide (CO$_2$)**, **methane (CH$_4$)**, and CFC-12 (see **ozone**). The current **radiative forcing** of N$_2$O is 0.16 Wm^{-2}, approximately 6% of the total positive radiative forcing (a warming effect) from the long-lived **greenhouse gases**.[4] Although N$_2$O is a far less important contributor to global warming than CO$_2$ and methane, emissions and concentrations have, nonetheless, been increasing at a relatively constant rate over recent decades and are expected to maintain this trend for at least the next few decades.

Sources of nitrous oxide

Approximately 60% of N$_2$O emissions arise from anthropogenic sources and 40% from natural sources.

Natural sources

Natural emissions of N$_2$O arise primarily from a series of complex biological processes in the nitrogen cycle. Nitrogen-fixing bacteria in the soil and/or root nodules in legume plants (such as clover, alfalfa, and soybeans) reduce free N$_2$ to ammonia (NH$_3$), which is then taken up by plants to support amino acid production and several other plant processes. NH$_3$ can also be oxidized by bacteria to form nitrites (NO$_2^+$) and then nitrates (NO$_3^{2+}$) in a process called *nitrification*. Other bacteria and fungi convert nitrates back into N$_2$, N$_2$O, and other oxides of nitrogen (referred to as NO$_x$: nitric oxide [NO] and nitrogen dioxide [NO$_2$]) in a process called *denitrification*. Aquatic environments such as lakes, wetlands, and oceans, where excess accumulations of dissolved nutrients in water occur through high sedimentation rates and/or thermal stratification (termed *eutrophication*) results in depleted oxygen levels, creating anoxic, or anaerobic (low or without oxygen) conditions. This enables the denitrification process, which results in the release of N$_2$, N$_2$O, and NO$_x$. Dentrification in soils and aquatic environments account for more than 95% of N$_2$O emissions from natural sources.[5] The remaining natural N$_2$O emissions emanate from atmospheric reactions via the oxidation of NH$_3$, which is produced from the digestion processes of ruminant animals and animal manure.

N$_2$O emissions from natural sources are difficult to quantify accurately and are subject to considerable uncertainty. This is due to the large number of diffuse sources and the significant variation in ambient environmental conditions that govern emissions from biological processes.[6] Nonetheless, best estimates of the natural sources of N$_2$O from the nitrogen cycle are approximately 11 TgN/year (5.2 billion tonnes **carbon dioxide equivalent (CO$_2$e)** per year), about 60% from soils and 35% from aquatic sources.[7] Emissions from natural sources have fallen since preindustrial times by as much as 9% as a result of the 30% decline in forest cover from deforestation (see **biosequestration**).[8]

Anthropogenic sources

Eighty-five percent of anthropogenic N$_2$O emissions arise from agricultural activities, 10% from the energy sector, and the remaining 5% from industrial processes and human sewage.[9]

The rapid expansion and intensification of agricultural production since 1950 has been the primary driver of anthropogenic emissions, largely due to fertilizer use (fertilizer consumption has grown 20-fold since 1950). Fertilizers increase soil biological productivity and hence dentrification. They are the single most important source of N$_2$O, accounting for 75% of all anthropogenic N$_2$O emissions. Human activities have added three to five times more reactive nitrogen to the biosphere than what would normally be available through the natural nitrogen cycle. Much of this additional nitrogen is taken up through increased plant growth but, as farmers generally "overdose" soils with nitrogen to ensure crop production is maximized, as much as 5% of available NH$_3$ is not taken up by plants.[10] The excess NH$_3$ either promotes microbial denitrification in soils or is washed out through surface and ground waters to terrestrial aquatic systems or the ocean where denitrification may take place: in either case, N$_2$O and NO$_x$ emissions are generated. The remaining N$_2$O emissions from agriculture arise primarily from livestock manure management (accounting for around 6% of anthropogenic N$_2$O emissions) and the combustion of residues and savanna burning (around 4% of N$_2$O emissions).[11]

Energy sector N_2O emissions arise primarily from the combustion of **fossil fuels** and **biofuels**. When these are combusted at high temperature and pressure, such as in an internal combustion engine, N_2O and NO_x emissions are generated. Fossil fuel combustion accounts for around 9% of anthropogenic N_2O emissions and biofuels around 1%. Although **biofuels** account for only a very small proportion of energy consumption, they have a much higher nitrogen content than fossil fuels and a significantly higher rate of N_2O emissions per unit of energy output.

The remaining 5% of anthropogenic N_2O emissions come mainly from industrial sources, particularly the manufacture of fertilizers, the manufacture of nylon, the use of ammonia as a solvent, and in nitric and adepic acid production. Landfills and sewage in waste treatment plants are also small sources of N_2O.

Indirect effects of nitrogen compounds

Although N_2O is the only nitrogen gas compound that directly contributes to **global warming**, several other nitrogen compounds indirectly influence global temperatures.

NO_x emissions from combustion are precursors to the formation of short-lived tropospheric **ozone** and so have a strong indirect warming effect. NO_x also has an indirect cooling effect by shortening the life of **methane**, though this effect is small.[12] Ammonia (NH_3) also has both an indirect warming effect (N_2O is produced when it oxidizes) and a cooling effect (it contributes to the formation of sulfate and nitrate **aerosols**). Overall indirect effects of nitrogen compounds result in a significant net warming, though the magnitude of this indirect effect cannot be separated from other precursors (see **ozone** and **aerosols**).

Changes in atmospheric N_2O concentrations

As with **methane**, direct measurements of atmospheric concentrations of N_2O only commenced in 1978. However, reliable historic data reconstructed from ice cores and other proxies show that, prior to 1750, atmospheric N_2O concentrations had not varied outside the range of 200–280 ppb in at least the last 650,000 years.

From 1750 to 1900, N_2O concentrations grew only slowly, by around 8 ppb (0.2% per decade). The rate of increase accelerated between 1900 and 1950 to 0.6% per decade (reaching 285 ppb by 1950), mostly as a result of fossil fuel combustion emissions. However, since 1950, N_2O concentrations have grown much more rapidly, averaging nearly 3% per decade, due primarily to the expansion of agriculture and fertilizer use.[13] The current rate of increase in N_2O concentrations due to anthropogenic sources is nearly 30 times that emanating from natural causes since the end of the last ice age around 17,000 years ago.[14]

Future trends in nitrous oxide concentrations

Future projections of N_2O concentrations are subject to uncertainty, particularly in relation to the longer term impacts of climate change on the nitrogen cycle, and hence emissions from natural sources. Although projections of anthropogenic emissions are relatively robust up until 2020, when they are expected to be 20% higher than 2000 levels, projections beyond this time are much less certain.[15]

Agriculture will remain the principal driver of anthropogenic N_2O emission growth. The combination of ongoing population growth, increases in per capita incomes, and possibly increased **biofuels** production will necessitate an expansion in agricultural production and fertilizer use, increasing N_2O emissions. However, improvements in fertilizer application, through more careful and accurate dosing to improve plant nitrogen uptake efficiency, may curtail emissions growth to some extent, as well as reduce nutrient runoff and watercourse eutrophication. Improved livestock manure management may also help temper the rise in agricultural N_2O emissions though available N_2O mitigation options in agriculture are relatively limited and emissions difficult to control. It is also unlikely, mainly for political reasons, that governments will limit or regulate the use or application rates of fertilizer or include agricultural N_2O emission sources in **emissions trading** schemes.[16]

Anthropogenic N_2O emissions from the energy sector are expected to rise considerably over the period to 2030, in line with projected increases in fossil fuel use (see **future emissions trends**). Furthermore, if the consumption of **biofuels** increases significantly, as is expected, this may increase energy sector N_2O **emissions intensity**.[17]

The only area where there is likely to be significant reductions in absolute N_2O emissions is from industrial sources (such as nylon or adepic and nitric acid production), due to regulatory controls and/or abatement incentives. For example, several successful **mitigation** projects are underway through the **Clean Development Mechanism**. Nonetheless, industrial source reductions will have little impact on N_2O emission trends as they account for such a small proportion of aggregate emissions.

Trends in N_2O emissions from natural sources remain much less certain, particularly in relation to the impacts of climate change on the nitrogen cycle and, to a lesser extent, future trends in deforestation. Overall, it is not clear whether natural source emissions will increase or decrease over the coming decades, but the net effect is likely to be small relative to the projected increase in anthropogenic emissions. Warmer temperatures and increased precipitation of nitrogen compounds from atmospheric reactions could stimulate plant activity and possibly enhance **biosequestration** of CO_2, a negative **climate change feedback**. However, increased atmospheric concentrations of N_2O and NO_x could also increase soil biological dentrification emission rates in mature forests, particularly deciduous forests of the Northern Hemisphere.[18] As yet it not clear which effect will dominate.

The amount of land under forest has a major bearing on the amount of natural N_2O emissions. Should deforestation cease, or net reforestation occur, N_2O emissions from forest soils could stabilize or even increase slightly.[19] This increase in N_2O emissions would be tiny compared with the benefits of reduced CO_2 emissions from halting deforestation.

Overall, N_2O emissions and concentration levels are likely to continue to increase for at least the next few decades, and possibly longer. As long as global population and fossil fuel combustion continues to grow, so too will N_2O emissions. Furthermore, due to the long lifetime of N_2O, its contribution to positive **radiative forcing** will persist well beyond 2100 even if anthropogenic emissions were to cease today.

See also: aerosols, anthropogenic greenhouse gas emissions, biofuels, climate change feedbacks, fossil fuels, future emissions trends, greenhouse gases, methane (CH_4), ozone, radiative forcing.

Notes

1 Under the **Kyoto Protocol,** N_2O is assigned an atmospheric lifetime of 120 years and GWP of 310.
2 IPCC 2007
3 Scheehle and Kruger 2007
4 IPCC 2007
5 IPCC 1997
6 IPCC TAR 2001
7 IPCC 2007
8 Klein Goldewijk 2001
9 Scheehle and Kruger 2007
10 Crutzen *et al.* 2007
11 Scheehle and Kruger 2007
12 IPCC 2007
13 Ibid.
14 Ibid.
15 Scheehle and Kruger 2007
16 Hooper 2006
17 Crutzen *et al.* 2007
18 Ambus *et al.* 2006
19 Klein Goldewijk 2001

Further reading

Crutzen *et al.* 2007, IPCC 2007, Scheele and Kruger 2007.

NORTH ATLANTIC OSCILLATION (NAO)

The North Atlantic Oscillation (NAO) is the second most important ocean–atmosphere coupled circulation system that influences the earth's climate after the **El Nino Southern Oscillation (ENSO).**[1] The NAO contributes to multiyear climate variations in Europe, North America, and North Africa. The NAO index is a measure of westerly wind strength over the North Atlantic in winter.

The NAO exerts a dominant influence on winter surface temperatures across much of the Northern Hemisphere. When the NAO index is positive, strong westerly winds, interacting with the **thermohaline**, push warmth toward northern Europe. This results in more frequent Atlantic storms and higher precipitation in northern Europe but drier conditions in central and southern Europe and north Africa. It is also associated with stronger northerly winds over Greenland and northeastern Canada, carrying cold air southward, which cools land and sea surfaces over the northwest Atlantic, North Africa, and the Middle East, but leads to warmer conditions in southeastern United States.[2] The reverse effects occur with a negative NAO index.

Since the 1960s, there has been a trend toward a stronger positive NAO index and a weaker and less frequent negative NAO index. This helps explain why wetter and stormier conditions have been observed in northern Europe over recent decades, while the Mediterranean and North Africa have experienced drier conditions. Recent severe droughts in Spain, Portugal, and Greece have been largely attributed to a stronger NAO.[3]

Global warming may have contributed to this trend, but as with the ENSO, the recent trend is still within the bounds of natural variability. Insufficient trend data is available to conclude that recent global warming has caused the trend toward more frequent positive NAO index readings. The NAO index has, on several occasions since the 1500s, been in the positive range for extended periods.[4] Since 2003 the NAO index has returned to near its long-term mean.[5]

If global warming does in fact lead to more frequent positive NAO index readings, it may result in more frequent drought conditions in Mediterranean and North Africa, which would have significant implications for agriculture and water security in these areas. It may also result in more winter flooding and severe storms in northern Europe. Such trends would be consistent with IPCC projections of climate trends for these regions over this century.[6] Whether these projected trends would correlate with more frequent positive NAO index readings is not yet known.

See also: carbon cycle, climate change impacts, extreme weather events, thermohaline.

Notes

1 IPCC 2007
2 Ibid.
3 Thompson *et al.* 2003
4 Jones *et al.* 2001
5 IPCC 2007
6 Ibid.

Further reading

IPCC 2007; McPhaden *et al.* 2006; Thompson *et al.* 2003.

NUCLEAR POWER

Nuclear energy can be derived from two types of reaction – *nuclear fission*, which involves splitting the nuclei of heavy metals (mainly uranium), and *nuclear fusion*, which involves joining together, or fusing, the nuclei of some forms of hydrogen. These reactions produce large quantities of heat, which can then be used to generate steam and produce electricity.

Nuclear fission is a proven electricity-generating technology and has been contributing to the world's electricity supply for more than 50 years. By 2006, there were 443 nuclear power plants operating in 31 countries, providing 15% of global electricity supplies (approximately the same contribution as hydroelectricity).[1] Total installed capacity is around 370 gigawatts (GW) – equivalent to 1.5 times the total generating capacity of Japan, the world's fourth-largest consumer of electricity – and thus represents an important source of electricity. A gigawatt of generating capacity is sufficient to power a small- to medium-sized city.

Nuclear fusion is still at the experimental stage, and it is considered unlikely that it will become a commercial source of electricity before 2050, if at all. However, should it ever prove to be reliable and cost-effective, it offers a near-limitless supply of zero emissions/low-waste energy.

Although nuclear power went through a rapid expansion phase from 1960 to 1990, the number of new plants constructed has since fallen dramatically, mainly due to cost competitiveness (relative to fossil fuels) but also due to public concerns over safety, the disposal of radioactive wastes and nuclear weapons proliferation. In particular, the nuclear incidents at Three Mile Island (1979) and Chernobyl (1986) were a major blow to public perceptions and acceptance of nuclear power and led several countries to adopt nuclear power phase-out plans.

The emergence of climate change as a major policy issue, heightened concerns over energy security, and improvements in reactor efficiency and safety have rekindled government and industry interest in nuclear power. Nuclear power is presently being characterized by some as a potential large-scale greenhouse gas emissions reduction option for the coming decades. Some proponents suggest that generating capacity could potentially be doubled or tripled by 2050 and could contribute to a reduction of 3–7 billion tonnes of CO_2 per year.[2] However, nuclear power remains a controversial source of energy and faces several major constraints, particularly cost and nuclear waste disposal.

The main attractions of nuclear power are the following:

- Large amounts of base-load electricity (available 24 hours per day) can be produced virtually free of CO_2 emissions.
- It is a well-known, relatively mature technology that has already been commercially deployed on a large scale.
- The costs and reliability of production are well known – nuclear is less sensitive to fuel price changes than fossil fuel plants, as uranium fuel accounts for less than 20% of electricity-generating costs, neither is it subject to the same supply variability issues as are many **renewable energy** technologies.
- Sufficient uranium resources exist to support a moderately expanded nuclear industry for most of this century.

The main disadvantages of nuclear power are the following:

- Electricity production costs are usually higher than for fossil fuel power plants (nuclear power is a very capital-intensive technology).
- Production costs are not markedly lower than several other low-emission technologies (notably **carbon capture and storage** and **renewable energy** sources such as **wind power** and geothermal, though it is generally cheaper than most forms of **solar power**).
- It produces high-level radioactive wastes that require safe storage for thousands of years.
- The prevailing public perception in many countries is that nuclear power still has operational safety issues.
- Nuclear fuel residues can be used to develop weapons-grade material and, therefore, a risk of nuclear weapons proliferation.

- Nuclear power based on fission technology is not renewable energy as uranium resources (the primary fuel source) are finite.

Technology status

Nuclear fission technology can be classified according to three distinct vintages (or generations of technology). The first nuclear-generating facilities (Generation 1 reactors) were commissioned in the 1950s and 1960s (mainly in the United States and the United Kingdom), utilizing several different nuclear fission reactor types. These reactors have relatively low-operating efficiencies (compared with today's technology) and most are decommissioned, in the process of being decommissioned, or are nearing the end of their operating lives.

During the 1970s and 1980s, the nuclear power industry went through a rapid expansion phase, and it was in this period that most of the reactors operating today were constructed. The technology deployed during this period (Generation II reactors) had several improvements over the initial designs (mainly in terms of efficiency and scale). While several different types were deployed, the water-cooled reactors, particularly the pressurized Light Water Reactor (LWR), dominated the technology mix, and today water-cooled reactors account for 90% of installed capacity.[3]

Most of these plants use an open fuel cycle, where uranium fuel is utilized only once and the spent fuel is stored for later disposal or reprocessing. Some countries, notably France, have utilized a closed fuel cycle, where some of the waste fuel components (particularly plutonium) are extracted and mixed with new uranium for reuse. The closed fuel cycle has the advantage of reducing the amount of waste that must be disposed of and substantially increases the amount of energy that can be derived from each kilogram of uranium (30- to 60-fold). Cost and safety issues, such as handling plutonium (a highly toxic, radioactive substance), has meant that the open fuel cycle remains dominant. Several fast breeder reactors (which utilize a closed fuel cycle) have also been constructed and are presently operating (in France, Russia, and Japan). These reactors are highly efficient at extracting energy from uranium fuel, but they have faced a range of technical and economic constraints – generation costs are around twice those of standard reactors.[4]

During the 1990s, the global nuclear industry largely stagnated and interest in nuclear power in most OECD countries waned. New reactor commissioning rates fell to a quarter of what they had been in the previous decade. This was largely due to economics (cost and reliability) and also partly due to heightened public concerns over safety and to frequent regulatory changes (particularly in the United States). Only a few nuclear reactors have been commissioned since the 1980s in Europe, and none in North America. As a result, the contribution of nuclear to global electricity supply has fallen from a peak of 18% in the mid-1990s to 15% by 2006.[5]

Against this trend, Japan has maintained an active nuclear programme and in 1996 introduced Generation III reactors – these are more efficient, inherently safer (incorporating several passive safety features), and more standardized in design (thereby reducing construction costs). They also possess longer operating lives and produce less waste. Growth in installed capacity utilizing Generation III reactors has been largely confined to Japan, South Korea, and more recently, China and India. Approximately 20 GW of new nuclear capacity has been commissioned since 2000.[6]

Of the 443 operating plants, just three countries accounted for approximately half of global capacity: the United States (104 reactors), France (59), and Japan (55).[7] However, several other countries have sizeable nuclear programs, including the United Kingdom, Germany, Sweden, Belgium, South Korea, Canada, and Russia. China and India account for most new nuclear plant construction. Several countries source a significant proportion of their electricity from nuclear facilities, especially France (78%), Lithuania (68%), Slovakia (57%), Belgium (55%), Sweden (45%), and Ukraine (45%).[8]

Overall, nuclear power has delivered a significant volume of electricity to the grid since the 1960s. This, in turn, has helped to keep **anthropogenic greenhouse gas emissions** at a lower level than would have been the case had the electricity been generated by fossil fuel facilities – approximately 3 billion tonnes of **carbon dioxide (CO_2)**/year less. However, it has also generated significant volumes of high-level radioactive waste.

Research into improved reactors continues and advanced reactor technologies are under development, such as Generation IV pebble bed reactors. Most of these technologies are based on a closed fuel cycle with additional efficiency and cost improvements but are not expected to be ready for commercial applications until 2025.[9]

Cost

In some countries, especially those heavily reliant on imported fossil fuels, nuclear power can be a cost-competitive source of electricity, but generally nuclear remains an expensive option compared with coal. Liberalization and deregulation of electricity markets have also not worked in nuclear's favor, due to its high capital costs and low flexibility in power production, which mean that plants must operate continuously to keep unit costs competitive.

The actual costs of nuclear-generated electricity has always been the subject of some controversy. During the 1960s and 1970s, many politicians and proponents of nuclear power asserted that it was "too cheap to meter." As it turned out, nuclear was far more expensive than anticipated. Current generating cost estimates for existing plants vary from 3 to 7 US cents per kilowatt hour (kWh),[10] and costs from new nuclear power plants varies between 4 and 7 US cents/kWh,[11] although some analysts estimate higher cost figures.[12] This is 30–40% higher than the 3–5 US cents/kWh for electricity generated from new coal-fired power plants in most countries. Although nuclear cost estimates include an allowance for the cost of plant decommissioning and radioactive waste disposal (usually less than 0.5 US cent/kWh), the actual end-costs of fuel cycle operation are not yet known with any certainty. To date, most of the plants decommissioned have been relatively small in scale, but many larger commercial plants are soon due for decommissioning. Decommissioning cost estimates vary widely, from around US$ 300 million to over US$ 1 billion for an average-sized reactor. Assuming a 40-year life of a 1,000 MW plant with 95% availability, this puts decommissioning at 0.1 to 0.3 US cents/kWh over the lifetime of the plant. As permanent, high-level radioactive waste disposal has yet to commence, waste storage costs are not yet known with any certainty.

Opponents of nuclear power assert that the industry has received significant direct or indirect public subsidies that have enabled nuclear to be more cost-competitive than it otherwise would have been.[13] Transparent cost data remain limited, and it is difficult to confirm how significant public subsidies have been to the nuclear power industry. Nonetheless, it is widely known that the nuclear industry has received, and continues to receive, public financial assistance and incentives (partly for energy security reasons, but also for military and technological reasons). This has helped reduce the cost of establishing and operating the nuclear fuel cycle.

While other energy technologies also benefit from public subsidies and incentives, in many countries the amount of public funding to the nuclear industry dwarfs that provided to other nonfossil fuel technologies. For example, between 1973 and 2002, the US Government invested US$ 100 billion in energy research and development, and around half of this went to nuclear.[14] The US government, in the 2005 Energy Policy Act, introduced a tax credit of 1.8 cents/kWh for eight years for the first 6,000 MW of nuclear capacity installed and provides loan guarantees to help cover a proportion of any cost overruns. A nuclear reactor currently under construction in Europe (the Olkiluoto plant in Finland) has also reputedly received subsidized low interest loans from French and German state-owned organizations.[15] Irrespective of the extent of subsidies, the fact is that, in most countries, nuclear remains a more expensive source of electricity than fossil fuel and large-scale hydro alternatives (where they are available).

Construction costs vary between countries, but the most recent IEA data estimate that nuclear capacity costs around US$ 1,500–1,800/kilowatt (compared with US$ 1,000–1,200 for coal-fired plants) and takes five or more years to build.[16] Recent estimates put the cost of constructing a new 1,600 MW plant constructed in Europe at US$ 3.5–4.0 billion ($ 2,200–2,500/kW).[17]

Expansion constraints

The main constraints are cost competitiveness, public acceptance of safety, radioactive waste disposal, nuclear weapons proliferation, and, in the longer term, possibly inadequate supplies of uranium. Each of these is discussed in turn below.

Cost competitiveness

Cost (unit generation costs, up-front financing, commercial risks) is the primary constraint to large-scale expansion of nuclear power. For significant expansion over the next few decades, the cost differential between electricity generated from **fossil fuels** (mainly coal) and nuclear power needs to be significantly reduced. Based on current operating costs, reductions of around 1.5–2.5 US cents/kWh (20–60% cost reduction) is required to make nuclear a cost-competitive source of electricity globally. In countries that are major producers of low-cost electricity from coal (e.g. Australia), the cost of nuclear would need to be more than halved to be competitive.

The cost differential between nuclear and coal-fired generation can only be achieved through either a significant reduction in nuclear capital costs or through increasing the costs of electricity from fossil fuels (such as through **emissions trading** or **carbon taxes**). In the absence of either of these, nuclear power is unlikely to substantially increase its share of global electricity production – and thereby reduce the growth of greenhouse gas emissions – over the period up to 2050.

Advances in nuclear technology, standardization of reactor designs, and reduced construction times are expected to deliver some capital cost savings over the next two decades. Estimates of capital cost reductions vary, but some analysts suggest reductions of up to 25% are plausible.[18] Capital cost reductions of this magnitude would improve the cost competitiveness of nuclear versus coal by around 1.0–1.5 US cent/kWh (assuming no cost reductions in fossil power plants) but is unlikely to substantially alter the competitiveness of nuclear on a global scale. However, if a carbon dioxide penalty of US\$ 40–50/tonne CO_2e were also introduced (increasing the cost of coal-fired electricity by approximately 1–1.2 cents/kWh), nuclear power would become much more cost competitive. This could stimulate a significant expansion in global nuclear capacity though this scenario would also make a number of other renewable energy or low-emission energy supply options also more competitive.[19] In the absence of any cost savings in the nuclear fuel cycle, carbon dioxide penalties would have to be higher (US\$ 70–100/tonne CO_2e). Furthermore, the presently unmonetized societal costs of permanent waste disposal, operational safety, and health also need to be fully reflected in power costs (see below).

Radioactive waste disposal

The lack of an effective, long-term disposal method for high-level radioactive wastes is a major unresolved issue facing the nuclear industry. Over a 40-year operating period, a standard 1,000 MW thermal reactor generates around 1,000 tonnes of spent fuel.[20] While reprocessing can reduce this to approximately 40–50 tonnes of concentrated high-level radioactive residues, these wastes are highly toxic (lethal after minutes of exposure) and must be safely stored – without risk of leakage into the environment – for 10,000 years or more. That's more than twice as long as the pyramids of Giza have been standing and 100 times a typical engineering design life of "permanent" major modern structures such as bridges, dams, and wharves. Opponents of the nuclear cycle question whether it is responsible to store large quantities of highly toxic materials together in one place as this represents a concentration of risk and a liability to descendents for thousands of years into the future.

To date, no country has commenced the permanent long-term storage of high-level radioactive waste, and an ever-increasing quantity of spent fuel is accumulating in temporary storage – awaiting reprocessing (an expensive process) or long-term disposal. At present, it is estimated that 45,000 tonnes of spent fuel are being held in temporary storage ponds in the United States alone. In addition to these fuel wastes, the radioactive nuclear reactor cores must also be permanently isolated from the environment after plant decommissioning.

The preferred disposal option is permanent storage several hundred meters below the surface in seismically stable, geological structures isolated from groundwater flows. The United States government is presently constructing a waste storage facility at Yucca Mountain, Nevada, but the project has encountered several technical problems (particularly higher-than-expected groundwater flows). There are doubts over whether this site can be completed to its design criteria and whether it will actually receive license approval. Even if approval is granted, waste storage is unlikely to commence until 2015 at the earliest.[21] A waste storage option is also being evaluated in Finland in conjunction with the Olikiluoto nuclear plant, but it is not yet known whether the facility will proceed. Even if storage facilities are available, the waste management, disposal, and security costs remain largely unknown but would be incurred for a very long time. It is evident that long-term safe waste disposal will remain a major issue for the industry and may constrain its future growth.

Public perceptions of nuclear safety

Although the new, generation III and IV reactor designs are inherently safer than earlier reactors, negative public perceptions of the safety of nuclear facilities persist. Heightened public concern following the Chernobyl nuclear disaster led several countries (Sweden, Germany, and Belgium) to adopt policies to phase out nuclear power entirely. Securing local public acceptance for new nuclear power plants remains a challenge in most countries and can result in lengthy delays.

While public perceptions could change over the coming years, particularly if the public concern over climate change continues to grow, considerable effort on the part of government and industry will be required to ensure that the risks from nuclear power are sufficiently low and that the industry operates in a sufficiently transparent manner to convince the public that nuclear power does not compromise health and safety.

Nuclear weapons proliferation

The threat of weapons-grade nuclear material being produced by a large number of countries, or falling into the wrong hands, remains an ongoing international political issue. Recent events in North Korea, Iran, and Iraq, and previously in India and Pakistan, highlight the international political concerns over nuclear proliferation. These issues could be expected to increase with a large-scale expansion in the number of countries constructing commercial nuclear power plants. Limiting who is able to fabricate or reprocess nuclear fuel could help limit the risk of nuclear proliferation, but it is by no means certain all existing, or potential, nuclear states would agree to this option. The use of nuclear weapons or irradiation of civilian targets is a threat that cannot be dismissed.

Long-term fuel supplies

While known economic reserves of uranium are considered adequate to support an expanded contribution of nuclear to global energy supplies for at least the next 50–70 years (if based on the open nuclear fuel cycle), uranium is not an abundant element and reserves will eventually be depleted. Just under half of low-cost uranium resources have already been consumed (around 2 million tonnes) and an additional 2.5 million tonnes of low-cost uranium (less than US\$ 40/kg) remains, plus an additional 2 Mt at US\$ 40–130.[22] There are also known to be another 2 million tonnes of uranium at costs greater than \$ 130/kg, and the total resource could be as much as

10–15 million tonnes, but this is highly speculative.[23] Thorium is also another potential nuclear fuel, and resources are estimated at 2–4 million tonnes. Since 2000 uranium prices have more than doubled and in 2007 reached US$ 90/kg. Although nuclear power is relatively insensitive to uranium prices (fuel cost represents less than 20% of nuclear generating costs), beyond US$ 100/kg they become more important.

If present nuclear capacity were to continue operating until all economic resources (less than $ 80/tonne) were depleted, these plants could continue to generate power for another 80–100 years, or around 40–50 years if nuclear capacity doubled. If all known resources were used, then it could last up to two centuries or more – and possibly more than 1,000 years if all uranium fuel was used in fast breeder reactors.[24] While limited economic uranium resources exist, this is not expected to be a significant constraint on nuclear power as a mitigation option, at least not until later this century.

Nuclear fusion?

Some consider nuclear fusion to be a near limitless source of energy. Fusion mimics the process that powers the sun, and it is estimated that just a few kilograms of fuel can keep a 1,000 MW plant (1 GW) – the size of a typical large-scale coal or nuclear power facility – operating for a day. Fusion also produces only a tiny fraction of the radioactive wastes associated with nuclear fission, and these wastes are relatively short-lived. Greenhouse gas emissions are negligible (although some emissions are created producing the raw materials and in construction). However, to date, a stable and sustained nuclear fusion reaction has not been demonstrated for more than a few seconds. Research into nuclear fusion has been underway since the 1970s and has recently gathered pace with the agreement of an international consortium to proceed with the development of a 500 MW demonstration commercial-scale reactor (the International Thermonuclear Experimental Reactor) at a cost of US$ 10 billion.

The main problem with fusion is the technical complexity of containing the fusion reaction (which operates around 100 million°C) and maintaining a reaction that has a net energy yield 24 hours per day. Furthermore, the costs are unknown at this stage, and many expensive materials are required by the reactors. It is estimated that at least 30 years will be required before engineers can finalize full-scale commercial reactor designs and possibly 50 years before large-scale deployment could commence. Nuclear fusion is more likely to be an energy option beyond 2050, assuming it can be proven to be commercial.

Potential mitigation contribution of nuclear power

The nuclear industry appears to be entering a possible renaissance period as the number of plants under construction or planned has increased significantly in recent years. As of 2006, 22 new nuclear plants (with a total installed capacity of nearly 19 GW) were under construction, and another hundred are at the planning and approval stage (mainly in Asia and Russia).

Several Asian countries have ambitious nuclear power capacity expansion plans. China plans to add another 34 GW and India 17 GW by 2020 – though both targets appear ambitious based on past nuclear construction performance.[25] Japan and South Korea also have active nuclear expansion plans and other countries are reevaluating the nuclear option (e.g. several European countries and the United States).[26] Any reversal in nuclear phase-out policies in Europe (or a tightening of the EU Emissions Trading Scheme – see **emissions trading**), combined with the introduction of financial incentives for nuclear in the United States, could lead to additional new reactor orders over the next decade.

While new nuclear plant construction is expected to increase over the next decade, this needs to be balanced against the anticipated declines in nuclear-generating capacity in Europe and the United States, as many existing nuclear reactors are approaching the end of their useful operating lives. More than half of the present reactor stock is due to be retired by 2025. Of the United Kingdom's 23 reactors currently in operation, for example, only one will still be in operation by 2025.[27] Based on current International Energy Agency projections, installed nuclear capacity is expected to increase only marginally over the period to 2030 and its contribution to global electricity supplies is expected to continue to decline from 15% to 10% by 2030 (assuming no major policy changes).

Despite this trend, several studies have estimated that the contribution of nuclear could be increased significantly by 2050 if the right policy frameworks and incentives were established. One modeling study concluded that nuclear's contribution could supply 16–19% of global electricity production by 2050 under favorable circumstances (including rapid advances in nuclear technology and a relatively high cost penalty on carbon dioxide emissions), and possibly provide 5–10% of required emission reductions to stabilize atmospheric CO_2 concentrations.[28] In the business-as-usual case, where there are no major policy shifts, nuclear is projected to supply only 6–7% of global electricity supplies by 2050 and provide only 2% of the required emission reductions. If nuclear's share were to triple by 2050 (assuming the safety and waste management issues can be adequately resolved), nuclear power could reduce annual global CO_2 emissions by 3–7 billion tonnes per year below what would otherwise be the case, depending on which fossil fuel is displaced. This would require a new 1,000 MW nuclear plant to be commissioned every two weeks for several decades and a total investment of over US$ 2 trillion.[29]

Whether a tripling, or even a doubling, of nuclear capacity is achievable by 2050 remains to be seen – even if conducive policy settings are in place. Under existing policy settings, and considering other constraints facing the nuclear industry, a doubling of capacity appears rather unlikely. Overall nuclear power could plausibly contribute up to 5–10% of the greenhouse gas emission reductions required to stabilize global emissions by 2050, but to achieve this would require a significant shift in present energy and climate policies.

See also: carbon tax, carbon capture and storage, emissions trading, fossil fuels, mitigation, renewable energy.

Notes

1 IEA 2006a
2 Deutch and Moinz 2006

3 IEA 2006b
4 Ibid.
5 IEA 2006a
6 Deutch and Moinz 2006
7 As of end 2006.
8 IEA 2006a
9 IEA 2006b
10 See Nuclear Energy Agency 2005 and Deutch and Moinz 2006
11 Deutch and Moinz 2006 and IEA 2006a
12 *NewScientist*, April 2006
13 See Lovins 2005a
14 Eckhart and Wierich 2006
15 *NewScientist*, April 2006
16 IEA 2006b
17 Wilcox 2006
18 Deutch and Moinz 2006
19 Ibid.
20 IEA 2006b
21 Deutch and Moinz 2006
22 IEA 2006b
23 Ibid.
24 Ibid.
25 Ibid.
26 Wilcox 2006
27 Ibid.
28 IEA 2006b
29 Deutch and Moinz 2006

Further reading

IEA 2006a; Deutch and Moinz 2006; Lovins 2005.

OCEAN CARBON SINKS

The oceans are the ultimate **carbon sink**, and they have the largest role to play in bringing the **carbon cycle** back into balance over the coming centuries. The oceans help to regulate the balance of carbon between the three active carbon reservoirs (the atmosphere, the land, and the oceans). As the oceans contain nearly all (93%) of the exchangeable carbon, they will, in the long term, largely determine the equilibrium **carbon dioxide (CO_2)** concentration in the atmosphere. Each year, large quantities of carbon (in the form of CO_2) are exchanged between the atmosphere and the oceans, with annual carbon *fluxes* (exchanges) in the order of 90–100 gigatonnes (Gt).

The oceans absorb atmospheric carbon through two primary carbon cycle mechanisms: the *solubility pump*, which accounts for around 30% of the ocean carbon sink effect, and the *biological pump*, which accounts for the remainder (see **carbon cycle**). At present, the ocean absorbs between 1.6 and 2.4 Gt more carbon from the atmosphere each year than it emits to it and, as a result, is a net **carbon sink**. While this range reflects the level of uncertainty surrounding net ocean–atmosphere fluxes, it is generally considered that the annual net uptake of carbon is around 2 Gt/year (7.5 Gt CO_2e), although some methods used to calculate net fluxes (like those calibrated for variations in wind velocities) have yielded uptake estimates of more than 3 Gt/year.[1] It is estimated that, over the past two centuries, the oceans have absorbed around a third of all anthropogenic (human-created) emissions of CO_2 to the atmosphere.

Ocean–atmosphere carbon fluxes are usually in balance, apart from small fluctuations associated with regular cycles of global cooling and warming (see **Milankovich cycles**) or some other perturbation like volcanic eruptions. Imbalances are generally short lived, and changes usually occur relatively slowly over thousands of years. However, occasionally there are major, and sometimes catastrophic, events that cause major imbalances in the carbon cycle. In the past, such events have included major meteor impacts, intense volcanic activity, large releases of methane gas from methane hydrate deposits, and sudden changes in ocean salinity that shut down the **thermohaline**.

The oceans have always been the primary means of restoring balance to the carbon cycle, even though this can, at times, take several thousand years. For example, scientists believe that there was a massive release of CO_2 to the atmosphere 55 million years ago when subsea volcanic eruption ignited frozen methane hydrates locked up in ocean floor sediments.[2] The release of CO_2 was so large (about 10 times larger than all anthropogenic CO_2 emissions to date) that it caused the oceans to become much more acidic and is believed to have caused a large-scale extinction of marine species. The oceans eventually restored equilibrium to the carbon cycle, but it is believed that it took 100,000 years before the ocean eventually returned to normal alkalinity levels (see **marine impacts**).[3]

While not as dramatic as the event 55 million years ago, the rise in atmospheric CO_2 concentrations over the past century has been so rapid, in an historical sense, that it too may be classified as a major perturbation to the carbon cycle. In response to increased atmospheric CO_2 concentrations, the oceans have again been working to restore balance. For example, ocean–atmosphere fluxes were roughly in balance until around 1850, but by 1900, the size of the annual negative flux (sink effect) had reached around 300 million tons and increased further to around one billion tons (1 Gt) by the 1960s. Since then it has doubled again and is now estimated to be around 2 Gt per year.[4]

The oceans will remain a net sink so long as atmospheric CO_2 concentrations are higher than concentrations in the surface oceans. But the ocean carbon sink is time constrained. If anthropogenic CO_2 emissions were to cease today, the oceans would eventually absorb 80–85% of the excess CO_2 in the atmosphere.[5] However, the process operates slowly, and it would take as long as 500–1,000 years before atmospheric CO_2 concentrations stabilized at a new equilibrium level (all other things being equal). The new equilibrium CO_2 concentration would be higher than that which prevailed in preindustrial levels of 280 parts per million (ppm), but only by as little as 10–15 ppm, and much lower than the current level (382 ppm). One may conclude that this is not much and the earth's living systems could easily accommodate a change in equilibrium concentrations of this magnitude. Unfortunately, due to the significant time lag before the new equilibrium was reached, there would be significant ongoing climate change in the interim period. Furthermore, other positive **climate change feedback** effects that amplify **global warming** may be set in motion (like releases of CO_2 and **methane (CH_4)** from the permafrost), which would contribute to atmospheric CO_2 loading and accelerate climate change even further.

How will global warming affect the ocean uptake of carbon dioxide?

There are four main factors that are likely to influence the magnitude of ocean–atmosphere net CO_2 fluxes over this century and beyond. Three of these are likely to reduce the uptake of CO_2, while one may increase CO_2 uptake. CO_2 uptake could fall due to increased surface ocean temperatures; reduced flow rate of the **thermohaline**; and reduced ocean alkalinity. The one factor that may increase CO_2 uptake is elevated levels of biological activity due to CO_2 fertilization and temperature increase (see **carbon cycle**).

Ocean temperature changes

Global warming will increase the temperature of the ocean surface, which will, in turn, tend to decrease its ability to absorb CO_2. This is because the solubility of CO_2 in water falls as water temperature rises (see **carbon cycle**). The mass of the ocean is so large that it adjusts only very slowly to changes in atmospheric temperature. Scientists estimate that it takes several decades before a given change in atmospheric temperature is fully reflected in surface ocean temperatures and several centuries for the temperature of the deep ocean to adjust. So, even if global temperatures remained constant at today's level, which is unlikely, it would take until at least 2040 before the surface oceans adjusted to the global temperature increase experienced to date. This means we are locked into a continued ocean temperature increase for some time, with or without any additional warming, and CO_2 solubility will decrease accordingly. The rate at which atmospheric temperature will rise this century is highly uncertain (see **global warming**, **stabilization targets**, and **climate sensitivity**). What is clear is that surface ocean temperatures will continue to rise throughout this century, and possibly at an accelerating rate. This is expected to reduce the annual net uptake of CO_2 by the ocean. The reduced capacity for the oceans to absorb has already been observed in the Southern Ocean where temperature and wind regime changes (due to global warming already experienced) has reduced annual CO_2 uptake by about 3 million tonnes/decade since 1980.[6]

Slowing of the thermohaline

Increased global temperatures could reduce the flow rate of the **thermohaline**. The thermohaline is one of the primary mechanisms of mixing surface ocean water with deep ocean water and operates like a giant conveyor belt that transports CO_2 to and from the deep ocean. Each year it accounts for about one-third of the export of CO_2 (in the form of dissolved inorganic carbon) to the deep (see **thermohaline** and **carbon cycle**). But the efficiency at which it functions is largely dependent on differences in water temperature and salinity in the North Atlantic Ocean. Global warming is likely to increase the flow of freshwater (from melting glaciers and ice caps and increased river discharges) into the north Atlantic and reduce the salinity of sea water in the higher latitudes. This is likely to slow the flow rate of the thermohaline or even possibly shut down the flow completely (see **thermohaline**). If this happens, it would significantly reduce the rate of turnover of the oceans and the export of carbon to the deep oceans and, thereby, reduce the net uptake of CO_2 by the oceans. If the thermohaline were to shutdown completely, it would represent a significant positive **climate change feedback**, and the consequences of this could be far reaching (see **thermohaline**).

Ocean acidification

Over time, as the oceans absorb more CO_2 from the atmosphere they become less alkaline (CO_2 is a mildly acidic gas, and it forms carbonic acid when it dissolves in seawater). Scientists often refer to this process as *ocean acidification* (see **marine impacts**). Ocean pH has already fallen by 0.1 pH units and could fall another 0.3–0.4 pH units by 2100. As pH falls, the oceans become more CO_2 saturated and, as a result, their ability to absorb CO_2 also falls. However, the most significant impact of ocean acidification is not so much the reduction in the oceans' ability to absorb CO_2 but the impact it will have on marine ecosystems (see **marine impacts**).

Ocean biological activity

The final major influence of climate change on the ocean carbon sink is through its impact on the amount of biological activity in the oceans. Increases in primary production through photosynthesis means more CO_2 is sequestered from the atmosphere. Through the action of the biological pump (see **carbon cycle**), large quantities of organic carbon are exported to the deep ocean. This process accounts for about 70% of the export of carbon to the deep ocean.

The factors that influence ocean net primary production (through phytoplankton photosynthesis) are complex, and the scientific understanding of ocean biological processes is not as well developed as the understanding of ocean chemistry. At present, models of the oceans have not yet been able to fully capture the processes at work, and there is little certainty about how the ocean ecosystem will respond to climate change. Higher water temperatures and higher levels of CO_2 availability are expected to stimulate primary production: a net positive gain in CO_2 uptake. However, other nutrient shortages (like the macronutrients nitrogen and phosphorous and the micronutrient iron) are likely to be a major limiting factor on primary production increases. Biological activity is also influenced by climate variations caused by regular **ENSO** cycles, as well as variations in wind and rainfall over the ocean.

While uncertainty exists, most models predict that net CO_2 uptake through biological activity will increase in the oceans over this century.[7] Even if climate change does cause net primary production to increase, for it to play a significant role in the ocean carbon sink effect the carbon must be exported to the deep oceans and remain there for extended periods of time. Knowledge of carbon export efficiency to the deep oceans remains limited, but it is known that it varies considerably between locations and at different times of the year.

On balance, the enhancements to biological activity are likely to increase CO_2 uptake, but this increase is unlikely to compensate for reduced CO_2 solubility, a weakened thermohaline and increased ocean acidity. Overall, the ocean effectiveness as a net carbon sink will reduce over time, and possibly by as much as 10–15% over this century.[8] Ongoing research and improved data collection and monitoring over the next few decades should enable us to more precisely determine how the oceans will perform as a carbon sink in future.

Can humans enhance the ocean carbon sinks?

While humans can influence the amount of CO_2 sequestered by land systems through actions such as afforestation, reforestation, and changed land management practices (see **land carbon sinks**), fewer opportunities exist to influence CO_2 uptake by the oceans. However, several studies since the early 1990s have found that ocean primary production can be enhanced with the addition of the micronutrient iron, a major determinant of biological activity. Iron fertilization experiments in the equatorial Pacific have demonstrated that significant increases in ocean primary production can be achieved.[9] These results have been confirmed by experiments in the Southern Ocean, which showed not only that primary productivity did increase with the addition of iron but also that there was generally low carbon export efficiency to the deeper waters.[10] This study also calculated that sequestration of large quantities of CO_2 would require iron fertilization activities to be undertaken over very large areas of ocean.

Whether large-scale iron fertilization is practical and cost-effective remains to be proven. Furthermore, concerns have been raised about possible side effects of iron fertilization on marine ecosystems (which are complex and not well understood). Adverse side effects could include the growth of toxic algae and increased emissions of dimethylsulphide (which affects cloud formation) and methyl halides (which cause ozone depletion). Much more detailed environmental impact assessments would be needed before any large-scale iron fertilization measures could be implemented.

Some entrepreneurs have viewed iron fertilization activities as a possible means of generating **carbon credits** for use in the carbon trading market or as contribution to meeting national targets under the **Kyoto Protocol**. Even if this were allowed under Kyoto Rules (which it presently is not), and it proved to be a cost-effective measure to undertake and if the side effects were considered acceptable, it would be still too difficult to verify how much carbon was actually exported to the deep oceans. At present, we do not have the monitoring capabilities to substantiate how much carbon remains stored in the ocean.

While it may seem somewhat comforting that the oceans will eventually correct the carbon cycle imbalance humans are creating, it also needs to be recognized that this is a very slow and long-term process. Between now and ultimate steady-state equilibrium, it is likely that the oceans will experience significant changes, as will their ability to absorb carbon dioxide from the atmosphere.

See also: albedo, anthropogenic greenhouse gas emissions, carbon cycle, carbon dioxide (CO_2), climate change feedbacks, climate sensitivity, El Nino Southern Oscillation (ENSO), global warming, Intergovernmental Panel on Climate Change (IPCC), Kyoto Protocol, thermohaline.

Notes

1 Feely *et al.* 2001
2 Malthe-Sorensson *et al.* 2004
3 Hendersen 2006
4 Matear and Hirst 2001
5 Feely *et al.* 2001
6 Le Quere 2007
7 Matear and Hirst 2001
8 Flannery 2005
9 Coale *et al.* 1996
10 Buesseler *et al.* 2004

Further reading

Feely *et al.* 2001; Matear and Hirst 2001; IPCC 2007.

OZONE

Ozone is a colorless greenhouse gas with a pungent odor. Ozone molecules are each composed of three oxygen atoms, denoted chemically O_3. It is solid at temperatures below $-193°C$ and a gas above $-112°C$. Ozone molecules are inherently unstable, resulting in a powerful oxidizing ability and high chemical reactivity.

Ozone occurs naturally in small quantities for short periods of time at ground level as a result of electrical discharges from lightning and the breakdown of naturally occurring Volatile Organic Compound (VOC) gases. Naturally occurring levels of ozone are only a trace gas, making up less than 0.00003% of the atmosphere by volume, and do not generally harm plants or animals. However, in higher concentrations, surface ozone can be dangerous to both plants and animal life.

Stratospheric ozone

It is important to distinguish between tropospheric ozone (also referred to as low-altitude or ground-level ozone) – which is dangerous to human health – and stratospheric (high-altitude) ozone, which is beneficial to human health. Approximately 97% of ozone occurs in the stratosphere, at altitudes of 15–55 km, and this constitutes the "ozone layer." The ozone layer is very important for life on earth, as it filters out nearly all (99%) of UV-B radiation, which causes mutations in plant and animal cells, leading to death at high levels of exposure. In humans, even relatively low exposure to UV-B can result in skin cancer and cataracts.[1]

From the 1930s onward, humans have produced ever-increasing quantities of **synthetic gases**, attractive for their properties of chemical stability, low toxicity, and manipulable physical characteristics. Of these **synthetic gases**, chlorofluorocarbons (CFCs) – molecules containing only carbon, fluorine, and chlorine atoms – were widely produced (and still are, to a lesser extent) for use as refrigerant gases, industrial cleaning solvents, aerosol spray-can propellants, and in the production of foams. Unfortunately, the low reactivity of CFCs means that they break down extremely slowly in the troposphere. Over time, these CFCs drift upward and reach the stratosphere; there, ultraviolet radiation catalyzes the breakup of the CFC molecules to release chlorine atoms, which themselves then react with ozone, creating a chain reaction, in which a single free chlorine atom can destroy thousands of ozone molecules. CFCs, along with several other **synthetic gases**, are Ozone Depleting Substances (ODSs), and their production and eventual phase-out are mandated under the Montreal Protocol (see **synthetic gases**).[2]

Since the 1930s, concentrations of CFCs in the atmosphere have increased from zero to more than 100 ppt[3] (parts per trillion), and the process of ozone destruction has resulted in a 4% decrease in average stratospheric ozone levels relative to the 1964–1980 period.[4] This decrease varies greatly with latitude, with most ozone loss being concentrated over the poles. Satellite observations indicate vast areas over Antarctica, and somewhat smaller areas over the Arctic, with greatly reduced densities of ozone. These mostly appear during summer months and can cover areas the size of the continental USA and exhibit densities up to 40% below "natural" levels.[5] These areas of lesser ozone are commonly known as the "ozone holes," though they are not in fact "holes."

Since ozone is a greenhouse gas, the depletion of the ozone layer means there are lower concentrations of **greenhouse gases** in the atmosphere, which should result in a cooling effect. Some cooling has been observed, and models indicate stratospheric cooling of up to 1°C since the 1970s. However, it is not clear that stratospheric cooling necessarily results in cooling of the troposphere or the surface, and considerable uncertainties concerning the magnitude of this effect remain. Countering this cooling, to some extent at least, is the increase in solar energy reaching the surface as a result of a "thinner" ozone layer, though this effect is believed to be quite small.[6]

Much more important is the contribution of CFCs and other ODSs to global warming. They are potent **greenhouse gases** with high **global warming potentials (GWPs)**, though not controlled under the **Kyoto Protocol**. CFC-11 and CFC-12 are the most abundant of these gases and have **GWPs** of 4,750 and 10,900, respectively.[7] In fact, CFC-12 is the third most important anthropogenic greenhouse gas (after **carbon dioxide** and **methane**) in the atmosphere. ODSs contribute approximately 12% of the total **radiative forcing** due to **anthropogenic greenhouse gas emissions** (excluding transient – short-lived – **greenhouse gases**).[8]

Overall, the **radiative forcing** resulting from ozone depletion in the atmosphere is estimated to be −0.05 (± 0.10) Wm^{-2} (a cooling effect), compared with the effect of ODSs as greenhouse gases of 0.32 (± 0.03) Wm^{-2} (a warming effect).[9] That is, even assuming that cooling in the stratosphere propagates to the troposphere (which has not been clearly demonstrated), the warming effect associated with ODSs is more than six times the cooling effect associated with a loss of stratospheric ozone. Best estimates for all **radiative forcing** sources, including positive and negative agents (see **albedo**, **aerosols**, **climate change feedbacks**), result in a total net anthropogenic **radiative forcing** of 1.6 Wm^{-2}: the net impact of ODSs (warming and cooling) represents almost 17% of the total contribution to **global warming** experienced to date.[10] As a proportion of the total anthropogenic positive **radiative forcing** of 2.64 Wm^{-2} (i.e. direct warming influence only, see **anthropogenic greenhouse gas emissions**), ODSs contribute slightly more than 10%.

Since the early 1990s, the successful implementation of the Montreal Protocol has resulted in a substantial reduction in ODS emissions: emissions are now only 20% of what they were in 1990,[11] and production of ODSs are due to be phased-out completely by 2030. Given the long atmospheric lifetime of most ODSs, it may take 100 years or more for the ozone layer to reach a new equilibrium. The ozone layer is expected to slowly recover throughout the twenty-first century, but unfortunately, this is expected to contribute a further 0.16 Wm^{-2} to global **radiative forcing**, thus contributing to additional warming.[12]

Tropospheric ozone

In addition to natural occurrence, ozone is produced commercially. High voltages of electricity are discharged in the presence of oxygen to produce ozone for use in water purification, soil decontamination, the fumigation of fruits and vegetables, and in medical applications. Ozone is considered superior to other chemicals since it leaves no environmentally persistent chemical residuals that can have carcinogenic or teratogenic (embryo-harming) effects. Aside from these relatively small sources, ozone arises as the indirect result of photochemical reactions between the precursors (see **greenhouse gases**) of oxides of nitrogen (NO, NO_2, and NO_3, collectively referred as NO_x) and VOCs (such as **methane**, butane, octane) in the presence of catalytic ultraviolet energy from sunlight. Emissions from industrial facilities, motor vehicle exhausts, gasoline vapors, and chemical solvents are the major sources of NO_x and VOCs. Ground-level ozone is the primary constituent of "smog" and is temporally and spatially determined by the concentration and distribution of these precursors, usually forming in large cities or in agricultural areas with intensive chemical fertilizer application or large-scale burning of biomass (with associated NO_x emissions). Tropospheric ozone levels have more than doubled since 1900 as a result of human activities.[13]

Although ozone is transient in the lower atmosphere, with a half-life of 12 hours at ground level, dangerous concentrations of more than 100 µg/m^3 (about 0.8 ppm at sea level)[14] can build up on hot, still sunny days, peaking in the afternoon when sunlight has fully reacted with vehicle exhaust emissions. Humans exposed to these levels, particularly the young, the old, and those exercising vigorously, may experience a range of health problems, such as eye, nose, and throat irritation; chest pain, coughing, and congestion; and reduced lung function. Breathing ozone can exacerbate respiratory illnesses, such as pneumonia and bronchitis, and chronic conditions such as emphysema and asthma. Repeated exposure may permanently scar lung tissue.

In addition to human health impacts, ozone has detrimental effects on plants. Exposure to ozone damages leaf cells when taken up through their stomata (cells on the underside of leaves that allow CO_2 and water to diffuse into plant tissue). In response, plants close their stomata to minimize damage, which also slows photosynthesis and hence plant growth (see **agriculture and food supply impacts**). High levels of exposure damage leaves and increase plant susceptibility to other stresses. In the United States alone, ozone is responsible for an estimated US$ 500 million in reduced crop production each year.[15]

Ozone as a greenhouse gas

The chemical formation of tropospheric ozone from precursors is relatively well understood, but emissions of precursors vary over time, are spatially highly variable, are short-lived when emitted, and the availability of sufficient ultraviolet radiation varies hourly. Furthermore, even with optimal formation conditions, the relationship between precursors and ozone formation can be nonlinear, such that known atmospheric concentrations of precursors do not necessarily result in predictable ozone formation.[16] These factors make modeling or measuring ozone levels very difficult, and the uncertainty relating to ozone's contribution to climate change is high.

Since ozone's tropospheric residence time is so short, comparing it with other, long-lived **greenhouse gases** through 100-year **global warming potentials** is meaningless. Reductions in emissions of precursors would result in almost immediate reductions in the formation of ozone, and thus reductions in its contribution to climate change. This is in stark contrast with the "Kyoto" **greenhouse gases**, whose emissions today commit the planet to warming for decades to come. For these reasons, tropospheric ozone is not a Kyoto gas, is not controlled under the **Kyoto Protocol** or **UNFCCC**, nor required for reporting in national **greenhouse gas inventories**. Ozone is usually not included in greenhouse gas emission analyses.

What is known with high certainty is that tropospheric ozone is an important contributor to **global warming** with an estimated **radiative forcing** of +0.35 Wm^{-2}, with an uncertainty range of +0.25 to +0.65 Wm^{-2}. This represents an estimated 13% of total positive anthropogenic **radiative forcing**, making tropospheric ozone the third most important greenhouse gas – slightly greater than the contribution of ODSs, but with a much higher upper bound of uncertainty.[17]

In addition to the direct **radiative forcing** of tropospheric ozone, it also constrains plant photosynthesis. Modeling and field experiments with elevated ozone levels have shown decreases in plant productivity (see **land carbon sinks**), resulting in reduced removal of CO_2 from the atmosphere (see **carbon cycle**), and thus higher CO_2 concentrations. This indirect effect is a positive **climate change feedback** that could result in as much, or a potentially greater, contribution to **global warming** than the direct **radiative forcing** of tropospheric ozone.[18]

Trends in tropospheric ozone concentrations

The assessment of long-term trends of tropospheric ozone is complicated by the sparsity of data prior to 1980 and the lack of spatially representative observation sites. Most models assume 90% lower emissions of precursors prior to 1750, implying a tenfold increase since the Industrial Revolution. However, given that ozone forms only under certain conditions, a tenfold increase in precursors does not necessarily equate to a tenfold increase in ozone. Nonetheless, concentrations of tropospheric ozone are known to have increased substantially since 1750. Recent trends are known to have varied regionally. Over Europe, concentrations increased from the early twentieth century to the late 1980s, after which the trend leveled off. Similar trends have been observed in North America. Over Japan, concentrations increased by approximately 12–15% from the 1970s to 2002, with the higher concentrations more recently recorded in southern Japan attributed to precursors from China. Since tropospheric ozone is dangerous to human health, emission controls of precursors (to reduce ozone formation) have been implemented in many high-income developed countries, which have reduced emissions of precursors in these countries since the late 1980s. However, as yet, very few developing countries have introduced emission controls, and as a result, precursor emissions in industrialized regions of developing countries continue to exhibit significant growth.[19] Globally, tropospheric ozone concentration, and its cumulative contribution to climate change, continues to increase.

While downward trends in tropospheric ozone may be expected in most high-income countries, and possible stabilization in some medium-income countries, this is unlikely to counter the ongoing growth of precursor emissions from developing countries in the medium term. Furthermore, the conditions for peak ozone formation require hot, sunny conditions, and as the globe warms, these conditions will be more frequently met. This represents a positive **climate change feedback** that can be expected to increase global concentrations of tropospheric ozone in the future. Atmospheric **methane** concentrations, an important precursor to ozone, are also increasing, albeit at a much lower rate since the late 1990s, which could also increase the contribution of tropospheric ozone to global warming (see **methane**). However, the additional radiative forcing from these effects may be partly offset by the expected increase in atmospheric water vapor, which has the effect of slowing the formation, and accelerating the destruction, of ozone.[20]

The expected trends in tropospheric ozone concentration are highly dependent on the modeling of broader emission trends (see **future emissions trends**), with a range of -27% (decrease) to $+55\%$ (increase) predicted by different models by 2030.[21] Better emission controls, particularly from the transport sector, are likely to help reduce precursor emissions, but unless substantive reductions are made in all sources of precursors, the upward trend in tropospheric ozone concentrations is expected to continue.

By 2100, models predict an increase in the **radiative forcing** associated with tropospheric ozone, with the estimates ranging from $+0.40$ to $+0.78$ Wm^{-2} compared with the current $+0.35$ Wm^{-2}.[22] This translates as an increase of 14–123% above current **radiative forcing** from tropospheric ozone, indicating that it will remain an important driver of global warming.

See also: aerosols, agriculture and food supply impacts, anthropogenic greenhouse gas emissions, climate change feedbacks, future emissions trends, global warming, global warming potentials, greenhouse gases, radiative forcing, synthetic gases.

Notes

1 US EPA 2007
2 UNEP 2006
3 IPCC 2007
4 Ibid.
5 NOAA 2006

6 IPCC 2007
7 Ibid.
8 Ibid.
9 Ibid.
10 Ibid.
11 Ibid.
12 Ibid.
13 Gauss *et al.* 2006
14 WHO 2005a
15 US EPA 2007
16 IPCC 2007
17 Ibid.
18 Sitch *et al.* 2007
19 IPCC 2007
20 Ibid.
21 Ibid.
22 Ibid.

Further reading

IPCC 2006a, 2007.

POLAR IMPACTS

The polar regions span the Arctic and Antarctic, as well as the high latitude areas bordering the north and south polar circles (above 60° north and below 60° south latitude). The Arctic and Antarctic together account for approximately 20% of the earth's land surface, and both are particularly susceptible to climate change. Temperature increases are expected to be greater in the polar regions than in other regions of the planet, particularly in the northern polar region, which has already warmed at more than twice the global mean rate over the past 100 years.[1] Climate-related changes in these regions will have a range of important geophysical and biophysical flow-on effects for other areas of the planet, including a slowing or shutdown of the **thermohaline**, releases to the atmosphere of large amounts of organic carbon trapped in permafrost, and significant impacts on the marine food chain.

In important respects, the northern and southern polar regions differ markedly. The Arctic is principally ocean (the Arctic Ocean) surrounded by land, while the Antarctic is continental land mass (Antarctica) surrounded by ocean. The Antarctic is colder than the Arctic. The Arctic contains most of earth's sea ice and the Antarctic most of its land-based ice. A warmer climate in these regions has important implications for long-term sea level rise. If the Arctic were to become completely ice-free, global mean sea levels would rise by 7–8 meters (due primarily to melting of the Greenland ice cap); if the Antarctic were to become ice-free, the seas would rise by around 60 meters (see **sea level rise**).

The different characteristics of the Artic and the Antarctic mean that the climate change impacts and processes will vary. Due to its geography, its land–sea characteristics and ice **albedo** feedbacks, the Arctic is warming faster than the Antarctic and is expected to maintain this trend.[2] The land ice mass of the Arctic is also expected to shrink at a faster rate than the Antarctic ice mass this century – which, by contrast, may remain largely unchanged, or even increase due to changes in precipitation patterns.

The polar regions are sparsely populated, and vegetation is limited: indeed, Antarctica is completely devoid of any trees or shrubs. While the Arctic and Antarctic terrestrial systems contribute very little to global terrestrial primary production, they do support important habitats for marine dependents such as birds and mammals (whales, polar bears, seals, and penguins). But most important, the oceans in the higher latitudes support very productive fisheries, such as cod and salmon, and climate change has already adversely affected the location and productivity of these species. In Antarctica, the vast krill stocks form an important base of the food chain, but they also have been declining in productivity in recent decades, partly due to retreating sea ice (see **marine impacts**). Unlike many other species on the planet, those that inhabit the polar regions have few, if any, options to adapt to a changing climate – many will simply become extinct.

Polar warming

There has already been a more rapid warming in the northern latitudes than the rest of the planet, particularly in recent decades. In some areas of Alaska, western Canada, and parts of Siberia, average winter temperatures have increased by 3–4°C since the 1950s – more than triple the global average increase.[3] Night time temperature increases appear to have been greater than day time changes (in part due to increased night time cloud cover), as have increases in winter temperatures relative to summer. From 1970 to 2003, the Arctic warmed by an average of 0.46°C/decade,[4] with Southern Alaska experiencing a temperature increase of 2–2.5°C since the 1970s.[5]

This accelerated warming is attributable to several factors. One contributing factor has been changes to the earth's **albedo** (surface reflectivity). Increased global temperatures have caused snow and ice cover to retreat, and as this occurs it exposes the underlying land and sea surfaces, which are darker and less reflective – darker surfaces absorb more heat than lighter surfaces such as ice (see **albedo**). Another factor has been changes in ocean currents and a significant inflow of warmer, more saline waters from the Atlantic in the past few decades. This has altered Arctic Ocean circulation patterns, accelerated sea ice melt, and contributed to decreased albedo and regional warming.[6] Decreased albedo is a positive **climate change feedback** effect (see **albedo**).

The rapid Arctic warming over past decades is much greater than climate change models predicted given the amount of greenhouse gas **radiative forcing** in the atmosphere. It is known that temperatures in the Arctic over the past century have experienced considerable variability, with alternating periods of warm and cold: for example, the Arctic went through a relatively warm period from 1925 to 1945 before cooling again. The recent rapid rate of change is believed, therefore, to be partly attributable to natural variability, arising mainly from changes in solar radiation.[7] Some scientists believe that in a few years the Arctic may enter another cooling phase. While this may indeed eventuate, it is unlikely to reverse the underlying warming trend caused by the buildup of greenhouse gases in the atmosphere.[8] Even a reversal of the current natural variation in solar irradiance may not result in an equivalent return to colder conditions. It is estimated that the Arctic will continue to warm on average by 0.45–0.75°C/decade this century, with the highest increases occurring in the winter.[9] This is more than twice the expected global average rate of change.

The higher-than-average Arctic warming forecast is consistent with previous trends in the earth's past. For example, about 50–55 million years ago, the earth experienced a very warm period (the Late Paleocene Thermal Maximum): average global temperatures were 5–6°C above present levels, and the Artic was as much as 12–15°C higher.[10] The most recent warm period occurred 125,000 years ago during the last interglacial peak, when Arctic temperatures were 3–5°C higher than present, mainly due to differences in the earth's orbit (see **Milankovich cycles**).[11] This level of warming resulted in a 4–6 m rise in global sea levels above current levels.[12] Based on mid-range IPCC global warming estimates, Arctic temperatures could increase by 4–7°C by 2100, which would exceed the temperature levels experienced 125,000 years ago.[13] Temperature increases of this magnitude will most likely result in complete deglaciation of Greenland, and at least an eventual 7 m sea level rise.

Although the Antarctic Peninsula and areas of the Southern Ocean have also experienced similar levels of warming to Alaska over recent decades, the average temperature of Antarctica as a whole has remained largely unchanged and there has been no discernible trend in temperatures.[14] Antarctica as a whole is expected to warm this century, but less rapidly than the Arctic.

Polar impacts

The impacts in the Arctic over the coming years will be more pronounced than in other regions due to the higher rate of warming. Key geophysical impacts will include continued loss of sea- and land-based ice, increased precipitation and coastal erosion, and changes in ocean circulation. Global warming will also lead to thawing of the permafrost, releasing more emissions of **carbon dioxide (CO_2)** and **methane (CH_4)** into the atmosphere and thereby generating further warming. These changes will affect polar ecosystems and biodiversity, as well as resulting in a range of **socioeconomic impacts** (particularly on infrastructure and fisheries).

Loss of ice coverage

The loss of sea- and land-based ice will have important implications for polar ecosystems and the earth's **albedo**. Between 1975 and 2004, Arctic terrestrial snow cover reduced by 10%, late summer sea ice coverage by nearly 20%, and ice thickness by 10–15%.[15] However, since 2004, the rate of loss of sea ice cover has accelerated appreciably, at a faster rate than predicted by most models. There has also been a contraction in sea ice coverage in Antarctica, although it has been difficult to detect a clear underlying trend due to significant interannual variability, but the area of sea ice has fallen by more than a million square kilometers over the past 50 years.[16] There has also been a noticeable acceleration in the loss of glaciers and increased ice cap melting, particularly in Greenland. Sea ice is expected to continue to retreat during the twenty-first century, particularly in the Arctic (see **glaciers and ice sheets**).[17]

Reductions in sea ice cover will have significant impacts on communities and local ice-dwelling wildlife. Sea ice provides an important habitat for polar bears and seals, and the length of their hunting season has been compressed in recent decades due to early disappearance of winter ice. The average number of Canadian grey seal pups born each year has fallen by 15% since 1980, as has the average adult weight of polar bears.[18]

Disappearing sea ice will also lead to enhanced coastal erosion, affecting coastal infrastructure and communities and reducing food supplies to the indigenous Inuit people of the Arctic. The Inuits have lived and hunted on the fringes of the Arctic for 5,000 years, using sea ice as a hunting platform. There are approximately 150,000 Inuits living in Russia, Canada, Greenland, and Alaska. The shrinking ice cover means that their hunting grounds, and the abundance and weight of the animals they hunt (particularly seals), is diminishing.[19] The loss of tundra will also reduce caribou and reindeer numbers and further limit food supplies for the Inuit and other communities.

The reduction in sea ice may, however, offer some benefits. Reduced ice coverage in the Arctic Ocean will enable new seasonal shipping routes to open up (the fabled Northwest Passage) and shorten transport time and cost between Asia, Europe, and North America. It will also facilitate greater access to the valuable mineral and energy resources of the Arctic, which is believed to contain up to one-quarter of the world's undiscovered oil.[20]

Precipitation changes

The poles presently receive little precipitation and are, in effect, cold deserts (Antarctica is the driest and coldest continent on earth). A warmer atmosphere will enable the air to hold more water and will consequently deliver more precipitation to the polar regions, particularly in the northern hemisphere where the warming is greatest. Air temperature also determines whether precipitation falls as rain or snow – as the atmosphere becomes warmer the proportion falling as rain will increase.

Changes in the quantity and type of precipitation have already been observed for the Arctic, and, to a lesser extent, the Antarctic. Since 1975, Arctic precipitation has increased by 8% overall and the fraction falling as rain has also increased. Rain-on-snow events, for example, have increased by 50% in western Russia over the past three decades.[21] Precipitation in the Artic is expected to increase by another 5–25% this century, depending on the extent of warming.[22]

In Antarctica, precipitation is also projected to increase – but the region is so cold that most precipitation will continue to fall as snow. Accumulating snow may equal or exceed the amount lost from melting and glacier flow to the ocean, and, consequently, scientists generally expect that the mass of the Antarctic ice sheet will remain unchanged, or even possibly increase, this century. If snow accumulations exceed losses, then Antarctica would tend to reduce the rate of sea level rise – a negative **climate change feedback**. However, uncertainty remains over whether Antarctica will gain or lose mass this century as recent evidence indicates that the rate of glacier flow to the oceans has accelerated and higher than models have predicted (see **ice sheets and glaciers**).

The type and intensity of precipitation has important implications for the amount of water retained on land (accumulated snow and ice) and that flows to the sea. It also determines the annual and seasonal characteristics of the hydrological cycle. The hydrological cycle in the Arctic has changed significantly over the past 50 years due to changes in sea ice coverage, precipitation patterns, river discharges, and glacier melt. These changes in turn have implications for ocean salinity and currents, flood risks, and terrestrial and marine ecosystems.

Greater inflows of freshwater to the Arctic Ocean and North Atlantic could lower salinity levels and could thereby possibly reduce the rate of flow of the oceans' most important circulating current, the **thermohaline**. Since the 1960s, an additional 37,000 cubic kilometers of fresh water (half from river discharge and half from melting sea ice) have flowed into the Arctic Ocean, which has led to a noticeable freshening in the Nordic sea and subpolar basins.[23]

While the overall salinity of the entire Arctic Ocean does not appear to have changed significantly in recent decades (as a majority of the freshwater is believed to have been exported to the Atlantic and Pacific oceans), it is considered likely that this could change in coming decades as freshwater additions increase.[24] Greater high-latitude precipitation, continued loss of sea ice, and accelerated glacier and Greenland ice cap melting are all expected to lead to ever-greater freshwater additions. The IPCC projects that the flow rate of the **thermohaline** could fall by 25% this century, while some scientists believe that it might shut down completely at some stage in the next 200 years.

The increased incidence of rain falling on snow will fundamentally affect Arctic ecosystems. Caribou and reindeer herds depend on the Arctic tundra for food, but increased rain and temperature will alter the freeze–thaw cycle. This will reduce the herds' ability to access food beneath the snow (by forming a hard ice layer) and will adversely affect the health, breeding success, and abundance of these species. Freshwater and marine fish species are also sensitive to changes in salinity, water temperature, and hydrological regimes. In recent decades, the shelf waters in several localized areas of the Arctic have freshened and water temperatures have increased, which has affected the abundance and seasonal cycles of phytoplankton, zooplankton, and fish species.[25] These impacts are expected to become more pronounced over the course of this century.

Permafrost thawing

Most of the land surface within and bordering the polar circles consists of permanently frozen ground (termed "permafrost"). Around 20% of Russia's land mass, for example, lies within the Arctic Circle, and 65% of its land surface is subject to permafrost. Global warming has already caused widespread thawing of the permafrost, and the amount of land subject to thawing can be expected to increase significantly over the course of this century. Arctic soil temperatures have already risen by up to 3°C since 1980, and the area of permanently frozen ground has fallen by 7%.[26] Recent studies have estimated that near-surface permafrost area could be reduced by up to 90% by 2100.[27] This will have important implications for economic infrastructure, the incidence of flooding and erosion, and the large quantities of organic carbon stored in frozen peat bogs and soils.

The high latitude regions are already experiencing infrastructure problems associated with thawing permafrost.[28] Thawing makes the soil unstable (it turns to slush) and causes infrastructure such as roads, railways, and buildings to subside. The Arctic region is, and will continue to be, an important petroleum production area. Many existing oil and gas production and pipeline facilities are built on permafrost and will need to be replaced or altered. Although reduced sea ice coverage will increase access to Arctic Ocean resources, permafrost thawing will reduce the ability to access land areas (where most of the processing and distribution infrastructure is located) and will incur considerable additional costs.

In terms of climate change of even greater importance is the thawing of the vast peat bogs and soil organic matter. The frozen soils contain around 400 billion tonnes of carbon (or approximately 1,500 billion tonnes of **carbon dioxide equivalent – CO_2e**), with an additional 500 billion tonnes of carbon estimated to be stored in frozen periglacial wind-blown sediments (loess), mainly on the Siberian plains.[29] There are also significant quantities of carbon contained in frozen **methane hydrates** located in seabed sediments in the northern latitudes.

As the soils and peat bogs thaw, they expose frozen organic matter to oxidation and accelerated aerobic decomposition. This releases both methane and CO_2 to the atmosphere (or to rivers and lakes in the form of dissolved organic carbon). A considerable increase in methane and carbon dioxide emissions from the marshes of Russia, Canada, and Alaska has already been observed (see **methane**), and this could continue to grow significantly over the course of the century. Preliminary estimates suggest that up to 18 billion tonnes CO_2e (equivalent to around one-third 2005 global emissions) could be released to the atmosphere by 2025, and approximately 360 billion tonnes CO_2e (more than seven times 2005 global emissions) by 2100, and possibly 850 billion tonnes CO_2e if loess sediments are included.[30] It is uncertain how much will be released as carbon dioxide or methane or carried to the ocean as dissolved organic carbon (most of which will be released to the atmosphere before it reaches the ocean). If a significant fraction is released as methane, this would deliver a much greater warming effect to the atmosphere than if it were released as CO_2 (see **methane**). These additional global warming-induced emissions represent another major positive **climate change feedback**.

Vegetation and biodiversity

As the climate warms and precipitation increases, the higher latitudes will be able to support more vegetation, and this will alter the types and structure of ecosystems. Tundra will progressively disappear, to be replaced by boreal forests. It is estimated that forest area will increase by 55% and tundra area will reduce by 40% by 2100: trees may move northward by up to 500 km.[31] It has been generally believed that the northern march of the forests will enable greater quantities of CO_2 to be sequestered from the atmosphere. While this will indeed occur, recent studies of the effects of temperature and CO_2 fertilization on forests suggest that the carbon gains in above-ground vegetation are likely to be largely offset by increased losses from soils. Increased forest coverage in the north will also reduce surface **albedo** (forests reflect less solar energy than ice and snow or open tundra). Some studies conclude that the reduced albedo effect alone will outweigh the gains from CO_2 absorption.[32] When the albedo effects are combined with the emissions from thawing permafrost and enhanced levels of soil respiration, the overall result will be that Arctic vegetation and permafrost changes will provide a positive **climate change feedback** and will accelerate global warming even further.

Climate change will induce significant changes for a range of migratory birds. By the end of the century, up to 50% of the Arctic's migratory bird breeding habitats are likely to disappear under mid-range global warming projections.[33] Some species, such as the spoon billed sandpiper, dunlin, and white fronted goose, could lose 50% of their breeding grounds with a temperature increase of only 2°C.[34] Commercially harvested fish and crustacean populations have already undergone substantial changes. There has been a large reduction in cod abundance, while the abundance of crustaceans such as shrimp and snow crab has increased (see **marine impacts**).[35] The US Environmental Protection Agency recently placed the polar bear on the endangered species list and indicated that the significant reduction in polar bear numbers can be largely attributed to disappearing sea ice as a result of climate change. In Antarctica, there will also be ecosystem effects. Some species of penguins, birds, and a large number of marine species are likely to suffer reductions in habitat zones and changes in ecosystem composition. As a general rule, those species that depend on the ice and tundra for their existence will be adversely affected – essentially they have no adaptation options available to them.

A general warming in the higher latitudes will, however, bring some benefits to humans and other species. The northward progression of the forests is expected to provide expanded habitat ranges for many forest species (but also for pests such as the Spruce Bark Beatle – see **biodiversity impacts**). It will also enable a spatial expansion of agriculture and forestry production, as well as increase the number of settlements and population in areas that were previously considered too inhospitable for habitation.

See also: albedo, biodiversity impacts, climate change feedbacks, glaciers and ice sheets, marine impacts, ocean carbon sinks, sea level rise, thermohaline.

Notes

1 IPCC 2007
2 Folkestad *et al.* 2006
3 Hassol and Corell 2006
4 Folkestad *et al.* 2006
5 Ibid.
6 Greene and Pershing 2007
7 Kerr 2007
8 Ibid.
9 Folkestad *et al.* 2006
10 Sluijs *et al.* 2006
11 IPCC 2007
12 Ibid.
13 Hassol and Corell 2006

14 IPCC 2007
15 Hassol and Corell 2006
16 Stern 2006
17 IPCC 2007
18 Hassol and Corell 2006
19 ACIAR 2004
20 US Geological Survey, www.usgs.gov
21 Hassol and Corell 2006
22 Folkestad *et al.* 2006
23 Peterson *et al.* 2006
24 Ibid.
25 Greene and Pershing 2007
26 IPCC 2007
27 See Canadell *et al.* 2007
28 Stern 2006
29 Zimov et al. 2006
30 See Canadell *et al.* 2007
31 Folkestad *et al.* 2006
32 Betts 2000
33 Hassol and Corell 2006
34 Folkestad *et al.* 2006
35 Greene and Pershing 2007

Further reading

ACIAR 2004; Folkestad *et al.* 2006; Hassol and Corell 2006; IPCC 2002, 2007.

RADIATIVE FORCING

The earth's climate is fundamentally driven by the balance between incoming solar radiation and outgoing infrared radiation (see **greenhouse effect**). Radiative forcing is derived from **anthropogenic greenhouse gas emissions** and other factors that force the radiation balance away from its natural state.[1] It is expressed in global annual average Watts per square meter (Wm^{-2}). In effect it is a measure of "how far out of balance" the atmosphere currently is. If net radiative forcing is positive, the earth's system is absorbing more energy than would normally be the case, leading to **global warming**. Conversely, if radiative forcing is negative, the earth absorbs less energy than normal, resulting in global cooling. At present, radiative forcing stands at $+1.6\ Wm^{-2}$, indicating a net warming effect.

Radiative forcing is a relatively simple, well-understood measure for comparing multiple influences on climate by quantifying the global average energy imbalance. It neither provides a measure of climate change, nor the magnitude of **climate change impacts**. Though climate change is closely correlated with radiative forcing, it is also influenced by other variables, such as **climate sensitivity**, and the strength of **climate change feedbacks** initiated by increases in global mean temperatures. The relationship is nonlinear – each additional $1\ Wm^{-2}$ change in radiative forcing does not yield the same amount of climate change.

Over paleoclimatic timescales, the earth–atmosphere system is constantly changing due to factors such as the **Milankovich cycles**, continental drift, and a range of other bio- and geochemical processes. These changes occur over thousands of years and are barely perceptible over timescales spanning a few centuries. Over these shorter time spans, the earth–atmosphere system is considered stable (in equilibrium). The radiative balance against which radiative forcing is compared are those that prevailed at the commencement of the Industrial Revolution, around 1750.

Components of radiative forcing

Annual radiative forcing depends on the status of different forcing agents, such as greenhouse gas concentrations, **albedo**, **aerosols**, and solar irradiance. These agents can have either a positive or negative influence on radiative forcing and can change due to natural or anthropogenic factors.

Natural factors

Changes in solar irradiance and volcanic activity are the two principal natural contributors to radiative forcing. The sun's energy output is variable, with regular 11-year cycles of up to 0.12% variation. These cycles average out over time, but since 1750, there has been a slight increase (0.05%) in total solar irradiance,[2] contributing a positive radiative forcing of 0.12 Wm^{-2}.

Volcanic eruptions eject enormous quantities of **aerosols** (such as sulphates) into the stratosphere, reflecting the sun's radiation back into space before it can reach the earth's surface, thus having a cooling effect. The most climatically important volcanic eruption in the past 150 years was Mount Pinatubo, Philippines, in 1991. This resulted in an estimated peak instantaneous (not annual) radiative forcing of -3 Wm^{-2}, similar in magnitude (but opposite in sign) to that attributable to the anthropogenic greenhouse gas effect. Pinatubo resulted in an annual global temperature decrease of 0.1°C from the decadal mean.[3] However, volcanic ash particles precipitate out of the atmosphere in about 3 months and sulphate **aerosols** in about a year.[4] Thus, while the effect of volcanic eruptions on radiative forcing can be substantial in the short term, the effect is transient and does not have a sustained long-term influence on the earth's climate.

Changes in cosmic radiation have also been suggested as another possible natural contributor to radiative forcing. It is postulated that recent changes in cosmic ray fluxes interacting with the atmosphere have caused an increase in low-level cloud formation, increasing **albedo** (a negative forcing).[5] However, this effect has yet to be scientifically substantiated and is not included in current radiative forcing estimates.[6]

Anthropogenic factors

By far the most important human contribution to positive radiative forcing is the rise in atmospheric concentrations resulting from **anthropogenic greenhouse gas emissions** since 1750. This is potentially the most significant sustained forcing in at least the past 16,000 years[7] and most likely the past several million years.

Table 16 2005 radiative forcing contributions[8]

	Radiative forcing value (Wm⁻²)	% of long-lived greenhouse gases
Positive radiative forcing components		
Carbon dioxide (CO_2)	1.66	63
Methane (CH_4)	0.48	18
Nitrous oxide (N_2O)	0.16	6
Synthetic gases	0.34	13
Sub-total for anthropogenic greenhouse gases	2.64	100
Tropospheric ozone	0.35	
Stratospheric water vapor from CH_4	0.07	
Surface albedo effect from black carbon on snow	0.1	
Aircraft contrails	0.01	
Solar irradiance (natural component)	0.12	
Negative radiative forcing components		
Depletion of stratospheric ozone	−0.05	
Increase in surface albedo from land use changes	−0.2	
Direct aerosol effect	−0.5	
Aerosol indirect effects on cloud albedo	−0.7	
Total net radiative forcing	1.6	

The long-lived anthropogenic **greenhouse gases** (**carbon dioxide**, **methane**, **nitrous oxide**, and the **synthetic gases**) contribute most to positive radiative forcing, with a substantial contribution from short-lived tropospheric **ozone**. Less important agents are stratospheric water vapor, resulting from the decomposition of **methane**, and decreased **albedo** from black carbon on snow. A minor contribution also comes from aircraft condensation trails (contrails) in the stratosphere. By 2005, anthropogenic activities had contributed approximately 3.2 Wm⁻² of positive radiative forcing (see Table 16).[9]

Aerosols (which affect **albedo**) account for about 80% of anthropogenic negative radiative forcing. The remaining 20% arises from increased **albedo** resulting from land-use changes and the destruction of stratospheric **ozone** (primarily from CFC emissions – see **synthetic gases**[10]). The aggregate negative radiative forcing is approximately −1.5 Wm⁻² (see Table 16).

The radiative forcing values of **greenhouse gases** have normally distributed probability density functions (uncertainty is evenly spread above and below the mean), and therefore, their mean values (shown in Table 16) can be directly added. The uncertainty of the other radiative forcing components are skewed (not normally distributed), and their means cannot be simply added. The total radiative forcing value of 1.6 Wm⁻² is a result of addition of probability density functions to account for different uncertainties, and not the simple addition of the mean values.

Note that the percentage of long-lived greenhouse gas radiative forcing attributed to **carbon dioxide** is 63%, in contrast to the much higher (77%) CO_2 contribution to atmospheric concentration of greenhouse gases in **carbon dioxide equivalent**

(CO_2e) terms (see **anthropogenic greenhouse gas emissions**). Conversely, **methane** represents only 14% of atmospheric greenhouse gas concentrations (in CO_2e terms) compared with an 18% share of radiative forcing. This apparent anomaly is due to radiative forcing being an annual measure, not accounting for the different atmospheric residence times of different gases. Methane has a relatively short atmospheric life of around a decade, while CO_2 has an assumed atmospheric life of 100 years (see **global warming potentials**). Consequently, when looking at the annual influence on climate, short-lived gases have a relatively greater impact.

The importance of atmospheric residence time on radiative forcing levels can be illustrated by the following example: if all anthropogenic emissions of **greenhouse gases** and **aerosols** ceased today, in the first few years radiative forcing would increase substantially. While there would also be an almost immediate fall in tropospheric **ozone** concentrations, reducing forcing by approximately 0.3 Wm^{-2}, the negative forcing of **aerosols** would decline to zero in under a year and add around 1 Wm^{-2} to radiative forcing, a net increase of around 0.7 Wm^{-2}. Over subsequent decades, the initial increase in radiative forcing would be eroded as long-lived greenhouse gases started to be removed from the atmosphere. By the end of the first decade, most of the anthropogenic **methane** would disappear, but the remaining greenhouse gases, which account for about 80% of the anthropogenic CO_2e loading, have atmospheric lives of 100 years or more. Some gases, such as the perfluorocarbons and sulphur hexafluoride, have atmospheric lives measured in thousands of years, and their concentrations in the atmosphere and contribution to radiative forcing effectively continue forever (see **synthetic gases**). Even if emissions of these gases ceased, they will still have a radiative forcing of around +0.3 Wm^{-2} a thousand years from now.

Future radiative forcing trends

Radiative forcing will not necessarily track **future emissions trends**, as it is in large part dependent on the rate of removal or breakdown of historic emissions in the atmosphere as well as current emissions of short-lived components of **aerosols** and tropospheric **ozone**. These factors in turn depend on nonlinear **climate change impacts** and **climate change feedbacks**. As such, projections on what level of radiative forcing will result from a given **carbon dioxide equivalent** concentration of **greenhouse gases** in the future are highly uncertain. Nonetheless, unless negative **climate change feedbacks** are substantially larger than current models estimate, higher greenhouse gas concentrations will result in higher radiative forcing, pushing the earth's climate system further away from its natural balance, with a commensurate increase in **global warming**.

See also: aerosols, albedo, anthropogenic greenhouse gas emissions, climate change feedbacks, climate change impacts, climate sensitivity, global warming, global warming potentials, greenhouse gases, stabilization targets.

Notes

1 IPCC 2006a
2 Wang *et al.* 2005
3 Ramaswamy *et al.* 2006

4 IPCC 2007
5 Marsh and Svensmark 2000
6 IPCC 2007
7 Ibid.
8 Adapted from IPCC 2007
9 IPCC 2007
10 Synthetic gases include those covered under the Montreal Protocol and the Kyoto Protocol.

Further reading

IPCC 2007.

RENEWABLE ENERGY

Renewable energy sources are those energy sources that can be constantly replenished. They include sources derived directly or indirectly from the sun (wind, solar, hydro power, wave power, and biomass), moon (tidal), or earth (geothermal).

Although other mitigation measures, such as **energy efficiency**, are expected to play a more significant role in reducing greenhouse gas emissions in the short to medium term, most consider renewable energy to be the principal means of developing a carbon-free energy system in the longer term. Renewable energy will be an important component of any global **mitigation** response.

Renewable energy sources first attracted the attention of governments after the 1970s oil supply crises (1973 and 1979), but interest waned in the latter half of the 1980s and early 1990s, primarily due to the higher cost of most renewable sources relative to fossil fuels. The emergence of climate change as a major policy issue has brought fresh interest in renewable energy and many governments have introduced policies (such as mandatory targets, subsidies, and tax incentives) aimed at increasing the contribution of renewable energy. These measures have stimulated considerable growth in the application of several renewable energy technologies, particularly **wind power** and **solar power**, and several other nonhydro renewable sources (like biofuels and geothermal energy).

Present contribution of renewables

Renewable sources of energy already make a significant contribution to global commercial energy supplies, accounting for an estimated 17%. Hydropower accounts for nearly all of renewable energy's 18% contribution to global electricity supplies, with other renewable sources (biomass, wind, geothermal, and solar) contributing less than 2%.[1] Since 1990, the contribution of renewable energy sources to global electricity production has remained largely unchanged.

Renewable energy sources also make a small contribution to direct thermal (heat) energy requirements (e.g. solar hot water systems, geothermal heat pumps, and process steam from biomass).

In the traditional energy sector (often referred to as noncommercial energy and usually not included in energy statistics), renewable sources are principally used for generating heat for cooking (biomass, plus agricultural and animal wastes) and for drying crops and pumping water (using solar and wind). In many developing countries, renewable energy accounts for more than half of household energy supplies, and in some countries, as much as 90%. At present, 2.5 billion people rely on renewable energy for a majority of their primary energy needs.[2]

It is in the electricity sector where renewable energy sources have the largest greenhouse gas mitigation potential, although opportunities also exist for **biofuels** in the transport sector. Renewable energy sources currently serve only about 2% of transport energy demand (mainly in the form of ethanol), but their share could increase over the next few decades.[3]

Advantages and disadvantages

As a greenhouse gas mitigation option, renewable energy sources have several advantages and disadvantages. The principal advantages are that they:

- are plentiful, inexhaustible, and widely distributed geographically;
- produce virtually no greenhouse gas emissions (although small quantities are produced in manufacturing and construction);
- have very limited health and safety risks and produce few wastes;
- are suitable for widespread application on a small to medium scale, sited near to the energy demand source (often termed *distributed generation*); and
- are not subject to volatile energy price changes.

The main disadvantages (excluding large hydropower and, to some extent, geothermal) are that:

- renewable sources are generally more expensive than fossil fuels;
- many renewable technologies provide energy only on an intermittent basis (e.g. solar and wind are not available 24 hours per day) and may be subject to variations in annual and seasonal availability (although hydro, geothermal, and some biomass technologies can supply power on a constant basis);
- due to their intermittent and variable nature, their value to the electricity system is usually lower than fossil fuel and nuclear (usually only the avoided cost of fuel);
- there are environmental and social issues that may constrain the growth of some renewable technologies, including the siting of renewable energy facilities, the impacts of large-scale conversion of forest and agriculture to biofuel production, and the impacts of large-scale hydropower development; and
- currently only a few renewable energy technologies (mainly wind turbines and solar water heaters) are approaching the mass production stage, and this limits the potential contribution to renewable global energy supplies in the short term.

Technology status

A wide range of renewable energy technologies are currently being deployed or under development. The principal renewable technology options include small- and large-scale hydropower, **wind power**, **solar power**, geothermal, ocean energy (waves/tidal power), biomass, and **biofuels**. The present technological and economic status of wind, solar, and biofuels and the contribution they could make to reducing global greenhouse gas emissions in the future are discussed in other sections (see **wind power**, **solar power**, and **biofuels**). This section reviews the potential contribution of three other renewable sources that have already been deployed on a commercial basis – hydropower, biomass, and geothermal energy – and several other renewable technologies that could contribute to global energy supplies in the future. These include ocean energy technologies (wave, tidal, and ocean thermal energy) and technologies based on hydrogen (nuclear fusion and hydrogen fuel cells).

Hydropower

Large-scale hydroelectric power production is a well-proven technology that has been deployed for nearly a century. Installed hydroelectric generation capacity is currently 850 gigawatts (GW), accounting for 16% of global electricity supplies in 2004.[4] Hydropower is the principal source of electricity for several countries: Canada, for example, sources 57% of its electricity from hydro, and the equivalent figure in Latin America is 66%.[5] Hydro has several advantages as an electricity-generation technology, including the ability to respond to sudden fluctuations in electricity demand; the ability to "cold-dark start" where unlike most fossil plants, hydropower can generate electricity without having initial energy input; and, where based on large-scale storage reservoirs, the ability to supply energy on a constant (24 hour) basis.

Hydro can be subdivided into two main categories: large-scale (greater than 30 megawatts [MW]) and small-scale (less than 30 MW), with further divisions of mini-hydro (1–5 MW) and micro-hydro (<1 MW).

Large-scale hydro generally uses storage reservoirs (large dams) and can be a controversial energy source. Reservoirs can inundate large areas of land (often involving the loss of pristine environments), sometimes require large-scale relocation of people, and can have adverse socioeconomic and ecosystem impacts downstream (such as reduced water flows, loss of wetlands, and diminished fisheries). Large-scale hydro is, however, usually a very cost-competitive source of electricity, and most existing facilities can generate power at 2–4 US cents per kilowatt hour (kWh) – very competitive with coal. Generation costs of new facilities are estimated at 3–5 US cents/kWh but vary considerably depending on the characteristics of the undeveloped hydro resources. Although hydro is a renewable energy source, the availability of economic undeveloped resources is finite and costs will rise over time as more marginal hydro resources are progressively developed.[6]

Small-scale hydro is generally based on run-of-river flows that have little or no water storage capabilities and are therefore subject to greater seasonal and annual supply variability. However, they are more environmentally benign than large-scale hydro facilities. Small-scale hydro generates electricity at costs ranging from 2 to 8 US cents/kWh but can be economic at higher costs in some locations. Similarly, mini- and micro-hydro are generally run-of-river with supply variability and cost implications but attractive for remote, village-level electricity supply.

Considerable undeveloped economic hydro potential exists, and less than one-third of the estimated economic resource has so far been developed.[7] However, the remaining undeveloped economic resources are unevenly distributed, and most are located in Africa, Asia, Latin America, and Russia. Many of these potential sites face a range of social and environmental constraints that may constrain their future development.

Hydropower will remain a major source of energy in the coming decades and will also help to reduce the future rate of growth of global greenhouse gas emissions. The International Energy Agency expects global installed hydropower capacity, under its business-as-usual "reference scenario," to grow by more than 500 GW over the period 2004–2030, mostly in developing countries where most of the remaining undeveloped hydro potential exists.[8] This growth is not, however, sufficient to maintain hydro's projected share of global electricity production, which is expected to fall to 14% over the period to 2030. But the IEA also estimates that hydropower could make a more significant contribution to global electricity supply under their alternative policy scenario, where governments adopt more aggressive energy policies driven by energy security and climate change considerations (see **future emissions trends**). Under this alternative scenario, more hydro plants are constructed and hydro is then able to maintain its present share (16%) of global production by 2030.[9] Long-term IEA projections of electricity production in 2050 suggest that hydro could provide an additional 1800–2400 terrawatt hours (TWh) above today's production levels. If it is assumed that 80% of this electricity would otherwise have been generated by fossil fuels (presently the most cost-competitive alternative), then hydro could deliver emission reductions of between 1 and 2 Gt carbon dioxide per year by 2050, depending on the type of fossil fuel displaced.[10] This represents a quite modest contribution to global mitigation efforts to 2050.

Biomass

The utilization of solid biomass to generate heat and electricity has been deployed on a commercial basis for more than a century (particularly in the timber, sugar, and other agricultural industries). Many forms of solid biomass feedstocks can be utilized, notably agricultural, municipal, and industrial wastes. Biomass can be combusted using standard steam boiler systems or can be gasified for direct use as a gaseous fuel or gasified for indirect use in turbines to generate electricity. Biomass can also be used to produce a range of **biofuels** for use in the transport sector. The main advantage of biomass is that the CO_2 produced is recycled through the operation of the **carbon cycle** and, if operated on a sustainable basis, has no net impact on atmospheric CO_2 concentrations. Put simply, the CO_2 emitted when the biomass is burnt is equivalent to the amount of CO_2 absorbed from the atmosphere during plant growth.

Biomass combustion efficiencies are generally lower (around 25–30%) than those of fossil fuel combustion technologies, and the scale of application is also usually smaller (in the 20–50 MW range). This increases unit costs relative to fossil fuels, with the result that the main economic attraction of biomass combustion lies in combined heat and power (CHP) applications in industry and urban district heating systems. However, depending on the cost of the biomass fuel source (it is often a free waste product), biomass electricity facilities can be economically competitive with fossil fuels. A wide range of factors influence generating costs (including feedstock quality and supply reliability, scale, and fuel transport and processing costs), and consequently costs show wide variation between commercial facilities – typically in the US 5–13 cents/kWh range.[11]

Potentially the most significant contribution of biomass to electricity generation is cofiring in existing coal-fired facilities. Coal/biomass blends of up to 10% can be used without any significant modifications to the coal boiler system, although overall combustion efficiency is lower (relative to coal alone). However, the combustion efficiency of the biomass component is higher (relative to biomass-only systems) and the costs are lower. Cofiring is already being practiced at over 150 installations in Europe, North America, and Australia. Utilizing 10% biomass to cofire coal plants delivers electricity at around US 5–6 cents/kWh (competitive with **nuclear power**) and is considered to be the least-cost biomass utilization option. It also offers additional environmental benefits in the form of lower emissions of dust, sulphur dioxide, and nitrogen oxides. Using biomass in Integrated Gasification Combined Cycle (IGCC) facilities is also being investigated, but costs are currently estimated at more than US 10 cents/kWh, making it a relatively expensive mitigation option.

A potential negative consequence of significant uptake of biomass usage is that the sourcing of biomass may not be sustainable. If old-growth forests are being cleared to source biomass or complex ecosystems are being replaced with monocultures for the more efficient production of biomass, then the loss of biodiversity may not justify the greenhouse gas benefits. If biomass is produced on arable land previously used for food production, there may be competition for resources, where the poor and the subsistence farmers could lose. There have also been concerns raised where biomass sources have been treated with insecticides and/or pesticides (such as cotton stalks) are incompletely combusted, or combusted at lower temperatures, they can form carcinogens and mutagens. Nonetheless, when biomass sources are waste products that are burned under carefully controlled conditions, they are a relatively safe fuel source.

Overall, biomass offers a reasonably cost-effective greenhouse gas mitigation option, particularly in developing countries (where significant biomass feedstock potential exists). It is expected to increase its contribution to global energy supplies over the coming decades, with an expected two- to three-fold increase in its share of global electricity supplies to 2030 (2–3% of electricity generated) and possibly account for 5% of global electricity supplies by 2050.[12]

Geothermal energy

Geothermal energy is obtained by extracting heat energy from the earth to produce electricity and is already deployed as a commercial electricity production technology in more than 20 countries. Like hydropower, geothermal power can produce reliable power on a continuous basis and, at most existing facilities, at a reasonably cost-competitive price. However, the most economically viable geothermal resources are restricted to just a few geographical regions, located in active volcanic areas (close to the boundaries of the major tectonic plates). Major geothermal development is currently limited to Central America and the United States, Japan, the Philippines, Indonesia, New Zealand, Italy, Iceland, and East Africa. However, the development of new technologies could expand the economic resource base over the coming decades.

327

Existing and prospective sites are generally based on steam and hot water, although "hot dry rock" technology could also prove economic in some locations. The most attractive geothermal resource is direct extraction of high temperature steam, but this is rare and only five sites have so far been identified. High temperature steam resources can produce electricity at a price as low as US 2–3 cents/kWh. Most other geothermal resources involve flash steam (using water hotter than 173°C) and binary systems (which can utilize water with temperatures as low as 85°C). Generally, the lower the water temperature, the more expensive the electricity. Most existing commercial facilities produce electricity in the 3–5 US cents/kWh range, although the cost from some European facilities is in the 6–11 US cents/kWh range. The majority of the attractive economic resources have already been exploited, and the generation costs associated with most future facilities are expected to be in the 5–8 US cents/kwh range (marginally economic in most locations).[13]

The future contribution of geothermal, based on existing technology, is considered to be relatively limited, although some increases in installed capacity are expected over the period to 2030. Further drilling could open up more resources, particularly in developing countries where exploration has so far been limited – but geothermal as a source of electricity is expected to remain a niche market. Geothermal investments are also a risky commercial venture as capital costs are high, the heat reservoir characteristics are difficult to determine precisely from drilling alone, and actual plant performance (and therefore cost) is often only known once the plant has commenced operation.

Hot dry rock technology (where water is pumped into hot rock formations and then brought back to the surface to generate steam) offers considerable longer-term potential due to the magnitude of the resource. Costs are currently similar to, or marginally higher than, solar power based on photovoltaic cells (see **solar power**). Some estimates put the cost of electricity at more than 20 US cents/kWh, although several plants in favorable locations are expected to produce electricity at much lower cost.[14] One area where considerable potential exists is in the provision of low-grade heat (40–60°C) to homes and buildings, and already more than 2 million geothermal heat pumps have been installed worldwide.[15] Where these systems displace electricity generated from fossil fuels, or oil and gas heating, they reduce greenhouse gas emissions.

Apart from local site disturbance, and in some places emissions of sulphur, there are no significant other environmental impacts from geothermal power, and because of this, it is often seen as a promising long-term energy supply option. Overall, geothermal energy will continue to contribute to global energy supplies, but its long-term contribution, especially as a source of cost-effective electricity, is expected to be much smaller than other renewable sources such as hydro, solar power, and wind power. However, up to 2050, up to 15% of the projected increase in renewable energy production (excluding hydro and biomass) could come from geothermal sources.[16]

Ocean energy

Several technologies can extract energy from the oceans. The principal forms of ocean energy currently deployed on a commercial basis, or at the demonstration stage, are wave power, to harness surface kinetic energy embodied in waves; tidal power, to harness the kinetic energy of ocean currents and tides; and ocean thermal energy conversion (OTEC), to utilize the heat gradients of the oceans to generate electricity. Several other ocean technologies, such as utilizing salinity gradients or ocean biomass production, have also been suggested, but these are only at the conceptual or laboratory stage.

The two ocean technologies that offer the most potential are wave energy and tidal/current energy. Wave energy is close to commercial-scale development, and several demonstration projects are operating: the first commercial facility is expected to be constructed by 2010.[17] Costs below US 10 cents/kWh are expected at favorable locations. Tidal energy involves similar technology to hydropower, using barriers and turbines to extract the energy of moving water. Tidal involves the construction of a barrier that straddles a bay or estuary, with turbines built into the barrier. As the tide advances and recedes, the water turns the turbines and generates electricity. Three tidal facilities are already operating – a 240 MW facility at La Rance, France (built in the 1960s), and two smaller facilities built in Russia and Canada. Tidal power can provide reasonably firm electricity supplies (can generate power for a considerable proportion of the day), but output varies over different stages of the tidal movement. Tidal power is very site specific and requires a very large tidal range. They also need to be located close to a major electricity grid in order to be economically viable. Many potential sites exist, but tidal energy is expensive – around 10–12 US cents/kWh at favorable locations.[18] OTEC offers considerable energy potential in the longer term but is far from being a proven commercial technology. A small-scale demonstration facility is currently operating in Hawaii, and a larger circulation system from Lake Ontario serves downtown Toronto, but the technology is likely to remain limited to niche markets and is unlikely to be deployed on a large scale until after 2050, if at all.

Overall, ocean energy systems are expected to remain only very small contributors to global energy supplies up to 2050 and will therefore remain a minor contributor to greenhouse gas mitigation efforts.

What contribution can renewable energy make to greenhouse gas mitigation efforts?

It is evident that renewable energy sources play an important role in the context of greenhouse gas mitigation. However, these sources of energy are unlikely, based on current projections, to deliver more than 10–20% of the carbon dioxide emission reductions required by 2050 to stabilize emissions at 2005 levels, let alone stabilize atmospheric concentrations. In the absence of supportive government policies, the contribution could be lower. Renewable energy sources are likely to make only modest contibution in reducing greenhouse gas emissions in the short term – they are likely to be a more important mitigation option in the medium to long term.

An International Energy Agency report, published in 2006, assessed the potential contribution of different technology options for stabilizing or reducing emissions up to 2050.[19] This study developed six different scenarios, characterized by different assumptions about policy changes, research, and development; technology advances; and technology commercialization efforts. Under the most optimistic scenario for renewables (termed the "Tech Plus" scenario), all forms of renewable energy (including hydro) are estimated to contribute approximately 20% of the emission reduction potential; under the more realistic scenario (the MAP scenario), the contribution of renewables is estimated to be approximately 15% (mostly from wind and biomass).

For renewable energy to achieve a 10–15% contribution to stabilizing emissions by 2050 will require a massive commitment by the global community. Even with strong government policies and incentive schemes, just the physical manufacturing and construction constraints of delivering a 30- to 40-fold expansion in nonhydro renewable energy output by 2050 (the amount required for these sources to deliver a 10% share of global commercial energy supplies by 2050) represent a significant challenge. Conceivably, renewable sources could make a greater contribution, but this would require a more radical change in policies, including the introduction of a significant cost penalty on CO_2 emissions (see **emissions trading** and **carbon tax**) and much more rapid progress in technology development. The resources devoted to research and development (R&D) of renewable energy technologies may need to undergo a ten-fold increase to achieve more than a 10% contribution to emission reductions.[20] This needs to be viewed against the long-term decline in R&D expenditures devoted to renewable technologies since the 1980s.[21] Nonetheless, investments in renewable energy production facilities have grown considerably in recent years, reaching US$150 billion in 2007 and is expected to exceed more than $US600/year by 2020.[22]

It is clear that renewable energy sources will play an increasingly important role in meeting the world's demand for energy over the coming decades, but it will be many decades before they surpass fossil fuels as the world's principal source of energy.

See also: biofuels, future emissions trends, solar power, stabilization targets, wind power.

Notes

1 Renewables 2005
2 IEA 2006a
3 Ibid.
4 Ibid.
5 IEA 2006b
6 Ibid.
7 IEA 2006a
8 Ibid.
9 Ibid.
10 Based on projections data in IEA 2006b
11 IEA 2006b
12 IEA 2006a and 2006b
13 IEA 2006b
14 Ibid.
15 Renewables 2005
16 IEA 2006b
17 Ibid.
18 Ibid.
19 Ibid.

20 Kammen 2006
21 Ibid.
22 UNEP 2008a

Further reading

IEA 2006a, 2006b; Kammen 2006; Renewables 2005; Ogden 2006.

SEA LEVEL RISE

Rising sea levels are frequently portrayed by the media as the major near-term impact of global warming. Images of disappearing sea ice, melting glaciers, collapsing polar ice shelves, and retreating mountain snow lines certainly capture the imagination and instill a sense of drama. That these things are happening and that human-induced climate change is contributing to these events are undeniable. It is also known that even if **anthropogenic greenhouse gas emissions** were to cease today, the sea level would go on rising for many decades, and possibly centuries. This is due to the long time lags between changes in atmospheric temperature and the resulting changes in ocean temperature and melting ice. The world is already committed to future sea level increase, regardless of our current **mitigation** efforts.

Despite the excited media coverage, the rate of sea level rise is actually likely to be quite slow. Sea level change over the next 100 years or so will be relatively modest, most probably up to 1.0 m and possibly only 0.5 m (although some estimates range as high as 1.5 m).[1] Rises of this magnitude would have coastal and ecosystem impacts, but it is expected that humans would be able to cope with, and adapt to, such increases. However, the response of the oceans and ice caps is so slow to changes in atmospheric temperature that the full impact of current warming on sea levels will not be realized for centuries, and possibly a thousand years or more. It is this long-term impact that is sometimes described as the *commitment to sea level rise*.

There is a high degree of uncertainty surrounding the timing and extent of sea level change. Scientific understanding of the complex processes and feedback mechanisms at work is limited. Much of the uncertainty arises because it is not clear how sensitive the earth's temperature is to changes in atmospheric concentrations of **greenhouse gases** (see **climate sensitivity**) and how sensitive the oceans and polar ice caps are to increased global temperatures. But it is known that the sea level will continue to rise, possibly at an accelerated rate, during the course of this century. Furthermore, recent evidence suggests that sea levels are rising faster than most models predict and sea level rise could be considerably higher than the 2007 **Intergovernmental Panel on Climate Change (IPCC)** estimates.

Measuring sea levels

The sea level changes constantly, according to air pressure, wind, and tectonic movements. For example, the prevailing southeast trade winds in the Equatorial Pacific constantly push warm surface waters westward across the ocean, with the result that the sea level is generally higher in the Western Pacific than it is in the Eastern Pacific. Air pressure also has a major impact on sea levels. During intense tropical cyclones, when air pressure is very low, the seas can rise a meter or more and completely cover some low-lying island atolls. These changes in sea level are temporary and can occur even if the average level of the ocean surface is unchanged.

Land is also constantly rising from, and subsiding into, the sea in different areas of the world due to natural land movements. For example, the northern coastline of Sweden is currently rising from the sea at a faster rate than the sea itself is rising. This is largely due to the huge weight that has been removed from northern Scandinavia through deglaciation since the last ice age.[2] There are many locations where the coast is rising and subsiding, and hence, the impact of sea level rise will vary by location.

To isolate the influence of such tectonic effects on surface-based measurement points, a variety of different measuring techniques are used to determine the *global mean sea level*. It is the change in global mean sea level that is generally referred to when assessing the impact of global warming. While ground-based data is still collected and used extensively, observations from satellites (like the Topex/Poseidon and Jason-1 satellites) have now become the primary means of recording changes in sea levels.[3]

How does global warming cause a rise in sea level?

There are two ways in which global warming can cause the sea levels to rise. First, sea level can rise due to the thermal expansion of the oceans. As water warms, it expands and occupies a greater volume; when it cools, it shrinks and occupies a smaller volume. The same principle applies to the oceans but on a much larger scale. It is not known precisely how much oceanic thermal expansion will occur in the coming decades and centuries as it depends on what happens to ocean circulation systems, the amount of global warming and variations in regional warming, and changes in atmospheric surface pressure. There is a large range of uncertainty, and estimated thermal expansion can be as much as ±100% of the mean.[4] Over the course of this century, possibly more than one-third of the predicted sea level rise is likely to come from the thermal expansion of the oceans (probably around 20–30 cm).

The second, and more visible, contribution to sea level rise is through the addition of more water from melting land-based ice. The principal source of this water will, at least initially, come from receding glaciers and mountain snow packs. However, in the long term (beyond 2050), it is the Greenland and Antarctic ice caps that will be the dominant forces determining sea level rise. If these two ice caps were to melt completely, the sea level would be 65 m above current levels. According to the **IPCC** (2007), the contribution from Greenland and Antarctica to sea level rise is not expected to be significant this century but could become so in the future. However, other studies suggest that based on the significant acceleration in melting and glacier discharge observed in recent years (which are not yet fully reflected in the most recent IPCC estimates), the contribution from Greenland and Antarctica could be greater than expected.[5] Considerable uncertainty still remains (see **ice sheets and glaciers**).

The large expanses of sea ice that cover the North Pole and some areas of the sea off the coast of Antarctica will not, in fact, contribute much to sea level rise. The reason for this is that they are already floating and have, therefore, displaced their own volume in the water. To illustrate why this is the case, imagine that a handful of ice cubes were added to a jug of water and corresponding water level marked. After a period of time, the ice would melt but the water level in the jug would remain at the same mark.

Observed rises to date

Toward the end of the last ice age, about 20,000 years ago, the sea level was about 120 m lower than it is today and ice covered a third of the earth's land surface (compared with around 10% surface coverage today). Most of the water was held in vast sheets of ice that covered much of North America, Russia, and Europe and some smaller areas of land in the southern hemisphere. As the ice age waned and the massive ice caps melted, vast quantities of water were released back to the oceans. Sea levels rose considerably until about 8,000 years ago but have since risen only very slowly. In the northern hemisphere, most of what remains of the land ice from the last ice age is contained in the Greenland ice cap, though the remaining mountain glaciers also contain a considerable amount of water.

Over the last 3,000 years, mean sea levels only rose on average by 1–2 cm/century, mainly from ongoing melting of residual glaciers from the last ice age.[6] However, during the twentieth century, the rise in sea levels accelerated considerably, to average 1.7 cm/decade (1.7 mm/year).[7] About half of the 17 cm rise since 1900 has been attributed to thermal expansion and the remainder through additions of water from melting glaciers with small amounts from the ice caps.[8] Since the early 1990s, there has been a discernible increase in the rate of sea level rise – to about 3.1 mm/year.[9] If this rate were to be maintained throughout this century, then the sea would rise by around 30 cm. While the observed rate has increased, it is too early to ascertain whether this represents a sustained acceleration, but indications suggest that it might be. It may require another 10–15 years of measurement and observation to determine the actual rate of change and the emergence of any accelerating trend.

Projected future rises?

Sea level rise can be viewed over two time horizons: what could occur this century and what could occur over the next 500–1,000 years. The 100-year time horizon usually occupies most policy makers' thoughts, but it must be realized that global warming will initiate a sea rise process that is unlikely to stabilize for more than 1,000 years.

The principal contributor to sea level rise will be the ongoing melting of **ice sheets and glaciers**. Of course, there is an upper limit to this rise (as there is only a finite quantity of land-based ice): if the earth were to become completely ice free, sea levels would rise by 65–70 m above their present levels. Thermal expansion of the oceans could, over the coming centuries, deliver up to an additional 5 m, but this is a relatively slow process. It takes several decades for the surface oceans to catch up with atmospheric temperature change and centuries to be reflected in deep ocean temperature. In total, the theoretical maximum amount the seas could rise is 70–75 m above present levels.

Projected increases to 2100

Continued increases in global temperature will accelerate the rate of glacier melt, sea ice retreat, and melting of the ice caps, though whether there will be a net loss of ice from Antarctica remains uncertain (see **ice sheets and glaciers**). Estimating future sea level rise involves an understanding of several complex processes that could contribute to higher sea levels. These include the rate at which the oceans absorb heat (which will determine the amount of thermal expansion), the amount of water that flows to the sea from melting glaciers and ice caps, and any changes in the amount of water stored on land. Most of the projections of future sea level changes are estimated on physics-based models of ocean heat absorption and glacier/ice cap responses, although some estimates are based on empirical data from what has already been observed.

The **IPCC** (2007) estimates that sea levels will rise this century by between 28 and 58 cm – up to half due to the thermal expansion of the oceans and the remainder will come from melting of glaciers and small ice caps and a small contribution from Greenland.[10] The IPCC estimates that the Antarctic ice mass would be neutral or gain in mass due to snow accumulation from increased precipitation, which could equal or exceed losses to the sea.

Some scientists believe that IPCC estimates are too conservative and that the models they are based on have systematically underestimated the actual rise in sea levels that have been observed. Since the late 1990s, the actual observed rate of increase has been at the upper end of the IPCC's range of uncertainty.[11] Projections based on observed data tend to produce higher estimates of future sea level rise this century. One recent study, based on observed data over the period 1880–2001, estimates that every 1°C rise in global mean temperature adds 3.4 mm/year to the rate of sea level increase.[12] Using IPCC projections of temperature increases this century (2–5°C) this would translate to 68–170 cm, much higher than the current IPCC estimate of 28–58 cm. This suggests that the uncertainty range for sea level rise may be higher than the IPCC estimates. The IPCC recognizes its projections may not fully incorporate recent changes in ice sheet and glacier dynamics and concludes there is some chance that sea level rises could be 10–20 cm higher and that even higher increases cannot be ruled out at this stage. Much depends on the rate of glacier flows and melting, the stability of the Western Antarctic Ice Shelf (WAIS), the rate of snow accumulation in Eastern Antarctica, the rate of melting of the Greenland ice cap, and the rate of absorption of heat by the oceans.

Projected rises beyond 2100

In the coming centuries, the melting of the Greenland and Antarctic ice sheets will be the dominant source of sea level rise, far outweighing any increases from thermal expansion. The timing and magnitude of change will depend on when trigger points are passed for large-scale Greenland deglaciation (estimated to lie in the 2.5–3°C global warming range) and collapse of the WAIS, which remains uncertain (see **ice sheets and glaciers**). When these thresholds are crossed (must probably during the latter part of this century), the earth would be committed to a global sea level rise of at least 13 m.[13]

If global temperature rise exceeds 4°C (which is certainly possible in the next 100 years), then thermal expansion could contribute an additional 0.3–0.8 m over the next two centuries and 1.5–2.0 m over the next 500 years.[14] Over 500–1,000 years, thermal expansion could contribute as much as 4–5 m, depending on assumptions about climate sensitivity, future greenhouse gas concentrations, and ocean mixing rates.[15] The actual amount of thermal expansion remains uncertain as it is determined by the physical transfer of heat from the atmosphere to the surface of the oceans and the rate at which the surface ocean waters mix with the deep ocean waters (see **thermohaline**).

Overall, the impact of global warming, based on mid-range temperature increases by 2100 of 3–4°C, would be 6–15 m over the next 500 or so years, depending on when the WAIS breaks up completely. If global temperature rise exceeds 4°C, then it is likely that the seas could rise by up to 20–40 m over the next 500–1,000 years. A key question is whether this would set in motion forces that could lead to an ice-free world. The chances of this are not, as yet, considered very high, based on current and near-term predictions of temperature rise. Nonetheless, there is a temperature point (currently unknown) that would trigger the move to an eventual ice-free world, even though it may take several thousand years. Once reached, the planet would be committed to a 70–75 m above current levels.

Can we stop the sea from rising?

If humans can constrain global warming to less than 2.5°C (below the Greenland deglaciation threshold), it is possible that significant sea level rises can be partially avoided. Nonetheless, even a 2°C rise would deliver substantial long-term sea level rise. It will not be possible to avoid sea level rise completely, and centuries, and possibly more than a thousand years, are likely to elapse before the sea level stops rising.

While the actual physical rise in sea levels and inundation of low-lying areas is relatively easy to visualize, there are a range of less obvious biophysical and socioeconomic impacts that will also affect coastal regions. These include loss of wetlands and wildlife breeding grounds; changes in storm surges, wave action, and coastal erosion/deposition patterns; increased agriculture/aquaculture loss and damage; salt water intrusion to freshwater aquifers; damage to coastal infrastructure (ports, sea walls, and buildings); and reef loss and tourism impacts (see **coastal zone impacts**). Given the timescales involved, and the fact that sea levels may rise by 0.5–1.5 m by 2100, it is evident that a range of adaptations will need to be implemented. It would be prudent to plan for at least a 1 m rise in sea level this century, and possibly more.

See also: climate change feedbacks, climate sensitivity, coastal zone impacts, ice sheets and glaciers, thermohaline.

Notes

1 Rahmstorf 2007
2 Bernes 2003

3 See NASA (http://sealevel.jpl.nasa.gov/)

4 Nicholls and Lowe 2006

5 Rahmstorf 2007

6 Bernes 2003

7 IPCC 2007

8 Ibid.

9 Ibid.

10 Ibid.

11 Rahmstorf 2007

12 Ibid.

13 Nicholls and Lowe 2006

14 IPCC 2007

15 Lenton *et al.* 2006

Further reading

Rapley 2006; Nicholls and Lowe 2006; Lenton *et al.* 2006; IPCC 2007; ACIAR 2004; Lowe *et al.* 2006; Rahmstorf 2007.

SOCIOECONOMIC IMPACTS

Anthropogenic greenhouse gas emissions are expected to continue to rise in the medium term and so too are atmospheric greenhouse gas concentrations. The amount of warming this century will depend on how quickly, and by how much concentrations rise, the sensitivity of global temperatures to different concentrations, and the extent of any climate change feedback effects that the warming initiates. Based on the most recent (2007) IPCC estimates, temperatures this century are likely to rise by between 2 and 5°C (see **global warming**).

Even with low levels of global warming (up to 2°C), there are likely to be significant climatic changes in all regions of the world. Temperature and precipitation regimes would evolve, the frequency and intensity of **extreme weather events** could be expected to increase, and many marine and terrestrial species would migrate toward the poles, reduce in abundance, or possibly become extinct (see **biodiversity impacts** and **agriculture and food supply impacts**). These changes would have significant impacts on both human and natural systems and could adversely affect the livelihoods of billions of people. Any course of action chosen by the international community to combat global warming will, implicitly or explicitly, have to weigh the expected risks of climate change against the costs of stabilizing atmospheric greenhouse gas concentrations at "satisfactory" levels (see **stabilization targets** and **dangerous climate change**).

Moderate levels of global warming (2–3°C) are expected to generate both positive and negative socioeconomic impacts, according to geographical location. The positive benefits will be relatively limited and largely confined to temperate regions of the mid- and high latitudes. For the lower latitudes, especially subtropical regions, socioeconomic impacts are expected to be overwhelmingly negative, even for low levels of warming. At high levels of warming (above 3°C), the socioeconomic impacts are expected to be negative for all regions: human society would be unambiguously worse off compared with a situation where the climate remained unchanged.

At temperature increases above 3–4°C, the risk of breaching critical thresholds capable of initiating major geophysical impacts increases significantly (see **climate change impacts**). Such impacts, such as the deglaciation of Greenland, the disintegration of the Western Antarctic Ice Shelf (see **polar impacts**), or the collapse of the **thermohaline**, would have major long-term repercussions for human civilization and could further accelerate global warming (see **climate change feedbacks**). The socioeconomic implications of these geophysical impacts are generally not fully accounted for in impact studies, although they remain a real risk. Most major studies also conclude that the socioeconomic burden of climate change will not be evenly distributed. The billions of poor that live in the developing world, and particularly those in the least developed countries, are expected to face disproportionately greater socioeconomic costs than the citizens of high-income, industrialized countries.[1]

Human welfare impacts

Climatic conditions have a major bearing on global economic output, human health, and welfare. Significant deviations from normal climatic conditions (such as droughts, floods, and extreme heat or cold) can have major socioeconomic repercussions, particularly through their effects on agricultural production. Global warming is expected to increase climatic variability and the incidence and severity of extreme weather events, lead to permanent long-term changes in average temperature and precipitation patterns, result in losses of biodiversity and shifts in the geographical distribution of species, and to alter the geographical range of pests and diseases. All of these changes will have socioeconomic implications.

The socioeconomic impacts of climate change will manifest themselves through a range of direct and indirect channels, many of which are subject to uncertainty. The key direct human impacts of climate change will be on human health; agricultural production and global food supplies (including changes to biodiversity); energy and water supplies; and loss and damage to infrastructure through extreme weather events, permafrost thawing, and sea level rise. The flow-on effects for household incomes and livelihoods, global capital markets, social cohesion, and political stability could be profound, even if global temperature rise can be constrained to 2°C (a temperature change the world may already be committed to).

Health

Climate change is likely to result in both positive and negative impacts on human health (see **health impacts**). Positive impacts are expected to be relatively limited and may include declines in cold-related morbidity and mortality, mainly in the higher-latitude regions of Russia, Europe, North Asia, and North America and possible reductions in the prevalence of some diseases in specific geographical areas (e.g. the incidence of malaria in some West African countries could decline as they become drier).

Negative impacts include potentially higher levels of malnutrition, mainly in developing countries, due to declines in agricultural productivity; reductions in national food supplies and increases in global food prices; increased mortality and morbidity from heat stress; an overall increase in the incidence and geographical range of vector-borne diseases (such as malaria, dengue fever, West Nile virus, and Lyme disease); greater incidence of water-borne diseases and parasites – malaria and diarrhea already account for one-quarter of child mortality in the developing world;[2] bacterial contamination of food; and higher levels of mortality and morbidity from extreme weather events, such as destructive storms, heat waves, and flooding.

Malnutrition is expected to generate the greatest health-related socioeconomic costs and will most severely affect low-income countries, particularly those in sub-Saharan Africa and South Asia.[3] With a 2°C rise in global mean temperature, an additional 30–200 million people could be at increased risk of hunger; at temperature increases of 3–4°C, the number increases to 500 million.[4] Malnutrition reduces worker productivity and increases the susceptibility of people to disease – both of which result in lost economic production.

Increases in the incidence of vector- and water-borne diseases are also expected to result in major socioeconomic losses in some regions. A 2°C rise in global mean temperature could result in an additional 40–60 million people being exposed to malaria in Africa alone.[5] Malaria is estimated to have reduced economic growth rates in some countries by more than 1% point over recent decades.[6]

Agriculture and food supplies

Changes to agricultural yields and world food supplies are likely to represent the single most significant economic impact of climate change. Agricultural output is highly susceptible to changes in climatic conditions. Crop yields can be substantially reduced by abnormal variations in temperature and/or water availability at critical stages of the growing cycle. All the following changes are expected to reduce crop yields: increases in the incidence of droughts, floods and heat waves; seasonal and long-term changes in temperature and precipitation patterns; changes to seasonal water flows due to reductions in mountain glaciers and snow packs; and changes in the geographical range of different pests and diseases (see **agriculture and food supply impacts** and **water impacts**).

At low to moderate levels of warming (up to 2°C), agricultural costs and benefits will coexist. In the mid- to high-latitude temperate regions (30–50° north and south of the equator), rising temperatures are expected to lengthen the duration of the growing season, and some regions are likely to experience elevated precipitation: both could result in increased crop yields. In addition, if the *carbon dioxide (CO$_2$) fertilization effect* is reasonably strong, this could also boost yields (see **agriculture and food supply impacts**). The higher-latitude regions of North America (especially Canada), Northern Europe, and some areas of Russia, Argentina, New Zealand, and Australia may experience some increase in agricultural productivity. This could make a positive contribution to the Gross Domestic Product (GDP) of these countries, particularly if accompanied by upward pressure on global food prices due to falling yields in lower-latitude countries. However, for global mean temperature increases beyond 3°C, the impacts on agricultural yields are expected to be predominantly negative in nearly all regions.

In the mid- to low latitudes, the impact on climate change is expected to be overwhelmingly negative, even with a relatively strong CO_2 fertilization effect. This is primarily because temperatures are already near optimum yield conditions for most crops (further temperature rises are likely to reduce yields) and the increased susceptibility of these latitudes to drought. India, Southern Europe, North Africa, the Middle East, South West United States, Southern Australia, and West Africa are all expected to become hotter and drier over the course of this century, and agricultural yields are expected to be adversely affected (see **water impacts**).

Overall, the low-income developing countries are likely to suffer disproportionately from changes in agricultural output. Agriculture contributes between 20% and 50% of GDP in many low-income countries and is the primary source of livelihoods for 40–70% of the population.[7] By comparison, agriculture generally accounts for only 1–3% of GDP in high-income, industrialized countries.

Climate-related events, particularly drought, can have devastating socioeconomic consequences, particularly in developing countries. In India, the failure of the 2002 seasonal monsoon rains resulted in a 3% reduction in GDP,[8] while the 1991–1992 drought in Zimbabwe resulted in a 9% fall in GDP and a 72% increase in food prices.[9] Reductions in GDP of this magnitude can have far-reaching socioeconomic consequences such as unemployment, civil unrest and crime, increased poverty, and negative economic impacts on other sectors.

Developing countries, especially communities that are dependent on hunting and gathering, are also more vulnerable to the impacts of climate change on biodiversity. Climate-induced changes to the geographical distribution of terrestrial and marine species, the abundance or health of specific animal species, and changes to migration and breeding patterns can result in major socioeconomic impacts for communities dependent on particular species (see **biodiversity impacts** and **coastal zone impacts**).

Water and energy

Water and energy are essential inputs to the global economy, and both are sensitive to climate change. In addition to the obvious impacts on agriculture, as discussed above, adequate and reliable water supplies are essential to industry, energy production, and the maintenance of urban populations. Any changes to water availability or reliability can have significant socioeconomic repercussions and may also result in conflicts over access to water resources.

Climate change is expected to induce major changes in the annual and seasonal availability of water across all regions of the planet (see **water impacts**). The socioeconomic impacts of these changes will vary according to geographical location and to the vulnerability of different communities to variations in supply. Developing countries are generally more vulnerable to variations in water availability than developed countries as they often have more rudimentary water storage and distribution infrastructure. Most developed countries have sufficient artificial storage capacity to meet water requirements during extended low-flow periods, often for several years, and a significant percentage of agricultural production has access to irrigation.[10] Nonetheless, some regions of high-income countries – notably South West United States, Southern Europe, and parts of Australia – are likely to face greater water stress arising from long-term declines in precipitation, increased frequency of droughts, and higher temperatures (see **water impacts**). Southern Europe is likely to see summer rainfall decline by one-quarter with a 2°C global temperature increase, and up to 40–50% with a 4°C increase.[11] Overextraction of groundwater resources will also exacerbate the socioeconomic impacts of declining water availability – as underground water stocks are depleted, climate risks and vulnerability will increase.

Even without climate change, accessing adequate water supplies is already a major issue in many regions due to growing populations and heightened demand from industry and agriculture. The continuing growth in urban populations and the emergence of very large urban centers, particularly in Asia, Africa, and Latin America, are expected to increasingly strain existing water infrastructure and create seasonal water availability problems.[12] Climate change could exacerbate these problems, particularly for cities that are dependent on seasonal water flows from mountain glaciers and snow packs in parts of India and China, South West United States, and several Andean countries (see **water impacts**). To accommodate increased variability in water availability, there will need to be significant improvements in water infrastructure (dams, pipelines, irrigation, recycling, and desalination facilities), which will incur economic, social, and environmental costs.

Diversion of water flows from rivers shared by several nations (such as the Mekong, Nile, Niger, and Tigris) may result in strained international relations in the coming decades, even in the absence of climate change. If climate change reduces water flows or increases seasonal variability, as is expected in many regions, this has the potential to exacerbate cross-boundary and interregional water issues.[13]

Climate change is also expected to result in strains on industry and energy production. While some countries are likely to experience a fall in energy demand for winter heating (mainly in the mid- and high latitudes), this will tend to be offset by heightened energy demand for summer cooling (air conditioning). On a global basis, the reductions in heating energy demand are expected to be more than outweighed by increased cooling demand.[14] Furthermore, drought reduced hydropower production and a greater incidence of heat waves (resulting in spikes in air-conditioning demand) will not only strain power production and distribution infrastructure (potentially leading to power outages) but may also actually reduce aggregate electricity production. During the European 2003 heat wave and the recent Australian drought, electricity output from some thermal and nuclear plants had to be reduced due to lack of cooling water.[15] The economic cost of interruptions to, or shortfalls in, electricity supplies can be significant. For example, the 1999–2000 drought in Kenya resulted in a 16% reduction in GDP, of which three-quarters was due to reductions in hydropower and industrial production.[16]

Extreme weather events

Each year, extreme weather events, such as droughts, floods, and destructive storms, inflict substantial loss of life, reductions in economic output, and infrastructure damage. In 2005, the direct insurance costs of extreme weather events (which generally underestimate the full cost impact of such events) reached an estimated $ 200 billion, or approximately 0.5% of global GDP.[17] As climate change increases the frequency and intensity of extreme weather events, their associated damage costs are also expected to increase (see **extreme weather events**).

For the developing world, the socioeconomic costs of extreme events will arise primarily from impacts on agriculture, health and water, and, to a lesser extent, infrastructure. The socioeconomic impacts on developed countries are likely to arise primarily from damage to infrastructure and disruptions to trade and industrial production. Developed countries have much greater quantities of capital invested in infrastructure, and many of the world's major industrial complexes and financial centers are located in coastal zones that are most vulnerable to extreme weather events (see **coastal zone impacts**). Destructive storms and precipitation events, and associated flooding and storm surges in coastal regions, can result in major infrastructure damage and loss. Hurricane Katrina (2005) is estimated to have caused $ 120 billion in direct damage to infrastructure and property, as well as interrupting oil and gas supplies from the Gulf of Mexico – leading to a spike in global oil prices and domestic gas prices.

For the period 1980–2004, the economic costs of weather-related natural disasters have been estimated at US$ 1.4 trillion.[18] A rise in global mean temperature of 3°C could double the damage costs associated with extreme weather events, with major implications for the insurance industry.[19] Given the importance of the insurance industry to global financial markets, this could, in turn, have major global economic repercussions (see **finance and insurance**).

Other impacts

The costs of coastal zone protection (construction of dykes, storm surge barriers, strengthening or relocating coastal industrial infrastructure) could become significant as the planet warms (see **coastal zone impacts**). Extensive thawing of the permafrost across large areas of the higher latitudes is expected to result in significant infrastructure damage and loss and may also adversely affect petroleum and mineral production and land-based transport (see **polar impacts**). The viability of the tourism industry may be threatened in many small island nations and coastal regions (see **marine impacts** and **coastal zone impacts**). Coral reefs support a large tourist industry: even at relatively low levels of warming (less than 2°C), the world's coral reefs are likely to suffer major damage and perhaps even permanent loss in some areas.

Sea level rise and heightened vulnerability in coastal zones could result in large numbers of people being displaced and forced to migrate. A 1–1.5 m rise in sea levels (which may occur this century – see **sea level rise**) could result in significant losses of land area for some countries without preventative measures (especially those in countries with large delta regions, such as Bangladesh and the Netherlands) and substantial loss of land in some island atoll countries (notably Tuvalu, the Maldives, and the Marshall Islands).

Estimated global socioeconomic costs

Overall, climate change is expected to result in a wide range of socioeconomic impacts, both positive and negative. The probabilistic nature of these impacts, and the uncertainty over the extent to which adaptation measures can reduce the negative impacts, makes estimating the future socioeconomic costs and benefits of climate change a difficult task.

Impact assessment models are used to calculate the potential economic costs of climate change. While the potential costs of some impacts – for instance, changes to agricultural and energy production and increases in infrastructure damage – may be estimated with some degree of certainty, many other (unpriced) socioeconomic costs – such as the loss of biodiversity or the deterioration in human health – are much more difficult to value. The complexity of the global socioeconomic system, combined with our limited understanding of how climate change will change key variables, means that impact assessment models inevitably provide only partial and indicative cost–benefit estimates of climate change.[20] Climate change impact cost estimates will undoubtedly underestimate the real human welfare impacts, let alone the welfare of other species.

Moreover, few impact assessment models consider the socioeconomic costs associated with major climate-related geophysical changes (such as the shutdown of the **thermohaline**) or the costs of political instability and regional conflict. By excluding many catastrophic and nonmarket effects, much of the literature tends to underestimate the true cost of climate change.[21]

Standard economic cost–benefit analysis tends to weight costs that occur decades in the future much less than costs incurred in the near term. Such "temporal discounting" serves to complicate the valuation process as climate change costs will tend to grow over time and the full costs may not arise for half a century or more (e.g. as with sea level rise). Impact assessment studies also report costs in terms of changes in GDP. However, due to the deficiencies of GDP as a measure of economic welfare (see **dangerous climate change**), some studies have begun to report results in terms of changes in consumption levels (e.g. Stern 2006). Impact studies also tend to attach higher values to losses incurred by high-income developed nations than those incurred by low-income countries (which have lower average income levels). To help overcome this problem, model results are often adjusted through a process called "equity weighting."

Several general conclusions emerge from a review of the literature: socioeconomic costs are likely to be proportionately greater for each additional degree Celsius rise in global average temperature (i.e. costs are expected to rise in a nonlinear fashion); negative socioeconomic impacts, on a global basis, are likely to become very significant for temperature increases above 3°C; and the burden of these costs is expected to be unevenly distributed geographically, with the poorer developing countries facing a greater socioeconomic burden than the developed world.

Estimates of the socioeconomic impacts associated with different global temperature changes vary considerably, depending on the model employed and the assumptions invoked concerning adaptation capabilities. But they can be broadly summarized as follows:

- For global warming of up to 1°C, most model results suggest that there is likely to be little or no change in global GDP and possibly an overall small increase, with a small net gain likely for most developed countries but losses likely for most developing countries.
- For temperature changes between 1 and 2°C, results in the literature indicate that the gains for most developed countries are still positive but the negative impacts

begin to increase significantly for developing countries, particularly in Africa and South Asia. The overall impact in a global sense ranges from a small benefit to a small loss. The impact assessment conducted by Tol (2002), for example, concludes that climate change leads to a net global GDP loss once temperatures exceed 1°C, while other studies indicate a higher break-even point.[22]

- At global temperature increases of 2–3°C, the results are also mixed but most models predict that the magnitude of global losses will increase significantly and that the impacts for developed countries also start to become increasingly negative. Most major studies indicate that Africa could expect a GDP reduction of at least 4% once temperature increases enter this range. Developed countries such as Europe, Australia, and possibly Japan are also likely to begin to experience overall net economic losses. The results for North America are mixed with some studies indicating that impacts may still be slightly positive, while others indicate a net loss. Global GDP losses are expected to be in the range of 1–5%, depending on the study.[23]

- At 3–4°C, climate change incurs negative socioeconomic effects for everyone, but particularly for the developing countries. Some models indicate global GDP losses of up to 11% for temperature increases for this magnitude, but others as low as 3–5% of global GDP.

- Above 4°C, all models predict significant losses on a global basis, with some in the order of 10–20% of global GDP, and possibly higher. However, our limited understanding of the repercussions of global warming of this magnitude means that predictions for such large temperature changes are highly uncertain.

One study that does warrant specific attention is Stern (2006) as it provides one of the most comprehensive assessments of the potential costs of climate change and it reviews many of the impact cost studies undertaken in recent years. The Review recognizes the complexities of estimating the costs of climate change but concludes that most past studies have tended to underestimate the costs. At moderate levels of warming (2–3°C), the Stern Review estimates that global GDP could be reduced by up to 3–5%. From 3–4°C, losses could reach 5–11% of global GDP. Stern (2006) acknowledges that if full account is taken of the entire range of potential impacts, and the higher burden faced by the developing countries, then losses of up to 20% of GDP over the period to 2100 are conceivable (see **stabilization targets**).[24] These figures have been criticized by many sources, often for contradictory reasons, particularly in relation to the discount rates used in economic modeling.[25] The debate continues and may only be resolved in hindsight, but the underlying thesis of the Stern Review – that the costs of limiting global warming to moderate levels are less than the costs of the impacts such warming would deliver – remains largely intact, even using some higher discount rates (see **mitigation**).

It is also important to put in context what the global GDP reduction estimates actually mean in terms of the future economic wealth. The GDP loss figures delivered by models and impact assessment studies refer to the percentage reduction in GDP compared with what GDP would otherwise have been at a given point of time in the future without climate change. For example, a 10% loss in global GDP by 2050 means that global GDP is 10% lower than it would otherwise have been but still likely to be much greater than what it is today – it is a relative loss in GDP and not an absolute loss in GDP. The world will, on average, still be richer than what it is today (in a purely monetary sense) but just not as rich. Nonetheless, as mentioned above, impact costs presented in terms of changes in global GDP may provide a broad indication of trends in global economic wealth, but they are likely to grossly undervalue the true human welfare costs and also mask some very important distributional and intergenerational equity issues.

Impact assessment studies conclude that there is likely to be significant cost variation across different countries and regions. Of particular concern are the impacts on developing countries, which will face a disproportionate burden. This is because:

- they are generally more dependent on climate-sensitive economic activities such as agriculture, forestry, fisheries, and tourism than high-income countries;
- most of the developing world's population live in the low- and mid-latitude regions, which are more likely to experience significant negative socioeconomic impacts at low to moderate levels of warming than the higher latitudes;
- the poor are generally more vulnerable to the impacts of extreme weather events; and
- they are less likely to be able to afford insurance cover, fund adaptation measures or obtain access to adequate emergency relief services.

Unless the international community can succeed in limiting the increase in mean global temperature to less than 3°C, and preferably to less than 2°C, the ability of a majority of the world's population to achieve their development aspirations is likely to be significantly constrained. Indeed, the number of people living in poverty is likely to increase.[26] The effects on the very poor in South Asia and Africa could be highly significant, and potentially devastating.[27] These are issues that decisions makers will need to consider when assessing appropriate climate change response strategies.

See also: agriculture and food supply impacts, biodiversity impacts, climate change impacts, coastal zone impacts, dangerous climate change, extreme weather events, future emissions trends, health impacts, marine impacts, mitigation, polar impacts, water impacts.

Notes

1 Tol *et al.* 2004
2 WHO 2005
3 IPCC 2007
4 Parry and Rozenweig 2004 [AQ: Please add to the Bibliography]
5 Warren *et al.* 2006
6 Stern 2006
7 Ibid.
8 Challinor *et al.* 2006b
9 IMF 2003
10 See Brown and Lall 2006

11 Schröter *et al.* 2005
12 Grey and Sadoff 2006
13 See World Bank 2003 and Niasse 2005
14 Stern 2006
15 Ibid.
16 World Bank 2006b
17 Muir-Wood *et al.* 2006
18 Epstein and Mills 2005
19 Stern 2006
20 See Tol 2002 and Nordhaus and Boyer 2000
21 See Kuik *et al.* 2006
22 See Stern 2006 for a review
23 See Tol and Yohe 2006 and Hare 2006
24 Stern 2006
25 See Nordhaus 2007 and Baer 2007
26 See UNDP 2007
27 Stern 2006

Further reading

Stern 2006; IPCC 2007; Tol *et al.* 2004; Tol 2002; Nordhaus and Boyer 2000.

SOLAR POWER

Most forms of renewable energy (wind, hydro, biomass, wave, and ocean thermal energy), as well as **fossil fuels**, are derived either directly or indirectly from the sun. *Solar power* refers specifically to those technologies that derive heat and electricity directly from sunlight. Along with **wind power**, solar power (also referred to as *solar energy*) is attracting considerable attention as a climate-friendly energy supply option.

Humans have used solar energy for thousands of years for activities such as heating buildings and drying crops, but, to date, very little of the sun's energy has been harnessed for commercial energy production. At present, solar energy contributes only a tiny percentage of global energy supply (less than 1%) and only 0.15% of global electricity production.[1] However, the sheer magnitude of the potential resource, its inexhaustible nature, and its environmentally friendly attributes suggest that solar energy is destined to become a much more important energy source over coming decades.

Each hour, more solar energy strikes the earth than the global economy consumes in an entire year.[2] However, due to its diffuse nature, the key challenge is how to capture this energy and convert it to other usable energy forms (such as electricity and heat) in a cost-effective manner. While solar power is already economic in some applications (e.g. solar water heaters and remote-area power systems), it remains a more expensive electricity-generating option than fossil fuels in the vast majority of applications. Nonetheless, technological progress over the past few decades has been rapid, and the medium- to long-term prospects for solar power have improved significantly.

Advantages and disadvantages of solar power

The main advantages of solar power are the following:

- The resource is large and renewable (can never be depleted through human consumption).
- Solar energy is more evenly distributed across the globe than fossil fuel stocks, providing greater energy independence and security benefits to nations.
- Virtually no greenhouse gas emissions are produced by the solar fuel cycle (though some are produced during manufacturing, construction, operation, and waste disposal) – it is therefore a climate-friendly technology.
- Solar technologies are well suited to *distributed generation* applications (energy production situated at or close to the source of energy demand), which can provide benefits through greater electricity system reliability and lower transmission and distribution losses. Currently, approximately 7% of global electricity production is lost during transmission and distribution.[3]
- Solar power attracts widespread public acceptance as an energy supply option, although the technology is still at an early stage of deployment.

The major disadvantages are as follows:

- Relatively high cost, particularly for grid-connected electricity generation – costs will need to be reduced at least three- or four-fold before solar power becomes cost-competitive with most fossil fuel-generating technologies.
- Solar power output (like wind) is intermittent and cannot be relied upon to supply electricity on a constant 24-hour-per-day basis (unless combined with energy storage systems, which are generally expensive).
- Most solar technologies are still in the relatively early stages of development and deployment, and their system-wide reliability as an energy source remains unproven (although solar hot water systems are relatively mature and deployed on a large scale).
- Large-scale centralized solar-generating facilities are relatively land intensive (requiring 10–20 hectares/megawatt[4]), which may limit their application in densely populated areas where opportunity costs for land are higher (though roof-mounted solar panels can occupy space that is otherwise unused).
- The existing highly centralized power generation and grid distribution systems of developed economies have large, capital investments with long engineering lives and are often in monopoly supply situations – this can be a disadvantage for solar power as there is often little political or financial incentive to incorporate distributed generation systems into their operations since this effectively erodes their market position and power.

Technology status and cost

Solar technologies can be broken down into three principal categories: *solar thermal*, where the heat energy is employed directly for end-use applications (e.g. solar hot water systems, solar passive buildings, and commercial and industrial heat); electricity generated by *photovoltaic cells* (PV), which convert sunlight directly to electricity; or *concentrating solar power* (CSP) systems, which heat water or other fluids to high temperatures suitable for driving standard steam turbines or Stirling engines. Each of these solar technologies has considerable long-term potential to supply energy and reduce greenhouse gas emissions.

Solar thermal

Solar thermal technologies can directly displace other energy sources, notably gas and electricity, used for water and space heating. They can also be used to provide low-grade heat for industrial and commercial agricultural processes. Thermal applications presently account for most of the solar energy contribution to global energy supply and, at least for the next 10–20 years, are expected to remain the principal means by which solar power can contribute to reducing emissions.

Solar water heating is the most commercially proven, and widely deployed, of the solar thermal technologies. By 2005, over 150 million square meters (m²) of solar thermal collectors had been deployed, mainly for domestic water heating, and in recent years installed capacity has been increasing by 10–13 million m² each year.[5] Solar thermal collectors can deliver a thermal energy equivalent of 400–1,000 kWh/m² per year: consequently, existing installations currently supply reasonably significant quantities of heat energy in some countries.[6]

Solar water heating technology is relatively mature, but further cost reductions and advances in collector efficiency are expected. In many countries, the life cycle cost of solar water heaters (the cost per unit of heat delivered over the life time of the unit) is already the lowest of the available water heating technologies. Although solar water heaters generally entail higher up-front capital and installation costs than other options, these costs are generally recovered in 3–8 years (payback periods depend on a country's available sunshine and its energy costs) and heaters normally last for 15–20 years.

As water heating usually accounts for 10–20% of a household's energy consumption, solar water heaters could make an important contribution to reducing greenhouse gas emissions. Solar water heaters are beginning to achieve considerable market penetration in some locations (such as Israel, parts of China, and some regions of Australia and the United States), but their overall market penetration to date has been quite low, and much lower than their economic potential would suggest. This is partly due to higher up-front costs, and also due to the inability, in most locations, for solar to supply all hot water needs throughout the year (it usually requires some form of backup supply, such as electricity) and a lack of public awareness of the financial benefits. In recognition of the economic and environmental benefits of solar water heating, many countries have introduced promotion and financial incentive schemes, which have assisted in overcoming some of these barriers. Nonetheless, it remains an underutilized technology, and the untapped market potential remains significant.

Solar thermal technologies can also provide cost-competitive space heating, particularly in combination with hot water heating. Buildings are the single largest consumer of electricity, accounting for 42% of global electricity consumption in 2004

(30–45% of this electricity is used for space and water heating, depending on the country), and account for approximately 15% of global greenhouse gas emissions.[7] Passive solar design uses building orientation and structural design to maximize the use of available solar energy to light and heat buildings, or to limit building heat gain in summer, are already cost-effective and can substantially reduce energy consumption. The use of high-efficiency windows, thermal walls and natural lighting can halve energy consumption for lighting and heating, at little or no additional cost.[8]

Lack of awareness amongst building designers, limited experience with the performance of new designs, and the existence of some regulatory barriers have constrained the uptake of solar passive building technologies to date, but they offer significant potential for reducing building energy consumption in the coming decades.[9] Overall, combined solar water and space heating/systems, in conjunction with solar-passive design features, can provide between 10% and 60% of water, lighting, and space heating requirements in a cost-effective manner for most buildings.[10]

Solar cooling technologies, based on absorption chillers, have also been developed, though the number of commercially operating systems is presently limited, mainly in Europe.[11] Solar cooling systems offer considerable potential in the medium term and, when integrated into the overall building design structure (rather than retrofitted to exiting buildings), these technologies are already cost-competitive on a full life-cycle cost basis.[12] They are just starting to be introduced on a commercial basis and so far account for only a tiny fraction of the building cooling market – but their medium-term potential is considered promising.

The utilization of solar thermal for large-scale industrial processes is much less developed and, in most industrial heat applications, is unable to compete with fossil fuels and biomass. Few large-scale applications of industrial solar heat have been deployed, but it is an area of active ongoing research and development and could become a more attractive option in the medium term, particularly for processes requiring low-grade heat.

Photovoltaic (PV) cells

Sunlight can be converted directly into electricity via PV cells, using silicon-based semiconductors. These cells are manufactured into modules (panels), each of which is capable of generating several hundred watts of electrical output. Panels can be mounted in arrays, generating 3–5 kilowatts (kW) for domestic applications (enough to power an average energy-efficient home), or into even larger commercial arrays consisting of hundreds of panels generating several hundred kilowatts, or even megawatts, for supply to electricity grids.

There are two principal types of PV cells: single crystal (monocrystalline) silicon cells and thin film (amorphous) silicon cells. Single crystal cells are twice as efficient at converting sunlight to electricity, but the panels cost 3–4 times more per m². Single crystal cells dominate the PV market at present, accounting for around 90% of the global sales, but recent advances in thin film cell manufacture, the use of less expensive silicon, and several other application and manufacturing advantages suggest that thin film technologies could double their market share over the next 5–10 years.[13]

Commercial single crystal PV panels currently have a sunlight-to-electricity conversion efficiency of 15–20% (6–9% for thin film), but conversion efficiencies of 25–30% have been achieved in laboratory situations. The efficiency limit of a single crystal cell is around 31%.[14] PV cells can be connected to the grid or used in off-grid applications (often combined with battery storage). They are already cost-effective in many off-grid applications.

There have been rapid advances in PV technology over the past two decades, and costs have fallen dramatically – from approximately $US10/peak watt (pW) in the early 1990s to $4–5/pW by 2008. Currently, they can supply electricity to the grid at around US 20–25 cents/kWh, compared with 3–7 cents/kWh for most other large-scale generating options based on **fossil fuels**, hydro or **nuclear power**.[15] Costs will need to be at least halved again, to around US$1–2/pW (or US 5–10 cents/kWh), before they can be considered cost-competitive with current mainstream technologies. Cost reductions of this magnitude are certainly possible in the next 15–20 years. Panel costs are expected to continue to decline at around 5% per year over the period to 2020.[16]

An important determinant of PV costs is the amount of silicon used per cell: PV-grade silicon is the single largest manufacturing cost component and can account for 20% of the panel cost.[17] As a result, panel costs are sensitive to high-grade silicon prices. Over the past decade, primary silicon costs have generally been around US$ 40–50/kg, but the rapid rise in PV demand since 2004 has led to a global shortage of high-grade silicon and prices have increased (to as high as US$60–70/kg) – this has, in turn, pushed up PV panel prices.[18] PV cells currently require approximately 13 g of silicon per watt of output capacity. Ongoing advances in manufacturing technology could reduce this to 7–8 g/watt over the next decade and would make an important contribution to reducing costs.

One recent promising technology advance that may help to reduce costs more quickly, possibly to as low as US$1–2/pW, is sliver technology. This is presently being developed by Australian scientists and involves slicing silicon into thousands of thin slivers, thereby substantially reducing the amount of monocrystalline silicon required per watt of output – to around one-tenth of current requirements. The technology, if it can be successfully rolled-out on a commercial scale, has the potential to reduce panel costs by up to 75%.[19] Another emerging technology that may also help to reduce costs is concentrator PV cells. These use mirrors to concentrate incoming solar energy on to a smaller surface area – this significantly increases the energy output per m^2 of cell area.[20]

A major advantage of PV technology is that it can be installed at the source of demand (*distributed generation*), which can reduce energy losses in transmission and improve system reliability. In terms of cost-competitiveness, the key consideration is the cost of delivered energy at the point of consumption and not necessarily the cost of energy entering the grid from other large-scale generating alternatives (such as coal-fired power stations). This is an important advantage for applications at the household level, where energy losses are highest, particularly those located at the end of the grid system. The system-wide benefits of PV systems vary between countries. In California, the system peak load (when electricity supply costs are the highest) usually occurs on hot summer days (due to air-conditioning), when PV production is at a maximum (thus providing maximum system stability and economic benefits). By contrast, in Northern Europe, the peak demand generally occurs on cold winter days, when PV power production is relatively low.

Concentrating solar power (CSP)

These technologies concentrate incoming sunlight to produce temperatures high enough to generate steam to drive turbines or Stirling engines to produce electricity. There are three main categories of CSP technologies: *parabolic trough*, *central tower*, and *reflective dish/Stirling engine*.

Parabolic trough technology involves rows of reflective parabolic dishes that heat fluid (usually synthetic oil) in tubes to temperatures of 250–400°C. The fluid is then passed through a heat exchanger to generate steam, which drives a steam turbine to generate electricity. This technology presently accounts for nearly all of the world's installed concentrating solar systems (around 350 MW, mostly in the United States). Although no new facilities have been commissioned for over a decade, recent advances in technology and the introduction of government incentives have renewed interest in CSP technologies. As of 2007, more than 500 MW of CSP capacity was under construction or had received development approval, mostly in Spain.[21]

Central tower technology also uses mirrors to concentrate sunlight to generate steam, but onto a receiver mounted on a central tower rather than a tube. Central tower facilities can heat fluids to 600°C, considerably higher than parabolic trough technology, and this offers higher thermal conversion efficiencies. Although a few central tower demonstration plants have been constructed, commercial-scale deployment only commenced in 2006 with the 11 MW Andalusia facility in Spain. Another 17 MW plant (Solar Tres) is expected in 2008–2009.[22]

Reflective dish/Stirling engine technology is similar to the central tower concept (in that it has a central receiver), but the heat is used to drive a Stirling engine rather than a conventional steam turbine. Stirling engines use the exchange of hot gases between two chambers to drive a piston connected to a generator. Only the United States (California) is presently moving forward with this technology and two plants, with a combined capacity of 800 MW, are at an advanced stage of planning.[23]

The ability to generate power after the sun has stopped shining is a major attraction of central tower and trough technologies. Several of the plants under construction in Spain are being built with energy storage systems. When the sun is shining, heat energy is stored in large thermal tanks of molten salts that can be tapped when sunlight is not available. This enables the plant to generate power for up to 8–10 hours without sunlight and, over the course of a year, could possibly achieve capacity factors as high as 75% (largely overcoming the intermittent power problem that many renewable technologies currently confront).[24] However, the addition of long-term storage facilities (greater than 2–3 hours' operation without sunlight) adds to initial capital costs, and it remains to be proven how effective they are in commercial operation. Some of the facilities under construction have opted for only two hours of storage due to lower capital costs. Nonetheless, the ability to generate power when the sun is not shining is an attractive aspect of CSP systems with combined storage.

Due to technical advances, all three CSP technologies have recently attracted government and industry interest. In terms of large-scale electricity generation, they could be cost-competitive before 2030. Unlike PV systems, CSP systems can be readily combined with standard fossil fuel generating facilities to provide primary heat energy (hybrid fossil fuel–solar facilities); when combined with new heat energy storage technologies, they can provide stable power for longer periods (and possibly even operate as base load suppliers in some locations); and they can provide reasonably large quantities of electricity (50–500 MW) at one location; and costs per kWh are presently lower than for PVs.

Finally, in terms of costs, some of the facilities under construction are expected to be able to generate power at 10–20 cents/kWh,[25] and the medium-term target is to lower CSP costs down to 5–8 cents/kWh by 2020, and below 5 cents/kWh by 2050, making the technology cost-competitive with **fossil fuels** and **nuclear power**.[26] Until that time, CSP technologies will continue to rely on government incentives and support schemes to be competitive. For example, the emergence of Spain as a significant CSP developer has been made possible by a new Spanish government regulation (Royal Decree 436, 2004), which guarantees the purchase of electricity produced from renewable energy sources and improves grid access provisions. In other jurisdictions (e.g. California, United States) stringent renewable energy targets have also led to considerable interest in CSP systems.

Solar's potential mitigation contribution

In the long term, solar energy is expected to become one of the principal means of generating climate-friendly electricity and underpinning less greenhouse-intensive economic development pathways. Thermal applications, such as water heating, passive solar designs, and solar heating and cooling buildings, offer the greatest cost-effective potential in the short to medium term (out to 2030). In many applications, they are already the most cost-effective options. The most significant constraint facing solar, in terms of mitigating power sector emissions, is the physical limits on how quickly capacity can be expanded.

The small size of the current solar industry means that scaling up to a point where solar power can make a meaningful contribution to greenhouse gas emission reductions presents a major challenge over the coming decades. Since the mid-1990s, installed solar generating capacity had been growing at an average annual rate of 25%: by 2006, it had exceeded 5,000 MW (5 GW) of capacity (approximately 0.15% of global electricity supply).[27] Capacity is expected to double again by 2010, when annual capacity additions could exceed 3 GW.[28] Currently, the greatest constraint is not lack of demand but the shortage of silicon and sufficient panel production capacity. To support a 5 GW/year capacity expansion would require a doubling of present silicon production capacity – but there are constraints on how quickly these facilities can be constructed.[29] To put this in perspective, the amount of capacity that was added in 2006 (around 1.5 GW) is roughly equivalent to the amount of coal-fired capacity added each week across the globe. If global solar generating capacity continues to grow at the rapid rate of the past decade (25% per year), then by 2020 it is still expected to account for less than 1% of global electricity supply.[30]

Recent modeling conducted by the International Energy Agency indicates that if technology progress continues on its present path, and governments adopt proactive policies to address climate change, then solar technologies could contribute approximately 2% of global electricity production by 2050.[31] Much depends on the magnitude, and rate of growth, of energy demand. If much lower energy demand growth rates can be achieved, then the contribution of solar could be greater, perhaps as much as 5% of global electricity supply. This would entail concerted climate policy action by the global community, a strong commitment to research and development, and active schemes to promote the uptake of solar. Of course, the development and deployment of many technologies is not gradual and linear (e.g. mobile phones, internet, etc.) as modeling suggests. Breakthroughs, rather than incremental improvements (as foreshadowed by thin film, flexible, and "paint-on" solar technologies), in combination with favorable policies, such as private investment incentives, could enable solar power to move beyond the incremental paradigm.

Overall, solar power is likely to make a relatively small, but increasingly important, contribution to emission reductions up to 2050. In the longer term, beyond 2050, solar is destined to become a much more important component of the global energy supply mix.

See also: energy efficiency, fossil fuels, mitigation, nuclear power, wind power.

Notes

1 Kammen 2006
2 Lewis 2006
3 IEA 2006b
4 Derived from data in Stirzaker 2006
5 IEA 2006b
6 Ibid.
7 WRI 2005
8 IEA 2006b
9 Ibid.
10 Ibid.
11 Ibid.
12 Ibid.
13 Cameron and Jones 2006
14 Lewis 2006
15 Ibid.
16 Cameron and Jones 2006
17 Ibid.
18 Ibid.
19 Normile 2007
20 Ibid.
21 Stirzaker 2006
22 Ibid.
23 Kammen 2006
24 Stirzaker 2006

25 Kammen 2006 and IEA 2006b
26 IEA 2006b
27 Kammen 2006
28 Cameron and Jones 2006
29 Ibid.
30 Morton 2006
31 IEA 2006b

Further reading

IEA 2006b; Stirzaker 2006; Kammen 2006; Morton 2006; Lewis 2006.

STABILIZATION TARGETS

Stabilization targets represent maximum mean global temperatures or greenhouse gas concentrations that the global community aims not to exceed. The principal objective of stabilization targets is to ensure that climate change is constrained to manageable levels. To date, no internationally agreed stabilization target has been formally adopted, but a range of potential targets have been proposed. Limits to temperature or concentration increases are also being considered in discussions concerning the successor agreement to the **Kyoto Protocol** (which expires in 2012). For example, the European Union has adopted a target of stabilizing global mean temperatures increase at 2°C or less.

It is widely recognized that to maintain **global warming** at manageable levels (those that avoid **dangerous climate change**), atmospheric **greenhouse gas** concentrations must be stabilized in the next few decades. By 2007, atmospheric greenhouse gas concentrations, when converted to their **carbon dioxide equivalence (CO_2e)**, had reached 455 parts per million (ppm), over 40% higher than preindustrial levels.[1] To date, this buildup of greenhouse gases in the atmosphere has caused the earth's average surface temperature to increase by 0.76°C.[2] However, due to the inertia of the climate system, global temperature is expected to increase by a further 0.5–1°C before it stabilizes (most likely somewhere between 1.5–2°C above preindustrial levels), even if concentrations were to remain at today's level (see **global warming**).

Greenhouse gas concentrations continue to increase by approximately 2.5 ppm/year: based on the projected growth in emissions over the next few decades, the annual rate of increase could accelerate to 4 ppm/year by 2040.[3] If this eventuates, atmospheric greenhouse gas concentrations could double their preindustrial levels before mid-century and would ultimately result in a 3°C or more rise in global mean temperature.[4] Temperature increases of this magnitude are likely to result in major **climate change impacts**.

If the international community considered a 3°C rise in global temperature to be unacceptable and agreed to limit greenhouse gas concentrations to a level that would constrain temperatures below this level (e.g. 500 ppm CO_2e), then this would constitute a stabilization target. Targets could also be set in terms of a specific increase in global mean temperature (e.g. 2°C above preindustrial levels) or even in terms of a maximum level of **radiative forcing** (measured in terms of watts/square meter).

Stabilizing total global greenhouse gas emissions is an urgent priority. However, this alone will not stabilize atmospheric greenhouse gas concentrations or the earth's mean temperature – it will simply slow the rate of increase. Thus, emissions stabilization in itself is not considered a meaningful target, unless adopted as a short-term interim measure until a global agreement can be reached on a specific concentration or temperature target.

Temperature versus concentration targets

A specific atmospheric greenhouse gas concentration can be associated with a range of different temperature outcomes. Although greenhouse gas concentrations are considered to be the main factor contributing to **global warming**, there are other factors, such as changes in the earth's **albedo** that also contribute to temperature change. As a result, the precise relationship between atmospheric greenhouse gas concentrations and global mean temperature changes (see **climate sensitivity**) remains uncertain, particularly in relation to the strength of possible **climate change feedback** effects.

The **Intergovernmental Panel on Climate Change (IPCC)** estimates that a doubling of preindustrial greenhouse gas concentrations would cause global mean temperature to rise anywhere between 2 and 5°C, with a best guess of approximately 3°C.[5] Consequently, any concentration target holds some risk that it may deliver a much higher temperature increase than is acceptable. For this reason, a concentration stabilization target should ideally be sufficiently flexible to be able to accommodate new scientific evidence about climate sensitivity as it emerges.

Much of the recent international discussion on stabilization targets has focused on setting a specific greenhouse gas concentration level, rather than a temperature-based target. While the ultimate objective is to minimize the rise in global temperature, it is likely that a temperature target would need to be converted to a specific concentration level to enable progress toward achieving the target to be monitored: greenhouse gas concentrations can be precisely monitored, even on a daily basis, whereas temperature increases tend to lag several decades behind increases in concentrations (see **global warming**). A specific atmospheric concentration level can also be linked to specific annual greenhouse gas emissions quotas (allowing for the uncertainties in **climate sensitivity** outlined above), which is important for apportioning emissions reduction obligations between different countries or regions. Nonetheless, any concentration target must be linked to a specific temperature rise for it to be meaningful.

Setting a stabilization target

There are a range of issues that need to be resolved in order to establish a stabilization target. The two key issues are (1) the degree of global warming that the international community is willing to accept (the maximum temperature threshold) and (2) the level of risk that the international community is willing to bear that a specific concentration target results in a higher temperature increase than anticipated. Other factors in the decision-making process would include the time frame in which emissions reductions need to be delivered; the costs and constraints of achieving these outcomes (see **mitigation**); and the apportionment of the task of reducing emissions (deciding who is responsible for reducing emissions). These are difficult and complex issues that are likely to take some time to resolve, assuming they can be resolved at all. The willingness to adopt a specific target will vary between countries and will be influenced by the type of climate change impacts different countries are likely to experience, and the relative costs of achieving different options.

While the international community has agreed, through ratification of the **United Nations Framework Convention on Climate Change (UNFCCC)**, that atmospheric greenhouse gas concentrations should be constrained to a level that avoids *"dangerous anthropogenic interference with the climate system,"* there is, as yet, no agreed temperature or atmospheric concentration threshold that would, if breached, result in **dangerous climate change**.

Given that the international community has yet to agree on a level of climate change that can be considered intolerable, establishing an internationally agreed stabilization target remains problematic. Nonetheless, since the Convention was signed in 1992, a scientific consensus has coalesced around the view that increases in global mean temperature of more than 2–2.5°C would most likely result in **dangerous climate change**. For the purposes of analyzing prospective stabilization targets, it is generally assumed that concentration levels exceeding 550 ppm CO_2e would most probably result in temperature increases of more than 2.5°C, though it could be higher or lower.

Identifying the likely temperature increase that would result from different concentration levels is important for establishing a stabilization target, and much depends on the **climate sensitivity** to a given increase in greenhouse gas concentrations.

Scientists have attached probabilities to the temperature outcomes associated with specific rises in greenhouse gas concentrations. For a doubling of CO_2e concentrations above preindustrial levels, the temperature increase is 3°C, but there is at least a 30% chance that temperature could be more than 3°C, a 10–15% chance that it could exceed 4°C, and a 5% chance that it could be greater than 5°C (see **climate sensitivity**).[6] That is, there is a 1 in 20 chance that a doubling of CO_2e concentrations will result in devastating temperature increases, rather than the relatively more manageable temperature increase of 3°C. Uncertainty increases with the greenhouse gas concentration level due to uncertainties surrounding the point at which higher temperatures initiate positive **climate change feedbacks** that amplify and accelerate global warming.

What is clear is that a doubling of greenhouse gas concentrations above preindustrial levels would be highly likely (almost certain) to exceed the 2°C threshold. If the international community were to agree that global mean temperature increases exceeding 2°C must be avoided, the maximum concentration target level would probably lie somewhere between 475 and 500 ppm CO_2e. However, there is a moderately significant chance (10–30%) that even 450ppm CO_2e (less than the present concentration level) could result in a 2°C increase.[7]

Table 17 lists the most probable (50% or greater chance of occurring) temperature ranges associated with different concentration stabilization levels and the expected

Table 17 Relationship between atmospheric concentrations and global mean temperature increases

Concentration level (ppm CO$_2$e)	Likely °C increase above preindustrial levels	Date reached with BAU emissions projections	Date reached with global emissions stabilized at 2005 levels
450	1.5–2.0	2008	2008
500	2.0–2.5	2025	2030
550	2.5–3.5	2035	2045
650*	3.0–5.0+	2060	2100

Source: Derived by authors based on data in Meinhausen (2006), Murphy et al. (2004), Stern (2006), IPCC (2007), IEA (2006a), and EIA (2005).

*Upper bound could be considerably higher due to uncertainty associated with climate change feedback mechanisms and concentrations at 650 ppm could result in temperatures that are well above 5°C.

dates at which the different concentration levels will be reached.[8] The estimates do not include the full range of potential temperature outcomes, which are up to one degree lower and often several degrees higher than shown.

Obviously, the level of emissions has a substantial bearing on the rapidity with which different concentration levels are reached. If we compare the concentrations that would result from "business-as-usual" (BAU) projections (which assume current emission trends persist) with those that would occur if global emissions were stabilized at today's levels, it is clear that a significant time gap emerges, particularly in the medium to long term. This has important implications for achieving the stated stabilization targets, as the longer emission reductions are delayed, the more difficult it becomes to avoid high concentration levels.

Emissions reductions required to stabilize concentrations

In order to stabilize atmospheric concentrations, global emissions must fall to a level equal to or below the natural ability of the earth to remove greenhouse gases from the atmosphere (through the operation of the **carbon cycle** and other atmospheric chemical processes). The longer emissions remain above the rate at which greenhouse gases are removed from the atmosphere, the higher the greenhouse gas concentration will rise, and the greater will be the corresponding rise in global temperature. An emissions rate lower than the rate of removal would result in reductions in atmospheric greenhouse gas concentrations.[9]

At present, the natural removal rate is of the order of 10–20 gigatonnes (Gt) CO$_2$e/year, well below the current global emissions rate of approximately 50 GtCO$_2$e/year (see **carbon sinks**). The rate of uptake of CO$_2$ by the land and oceans is expected to fall over the long run as these reservoirs become progressively more CO$_2$ saturated. Based on recent modeling, the ability of the terrestrial land system to absorb CO$_2$ could fall by at least 20–30% by 2100, and possibly more.[10] There is even a possibility that the land could shift from being a carbon sink to a carbon source (see **land carbon sinks**).[11]

By next century, the natural rate of removal may decline below the present level of 10–20 GtCO$_2$e/year – possibly to below 5 GtCO$_2$e/year, less than 10% of current emissions levels.[12] If this occurs, global emissions would also need to fall to match the decline in the natural removal rate. Should the rate of removal fall below global emissions, atmospheric greenhouse gas concentrations would begin to climb once again.

Much of the recent international debate has tended to focus on reducing emissions to a level that would stabilize concentrations at or below 550 ppm CO_2e. Stern (2006), for example, concluded that concentrations need to be stabilized at 550 ppm CO_2e or lower in order to avoid serious climate change impacts; others have suggested that concentrations may need to be stabilized at between 450 and 475ppm CO_2e (around present levels) to be reasonably confident that global mean temperatures do not exceed 2°C.[13] This is not considered to be realistically achievable unless a concentration "overshoot" is accepted in the short to medium term, until concentrations are brought back to the threshold level at some time in future.

Since the late 1990s, a large number of studies have investigated emissions reduction options. While it is not practical to list the full range of possible outcomes, most studies come to broadly similar conclusions.[14]

First, in order to stabilize atmospheric greenhouse gas concentrations, emissions will have to eventually fall to at most one-fifth of present levels (i.e. at least 80% below current emissions). Depending on the rate at which greenhouse gases continue to be removed from the atmosphere, the emissions level may need to fall even further, potentially to less than one-tenth of the current level.

Second, the ability to achieve a specific concentration target is highly dependent on when global emissions peak and on the magnitude of the peak. The longer the delay in stabilizing global emission levels, the more rapid the required rate of annual emissions reductions. If, for example, emissions peak in 2030, rather than 2020, the annual rate of reduction in emissions required to achieve an atmospheric concentration of 550 ppm CO_2e doubles.[15]

Third, the lower the concentration target, the more rapidly emissions must decline to achieve the target. For example, if a 500 ppm CO_2e target were to be adopted, and global emissions were to peak before 2020, then emissions would need to decline by 4–6% per year to 2050 (to be 60–70% below present levels). If global emissions did not peak until 2030, then a 500 ppm target would be unlikely to be achievable. A 550 ppm CO_2e target would necessitate reductions in emissions of 1.5–2% per year, bringing emissions to at least 25–30% below current emissions by 2050 (also assuming that global emissions peak before 2020).[16]

The higher the concentration target, the easier it is to achieve – but this must be traded off against associated increases in impacts. Eventually, once thresholds for major positive climate change feedbacks are exceeded – for example, the trigger point for major releases of methane from the **methane hydrates** – concentration targets would no longer be necessary: the world would most likely enter a catastrophic non-equilibrium climate and anthropogenic emissions levels may be of little consequence.

Finally, a range of different emissions reduction options are available for achieving a specific concentration target. Emissions can peak soon and decline at a modest rate or reductions can be delayed until cheaper mitigation technologies become available, which would allow emissions to continue growing for some time but would then entail greater annual emission cuts later in order to achieve a specific target. The latter option underpins the approach adopted by the six countries that signed the

Table 18 Emissions pathways to 500 and 550 ppm stabilization

Concentration target (ppm CO_2e)	Year global emissions peak	% yearly reductions to achieve target	% reduction relative to 2005 by 2050
450*	2010	7.0	90
500	2010	3.0	50
500	2020	4.0–6.0	60–70
550	2010	1.0	25–30
550	2020	1.5–2.5	25–30

Source: Stern 2006.

2005 Asia-Pacific Partnership on Clean Development and Climate, which aims to invest in technologies (such as **carbon capture and storage**) that will reduce the cost of emissions abatement in the future.[17] However, this is a very risky approach and one that transfers the burden of emission reductions to future generations. Furthermore, without the necessary market signals (like a cost penalty on greenhouse gas emissions) or binding targets, it is less likely that these new technologies will emerge in a time frame that would deliver the required emission reductions.

Table 18 indicates the different emissions pathways needed to achieve selected stabilization targets and how they would vary according to the timing of peak emissions. These estimated emissions reduction requirements are illustrative only and will vary to some degree according to how much higher peak emissions are relative to 2005 levels. These examples assume either a 2010 emissions peak of 7% above 2005 levels or a 2020 emissions peak of 15% above 2005 levels.

Achievable targets

The adoption of a specific stabilization target will, to a large extent, be governed by the mitigation costs involved, the rate at which emissions reduction measures can be implemented, and the political will to do so. Adopting a stabilization target below 500 ppm is most likely to involve greater expense than a target between 500 and 550 ppm or 550–600 ppm, and so on.

While a high annual emissions reduction rate is more likely to incur greater expense than a lower one, due to the structural adjustment costs it would involve, there is considerable debate about whether reducing emissions will actually result in a net cost to the global economic system. There are, for example, significant unrealized economic gains available through **energy efficiency** measures that could deliver emissions reductions at negative cost. Nonetheless, any decision to set a stabilization target will undoubtedly involve some trade-off between the costs of **mitigation** and the costs associated with **climate change impacts** incurred.

To put into perspective the magnitude of the effort involved, it is useful to consider what the existing **Kyoto Protocol** agreement is likely to deliver and the experience of several countries whose emissions have fallen in recent decades. If, by 2012, the Annex I countries (that have binding emissions reduction commitments under the

Kyoto Protocol) honor their commitments and meet their reduction targets, this would be equivalent to an average annual rate of decline in emissions of 0.3% per year over the period 1990–2012 (excluding non-Annex I countries, whose emissions are expected to grow rapidly).[18] Achieving a 500 ppm stabilization target would involve a ten-fold greater effort than the Kyoto commitments, and these reductions would need to be achieved on a global basis, not just in Annex I countries.

Based on past experience, it is possible to achieve annual emissions reductions of up to 1% per year while maintaining economic growth. For example, between 1977 and 2003, France's emissions declined by an average of 0.6% per year (due to the large increase in nuclear-generating capacity), while between 1990 and 2000, the United Kingdom's emissions fell by an average of 1% a year (due primarily to the substitution of coal by natural gas).[19] It is generally considered that emissions reductions of up to 1–1.5% per year would be achievable without compromising economic growth for most countries and could possibly even lead to higher rates of growth due to efficiency gains.[20] However, annual reductions greater than 1.5% per year may be difficult to achieve without lowering economic growth trajectories, at least in the next 20–30 years.

A major constraint is the inherent inertia of the global economy, whose structure changes only relatively slowly (see **future emissions trends**). Much of the world's existing plant and equipment, responsible for the bulk of global CO_2 emissions, have operating lives ranging from 10–20 years (for appliances, cars, and manufacturing equipment) to 30–50+ years (for electricity-generating plants, heavy industry, and residential and commercial buildings). Replacing the existing capital stock prior to the end of its economic life is likely to incur economic cost, although this may be offset by economic gains elsewhere in the economy. Furthermore, developing, commercializing and rolling out low-emission technologies to the extent that they can make a significant contribution to emissions reductions takes time, often several decades (see **solar power** and **wind power**).

The complexities of domestic and international political processes are also likely to mean that it may take a decade, and most likely several decades, to implement the measures needed to stabilize global emissions, let alone stabilize atmospheric greenhouse gas concentrations. Any measures that could potentially result in lower rates of economic growth are likely to face considerable political resistance and may prove difficult to implement.

Given the expected future expansion of the global economy and population, and the structural inertia of the global economic system and political governance processes, emissions are unlikely to be stabilized before 2020 at the earliest, and possibly not before 2030, even if the international community agrees to a relatively ambitious concentration target. Even then, emissions reductions of greater than 2% per year would be very difficult to achieve.

It is still technically possible to reduce emissions to stabilize atmospheric greenhouse gas concentrations at 475–500 ppm CO_2e and thereby constrain global mean temperature increase to 2–2.5°C, but it would entail an enormous commitment by the international community and urgent and immediate action to reduce emissions. Without immediate and significant action to reduce emissions, stabilizing the atmospheric greenhouse gas concentration even below 550 ppm does not appear to be realistically achievable. Any legally binding target, if one is to be eventually adopted by the international community, is more likely to be in the 550–650 ppm range.

There is a real and growing likelihood that the international community will not be able to restrain global mean temperature increase to below 3°C. This has very serious implications for the climate change impacts the world can expect to endure over the course of this century and beyond.

See also: anthropogenic greenhouse gas emissions, climate change impacts, climate sensitivity, future emissions trends, global warming, mitigation.

Notes

1 IPCC 2007
2 Ibid.
3 Stern 2006
4 IPCC 2007
5 Ibid.
6 Meinhausen 2006
7 Ibid.
8 Concentration levels are for those greenhouse gases covered by the Kyoto Protocol and do not include CFCs and halons (covered by the Montreal Protocol) or ozone (see **anthropogenic greenhouse gas emissions**).
9 It is possible that the rate of removal of greenhouse gases from the atmosphere could be increased by utilizing biomass energy combined with carbon capture and storage.
10 Jones *et al.* 2006
11 IPCC 2007
12 Stern 2006
13 See Meinhausen 2006
14 See Meinhausen 2006 and Stern 2006 for a review of the literature on stabilization pathways.
15 Stern 2006
16 Ibid.
17 Kallbekkan and Rive 2006
18 Ibid.
19 Stern 2006
20 Ibid.

Further reading

Stern 2006; Meinhausen 2006; Murphy *et al.* 2004; Kallbekkan and Rive 2006.

Synthetic gases

Most **anthropogenic greenhouse gas emissions** are adding to naturally occurring greenhouse gas concentrations in the atmosphere, enhancing the **greenhouse effect** and leading to **global warming**. The synthetic gases, however, have been manufactured, cannot occur naturally (with the exception of negligibly small emissions of CF_4 from granite), and have terrestrial radiation absorption spectra (see **greenhouse effect**) that are usually different from naturally occurring **greenhouse gases**. This, along with their typically long atmospheric lifetimes, contributes to their high **global warming potentials (GWPs)**, making them important contributors to **global warming**.

The synthetic gases are manufactured with the deliberate characteristic of low reactivity and are used for specialized purposes such as air conditioning and refrigeration gases, propellants (in metered dose inhalers and aerosol delivery such as deodorants), foam blowing, cover gases, insulation, fire suppression, and solvents for cleaning. They are generally characterized by high atomic masses and fall into three main groups:

1   those included in Annex A of the **Kyoto Protocol**,[1] the hydrofluorocarbons (HFCs), perfluorocarbons (PFCs), and sulfur hexafluoride (SF_6) (see **global warming potentials** for a complete list);
2   the Ozone Depleting Substances (ODSs) covered under the Montreal Protocol,[2] consisting of chlorofluorocarbons (CFCs), hydrochlorofluorocarbons (HCFCs), halons, carbon tetrachloride (CCl_4), methyl chloroform (CH_3CCl_3) and methyl bromide (CH_3Br) (see **ozone**); and
3   others, including ethers and fluorinated ethers, that are not covered by any international agreements (see **greenhouse gases**).

Global warming contribution

Although synthetic gases make up a very small percentage of emissions by mass, due to their high GWPs, together they have contributed 0.337 Wm^{-2} (approximately 13%) to anthropogenic **radiative forcing** (or warming). This makes synthetic gases the fourth most important group of **greenhouse gases**, after **carbon dioxide**, **methane**, and tropospheric **ozone**.[3] The Montreal Protocol gases make the largest contribution to **radiative forcing** (0.32 Wm^{-2} or 20% of the total and almost 99% of the synthetic gas contribution) (see **ozone**). The "other" gases make a negligible contribution and are discussed under **greenhouse gases**. This section deals with synthetic gases covered under the **Kyoto Protocol**.

The synthetic gases covered by the **Kyoto Protocol** have, to date, made only a relatively small contribution (of 0.0172 Wm^{-2}, about 1% of anthropogenic **radiative forcing**) to **global warming**. However, since synthetic gases are manufactured specifically to be nonreactive and stable, most do not break down easily in the atmosphere and are not removed from the atmosphere through photosynthesis or oceanic absorption as is **carbon dioxide** (see **carbon cycle**). Thus, they are generally long-lived greenhouse gases, and some (PFCs in particular) have atmospheric residence times of up to 50,000 years, effectively making their radiative alteration to the atmosphere permanent. While the current impact of Kyoto synthetic gases is relatively small, emission rates have increased from effectively zero at the beginning of the twentieth century and steadily increased throughout the century. Their **radiative forcing** contribution grew at 10% per year between 1998 and 2005 period, the fastest rate of any of the anthropogenic greenhouse gases. It is thus important to reduce these emissions to avoid a long-term cumulative increase in their concentrations and, hence, their contribution to **global warming**.

Major emission sources

PFCs and SF_6 predominantly emanate from industrial sources and contribute 0.0042 Wm^{-2} and 0.0029 Wm^{-2} to global **radiative forcing**, respectively (24% and 17%, respectively, of the Kyoto synthetic gas contribution). HFCs from various industrial, commercial, and residential sources contribute the remaining 59%, 0.0101 Wm^{-2} to **radiative forcing**.[4]

Perfluorocarbons

Around 70% of PFCs are used as cover gases in the manufacture of aluminium. Due to their very low reactivity and density (heavier than air), they "cover" the liquid aluminium during smelting to prevent its oxidation. Since it is an expensive gas to produce, aluminium smelters try to avoid PFCs from escaping to the atmosphere (termed "fugitive" emissions). The use of PFCs in aluminium smelting began in the late nineteenth century, and PFC emissions grew steadily to around 80 $MtCO_2e$/year by 1970. Even though aluminium production has grown dramatically since 1970, ongoing manufacturing and process management improvements, particularly the elimination of reactive anodes, have resulted in a considerable reduction in PFC emissions per tonne of aluminium produced (they are now about a quarter of what they were in 1990).[5] These improvements have meant that aluminium industry PFC emissions have remained relatively steady through the 1970s and 1980s, and declined slightly during the 1990s.

Most of the remaining PFC emissions arise from the semiconductor industry where PFCs are used for etching circuits and as a solvent to clean chips and the chambers in which chips are manufactured. PFCs from fire suppressants (fire extinguisher use and testing) contribute only a very small amount. The rapid growth of the semiconductor industry since the early 1990s, driven largely by the demand for computers and electronics, has led to an increase in PFC emissions from the industry, from less than 5 $MtCO_2e$/ year in 1990 to around 40 $MtCO_2e$/year by 1998.[6] Since then emissions of PFC-116 have grown a further 22%,[7] with similar growth rates expected for PFC-14, the other most commonly used PFC. The increased PFC emissions from the semiconductor industry have far outweighed the decreases from the aluminium sector, resulting in total PFC emissions of about 120 $MtCO_2e$/year by 1998 (the last year of reliable global records).

Sulfur hexafluoride

SF_6 is the most radiatively potent of all greenhouse gases (with an atmospheric lifetime of 3,200 years and GWP of 23,900), and thus even small quantities of emissions can have a significant warming impact. About 90% of SF_6 emissions are related to the electricity sector, primarily electrical switchgear and transformers, where SF_6 is used as an insulator between high-voltage components to prevent arcing, a testament to the stability of the gas. Emissions result from leakage during the charging of equipment (during manufacture and maintenance), leakages through seals during equipment operation, and deliberate or accidental venting of the gas at the end of equipment life. The use of SF_6 in electrical equipment has grown with the global expansion of high-voltage electrical grids. SF_6 emissions also emanate from magnesium metal manufacturing, where it is used as a cover gas similar to aluminium production. The use of magnesium as a lightweight metal, particularly in automobiles, has grown in recent years.

While improvements in manufacture and design have reduced SF_6 leakage, SF_6 emissions are still growing. Since 1970, emissions have still grown at around 5% per year, and by 1998, global emissions had grown to 150 $MtCO_2e$/year and to 200 $MtCO_2e$ per year by 2006 (more than the annual greenhouse gas emissions from Sweden, Norway, Finland, and Denmark combined).

Hydrofluorocarbons

HFC emissions are closely linked to the stratospheric **ozone** depleting gases controlled under the Montreal Protocol. The most significant HFC from a global warming perspective is HFC-23 – a powerful and long-lived greenhouse gas. HFC-23 is a by-product of HCFC production (a refrigerant gas controlled under the Montreal Protocol gas) and is typically vented to the atmosphere. HFC-23 emissions have grown substantially since the 1970s, doubling from 40 to 80 $MtCO_2e$ in the 1980s. Since 1990, the growth in HFC-23 emissions has moderated slightly but had reached 120 $MtCO_2e$ by 2003.

Other important hydroflourocarbons that contribute to global warming are HFC-134a and HFC-152a, used as propellants, refrigerants, foam blowing agents, and insulants as substitutes for CFCs and HCFCs that are being phased out under the Montreal Protocol. Since the Montreal Protocol was signed, the production of HFC-134a, HFC-152a, and several other HFCs rapidly increased to replace the CFCs and HCFCs being phased out. From near-zero in 1990, HFC-134a emissions grew to around 70 $MtCO_2e$ by 1998, while other HFC emissions grew from near zero to around 41 $MtCO_2e$.[8]

In general, HFC emissions have increased as CFC and HCFC emissions have decreased. Since the Montreal Protocol gases have higher GWPs than HFCs, net **radiative forcing** from synthetic gases is likely to begin to fall over the next few decades (the concentration of CFCs only peaked in 2003). Both CFCs and HCFCs decay slowly (about 1% and 5% per year, respectively) in the atmosphere, so their **radiative forcing** will remain evident at least to mid-century, and in the case of HFC-23, for several centuries.

Future trends

Since synthetic gases are for specialized uses, their production is limited to a relatively small number of manufacturers and therefore a coordinated international approach may be easier to manage (see **future emissions trends**). Further, because these gases have such high GWPs, the introduction of a carbon cost penalty (see **carbon taxes** and **emissions trading**) would provide a financial incentive to reduce emissions.

PFCs

Even though the demand for aluminium continues to grow, production process improvements designed to reduce PFC emissions/tonne of aluminium produced are expected to counterbalance the aluminium production increases. In the medium term (to 2030), a gradual shift to "inert anode" processes should ensure a significant reduction in PFC emissions from the aluminium sector.[9]

The semiconductor industry is expected to continue to experience rapid growth, and with it PFC emissions are also expected to grow. Even though the industry has adopted specific sector emission targets (10% below 1997 levels by 2010),[10] industry-wide PFC emissions continue to grow. However, over the period to 2020, process improvements, including a shift from CF_4 (the most long-lived PFC) to other, lower-GWP PFCs emissions are projected to decrease by up to 50%.[11] Considerable potential for further reductions exists in the longer term with the possibility of complete cessation of PFC emissions beyond 2050.[12]

SF_6

Currently the commercially available alternatives to electrical switchgear are high-density fluids, most commonly heavy oil. This is not as effective an insulator as SF_6 and has significant safety concerns (explosion risks) with its use. Vacuum-sealed switchgear has been proposed and tested as a viable alternative but has yet to be proven at the commercial scale. As such, SF_6 emissions are expected to continue to increase in line with the expansion in electrical supply networks, at least for the next few decades. Emissions from magnesium production are also expected to grow substantially in the near term, but with decreases in SF_6 emissions per tonne of metal produced. Total SF_6 emissions in 2020 are projected to be around 10% higher than current levels, but thereafter emissions are likely to stabilize at around these levels over the period to 2050.[13]

Given the 3,200-year life of SF_6 and the 50,000- and 10,000-year lives of the two most common PFCs, respectively, even the complete cessation of emissions would not result in a reduction of atmospheric concentrations of these gases for thousands of years. Given that aggregate emissions of these gases are not expected to be reduced for many decades, atmospheric concentrations will continue to grow. In effect, these gases represent a near permanent change to the radiative budget of the atmosphere (a permanent warming effect).

HFC emissions are expected to continue increasing until at least 2020. While hydrocarbons (such as butane), ammonia, and carbon dioxide are commercially available for use as alternatives to HFCs (and CFCs) as refrigerants, because of their lower density, more energy is used in the operation of the air conditioners and refrigerators to achieve the same cooling performance. This has been a disincentive to broader uptake of these alternatives, and none has yet made significant market penetration, even though the high cost of production and handling of HFCs makes the alternatives more attractive on a life-cycle basis.[14] HFC emissions, particularly HFC-134a and HFC-152a, are projected to grow rapidly to reach 250 $MtCO_2e$ by 2020, roughly double current levels. Projections of emissions beyond 2020 remain uncertain, since technical alternatives already exist, and measures to increase their market penetration, such as instituting a multilateral agreement akin to the Montreal Protocol, could lead to significant reductions in HFC emissions.

HFC-23 emissions, as a by-product of HCFC-22 production, should decrease as HCFC-22 production is phased out by 2030 under the Montreal Protocol. In addition, the waste HFC-23 can be captured and destroyed by incineration at the manufacturing plant. However, there is no legal requirement to capture and destroy HFC-23, and it is generally just vented to the atmosphere. Recently, the **Clean Development Mechanism (CDM)**, which recognizes HFC-23 capture and destruction projects as eligible projects for crediting emission reductions, has provided a financial incentive to capture and destroy HFC-23 emissions in non-Annex I countries. To date, 18 CDM projects that capture and destroy HFC-23 have either been implemented or proposed, which may reduce emissions by around 80 MtCO$_2$e/year over the period to 2012 compared with the 120 MtCO$_2$e in 2003.[15] Provided these projects continue to operate after 2012, HFC-23 emissions are expected to be much lower than pre-2000 levels. HFC-23 replacement gases, such as HFC-134a and HFC-152a, have relatively short atmospheric lifetimes of 13.8 and 1.4 years, respectively, and lower global warming potentials than HFC-23 so are less of a problem from a climate change perspective – once emissions cease, their atmospheric concentrations will decline relatively quickly. HFC-23 on the other hand has a 260-year life (a concentrations decline of only 0.4% per year once emissions cease) and will remain a long-term problem gas.

Overall, Kyoto synthetic gas emissions are likely to continue to increase until at least 2020 but could decline thereafter. However, given the long atmospheric lifetimes and potency of these gases, their contribution to global warming will persist for many centuries.

See also: anthropogenic greenhouse gas emissions, Clean Development Mechanism (CDM), future emissions trends, global warming potentials, Kyoto Protocol, ozone, radiative forcing.

Notes

1 United Nations, 2005
2 UNEP 2006
3 IPCC 2007
4 Ibid.
5 WRI 2005
6 Olivier and Bakker 2003
7 IPCC 2007
8 Olivier and Bakker 2003
9 Harnisch *et al.* 1998
10 For example, Semiconductor Industry Association voluntary agreement hosted by US Environmental Protection Authority.
11 Raoux 2007
12 Schaefer 2007
13 Ibid.
14 European Union Directive 2006
15 UNEP Risoe Centre, http://cdmpipeline.org

Further reading

IPCC 2006.

TECHNOLOGY, STRUCTURAL CHANGE, AND ORGANIZATIONS

Climate change has emerged as a critical issue for the global economy, both in terms of its predicted **socioeconomic impacts** and the **mitigation** measures implemented to limit its effects. A crucial question that has attracted the attention of economists and climate policy analysts is whether the transition to a less emissions-intensive economic system will reduce the rate of growth of the global economy. This question has yet to be resolved, but several recent studies have concluded that the transition could be achieved with little change in long-term growth rates and may even stimulate growth (see **mitigation**). Much depends on the types of mitigation policies adopted by governments, the impacts these have on the development and deployment of low-emission technologies, and how organizations respond to these policy measures.

Given the projected increases in the two principal drivers of emissions, per capita income and population, over the same period, delivering the required emissions reductions to stabilize concentrations at a level that avoids **dangerous climate change** will prove to be a significant challenge (see **mitigation** and **stabilization targets**). It will require substantial structural changes to economies and, most important, a "decarbonization" of the energy system through a reduced reliance on fossil fuels. It will also entail technological innovation to reduce the **emissions intensity** of economic activities. These changes will have important implications for organizations and their long-term survival. For those organizations that compete in the market place (mainly profit making firms) climate change could have significant effects on their competitiveness and profitability.

Technology and innovation

Technology is defined as the processes by which an organization transforms labor, capital materials, and information into products and services of greater value. All firms have technologies. *Innovation* refers to a change in these technologies.

Joseph Schumpeter, a renowned growth economist, defined five types of innovation: (1) the introduction of a new good, (2) the introduction of a new method of production (*new equipment and processes*), (3) the opening of a new market(s), (4) accessing a new supply source for factor inputs (*the application of new materials and components*), (5) and the reorganization of an industry (*the introduction of new organizational forms*).[1]

Technological innovation has underpinned the process of industrialization and expansion of the global economy over the past two centuries. Schumpeter viewed innovations by *entrepreneurs* as the principal driver of economic growth. Entrepreneurs act as catalysts in the process of "*creative destruction*," a phrase used by Schumpeter to describe how innovations repeatedly change and replace established technologies, displace established firms, create new firms, increase demand and investment, and rejuvenate the economy.

Ongoing innovation, and the phasing out of old technologies and companies, is a key dynamic that propels an economy. Clusters of new innovations trigger technology transformations that subsequently establish new industries and commercial enterprises, creating waves of economic development. Innovation often concentrates

Table 19 Long waves of innovation and economic development

First wave 1785–60	Second wave 1845–55	Third wave 1900–50	Fourth wave 1950–80	Fifth wave 1980–2000
• Water power • Textiles • Iron	• Steam • Rail • Steel	• Electricity • Chemicals • The internal combustion engine	• Petrochemicals • Electronics • Aviation	• Digital networks • New media

in space and time, and considerable post-Schumpeter theoretical work has been undertaken to identify underlying cycles of technological innovation (notably so-called Kondratieff or Long Waves).[2] Since the mid-eighteenth century, five major waves of innovation are generally recognized, each one driven by the emergence of new industry clusters and technical advances.

The first wave is commonly referred to as the "*Industrial Revolution*," which began in Britain around 1750 when innovations in metallurgy, mechanized production, and energy led to a transformation of the economic system and the emergence of capitalism. The second wave commenced in the mid-nineteenth century, driven by the deployment of steam power technologies and the expansion of railways and transportation networks (which led to a significant expansion in the demand for steel). At the start of the twentieth century, a new wave of economic expansion commenced, driven by electricity production and the automobile, followed by the fourth wave (mid-twentieth century) driven by chemicals, aviation, and electronics. The latest wave commenced in the 1980s, driven primarily by information industries and telecommunications.

With each consecutive wave, the rate of industrial transformation (structural change) has increased, and the period between waves has shortened. While economic growth, and its potentially cyclical nature, remains an area of active research and lively academic debate, it is clear that certain technologies and industry clusters have been important contributors to economic growth. Table 19 lists the five major waves of economic expansion over the past two centuries and the industry clusters associated with each one.

With each subsequent wave of innovation, the demand for energy has increased, adding to the cumulative base energy demand created by previous waves of industrial development. With each wave, human and animal power has been progressively substituted by machinery and capital that use other forms of energy, principally **fossil fuels**.

Another characteristic of these successive waves of innovation has been changes to the scale of human organization. While the factory epitomized the industrial revolution, subsequent waves created levels of organization that moved well beyond the factory to encompass industrial zones and entire cities.[3] Today, approximately 80% of the population of high-income, industrialized countries lives in urban areas, and globally there are now more than 250 cities with populations greater than 1 million.[4]

Technological innovation, and the corresponding rise in economic output and per capita income, has underlain much of the rapid increase in greenhouse gas emissions over the past century. Although ongoing refinement of technologies has often led to reductions in the energy and emissions intensity of various production processes, it is the types of technologies that humans have deployed, and the sheer magnitude of the economic output associated with these technologies, that have driven greenhouse gas emissions. In the coming decades, technological innovation may lead to even greater levels of emissions or, alternatively, lead to significant reductions – it is both a potential threat and an opportunity. It is possible that climate change, and the need to wean the economic system off fossil fuels, could provide the impetus for a new wave of innovation and economic development.[5]

A range of factors will determine how quickly, and to what extent, new and/or improved technologies can contribute to reducing emissions. These include their cost competitiveness relative to existing and alternative technologies, which will be influenced by whether or not governments introduce a cost penalty on emissions, the level of research and development (R&D) funding devoted to low emission technologies, access to technologies and the rate of technology transfer between countries, and the technical capabilities and infrastructure to support new technologies.

Governments have an important role to play in promoting the development of low emissions technologies through their ability to establish enabling policy and regulatory environments and by supporting R&D programs. For example, the introduction of mandatory minimum energy performance standards for domestic appliances (such as refrigerators and washing machines) since the early 1990s has led to significant improvements in the efficiency of these technologies (see **energy efficiency**). The significant increase in energy R&D funding following the two oil price shocks of the 1970s produced significant advances in many nonfossil fuel technologies (**renewable energy** sources, **energy efficiency** and **nuclear power**). Governments are an important contributor to R&D, but since the late 1980s the level of government funding for alternative energy technologies has been declining.[6] This will need to be reversed if the rate of development of nonfossil fuel technologies is to be accelerated. Governments generally view technology development support as an attractive policy option, relative to less politically popular policy measures such as the introduction of a carbon tax on strict regulatory provisions. Technology-based policy approaches generally take the form of financially supporting research, development, and deployment (RD&D) of particular technologies and underpins the climate policy response by several governments.[7] However, governments have a mixed record in terms of supporting the right technologies, and therefore a heavy dependence on policies that support particular technological solutions to climate change may fail to deliver the required outcomes.

Nonetheless, even if the development of cost-competitive low or zero emissions technology is accelerated, their contribution to reducing future emission trends will depend on how quickly they can be deployed in the market place. For example, solar water heaters have been available for several decades and, on a life-cycle basis, are one of the most cost-competitive forms of hot water heating, yet their uptake in the market has been much lower than anticipated (see **solar power**). The transfer and uptake of low emission technologies (such as **renewable energy**, **carbon capture and storage**, and **energy efficiency** technologies) by the rapidly growing developing economies will be an important determinant of future trends in **anthropogenic**

greenhouse gas emissions, as this is where much of the future growth in global emissions will come from. As such, it is difficult to see a rapid uptake of new low emission technologies in the absence of strong market signals and incentives to do so (e.g. through the introduction of a significant cost penalty on greenhouse gas emissions).

The **UNFCCC** and the **Kyoto Protocol** both emphasize the importance of technology transfer and deployment to addressing climate change. One of the key arguments supporting the establishment of the **Clean Development Mechanism (CDM)** under the Kyoto Protocol was that it would be a means of promoting the transfer of new technologies. However, accelerating technology uptake and transfer faces many constraints, including the absence of a financial or regulatory incentive to reduce emissions; access barriers and licensing provisions (as most new technologies are owned and controlled by private entities); lack of the skills and support infrastructure to construct, operate, and maintain new technologies; and vested industry and political interests to protect a market niche and the existing infrastructure stock.[8]

Structural change

Structural change refers to changes in the relative importance of different economic activities, technologies, and production processes that generate economic output. Structural change reflects long-term changes in the *composition* of output and is often contrasted with the short-run fluctuations in aggregate *volume* of output during the business cycle.

Over time, as economies grow, the contribution of some sectors to economic output (e.g. agriculture and heavy industry) decline, while the contribution of the tertiary (service) sector (finance, education, leisure, and professional services) tends to rise. These changes are often accompanied by shifts in agriculture, manufacturing, and heavy industry to medium- and low-income countries that have a comparative advantage in these activities. Such structural shifts tend to result in a geographical redistribution of emission sources that can reduce the emissions intensity of high-income economies but which can also, in some circumstances, increase the emissions intensity of low-income countries. To limit future climate change, the rate of decline in emissions intensity will need to accelerate appreciably, both in developed and developing countries, and possibly triple or quadruple the current rate (see **emissions intensity**). This will require additional measures (above those generated by market forces) to reduce the dependence on **fossil fuels**, limit waste, and manage the land sustainably.

Major determinants of structural change include variations in relative prices of different goods and services, economic competitiveness, and technological innovation. Policies introduced to reduce greenhouse gas emissions (e.g. **carbon taxes**, standards and regulations, or subsidies) will result in changes to the structure of the economic system. Some activities and sectors may decline in importance (notably, the fossil fuel industries), while others may prosper and expand (such as the **renewable energy** industries).

Furthermore, a significant cost penalty on emissions may, over time, make countries, such as Iceland, Canada, and Brazil, that have less emissions-intensive energy systems based on low-cost renewable energy sources become more attractive locations for energy-intensive activities, such as aluminium smelting. This would result in a geographical redistribution of certain industries. There may, however, be instances of industry relocation that do not reduce aggregate emissions. For example, variations in the stringency of mitigation policy measures adopted by different countries may provide an incentive for some industries to relocate to countries with lax mitigation policies – merely resulting in a geographical redistribution of the same, or potentially higher, aggregate emission level. This is commonly referred to as "carbon leakage."[9]

Greenhouse gas mitigation measures will require significant new investment in low-emission technologies and will probably divert investment away from emission-intensive industries and/or accelerate innovation in these industries to reduce emission intensity. These shifts in investment patterns could drive higher levels of innovation and generate new forms of economic activity and sources of growth. The total value of economic activity could be higher or lower than would otherwise have occurred in the absence of measures to reduce emissions. The extent of any relative economic gains and/or losses will largely depend on what technologies are deployed and the stringency of the mitigation task (see **stabilization targets** and **mitigation**).

Cost of structural change

To stabilize atmospheric greenhouse gas concentrations, the global economic system will need to undergo significant structural adjustment. Energy and transport systems will need to become much more efficient and less carbon-intensive; methods of production and patterns of consumption will need to evolve, as will the strategies by which firms create value. There is ongoing debate amongst economists and policy analysts as to whether or not these structural changes will result in a reduction in the rate of economic growth (an economic cost) or stimulate higher levels of innovation and growth (an economic benefit). To ascertain the economic repercussions of different mitigation objectives, economists have developed a range of economic- and technology-based models, which attempt to simulate how an economic system may adjust to different policies, changes in the price of fossil fuels, the costs of low-emission technologies, and to a range of other key economic variables. The debate is, of course, largely confined to changes in economic measures of human welfare (principally GDP per capita) and does not encompass other important determinants of human welfare (especially nonmonetized components, such as biological diversity, community and cultural cohesion, maintenance of pristine landscapes, health and clean air). Decisions on the costs and benefits of climate change should not be based solely on the impacts of policies on GDP growth rates, although this is an important consideration (see **dangerous climate change** and **socioeconomic impacts**). Nonetheless, most studies tend to express climate change mitigation costs and benefits in terms of the impacts on future GDP growth rates.

Economic models primarily focus on how resources are allocated within an economic system and whether this varies from the optimum allocation for maximizing economic efficiency and output. A majority of mitigation studies, but not all, tend to suggest that stabilizing concentration levels below 550–600 ppm CO_2e is likely to result in a modest reduction in the rate of growth in GDP: as a general rule, the lower the target concentration level, the higher the cost in terms of reduced rates of GDP growth (see **mitigation**).

However, few models effectively simulate the underlying drivers of economic growth (such as new technology innovation), and therefore, their results may not provide an accurate reflection of actual structural adjustment costs and benefits. The economic cost, or benefit, of reducing emissions very much depends on how the models treat technological change. In many economic models, the rate of change of technology, and the costs of alternative technologies, are exogenously determined (in other words, imposed on the model from outside), with no consideration of how technological development may be coupled to economic growth, and vice versa.

In recent years, more sophisticated models have incorporated technology learning processes, termed "induced technological change," that mimic the impact of climate policy on technological innovation.[10] These models produce substantially lower mitigation cost estimates than those where technology change is exogenously determined. They also indicate that the cost structure of an economy based on low-emission technologies need not be higher than one dependent on fossil fuel. Under some circumstances, these models produce results that suggest that mitigation measures may even reduce systemic costs and increase economic output, particularly if the full economic potential of **energy efficiency** measures is taken into account.[11]

Increased economic output due to **energy efficiency** improvements often results in an offsetting "rebound effect": the economic resources freed up by energy efficiency gains tend to be spent on other goods and services, which also have emissions associated with their production. Considerable debate surrounds the precise magnitude of the rebound effect, but it is generally acknowledged that **energy efficiency** results in net emission reductions.[12]

Even though new technologies offer the opportunity to create a more efficient and less greenhouse gas-intensive economic system, their potential role is time constrained. It can often take several decades for new technologies to be deployed on a large scale in the market place. While some products and production processes are relatively short-lived (e.g. the capital stock of cars and domestic appliances is generally replaced every 10–20 years), significant capital equipment and infrastructure has relatively long lifetimes (up to 50 years in the case of fossil fuel power stations and even longer in the case of buildings). As a result, achieving a significant change in the structure and emissions intensity of the global economic system is likely to take at least 30 years, and possibly 50 years. More rapid structural change is technically possible but is likely to involve greater socioeconomic and political cost (see **stabilization targets**).

Organizational change

The response of private sector firms to climate change risks and mitigation policies will have an important influence on how effectively, and at what cost, human society can make the transition to a less greenhouse gas-intensive economic system. Private firms control vast sums of capital, account for a majority of capital investment decisions, and are the primary source of technological innovation: in short, they are important drivers of economic growth.

Ultimately, the objective of all firms is the creation of value, and from that, profit, for the owners of the firm. Value, however, has different meanings to different stakeholders. Firms manage resources to create value through their capabilities to deliver products or services that consumers desire; maintain relationships with resource providers and customers, and organize activities through governance, management systems, and processes.[13]

Climate change is a recent addition to the array of market and political issues that organizations need to consider in formulating their value-creation strategies. Climate change is likely to affect the cost and supply of raw materials, create finance market volatility due to **extreme weather events**, cause crop failures (see **agriculture and food supply impacts**), disrupt energy supplies, and, potentially, lead to political instability, such as climate-induced migration or resource access issues (see **climate change impacts**). These climate change risks will become increasingly important considerations for firms, and they will need to develop risk-reduction and hedging strategies to accommodate these risks (see **finance and insurance**). In particular, firms will need to respond to changing consumer sentiments toward the environment and the evolving political climate.

The multifaceted nature of climate change (environmental, economic, political, and social) means that systematic risks exist throughout the whole economy, many of which are likely to impinge on a firm's operations, market environment, profitability, and competitiveness. It is important that shareholders and managers are aware of the risks climate change presents, the opportunities for minimizing these risks, and the costs and benefits associated with different actions. Climate change will need to become a core consideration in decision making in order to ensure that shareholder value is not eroded.[14]

New business strategies

Climate change impacts and policies are likely to have a profound influence on how firms conduct their operations and the business strategies they adopt. In formulating an effective climate change response strategy, firms will need to identify their level of exposure to climate risks, including the impacts of mitigation and adaptation policies governments have, or could introduce to ameliorate climate change; quantify and track their greenhouse gas emissions, and the costs and benefits of measures to reduce emissions; be cognizant of the attitude and response of investors and consumers; and identify potential market and business opportunities that could emerge in a climate-constrained market place.

Climate change represents some degree of risk to all firms.[15] These can be broadly categorized under *sector-specific risk* or *company-specific risk*.

Sector-specific risks

These entail risks faced by a sector or industry, consisting of *regulatory risk* (where greenhouse gas emissions are regulated or priced, which can have a material financial impact on emission intensive sectors), and *physical risk*, where extreme weather events (agriculture, tourism, insurance, transport infrastructure) will have a direct impact on specific sectors of the economy.[16]

Company-specific risks

The risk impact on particular organizations within a sector or industry includes *competitive risk* (where some companies within an industry or sector are more exposed to policy changes or climate impacts than others), *litigation risk* (where organizations with large greenhouse gas footprints, or who undertake no actions to reduce emissions may face legal risk such as lawsuits from investors or the general public), and *reputational risk* (where organizations could encounter a negative public, investor, or media response due to their operations and products).

Most organizations already have strategies and processes to deal with climate variability. In the future, however, organizations will need to accommodate permanent changes in underlying conditions and the frequency and intensity of extreme events and a changed policy and market environment. Climate change has already had a noticeable impact on many companies, and some have begun to reshape their core business strategies in order to succeed in a climate-constrained market place. Climate change and the climate policies governments introduce will affect firms to varying degrees: some face greater climate change and policy risk than others. For example, energy companies and industries with a high level of fossil fuel dependence are likely to face higher climate policy risk than those that produce few greenhouse gas emissions, while some companies face greater risk exposure to climate change impacts (such as insurance companies) than others.

There is a rapidly growing number of examples of the financial benefits of proactive responses of companies to climate change. These stem mainly from identifying cost-effective efficiency measures and product redesign. For example, BP has reduced its greenhouse gas emissions by more than 10% since 1990 at a net financial benefit of $ 650 million, while the industrial conglomerate BASF reduced its emissions by 38% between 1990 and 2002 and saved more than 500 million euros.[17]

However, there is also some evidence to suggest that a significant proportion of companies have yet to pay much attention to climate change. Recent company surveys conducted in several countries have concluded not only that a majority of companies are aware of climate change as an emerging risk, and one that they need to manage and respond to, but also that most do not yet fully understand the climate change risks and policy exposure they face, or have introduced measures to reduce these risks. For example, a recent Australian survey of over 800 manufacturing companies found that three-quarters of companies recognized that climate change was an important issue and believed that they had a responsibility to reduce their greenhouse gas emissions, but less than half had implemented any measures to manage climate risks or reduce emissions, and only one in ten actually knew what their greenhouse gas emissions were, or understood how **emissions trading** schemes would affect their business operations.[18] Another study also found similar results across different companies but also that companies that adopt strong climate change business strategies are more likely to be able to maintain or strengthen their competitive position. They are also better able to identify and reap benefits from efficiency and process improvements, identify new and emerging market opportunities than those that have yet to adopt climate-sensitive business strategies.[19]

While it is clear that many companies are adopting more proactive positions on climate change, companies face a range of uncertainties in relation to future government policy action, consumer and investor response to the climate change actions they take, the costs and benefits of early action, and the timing and extent of climate change impacts. In short, it is difficult for them to decide how much action to take to reduce greenhouse gas emissions and hedge against climate change impacts that may or may not eventuate. While uncertainty in the market place exists, it is clear that over the coming decades the market and regulatory environment that companies face will change considerably.

At present, many firms lack the internal technical capacities to understand, identify, and manage climate change risks – many are aware of the issue but are unsure of what to do. Since the early 1990s, a wide range of tools, guidance documents, and specialist service providers have emerged that are aimed at assisting companies to identify and manage their climate change risks. The Greenhouse Gas Protocol, developed by the World Resources Institute and the World Business Council for Sustainable Development, is one document that has provided an important guide to business in terms of identifying their greenhouse gas footprint and climate change risks.[20] Other organizations, such as the Pew Center, as well as government initiatives, such as the Australian Government's Greenhouse Challenge Program and the United States' "Climate Leaders Program," have also assisted organizations to build their capacity to manage climate change issues.[21]

Climate strategies cannot be viewed as just an add-on to business as usual; they need to be integrated into a company's core business activities. Investing the necessary time and effort to come to grips with emerging climate change risks is likely to provide long-term dividends for organizations. It is becoming increasingly evident that those companies that take measures to minimize their climate change exposure and risks are likely be more competitive, better positioned to take advantage of emerging business opportunities, and more economically sustainable in the longer term than those that do not.[22]

Organizations, and private companies in particular, will play a pivotal role in the development and deployment of the technologies and operational strategies required to make the transition to a low-emissions society. To enable organizations to maximize their contribution to achieving this transition will require governments to put in place policies, regulatory frameworks, and financial incentives that provide appropriate long-term market and political signals; the integration of climate change as a core component of organizational business strategies; and an enabling response from key stakeholders, especially customers and investors.

See also: emissions intensity, finance and insurance, mitigation, socioeconomic impacts, stabilization targets.

Notes

1 McGraw 2007
2 See Kohler 2005

3 Steams 2007
4 Faber 2002
5 See Barker *et al.* 2006 and Stern 2006
5 IPCC 2007
6 Kammen 2006
7 See US DOE 2006 and IEA 2006b
8 EuroActive, February 2007
9 See Stern 2006 for further discussion
10 See Manne and Richels 2004
11 Barker *et al.* 2006
12 See Frondel 2004
13 Birkett 2001
14 See Hart and Milstein 2003
15 Australian Greenhouse Office 2006
16 Wellington and Sauer 2005
17 Stern 2006
18 Australian Industry Group 2007
19 Pew Center 2006
20 Wellington and Sauer 2005
21 See Greenhouse Challenge, www.greenhouse.gov.au/challenge
22 See Pew Center 2006

Further reading

Barker *et al.* 2006; Stern 2006; Manne and Richels 2004; Pew Center 2007; IPCC 2007.

THERMOHALINE

The Atlantic Thermohaline Circulation (ATHC), or often referred to simply as the *thermohaline*, plays an essential role in regulating the earth's climate. It is also an integral component of the **carbon cycle**. The name *thermohaline* is derived from the word temperature (thermo) and salt (haline). Temperature, salinity, and prevailing winds drive the thermohaline.

The thermohaline is the major driving force underlying the movement of water between the different oceans of the world and is an important mechanism for mixing the water of the surface and deep oceans. It operates like a giant conveyor belt, moving heat, salt, and dissolved inorganic carbon around the oceans. It is also an important means of transporting oxygen to the deep oceans, and without it the deep ocean would be depleted of oxygen and become anoxic.

The thermohaline transports very large quantities of heat in the surface oceans from the equator toward the polar region of the North Atlantic. The amount of heat it transports is very large (1 petawatt/year), equivalent to around 100 times the energy consumed by humans each year. This river of heat, combined with the heat contributions from other wind systems, helps maintain Europe 8°C warmer, on average, than it would normally be at that latitude, with the strongest warming effect evident in the winter. If you have experienced a winter in both Montreal and Paris (similar latitude), you will appreciate the difference the thermohaline makes to Europe's climate.

Due to its importance in regulating the earth's climate, and its role in transporting large quantities of **carbon dioxide (CO_2)** to the deep oceans, considerable scientific research has been devoted to assessing the impact global warming may have on the operation of the thermohaline in the future. Many scientists believe that **global warming** may cause the thermohaline to slow, or even eventually shut down completely. If this did happen, the repercussions for the earth's climate would be significant, particularly at a regional level. Not only would Europe experience a general cooling (which seems paradoxical given that the rest of the globe is becoming warmer), but it would generate several positive **climate change feedback** effects that might contribute to an acceleration of climate change, primarily through a reduced rate of transport of dissolved CO_2 to the deep ocean, thereby reducing the **carbon sink** potential of the oceans.

The thermohaline is driven by variations in the density of water. When water cools, it contracts and increases in density, and as water warms, it expands and decreases in density. The density of water also increases as it becomes more saline, and, as a result, is denser than freshwater at the same temperature. Thus, freshwater tends to float relative to colder and more saline sea water. These two basic factors drive the thermohaline.

Prevailing ocean currents and wind systems drive warm water from the equator toward the poles, and as it moves north, it begins releasing heat (heat loss generally starts around the Tropic of Cancer – 23.5°N). It becomes progressively cooler the further north it moves and also becomes saltier due to the evaporation effect of the winds. By the time it reaches the higher latitudes of the North Atlantic, it has become much denser and saltier than it was when it left the equator.

Sea ice in the higher northern latitudes increases the salinity levels yet further through a process termed *brine exclusion*. Sea ice is, in fact, freshwater: salt is excluded during the freezing process. The excluded brine increases the salinity of the surrounding water and reduces its freezing point below 0°C (which enables it to remain liquid even though it is below zero). The water becomes so cold and saline (heavy) that it begins to sink into the deep ocean.

It is at this point that the thermohaline plays its important role in the operation of the **ocean carbon sink**. As water cools, its propensity to absorb CO_2 from the atmosphere increases. On its journey northward, the cooling seawater not only becomes saltier but also absorbs more dissolved CO_2. When this water sinks in the far North Atlantic, it carries the salt and dissolved CO_2 with it into the deep ocean. This process accounts for nearly one-third of the **ocean carbon sink** effect, and hence it is critically important to understand how global warming could affect the thermohaline.

When the cold saline water sinks, it forms what is termed *deep water*. This deep water slowly migrates southward, driven by the force of gravity, to the deep regions of the Atlantic Basin. From there, it travels down the length of the Atlantic to the Southern Ocean and into the Indian and Pacific Oceans. There, the deep water separates and some moves north into the Indian Ocean, where it slowly resurfaces through *upwelling* in the Arabian Sea, while the remainder travels across the Southern Ocean below Australia and finally makes its way north across the Pacific Basin, where it eventually resurfaces

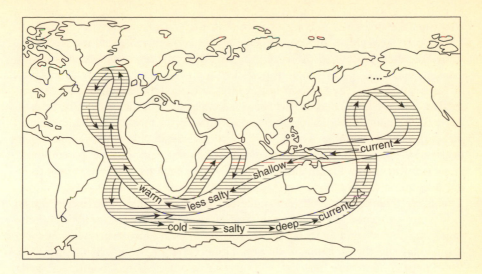

Figure 6 The thermohaline

Source: http://www.killerinourmidst.com/THC.html

through deep water upwelling in the Northern Pacific. The upwelling deep water is rich in nutrients and is a major contributor to biological activity in the surface oceans (see **carbon cycle**). The deep water is also rich in dissolved inorganic carbon: as it reaches the surface and warms, it releases CO_2 back into the atmosphere. This water eventually travels back in the reverse direction through the Indonesian Archipelago, rejoins the warm water flowing south in the Indian Ocean and back round into the Atlantic. It then flows northward up the length of the Atlantic and arrives back where it started and so completes the cycle (see Figure 6). The process of exchanging water between the deep ocean and the surface ocean, and vice versa, is called *overturning*.

The thermohaline is a very slow-moving but vast stream of water (about 100 times the quantity transported by the Amazon River). Once the seawater sinks into the North Atlantic, it may take up to a thousand years before it returns to the surface. It is this slow *overturning* via the thermohaline that ensures that the absorbed CO_2 remains buried in the deep ocean for long periods of time. Should climate change cause the thermohaline to slow, or to shut down completely, the uptake of atmospheric CO_2 by the oceans would be substantially reduced. This would result in a positive climate change feedback effect that could accelerate global warming by allowing more CO_2 to build up in the atmosphere.

Will present global warming cause the thermohaline to slow or shut down?

Paleoclimatic records indicate that the thermohaline has collapsed in the past, and modeling indicates that it may do so again in the future due to climate change. If

377

sufficient freshwater is added to polar oceans, it can reduce the salinity and, therefore, its density. This, in turn, reduces the rate at which the water sinks and thus slows the thermohaline. Records show that at the end of the last ice age, about 15,000 years ago, global surface temperatures began to increase and the ice sheets started to retreat northward. However, shortly afterward there was a brief return to ice age conditions, and evidence suggests that it may have been due to a shutdown of the thermohaline. The cause of this event is attributed to the collapse of a major ice dam (Lake Agassiz) that had held back large quantities of meltwater.[1] This freshwater flowed into the North Atlantic and reduced salinity to the point where the thermohaline is believed to have shut down. Records indicate that some shutdowns in the past were abrupt and persisted for relatively short periods of time, possibly just a few decades, before the flow resumed.[2] There has been no shutdown of the thermohaline detected over the past 8,000 years.

Once the thermohaline shuts down, the ocean current would most likely flow in the opposite direction. Eventually salinity would rise to sufficiently high levels to enable the thermohaline to recommence. However, uncertainty surrounds how long it would take the thermohaline to start up if global warming resulted in higher melt rates (and therefore more freshwater). This is an area of ongoing scientific research. In most models, **radiative forcing** from elevated atmospheric concentrations of **greenhouse gases** reaches a critical point, after which the only stable state is for the thermohaline to shut down. Even if atmospheric temperatures begin to fall, some models indicate that the thermohaline would remain dormant for extended periods of time. There is insufficient understanding at present to say whether the thermohaline is bistable (i.e. reversible) should it shut down.

It is not yet known whether expected temperature changes would be significant enough to create an abrupt change in salinity, one sufficient to shut down the thermohaline completely. Circumstances today are quite different from those in the early years following the last ice age, as are the likely feedback effects. For example, in the past there was much more ice cover, and therefore different **albedo** feedback effects were at play. Furthermore, the chances of a sudden massive one-off discharge of freshwater to the North Atlantic (e.g. due to sudden ruptures of land-based ice dams) is no longer a threat as these ice dams disappeared long ago.

These past events are more sudden than the current slow melting of the Greenland ice cap. Nonetheless, global warming is predicted to accelerate sea ice retreat (thereby reducing *brine exclusion*) and to promote enhanced in-flows of freshwater from melting land ice and rivers that empty into the Arctic Ocean (due to increased precipitation in the higher northern latitudes). These factors will reduce salinity levels in the North Atlantic. This trend has already been observed. Over the period 1950–1990, the Atlantic surface waters have become much more saline, due to increased evaporation, and the waters of the high northern and southern latitudes around the poles have become much fresher, due to melting ice and increased precipitation.[3] Most scientists consider that this will lead to a slowing of the thermohaline over the next century and beyond. If the salinity changes are rapid enough, it could lead to a complete shutdown.

In recent years, scientists, using a variety of models, have conducted research to try to establish the potential impact of future global warming on the thermohaline.

The chances of a shutdown this century are considered to be low, but there are significant variations between modeling results. This is due to variations in the dominance of different positive and negative feedback mechanisms between models. Some simulations show negative feedback effects that reduce the chances of a shutdown, while others show much weaker feedback influences. For example, some model simulations show that increased heat in the tropical Atlantic will lead to increased evaporation (the amount of water vapor emitted to the atmosphere) and that this water vapor is likely to be transported to the Pacific, where it results in increased precipitation. This loss of water leads to a saltier Atlantic, and as the winds move, this saltier water north of it offsets some of the freshwater-induced salinity changes.[4] It is not yet known to what extent this feedback could counteract the salinity change from freshwater inflow, but it would be expected to have some moderating influence.

However, under a scenario of a doubling of CO_2 concentrations by mid-century, leading to a 3°C temperature rise this century, nearly all models predict at least a slowing of the thermohaline. Modeling results suggest that there is less than a 40% chance of a shutdown by 2100 but that this rises to a 65% chance of shutdown by 2200.[5] At higher atmospheric greenhouse gas concentrations, several models indicate a complete thermohaline shutdown and the initiation of a reverse flow. For example, one simulation that assumed CO_2e concentration reached 1,000 ppm early next century resulted in a complete shutdown and remained shut down indefinitely.[6]

It is too early to know at this stage what the climate change threshold for a shutdown would be or how fast this threshold would be approached for different levels of increases in greenhouse gas concentrations. It is also possible that the rate of change in concentration levels, as well as the concentration level itself, is an important determinant of shutdown: the faster we approach the threshold, the more likely it is that the thermohaline will shut down.[7] Ongoing advances in modeling techniques are likely to reduce this level of uncertainty over the next decade, and we then may be in a position to more clearly define critical thresholds. A 2005 study of the flow of North Atlantic Deep Water (an integral component of the thermohaline) indicated that the rate of flow appears to have declined by up to 30% since the 1960s, but more research and time series data are required to substantiate these findings.[8] The IPCC predicts that the thermohaline flow rate is likely to decrease by around 25% by 2100.[9]

What would happen if the thermohaline did shut down?

The main impact would be to reduce the carbon sink potential of the oceans through the reduced transport of dissolved carbon dioxide into the deep oceans. A significant slowing or shutdown would tend to accelerate the rate of increase in CO_2 concentrations in the atmosphere and would provide a positive **climate change feedback**. It is also possible that a shutdown of the thermohaline would reduce the size of the **land carbon sink** in the northern latitudes due to cooler and drier conditions reducing plant growth. Some studies suggest that primary production could be reduced by 10% or more in some regions of the northern hemisphere, thus reducing CO_2 uptake by vegetation.[10]

Temperature changes are expected to be more significant in some regions than others. Greenland, Europe, the North Atlantic, and the US Mid West would most likely experience a general cooling. However, the Alaskan Peninsula may warm, possibly inducing the release of carbon and methane stored in the permafrost, thus generating another positive **climate change feedback**. The southern hemisphere winter would also most likely experience a warming, which may lead to greater instability of the West Antarctic Ice Sheet (see **ice sheets and glaciers**).[11] The resulting thermal expansion of the ocean could add up to 10 cm/century to **sea level rise**.[12]

If the thermohaline slows, it will reduce the rate of mixing of the ocean waters. This will tend to isolate the surface ocean (where the phytoplankton resides) from the deep ocean (where most of the nutrients are found). It will mean a reduction in oxygen supply to the deep oceans, which would eventually become anoxic (depleted of oxygen). The result of this stratification process would not only be to kill off many deep ocean aerobic marine species but also to promote anaerobic organisms, such as sulphate reducers, that produce hydrogen sulfide as a waste product. This, in turn, would deplete soluble iron, which is an important nutrient source for phytoplankton and could serve to reduce biological activity and primary production.

See also: carbon cycle, carbon dioxide (CO_2), carbon sinks, climate change feedback, global warming, greenhouse gases, ice sheets and glaciers, land carbon sink, ocean carbon sink, radiative forcing, sea level rise.

Notes

1 Teller *et al.* 2002
2 Wood *et al.* 2006
3 Curry *et al.* 2003
4 Wood *et al.* 2006
5 Schlesinger *et al.* 2006
6 Stocker and Schmitter 1997
7 Ibid.
8 Bryden *et al.* 2005
9 IPCC 2007
10 Wood *et al.* 2006
11 Schlesinger *et al.* 2006
12 Ibid.

Further reading

IPCC 2007; Schlesinger *et al.* 2006; Wood *et al.* 2006.

UNITED NATIONS FRAMEWORK CONVENTION ON CLIMATE CHANGE (UNFCCC)

The United Nations Framework Convention on Climate Change[1] (UNFCCC, "the Convention"), and its affiliated **Kyoto Protocol** is the most important international cooperation agreement on climate change and provides the basis for international action to curb **global warming**. The Convention was agreed in May 1992 and entered into force on March 21, 1994. The UNFCCC has been ratified by 192 countries (Parties to the Convention), and only Andorra, the Holy See (Vatican City), Iraq, and Somalia have yet to ratify. The Convention sets an overall framework for nations to cooperate to address climate change.

The overriding objective of the Convention, under Article 2, is "to stabilize greenhouse gas concentrations at a level that would prevent dangerous anthropogenic interference with the climate system." The Convention states that this objective "should be achieved within a time-frame sufficient to allow ecosystems to adapt naturally to climate change, to ensure that food production is not threatened and to enable economic development to proceed in a sustainable manner" (see **dangerous climate change**). The Convention is informed by interdisciplinary studies from the **Intergovernmental Panel on Climate Change (IPCC)**, which assist in determining what greenhouse gas concentrations would result in dangerous interference. The Convention does not set numerical targets, but, under Article 4.2, calls for Annex I Parties to "return their greenhouse gas emissions … to 1990 levels."

Ratification of the Convention requires Parties to recognize 14 important principles and sets an ambitious framework for international action. Of these principles, four of the most important are:

1   Recognition that the adverse effects of climate change are a common concern of mankind.
2   Agreement that the precautionary approach be applied. That is, where there are threats of serious or irreversible damage, lack of full scientific certainty shall not be used as a reason for postponing cost-effective measures to prevent environmental degradation.
3   Acknowledgment that developed countries account for the greatest share of historic and current emissions, that per capita emissions from developing countries are relatively low, and that developing countries need to grow economically to meet their social and development needs.
4   Acceptance that the global nature of climate change calls for cooperation by all countries, with common but differentiated responsibilities, and according to respective capabilities.

The two principles relating to greater responsibility for developed countries and creating "common but differentiated responsibilities," are grounded in considerations of equity, but they have led to important and contentious outcomes in the **Kyoto Protocol**: different targets for different developed countries, and no targets at all for developing countries.

Administrative structure

The Convention consists of a range of different administrative and coordinating bodies, including the Conference of the Parties (COP), the President and Bureau, Subsidiary Bodies, the Secretariat and other committees, working groups and expert bodies. Each Party (ratifying country) to the Convention may participate in any of these organizational bodies.

Conference of the Parties

The COP is the ultimate decision-making body of the Convention. The COP is made up of delegates from each of the Parties and has overall responsibility for the implementation of the Convention, including specifically:

1 Examining Parties' commitments to the objectives of the Convention;
2 Facilitating the exchange of information;
3 Facilitating the coordination of actions and measures by Parties;
4 Guiding the development of methodologies, particularly those related to preparing national **greenhouse gas inventories**;
5 Assessing the ongoing implementation of the Convention by Parties;
6 Reviewing and adopting reports by Parties;
7 Making recommendations;
8 Mobilizing financial resources;
9 Reviewing reports submitted by Subsidiary Bodies; and
10 Other functions to achieve the objectives of the Convention.

The core work of COP decision making occurs in public plenary sessions conducted in the six official UN languages by delegates – usually high-ranking government officials – from each of the Parties.

The rules for the decision-making process of the COP were drafted in the first COP held in Berlin in 1995, based on the UN practice that all decisions of substance be made by consensus, with voting implemented as a last resort. However, agreement could not be reached on the voting system to be employed (proposals included simple majority voting, two-thirds majority voting, and representation of a minimum number of votes from different regions or groups): consequently, all substantive decisions of the COP must be taken by consensus. Procedural matters (timing of meetings, venues, etc.) are, however, determined by a simple majority vote amongst the Parties that are present.

The COP meets for two weeks annually, usually in November or December. The COP attracts thousands of official delegates and observers, occasionally numbering 10,000 people. In addition to official government delegates, there are a large number of observers from international organizations and multilateral agencies, nongovernment organizations (NGOs), industry associations, and the media. While participation in plenary sessions is limited to those accredited by the Secretariat, the COPs have spawned a multitude of parallel meetings, both inside and outside the official COP meeting venue (see Table 20).

These parallel meetings take advantage of the confluence of the world's leading technical, business, and political figures and enable broader public discussion and debate of key climate change issues. Extraplenary meetings and activities include official and unofficial side events, stalls for information and promotions, business roundtables and meetings, launches of anything from Internet sites to new technologies, social

Table 20 Conference of the Parties (COP) meetings

COP	Year	Location
COP1	1995	Berlin, Germany
COP2	1996	Geneva, Switzerland
COP3	1997	Kyoto, Japan
COP4	1998	Buenos Aires, Argentina
COP5	1999	Bonn, Germany
COP6	2000	The Hague, The Netherlands
COP6 (Part II)	2001	Bonn, Germany
COP7	2001	Marrakesh, Morocco
COP8	2002	New Delhi, India
COP9	2003	Milan, Italy
COP10	2004	Buenos Aires, Argentina
COP11, CMP1*	2005	Montreal, Canada
COP12, CMP2*	2006	Nairobi, Kenya
COP13, CMP3*	2007	Bali, Indonesia
COP14, CMP4*	2008	Poznan, Poland
COP15, CMP5*	2009	Copenhagen, Denmark

*CMP = Conference of the Parties of UNFCCC held in conjunction with the Meeting of the Parties of the Kyoto Protocol.

gatherings and public demonstrations. What happens outside the COP plenary is often more influential and informative (and certainly more dynamic) than what happens in the official plenary sessions. Indeed, in recent years, nonplenary events have played an increasingly influential role in guiding the direction of the international discussions.

Most decisions of the COP are predrafted by the Subsidiary Bodies, the President, the Board, the Secretariat, or groups of Parties prior to a COP. While the mandate of the COP is the implementation of the Convention, COPs are more akin to international negotiations than to decision making by a company's board of directors. The COP is, in essence, a political forum, with all the alliances and enmities, lobbying and positioning, historic grievances, linked agendas, influencing, and posturing that such a forum entails. The real work of reaching consensus on draft decisions is mostly done in bilateral meetings, discussions "in the corridors," and special meetings often facilitated by the President.

At the end of the intensive two weeks of a COP, there is a high-level Ministerial session, in which Ministers (usually the Environment Minister) of the Parties make a set presentation. These usually provide an optimistic view of their government's implementation of the Convention, with an eye to domestic audiences as much as to those present at the COP. Often, Ministers will use the opportunity to criticize other Parties' actions or inactions (in particular, developing countries seeking more efforts from developed countries to reduce emissions), seek special exemptions, or present claims for greater funding for their country or project.

COP President and Bureau

The role of the President is to promote and facilitate the work of the COP. The office of the President is expected to rotate among the five UN regional groups (Africa,

Asia, Central and Eastern Europe, Latin America and Caribbean, Western Europe, and Others). In practice, the government that hosts the COP has a representative (usually the Environment Minister) elected after the opening of the COP, and this person remains President until the following year's COP. The President meets with Party representatives (usually at Ministerial level) to facilitate bilateral and multilateral agreements that can be taken to the COP with known support and sponsors. The President is usually a senior member of the ruling government and often a person with experience in dealing with international affairs and negotiations. The success or failure of a COP reflects very much on the President.

The Bureau deals primarily with procedural and organizational issues and is made up of seven elected Vice Presidents, the Chairs of the two Subsidiary Bodies, a Rapporteur, and the President. In contrast to the President, Bureau members usually serve for two years during and between sessions.

Subsidiary Bodies

The Convention established two permanent Subsidiary Bodies, the Subsidiary Body for Scientific and Technological Advice (SBSTA), and the Subsidiary Body for Implementation (SBI). Like the COP, they each have a Bureau consisting of a Chair, Vice Chair, and Rapporteur. These bodies constitute the principal working bodies of the Convention and meet twice per year (mid-year and in conjunction with the COP) for one or two weeks. Meetings are open to all Parties and are usually attended by government bureaucrats and technical specialists rather than political negotiators. Neither body can make decisions on the Convention, and the outcomes of their work consist primarily of recommendations and draft decisions referred to the COP for decision.

The SBSTA's stated role under the Convention is to provide "timely advice on scientific and technological matters relating to the Convention."[2] The COP further specified the SBSTA's roles to:

- provide assessment of the state of scientific knowledge, particularly with links to the **Intergovernmental Panel on Climate Change (IPCC)**;
- prepare scientific assessments of the effects of measures taken by Parties;
- identify new technologies and advise on promoting their rollout;
- advise on scientific programs, cooperation in research and development, and capacity building; and
- respond to scientific, technological, and methodological questions that the COP raises.

The SBI's stated role under the Convention is to assist the COP "in the assessment and review of the effective implementation of the Convention."[3] The COP further specified the SBI to:

- consider the information communicated by all Parties to assess the effect of actions taken to meet the objectives of the Convention;
- consider information provided by Annex I parties in reviewing the adequacy of their commitments;

- assist the COP in preparing and implementing decisions, including reviewing the financial mechanisms of the Convention.

Secretariat

The Secretariat, with a permanent staff based in Bonn, Germany, provides support services to all the bodies of the Convention, including:

- administrative arrangements and support to all sessions and negotiations;
- assistance to Parties (particularly developing countries) in meeting their commitments;
- coordination of work with the secretariats of other international bodies, particularly the Global Environment Facility, the IPCC, the United Nations Environment Program and United Nations Development Program, and other UN agencies.

Other Bodies

The COP has established a number of other groups to work on particular issues with set time frames, including *ad hoc* groups to facilitate negotiations, expert groups to work on technical issues, and groups to work on the issues of special Parties, such as the Least Developed Countries.[4]

Annex I and non-Annex I Parties

The Convention recognizes several groups of Parties, but the primary distinction is between the 41 Annex I countries, which consist primarily of developed industrialized countries with relatively high per capita emissions (see **emissions per capita**), and 151 "non-Annex I" Parties, which are mostly developing countries. Within Annex I, there is a further subdivision, which distinguishes members of the Organisation for Economic Cooperation and Development (OECD) in 1992 (Annex II), and Economies in Transition (EIT) countries, which were previously part of the Soviet Bloc.

Within non-Annex I Parties, several groups negotiate as blocks, but the Convention explicitly recognizes the 50 Least Developed Countries (LDCs), as identified by the United Nations[5] (see Table 21).

Different Parties have different obligations and responsibilities under the Convention. These responsibilities can be grouped under all Parties, Annex I, Annex II, non-Annex I, and Least Developed Countries.

All Parties

The Convention recognizes "common but differentiated responsibilities [of Parties] and their specific national and regional development priorities, objectives and circumstances." Nonetheless, all Parties commit to "formulate, implement, publish and regularly update national ... programs containing measures to mitigate climate change by addressing anthropogenic emissions by sources, and removals by sinks, of all **greenhouse gases** not controlled by the Montreal Protocol" (see **ozone**). With

Table 21 UNFCCC country categories

Annex I	Annex II	Least developed countries
Australia	Australia	Afghanistan
Austria	Austria	Angola
Belarus	—	Bangladesh
Belgium	Belgium	Benin
Bulgaria	—	Bhutan
Canada	Canada	Burkina Faso
Croatia	—	Burundi
Czech Republic	—	Cambodia
Denmark	Denmark	Cape Verde
European Economic Community	European Economic Community	Central African Republic
Estonia	—	Chad
Finland	Finland	Comoros
France	France	Democratic Republic of the Congo
Germany	Germany	Djibouti
Greece	Greece	Equatorial Guinea
Hungary	—	Eritrea
Iceland	Iceland	Ethiopia
Ireland	Ireland	Gambia
Italy	Italy	Guinea
Japan	Japan	Guinea-Bissau
Latvia	—	Haiti
Liechtenstein	—	Kiribati
Lithuania	—	Laos
Luxembourg	Luxembourg	Lesotho
Monaco	—	Liberia
The Netherlands	The Netherlands	Madagascar
New Zealand	New Zealand	Malawi
Norway	Norway	Maldives
Poland	—	Mali
Portugal	Portugal	Mauritania
Romania	—	Mozambique
Russian Federation	—	Myanmar
Slovakia	—	Nepal
Slovenia	—	Niger
Spain	Spain	Rwanda
Sweden	Sweden	Samoa
Switzerland	Switzerland	Sao Tome and Principe
Turkey	—	Senegal
Ukraine	—	Sierra Leone
United Kingdom	United Kingdom	Solomon Islands
United States of America	United States of America	Somalia (not ratified UNFCCC)
—	—	Sudan
—	—	Tanzania
—	—	Timor-Leste
—	—	Togo
—	—	Tuvalu
—	—	Uganda
—	—	Vanuatu
—	—	Yemen
—	—	Zambia

regard to sinks and reservoirs (see **carbon cycle**), Parties commit to "promote sustainable management, and … the conservation and enhancement … of sinks and reservoirs of all greenhouse gases … including biomass, forests and oceans as well as other terrestrial, coastal and marine ecosystems." More generally, all Parties agree to take climate change into account in their national development policies and cooperate in the exchange of scientific, technological, and other information relating to climate change.[6]

These goals are clearly desirable, but their qualitative nature does not easily lend itself to clear mitigation outcomes or to measurable success or failure. The non-Annex I countries were, at the time of drafting the Convention (and remain still), adamant that the richer, developed countries have been responsible for the vast majority of historic emissions and must, therefore, lead the way in addressing climate change. Furthermore, even if there is a recognition and desire to take greater actions, many non-Annex I countries simply cannot justify devoting resources to address climate change ahead of such fundamental development needs as health, education, and poverty reduction. This issue of "ability to pay" is addressed in the differing responsibilities of Annex I and non-Annex I countries. However, the ability to pay for non-Annex I countries will change over time and some, for example South Korea, Singapore, and several others, are already high-income countries that arguably should have equivalent obligations to Annex I countries.

Annex I Parties

Under the Convention, Annex I Parties make further commitments. These relate primarily to taking the lead in addressing climate change, with Article 4.2 requiring Annex I Parties to "adopt national policies and … corresponding measures on the mitigation of climate change." The Convention also states that Annex 1 emissions should aim to "return [emissions] by 2000 to earlier levels." While all Parties commit to reporting on national policies to address climate change, Annex I Parties "shall communicate detailed information"[7] on how their programs will reduce their emissions. These two Articles are, together, taken as a commitment by Annex I Parties to reduce emissions to 1990 levels by 2000, but from the nonspecificity of the language it is not clear how legally binding this commitment is.

Annex I Parties must submit regular "national communications," which include national **greenhouse gas inventories**, establish a 1990 baseline against which annual emissions can be compared, and summarize programs and measures in place to mitigate climate change. National communications and **greenhouse gas inventories** are important mechanisms for tracking emission trends as well as policies and measures to reduce greenhouse gas emissions.

Given the major political and institutional changes taking place in ex-Soviet states in the early 1990s, EIT Annex I Parties are permitted some flexibility in their commitments under the Convention. This has meant some delay in submitting national communications and has resulted in some EIT Parties selecting a base year other than 1990.

By 2008, all 41 Annex I Parties had submitted their first national communications, 39 have submitted their second (Turkey and Luxembourg not submitted),

38 their third (Turkey, Luxembourg, and Ukraine not submitted), and 38 have submitted their fourth national communications, which were due in January 2006 (Turkey, Luxembourg, and Ukraine not submitted).

Annex II Parties

In addition to the requirements of Annex I, the 24 Annex II Parties have an obligation to provide "new and additional financial resources"[8] to non-Annex I Parties to assist them in addressing climate change (see financial mechanisms discussion below), facilitate the transfer of clean technologies to non-Annex I and EIT countries, and to report on their actions in regard to these commitments in their national communications.

Non-Annex I Parties

As Parties to the Convention, non-Annex I Parties are required to report on their policies and programs to address climate change, but these reports are more general in nature. There are less stringent requirements for national inventories, and a 1990 base year need not be established. Moreover, preparation of national communications by non-Annex I Parties is contingent on receiving "new and additional" funding from Annex II Parties. As such, submission of national communications from non-Annex I countries has been delayed. As of early 2008, 135 of the 151 non-Annex I Parties that have ratified the Convention had submitted their first national communications, three (Argentina, South Korea, and Uruguay) have submitted their second national communications, and only one (Mexico) has submitted a third national communication. Non-Annex I national communications are essential for tracking global emissions, identifying emission trends, and as input to negotiations on future international climate change agreements.

Least Developed Countries

The 50 LDCs may submit their national communications at their own discretion. By early 2008, 43 of the 49 LDCs that have ratified the Convention had submitted their first national communications, and none had submitted subsequent national communications.

Convention Funding

Financing of the Convention falls into three main areas: core funding, new and additional funding from Annex II Parties, and voluntary contributions.

Core Funding

The Climate Change Secretariat forms part of the United Nations system and, as a consequence, uses the UN approach of two-year budgets and contributions, subject to the accounting and audit processes of the United Nations. The Secretariat proposes a two-year budget, which is reviewed by the SBI and submitted to the COP for approval by consensus. The core budget for the 2006–2007 biennial period was approximately US$60 million (US$30 million per year).

The level of each Party's payments to the core budget are based on dues assessed in the same manner as for the United Nations General Assembly, which reflects the capacity of countries to pay. The scale reflects the country's Gross National Product (GNP), adjusted to take into account factors such as per capita income and external debt: the result is that wealthier countries pay more.[9] However, no Party is required to pay more than 22%[10] and all parties must make a contribution (a minimum contribution of 0.001% is set for the poorest countries). Parties are required to pay their dues on January 1st of each year, and many have consistently paid late, which has caused problems for the functioning of the Convention.

The Convention is very much intertwined with the **Kyoto Protocol**, which has the same financial rules and procedures. In 2005, the COP agreed to apportion 63.2% of the budget to the Convention and 36.8% to the Protocol, with countries that are Parties to the Convention but not the Protocol (in particular the United States) only contributing to the Convention.

Annex II funding

The Convention established a financial mechanism for Annex II contributions, operated by the Global Environment Facility (GEF)[11] and implemented by the United Nations Development Program, the United Nations Environment Program, and the World Bank. The number of implementing agencies has recently expanded to include regional development banks and several other UN agencies. The fund remains accountable to the COP, which determines policies and program priorities, with reviews on implementation every four years.

The GEF is replenished every four years and includes contributions to environmental priorities other than climate change (biodiversity, international waters, land degradation, persistent organic pollutants (POPs), and protection of the ozone layer). The GEF is considered development aid over and above "normal" Official Development Assistance and is, therefore, eligible to be considered "new and additional financial resources" under the Convention. By 2008, US$3.1 billion in GEF funds, and an additional $ 14 billion in cofinancing, had been allocated to climate change activities.[12]

In addition to the GEF, COP7 in Marrakech (2001) established three other funds (the "Marrakech funds") to complement the GEF climate change activities. The Special Climate Change Fund (SCCF) predominately targets **adaptation** projects but can also be used to fund technology transfer activities. By mid-2008, the SCCF had received pledges totaling US$ 60 million for disbursement. Most of these funds have already been spent or committed to project activities.

The Least Developed Country Fund (LDCF), which assists LDCs to establish their own, country-specific, National Adaptation Programmes of Action (NAPA). The NAPA identifies priority policies and projects that may then apply for funding through normal GEF procedures. By mid-2008, the LDCF had received pledges totaling approximately US$120 million from Annex II Parties.[13] A majority of these funds have also either been expensed or committed to activities.

The Adaptation Fund is different from the other funds in that it is not regarded as "new and additional financial resources" in Annex II national communications. The main source of finance for the Adaptation fund comes from a 2% levy on **Clean Development Mechanism (CDM)** projects and not solely reliant on Annex I donor funds. The Adaptation Fund is to be administered by the GEF on an interim basis, but as of mid-2008, the fund had yet to be fully operationalized. It is expected that up to $ 300 million may be available over the 2008–2012 period to fund adaptation measures (see **adaptation**).

Relative to the scale of the mitigation and adaptation challenges facing the poorer developing countries, the level of resources provided to date for the three Marrakech funds is very small and will need to increase significantly over the coming years if the developing world is to be able to cope with the emerging climate change impacts.

Voluntary contributions

In addition to the required fees, some countries make voluntary additional payments, notably the Government of Germany's contribution of approximately US$ 1 million per year to support the Secretariat in Bonn. Annex II Parties can make voluntary contributions to any of the funds discussed above at any time, and this is encouraged by the COP. Annex II Parties can also fund programs, capacity development activities, and projects bilaterally or through mechanisms other than those facilitated by the COP.

There has been ongoing tension between recipient and donor countries, as some non-Annex I Parties believe that Annex II Parties provide funding to meet their Convention obligations at the expense of traditional development assistance (such as for poverty reduction, education, and health). Indeed, the magnitude of Official Development Assistance has been largely stagnant since the late 1990s, while funding for climate change-related activities has increased substantially. This would, of course, be counter to the "new and additional" Article of the Convention, which was included precisely to avoid this situation. The issue is particularly contentious since most developing countries do not have climate change mitigation at the top of their development priorities and insist that Annex I countries created the problem and therefore should address it without diminishing development assistance. There is, however, no international convention or treaty that specifies a required level of donor assistance, and it is, therefore, not possible to substantiate or deny the claim of ODA diversion. Further, projects that are financed through the Convention funding mechanisms and voluntary Annex II contributions are targeted toward addressing climate change but almost always have at least some development benefits associated with their implementation.

Can the Convention be viewed a success?

Only the Geneva Conventions and Montreal Protocol exceed the UNFCCC for number of countries ratified, and COPs regularly rival the UN General Assembly as the largest gathering of government leaders and senior officials during the year. The founding of the United Nations is the only intergovernmental agreement that can compare with the Convention with respect to the speed of identification of a global need, convening almost all nations and agreeing ambitious goals. While, at times, the COP process can seem painfully slow, compared with other global multilateral agreements the pace and scale of the COP decision making is actually quite impressive. Climate change is a global problem that requires a global solution. As a mechanism to bring nations together to recognize the need for cooperation in addressing climate change and providing a forum for discussion, the Convention must certainly be considered a success.

Unfortunately, recognizing a problem and agreeing that something needs to be done is only the first step toward the Convention's objective of stabilizing greenhouse gas concentrations to prevent **dangerous climate change**. In aggregate, Annex I Parties have met the goal of reducing emissions to 1990 levels by 2000 as actual emissions were 6% below 1990 emissions. Few seriously point to this as a sign of success of the Convention however, since this decrease has been almost entirely as a result of the major economic changes, and subsequent substantial emission reductions, that took place in the early 1990s in the Economies in Transition (EITs) countries. By 2000, aggregate EIT emissions were nearly 40% lower than 1990 emissions. In contrast, aggregate Annex II Party emissions steadily increased over the same period, to be 8.8% above 1990 levels in 2000.[14] However, since 2000, emissions in most Annex I parties (including EIT economies) have been growing steadily. By 2005, total Annex II Party emissions were 10.0% above 1990 levels (see **Kyoto Protocol** and **anthropogenic greenhouse gas emissions**), although overall aggregate Annex I emissions (including EITs) were still 4.6 % below 1990 levels.

Emission trends for non-Annex I Parties are harder to gauge since their national communication reporting is less comprehensive and often quite dated (see **greenhouse gas inventories**). Nonetheless, it is clear that emissions have been rising, particularly from the large, rapidly industrializing countries such as India and China. Annex II technology transfers have made no discernible impact on non-Annex I Party emission trends.

By 2006, global emissions were 35% above 1990 levels. It is clear that the various commitments of all Parties to date have not yet stabilized greenhouse gas concentrations in the atmosphere – in fact the rate of increase of atmospheric greenhouse gas concentrations is accelerating, and this trend is expected to continue for at least the next few decades (see **future emissions trends**).

The Convention foresaw this and explicitly required a review of the adequacy of the commitment of Annex I countries returning to 1990 emission levels by 2000 and subsequent regular reviews.[15] The initial review at COP1 resulted in the "Berlin mandate," which eventually led to the drafting of the **Kyoto Protocol** at COP3 in 1997.

Overall, the Convention provides an unprecedented platform for global cooperation on climate change and has led to the development of an effective international administrative framework that can support future international agreements to be negotiated, implemented, and monitored. However, the Convention, while supported almost unanimously by the international community, has not, as yet, delivered any meaningful reduction in emission trends: at best, the Convention's success can be viewed as mixed. Even though concentrations have probably not yet reached the **dangerous climate change** threshold, they are likely to pass this point before 2025 (see **stabilization targets** and **dangerous climate change**). To honor the objectives of the Convention, the international community will need to significantly increase the strength of its commitments post-2012 and ensure they are fully implemented.

See also: adaptation, anthropogenic greenhouse gas emissions, dangerous climate change, emissions per capita, future emissions trends, greenhouse gas inventories, Intergovernmental Panel on Climate Change (IPCC), Kyoto Protocol.

Notes

1 UN 1992
2 Ibid.
3 Ibid.
4 See Yamin and Depledge 2004
5 See UN Department of Economic and Social Affairs, http://www.un.org/esa/policy/dev-plan/profile/index.html
6 UNFCCC 1992
7 UNFCCC Article 4.2(b)
8 UNFCCC Article 4.3
9 UN Public Web site
10 UN General Assembly resolution 55/5
11 See Global Environment Facility, http://www.gefweb.org/ for most recent data.
12 UNFCCC Handbook 2006
13 See Global Environment Facility http://www.gefweb.org/interior.aspx?id=232 for most recent data.
14 UNFCCC/SBI, May 2003
15 UNFCCC Article 4.2(d)

Further reading

UNFCCC Handbook 2006; Yamin and Depledge 2004.

WATER IMPACTS

Climate change will have significant effects on the global water cycle and could have far-reaching impacts on humans and natural ecosystems. Expected changes include variations in the distribution, timing, and intensity of precipitation events (rainfall, snow, hail, and cloud mist) and changes in the timing of seasonal water flows. Although some modifications to the global water cycle have already been observed, the changes over this coming century are expected to be much more pronounced. The impact on humans and natural ecosystems are expected to be significant.

Access to water, both in terms of quantity and quality, have a major influence on people's livelihoods and on the structure and location of economic activities, particularly agriculture and industry. Water access and availability are already pressing issues for many countries and regions. Currently, more than 1 billion people lack access to clean, piped water for basic drinking and sanitation needs, and more than 2 billion live in water-stressed areas (those where the average allocation of available water is less than 1,000 cubic meters per capita per annum). With increasing population and higher levels of economic activity, these water resource problems are expected to worsen, with or without climate change. Climate change is expected to exacerbate such problems in some regions but may alleviate them in others.

Temperature effects

The water-retention capacity of air increases with temperature and is a key determinant of precipitation. The relationship is described by the *Clausius-Clapeyron equation*. Cold air is drier than warm air, and this is the reason why the poles experience such low levels of precipitation (they are essentially cold deserts) and why the equatorial regions have so much rain. As the planet warms, the atmosphere will tend to hold more water and the water cycle will intensify. For every 1°C rise in earth's mean surface temperature, global precipitation is expected to increase by approximately 1%.[1]

Increasing global temperatures will also result in more precipitation falling as rain, rather than snow, and more intense precipitation events. Warming will also change the seasonal melt rates of glaciers and mountain snow packs, which will, in turn, change the timing and magnitude of seasonal water flows. Higher temperatures will also increase evaporation rates, which can have significant impacts on the availability of soil moisture at different times of the year. All these changes are expected to result in increased frequency and intensity of both droughts and floods.

Rainfall distribution

While the impact of climate change on particular stages of the water cycle remains uncertain, aggregate global precipitation is expected to increase as the atmosphere warms, but it will not be evenly distributed across the planet. Some regions will become drier and others wetter. In some areas, average annual rainfall may remain unchanged, but rain may fall at different times of the year or in fewer, more intense, events.

While there is general agreement between most climate models concerning the broad regional trends in precipitation that are likely to occur with **global warming**, there remains considerable uncertainty about the precise magnitude and distribution of the changes.

Which regions become drier?

Models suggest that the subtropics and the mid-latitudes will generally become drier. In the southern hemisphere, the southern regions of Australia and Africa are likely to experience lower precipitation. In the northern hemisphere, the most pronounced drying will be in the Mediterranean region (both North Africa and Southern Europe), South West United States and Mexico, and probably West and Central Asia and Western China. In some areas, there could be reductions in available runoff of up to 30% by 2050 if temperatures rise by 2ºC, and possibly up to 50% if global mean temperature exceeds 4ºC.[2] Some areas of the tropics may also become drier. For example, the East Amazon region of Brazil is expected to experience a reduction in rainfall that could accelerate the level of forest dieback already being experienced in the region, ultimately leading to significant ecosystem deterioration.

Some regions are already experiencing drying trends. The southern regions of Western Australia have, since the mid-1970s, experienced a 15% decline in average annual rainfall, and winter rainfall is 25% below the long-term average. This has already adversely affected natural ecosystems, urban water supplies, and agricultural production.[3] Runoff has decreased more than rainfall: since 1975, the amount entering Perth's urban water catchments has fallen to half the long-term average.[4] The east coast of Australia has also experienced a fall in rainfall, due to a southward shift in the weather system and the more persistent El Nino that has prevailed in recent decades.

Since the 1970s, average precipitation has decreased by 2.4% per decade in African tropical rainforest regions and by 4% per decade in West Africa.[5] Falls in precipitation have been observed in some areas of Southern Africa but not in others – there is, as yet, no discernible trend.

Which regions become wetter?

The higher latitudes are expected to experience increased precipitation as atmospheric temperatures increase. In particular, Russia, northern Europe, Alaska, northern Canada, Greenland, and Antarctica will receive much more precipitation. Rainfall, runoff, and river discharge have already increased significantly in the higher northern latitudes, particularly in eastern Russia.[6] Precipitation (falling mainly as snow) has also risen in the past two decades in Antarctica and Greenland. Northeastern and northwestern North America are also likely to become wetter. Some areas of the tropics are expected to become much wetter, particularly in the equatorial central Pacific. Parts of East Asia are also expected to receive more rain. Even the Sahara Desert might receive elevated winter rainfall. For all regions that experience increased precipitation and rainfall intensity, there will be associated increased risks of flooding.

Overall, since the mid-1990s, there has been a general drying in the mid-latitudes in both the northern and southern hemispheres and a noticeably wetter trend in the high latitudes of the northern hemisphere and Antarctica.[7] While this seems to be consistent with what the models suggest will happen under climate change, it is too early to judge whether these are permanent shifts in the underlying average precipitation patterns or part of regular cycles, such as the **North Atlantic Oscillation (NAO)** and the **El Nino Southern Oscillation (ENSO)**. In the high latitudes of the northern hemisphere, precipitation is known to be influenced by the decadal variability of the NAO.

There are some regions where it is uncertain whether wetting or drying trends will prevail as the models yield inconsistent or conflicting results. For example, there is considerable uncertainty over the future trend in rainfall for the Sahel region of sub-Saharan Africa. From 1970 to 1990, the Sahel experienced a series of extended droughts, which led to widespread starvation. While precipitation became more regular during the 1990s, low rainfall and severe drought conditions have since returned.

Scientists cannot yet say whether the return to low-rainfall conditions is just a short-term trend or part of a long-term systemic shift. Computer models for the Sahel disagree, ranging in their forecasts from persistent drought, mild drought, and even wetter conditions. Although the most recent modeling predicts that the Sahel is likely to become drier, there is not yet sufficient data to conclude with any certainty that this will be the case.[8] There is also uncertainty over trends in the central plains region of North America and the southern half of South America.

Seasonal water availability

Although changes in the amount of annual rainfall are important in determining aggregate water availability for a given region, the timing and intensity of rainfall events is also critical. It is the changes in extreme dry and extreme wet periods, rather than changes in long-term averages, that can have the greatest impact. Aggregate rainfall may increase, but if it is compressed into a narrow time period much of the additional water may simply end up as increased runoff to the oceans, with little change in terrestrial water availability. If dry periods become even drier, this will tend to aggravate existing seasonal water shortages.

In many regions, seasonal water flows are influenced by the melt rates of mountain glaciers and snow packs. Over a billion people currently depend on dry season flows from glaciers and snow pack melt, including parts of southwest United States, the Andean region, but especially South Asia and Western China.[9] A full 70% of the summer flow of the Ganges is fed by Himalayan melt waters.[10]

The initial impact of global warming on glaciers is to increase the rate of summer melting and to increase water flows. In some areas, increased melt rates can cause rapid buildup of water in glacial lakes to the point where they burst and result in flash floods. This is already a serious risk in the Himalayan region. In Nepal, for example, several recent glacial lake bursts have resulted in major damage to infrastructure and loss of life.[11] There has also been increased water flow and flash flood activity in the Andes region, resulting in increased erosion and mudslides.

Over time, as the glaciers and snow packs disappear, dry season water flows will diminish. This could result in water shortages and adverse impacts on agricultural production and people's livelihoods. Glaciers and snow caps in the Andes region are an important source of water to many countries (Peru, Columbia, Ecuador, and Bolivia), and several major cities depend on them for their water supplies. Rapid Andean deglaciation may entail a need for substantial increases in water storage capacity to cover future supply needs. Over the past half century, the Andean glaciers have been contracting at an accelerating rate. By 2020, the Chacaltaya Glacier in Bolivia, which is an important source of water for La Paz, is expected to have melted completely, as will four of Ecuador's eight major feed water glaciers.[12] In Columbia, the El Cocuy glacier region is an important source of water, but the ice field has contracted by 75% since 1950 and could be largely gone by 2050. Peru is also heavily dependent on glacial melt water, but it has lost a quarter of its glacier water resource since 1970, and, in some areas, ice mass has nearly halved. Overall, climate change is expected to increase the vulnerability of many Andean countries to seasonal water shortages. Changes in water availability in South Asia are likely to become particularly severe in the next few decades as most of the major feed glaciers are disappearing rapidly, and this will have widespread repercussions for more than 1 billion people that depend on Himalayan meltwater flows.

The southwestern part of the United States is also likely to experience significant changes in water resource availability. The region is already semiarid, and much of California's agriculture depends on montane snow pack to provide a consistent flow of water over the summer months. Substantial reductions in snow pack in the Sierra Nevada Mountains, 25–40% by 2050 and up to 80% by 2100, and peak water flows will occur up to three weeks earlier each spring.[13] Summer water flows could be halved by mid-century, with significant implications for California's agricultural sector. Several major cities in this region are also likely to face increased water shortages.[14]

The human impacts

The impact of changes in the water cycle on human populations will depend on a range of demographic, economic, and physical changes over time. Water is an essential input to economic activity. Patterns of human settlement and types of agricultural systems have, to a large extent, evolved according to the availability of water resources. While humans only need around 50 liters per day to meet their basic drinking and sanitation requirements, much larger allocations are needed for agriculture and industry. At present, more than two-thirds of freshwater extractions by humans go to agricultural activities (mainly irrigation) and another quarter is used by industry and other economic activities. Less than 10% is used for drinking and sanitation purposes. Water consumption generally increases with income and the level of economic activity – as economies grow, they usually use more water. Per capita consumption is usually higher in developed than developing countries.

Current global water extraction by humans is well below the potential extraction limit – there is no shortage of freshwater in a global sense. Indeed, climate change is expected to increase global freshwater supplies. The main issues concern the geographical distribution and the timing of seasonal water flows. Climate change is likely to lead to changes in both the volume and timing of river flows and runoff, which will impact economic and social systems and could also lead to heightened regional tensions and conflicts over water resources (see **socioeconomic impacts**). While it is usually preferable to receive more water rather than less, too much water can also have detrimental effects, particularly if the extra precipitation falls at the wrong time of year (resulting in crop spoilage) or over very short time periods (resulting in flooding). Nonetheless, it is the risk of water shortage that poses the greatest threat to society.

Water stress is the most common indicator of the adequacy of water resources to meet human needs. Water stress can be expressed in terms of *availability* of water per person, *access* to safe water, or *exposure* to risks (flood, drought, and health effects). Generally, if water extraction exceeds 20% of available flows, then an area is considered water stressed.[15] Approximately 2 billion people currently live in watersheds that can be classified as water stressed.[16] Water stress is often calculated on the basis of average allocation of water per person (derived by dividing the annual available flows of renewable freshwater by the number of people in the watershed). If availability falls below 1,000 m^3/person per year, a region is considered to be *water stressed*, and below 500 m³/person per year, a region is considered *severely water stressed*.[17]

Climate change is expected to increase the number of people subject to water stress. Under mid-range IPCC estimates of global temperature increase, the number of people deemed to be living in water stressed areas could grow to 3 billion by 2025, and 4–5 billion by 2050.[18] For low levels of temperature increase (less than 1.5°C), the impact could be restricted to an additional 500 million people, though at 2°C the number jumps sharply to 1.5 billion and doubles again to 3 billion at 3°C.[19] The areas that will be most affected by increased water stress are Africa, the Middle East, the Mediterranean, Australia, South America, Central and West Asia, and Southwestern United States. Offsetting this to some extent will be a reduction in the number of people under water stress in East Asia, which could experience increased water availability. In general, the regions that will experience the largest increase in water availability are those that are not presently subject to water stress.

Climate change impacts on the water cycle are expected to have significant repercussions for urban populations. The world is rapidly urbanizing, particularly in the developing world, where a significant number of megacities are emerging. In 2000, 47% of the world's population lived in urban areas, surpassed 50% in 2007, and is expected to increase to 60% by 2030. Large cities tend to concentrate water demand in a confined area. They draw significant amounts of water, especially during droughts and heat waves, which usually coincide with low flow periods. This could amplify water scarcity issues in regions where climate change is expected to result in greater variability in precipitation or seasonal flow.

Many cities on the Pacific coast of the Americas depend on glaciers and snow melt for their water supplies. Over the next few decades, many of these glaciers will disappear, and this will put considerable stress on urban water supplies. Nearly one billion people live in slum areas surrounding these cities – they are particularly vulnerable to the water, sanitation, and food problems that arise during droughts and floods. Increased urbanization also increases the number of people who are vulnerable to flooding, as many large cities are located on major rivers and deltas.

With the expected increase in the number of people living in water stressed areas, there are likely to be increased pressures on agricultural production and food security, and possibly greater extractions from shared water resources. These factors could lead to regional conflicts over the allocation of water resources and periodic migrations due to food and water shortages. Major shared water resources like the Nile, Niger, Tigris-Euphrates, Mekong, Ganges, and Indus rivers will need to be carefully managed to ensure reliability of water flows and equitable transboundary water allocations. Nile flow could fall by up to 50–75% by 2100, and this could increase the vulnerability of Egypt to periodic water shortages.[20] Water resource conflicts may emerge with or without climate change, but in many regions climate change is likely to exacerbate these problems over the coming decades.

Human adaptation options

Humans have had to adapt to variations in water resource availability for centuries. In some cases, this has involved the construction of water transport and storage infrastructure (aqueducts, canals, pipelines, and reservoirs), adjusting cropping patterns, adopting water harvesting and irrigation techniques, tapping ground water resources, and even the abandonment of human settlements.

With greater variability in the patterns of precipitation, and changes in the timing of seasonal flows, there will be a need to construct more water storage reservoirs to smooth out seasonal fluctuations, build long-distance pipelines to transport water from regions with surplus water to water deficit regions, and to construct more irrigation facilities to ensure food security. Significant potential exists to increase water use efficiency, through improved technology, recycling, and conservation measures. Measures can include drip irrigation, reducing irrigation losses, changing crops, water recycling and reuse (particularly in industry), and improved water harvesting and soil moisture retention techniques. Israel, among other countries, has employed a wide range of these techniques to adapt to water scarcity problems. Increased international trade of water-intensive crops (e.g. rice and cotton) and products will be an important adaptation response. This is often referred to as "virtual water trade" as the products embody significant amounts of water.

Finally, desalination of brackish ground water or seawater is another adaptation option that has been suggested as a means of addressing emerging water shortage. While there are already more than 15,000 desalination plants operating around the world, producing around 30 million cubic meters of freshwater per day, it remains a relatively expensive and energy-intensive option. The most efficient state-of-the-art plant still requires 2–3 kilowatt hours to produce 1 m^3 of water.

It is evident that a range of potential options exist that can help humans adapt to changes in water availability. However, most require access to technology and sufficient financial resources to enable them to be implemented. This is not likely to be a major constraint in wealthier countries, but it could be a significant barrier to many of the poorest countries. It is these countries that are the most vulnerable to climate change impacts on water, but also the least able to implement suitable adaptation measures. They also tend to be most dependent on rain-fed agriculture, and where agriculture accounts for a much higher percentage of national economic output. It is here where the climate change water impacts will have the greatest impact on human welfare.

Ecosystem impacts

While humans will confront an increasing number of climate change-related water resource issues over the coming decades, the impacts on natural ecosystems will also be significant. Already many aquatic ecosystems are under stress due to human-related extractions and diversions for irrigation and industrial usage. River flows in dry seasons have already reached critical levels in many of the world's major river basins, freshwater lakes and wetlands, particularly in Africa, Australia, and Central and South Asia, with subsequent impacts on both aquatic and terrestrial ecosystems.

In some areas, climate change is expected to reduce water availability to these natural systems and further exacerbate ecosystem stress. Already changes in precipitation patterns and extreme events have had significant impacts on ecosystems and biodiversity, particularly in the Artic and montane systems (see **polar impacts** and **biodiversity impacts**). Changes in precipitation patterns will also change flood regimes and salinity levels in coastal systems, which can have significant flow on impacts to marine ecosystems (see **marine impacts** and **coastal zone impacts**).

See also: biodiversity impacts, coastal zone impacts, global warming, marine impacts, polar impacts.

Notes

1 Hadley Centre 2003
2 Arnell 2006
3 Flannery 2005
4 Ibid.
5 Nyong and Niang-Diop 2006
6 Curry *et al.* 2003
7 Peterson *et al.* 2006
8 Brahic 2006
9 Barnett *et al.* 2005
10 Stern 2006
11 Agrawala *et al.* 2005
12 Simms and Reid 2006
13 Hayhoe *et al.* 2006
14 Service 2003
15 Arnell 2006
16 Ibid.
17 Stern 2006
18 Arnell 2006
19 Hare 2006
20 Nyong and Niang-Diop 2006

Further reading

Arnell 2006; IPCC 2007.

WIND POWER

Wind power technologies harness the energy of the wind. The global wind resource is very large and is estimated at 72 terrawatts, or equivalent to 40 times annual electricity production.[1] This is a huge energy resource, but only a tiny fraction (less than 1%) could ever be realistically captured.

Wind energy has been used as an energy source for centuries, mainly as mechanical power for grinding grain, ocean transport, and pumping water. Attempts to harness wind energy to generate electricity on a commercial scale only really commenced in the 1970s following the oil supply shocks (1973 and 1979). While early turbines were expensive, and subject to a range of economic and technical problems, wind technology has since progressed rapidly. Reliability of supply has increased and costs have fallen to the point where wind is now cost competitive with fossil fuels in many locations.

Concern over climate change has significantly increased the interest of governments and industry in wind power as a potential greenhouse gas **mitigation** option. The principal attractions of wind power are as follows:

- It produces virtually no greenhouse gas emissions or other wastes.
- Wind resources are relatively abundant and widely distributed geographically, which has benefits in terms of national energy security (by reducing the need for imported energy).
- Employment generated per kilowatt of installed capacity is generally greater than for most other electricity generation options, thus making a positive contribution to economic activity.
- Wind power technology is reasonably mature, already deployed on a large-scale commercial basis and is cost competitive with **fossil fuels** and **nuclear power** in favorable locations.

The principal disadvantages of wind power are as follows:

- Wind energy supplies are intermittent and cannot be relied upon to contribute electricity 24 hours per day (at least, not without energy storage facilities) – thus requiring additional generating capacity to be installed to meet demand when the wind is not blowing.
- While power costs delivered to the grid can be competitive in many locations, the uncertain nature of supply means that the real value of wind is largely the avoided cost of fuel (or avoided carbon dioxide cost penalties) used for electricity generation – thus subsidies, regulatory provisions, or other financial incentives are often required to ensure wind is an attractive option for electricity buyers, except in very favorable locations.
- Reliability of supply due to mechanical outages has been a problem with some turbine designs, although the reliability of turbines has improved considerably since the early 1990s.
- The siting of wind turbines has often met with considerable public resistance due to concerns about noise, aesthetic impacts on the landscape, and potential dangers to local birdlife (although opinions on wind's environmental impact vary considerably).

Further, wind turbines are relatively small in scale and large numbers of turbines over a large area are needed to provide the same energy output as large centralized fossil fuel or nuclear facilities: 1,000 MW of wind capacity occupies 20,000–25,000 hectares, though a significant proportion of this area can coexist with farming activities. However, decentralized power sources can contribute to improving system reliability.

Technology status

The main variables affecting the economic viability of wind power are annual average wind speeds and consistency, location relative to electricity grids, and turbine size.

Energy is the cube of the wind speed – so a 10% increase in wind speed yields a 33% increase in energy output. Wind power facilities are normally confined to areas where average annual mean wind speeds are greater than 4–5 meters/second (m/s) – this limits the number of suitable locations and hence the fraction of the global resource that can be utilized. The consistency and seasonal characteristics of wind also influence the attractiveness of the resource. With consistent mean wind speeds of 8 m/s, wind is generally a very attractive generating option, if located close to electricity grids. However, if the majority of the wind energy is delivered in just a few months of the year, or compressed into short periods of high wind velocities, the wind resource may not be economically viable.

Wind resources are attractive in high latitudes and polar environments, as colder air is denser, and so more energy can be captured for the same wind speeds, or alternately, wind generation is viable with lower wind speed regimes. The greater wind resource availability is of course countered by higher maintenance costs and the need for de-icing. In areas subject to severe wind regimes such as tropical cyclones, wind turbines need more robust structures, which adds to construction and maintenance costs. Most wind turbines need to shut down when wind speeds exceed 25 m/s to avoid damage.

With the exception of small-scale, remote applications, wind generation is most economically attractive when located near large electricity grids. For example, the wind resource across the Southern Ocean (40–50°S of the equator – the "Roaring 40s") is enormous, but extracting all but a tiny fraction (mainly in Tasmania, New Zealand, and Southern Chile) is never likely to be financially viable.

Turbine size is an important determinant of wind power cost. The greater the swept area of the blades, the more wind energy that is captured – this significantly reduces costs per unit of energy. Engineers are well aware of the importance of wind turbine scale to production economics, but increasing turbine size has proved a major technical challenge. Initial attempts at building large-scale turbines, with capacities of greater than 1 megawatt (MW), generally failed, and many large prototype turbines could not cope with the forces and stresses involved. Technical problems such as vibration, gearbox breakdowns, blade loss, and even total destruction of the turbine structure have proved to be major technical constraints for turbine manufacturers. Maintenance and repair costs still represent approximately one-third of the delivered cost of electricity. For this reason, early turbine capacities, at least up until the 1990s, were relatively small, generally 150 kilowatt (kW) or less, with blade diameters of 10–20 m. As a result, production costs per kilowatt hour (kWh) were initially expensive, often more than US 20 cents/kWh (although less at very favorable wind sites).[2] To be cost competitive, generating costs must normally remain below US 6–8 cents/kWh.

By the early 1990s, engineers were successful in scaling up reliable turbine sizes to 250 kW, and to 500–750 kW by the mid-1990s. These machines are capable of producing power at less than US 7–8 cents/kWh. By 2000, commercially deployed turbine sizes had reached 1–1.5 MW (a ten-fold increase in capacity in one decade), with swept area diameters of 100–120 m. In many locations these machines can produce electricity at costs as low as US5 cents/kWh. By 2007, standard turbine size reached 1.5–2.5 MW, capable of delivering electricity to the grid at US 4–5 cents/kWh, and even US 3–4 cents at highly favorable locations (half the cost of a decade earlier).[3] The costs of installed capacity are presently around US$ 850–1,150 per kW for land-based turbines, but higher for offshore systems (US$ 1,100–2,000 per kW).[4]

Due to these significant advances in technology, wind is now a cost-competitive source of electricity in some locations. Further advances in wind turbine technology and reliability are expected over the period to 2020. At present, 5 MW turbines are under development and could become the industry standard by 2010, particularly in offshore applications.[5] Improvements in manufacturing processes, increased operational reliability, larger capacity turbines, and market competition from an increasing number of manufacturers (particularly in India and China) will undoubtedly deliver further cost reductions. However, wind technology is approaching maturity and wind power costs are expected to stabilize at between US 3–6 cents/kWh in the medium term, depending on location and turbine size.

Current contribution to global electricity supplies

The wind power industry has grown rapidly since 2000, and installed capacity has been growing at an annual average rate of 32%: global capacity increased by 42% in 2005 alone.[6] By 2008, global installed capacity had reached 100 gigawatts (GW).[7] This is equivalent to 75 large coal-fired power plants, or around 1.5 times the total generating capacity of Australia. Despite the fact that wind is the fastest growing renewable energy source, wind still only accounts for 0.5% of global electricity supplies.[8] However, this masks the importance of wind in some countries. In 2006, wind supplied 18% of Denmark's electricity, 9% of Spain's, and 7% of Germany's.[9]

Although the United States accounted for most of the installed capacity in the 1970s and 1980s, Europe has since become the dominant force in the global wind power industry and presently accounts for over 70% of global wind capacity. North America accounts for approximately 20%, with India and China accounting for much of the remainder. In 2006 Germany was the single largest wind power producer, with installed capacity of 21 GW, while Spain (12 GW), United States (12 GW), and Denmark (3 GW) were also large producers.[10] In Europe, wind power has developed into an important industry, with turnover of €6 billion (US$7.5 billion) in 2006.[11]

In the United States, the existence of a 1.9 cent/kWh tax credit has boosted investment in wind power.[12] US capacity increased by 2.5 GW in 2006, the single largest expansion of any country. India and China also have ambitious expansion plans: combined capacity additions of at least 2 GW/year are planned to 2010.[13] In Canada, the Wind Power Production Incentive Scheme is on track to stimulate 6 GW of new wind capacity by 2010.[14] Overall, global installed wind capacity is expected to reach between 160 GW by 2010, making wind the second most important source of renewable electricity after hydropower (see **renewable energy**).[15]

Future supply prospects

In the longer term, installed wind power capacity is expected to expand significantly, though its growth rate will depend on several key factors. These include the availability of economic wind resources, government energy and climate change policy, the extent to which costs can be further reduced and reliability improved, and the importance of social and environmental concerns.

Resource availability

Sufficient economic resources exist to support a substantial expansion in global wind generating capacity. In most major energy consuming countries wind power is capable of meeting at least 5–10% of electricity requirements and significantly more in countries with favorable wind resources and high electricity supply costs. For example, Denmark aims to meet 40% of national electricity supplies from wind by 2030, and it is estimated that the United States could meet up to 20% of national electricity requirements, if it fully developed known economic wind resources (although this is highly unlikely).[16]

Nonetheless, there is, like hydroelectricity, a limit to the economically viable resource base, and this will, in the longer term, place a ceiling on the amount of energy that can be sourced from the wind. Indeed, some analysts believe that Germany will soon approach the wind energy saturation point for its electricity system.[17] Furthermore, some regions have much more limited wind power potential as wind regimes are much less favorable in the tropics and subtropics (within 20° north and south of the equator).

The Policy environment

Energy and climate change policies will be an important determinant of future expansion of wind power. If the major energy consuming countries adopt aggressive greenhouse gas emission reduction policies, such as cost penalties on carbon dioxide emissions (either through a broad-based **emissions trading** or through **carbon taxes**), and energy security remains a concern to policy makers, then wind capacity could expand strongly over the next 10–20 years. The provision of subsidies, tax incentives, renewable energy targets, and green energy certificate schemes have all played an important role in the growth of wind power to date, particularly in the United States and Europe. Uncertainty over how long governments will maintain existing incentive schemes does, however, have an impact on investor confidence. For example, there is no certainty that the present wind power tax credit provisions in the United States will be maintained over the medium term.[18]

Despite its impressive growth over the past decade, the wind power industry still faces a range of technical and cost-competitiveness issues. While construction, maintenance, and operating costs are expected to continue to fall over the next decade, the ability of the turbine manufacturers to supply enough turbines to meet demand remains a constraint, at least in the short to medium term. The rapid growth in turbine demand over the 2004–2007 period strained manufacturing capacity, and during 2006–2007, a shortage of turbines and spare parts placed upward pressure on costs – installed capacity costs actually even rose slightly.[19] There are physical limits on how quickly manufacturing capacity can expand, though India and China have significantly expanded their domestic turbine manufacturing capacity.

Intermittent supply

The intermittent nature of the resource will remain a key constraint. The global average availability factor (the percentage of the time wind provides power to the grid) was 26.5% in 2005.[20] This means, on average, that wind can provide power for only 6–7 hours per day (although availability factors can be as high as 30–35% in favorable locations). This compares with 80–90% for most **fossil fuel** and **nuclear power** facilities.

Improving the wind energy capacity factor is a major focus of wind power producers, and several options for increasing availability exist. The expansion of offshore generating capacity is believed to offer considerable promise for raising capacity factors, possibly to as high as 40%, as offshore wind regimes are generally more consistent than on land. Although offshore installation costs are greater than land-based systems, superior wind consistency, combined with advances in offshore construction technologies, are expected to make offshore installations increasingly cost-competitive in the future. They also face fewer social and environmental constraints than onshore facilities. There are more than ten offshore facilities presently under construction, led by the United Kingdom, The Netherlands, and Denmark, and offshore wind is expected to be a major area of expansion in the coming decades.[21]

One option presently being considered is to establish an offshore grid system that will link wind-generating facilities in the Baltic, North Sea, Irish Sea, and the Atlantic coast – the so-called "Wind Super Grid." This grid would consist of two thousand 5 MW turbines (10 GW capacity) over an area of 3,000 square kilometers.[22] As the wind is generally blowing somewhere within this area, the reliability of wind power supplies would be improved and would possibly enable wind to claim a greater capacity credit. This would, in turn, significantly improve the attractiveness of wind as a source of electricity. Other options include combining wind with pumped hydro storage facilities that can generate power when the wind is not blowing, and other storage options, including the generation of hydrogen that can substitute for petroleum transport fuels (see **renewable energy**). All these options involve higher costs and most are not cost competitive at present.

In countries with large existing hydropower capacity (particularly Canada, Norway, and Sweden), the annual generation from hydro reservoirs is often constrained by limited water inflows. Where wind power is available in the same grid, when wind turbines are generating, less or no water need pass through the hydro turbines and so remains in the reservoir for use at a later time when the wind resource is not available. This is possible since hydro turbines can start almost instantaneously and operate effectively over a wide range of flow rates in contrast to coal or nuclear generators, which operate efficiently only within a narrow range and take several hours or even days to start up. The wind–hydro combination effectively provides a renewable energy "storage" option that helps overcome the problem of wind power's intermittent supply availability.

Social and environmental constraints

The siting of wind power facilities has often met with strong local resistance, mainly due to visual impacts on the landscape and to noise. Wind power facilities often require large areas of land (relative to traditional electricity-generating options), and many opponents have raised this as an issue of concern. However, in practice, wind has generally been able to coexist with other uses, such as farming, with little or no impact on production. While noise was a problem with some early turbine designs, modern turbines are much quieter, and generally quieter than the noise of the wind itself.

Possibly the most significant environmental issue faced by wind has been the impact on bird life. The wind turbines installed in Altamont Pass in California have been found to be responsible for the deaths of 800–1,300 birds of prey (Golden Eagles and Red Tailed Hawks) each year, attracting considerable public criticism from some environmental groups.[23] There is presently insufficient empirical data available to draw definitive conclusions regarding the extent of this problem, and only a few reliable data sets exist. Some suggest that the bird kill issue has been exaggerated and is less significant than other human causes of bird deaths. One recent study estimated that even if wind power were to expand to the point where it supplied 6% of US electricity needs (ten times current capacity), the number of bird deaths would still only account for 9 out of every 10,000 human-related bird deaths.[24] More recent wind turbines have much slower blade speeds (with higher turbine gear ratios), which greatly reduces the risk of bird-strikes. Developers have also used passive design approaches, such as avoiding major migration routes and planting trees attractive to bird species that may be at risk away from the wind towers to further avoid bird strikes.

Nonetheless, bird deaths (or the perception of this risk) could prove to be a barrier for wind energy at ecologically sensitive sites. The social and environmental impacts of wind power needs to be weighed against the social and environmental impacts of other energy alternatives. It should be borne in mind, for example, that fossil fuels are also responsible for a large number of human deaths each year and cause widespread damage to ecosystems.

Future supply potential

While constraints exist, it appears that the contribution of wind power is likely to grow significantly over the coming decades, albeit from a very low base. The International Energy Agency expects that wind power could increase its share of global electricity supply from around 0.5% at present to as much as 2–3% by 2030, and possibly 5–10% by 2050 (depending on which scenario is chosen).[25] By 2050, wind is expected to provide between 20% and 30% of total renewable electricity supplies (3–6% of total global commercial energy supplies). Projections beyond 2050 are speculative, though wind could, conceivably, provide 5–10% of total global energy supplies– a very significant contribution.

Mitigation potential

Wind power is one of the most promising low-emission renewable energy alternatives. The extent to which it can reduce future emission trajectories will depend on the growth in global electricity demand and the type of electricity generation facilities that wind power displaces. For example, if every kilowatt hour of wind power were to substitute for coal-fired electricity, then each kWh would reduce emissions by around 1 kg (or one tonne per megawatt hour). If it were to displace gas-fired electricity, the reduction would be 0.5–0.7 kg/kWh, or zero if it displaced hydro- or solar power. The type of energy wind displaces will be largely determined by the relative supply costs of different sources and the time of day wind energy feeds into the grid.

Out to 2030, the contribution of wind to CO_2 emission reductions is likely to be relatively small on a global basis, perhaps 1–2%. By 2050, modeling suggests that wind could provide as much as 4–8% of the reductions required to stabilize emissions, depending on the scenario.[26]

See also: carbon dioxide (CO_2), fossil fuels, future emissions trends, nuclear power, renewable energy, solar power.

Notes

1 Archer 2004
2 IEA 2006b
3 Ibid.
4 Ibid.
5 De Vries 2006
6 Douglas 2006
7 UNEP 2008
8 Douglas 2006
9 World Wind Energy Association 2007
10 Ibid.
11 CommoditiesNow 2006
12 Popisil 2006
13 Hays 2006
14 Ibid.
15 See World Wind Energy Association 2007 and Douglas 2006
16 Douglas 2006
17 Hays 2006
18 Popisil 2006
19 Ibid.
20 Douglas 2006
21 Hays 2006
22 Cleary 2006
23 Douglas 2006
24 Wood 2006
25 See IEA 2006a and 2006b
26 Based on data in IEA 2006b

Further reading

Douglas 2006; IEA 2006a, 2006b; World Wind Energy Association 2007; Hays 2006.

BIBLIOGRAPHY

AAAS, "Melting Faster," *Science*, 316(5827): 957, 2007.

Abysov, G. *et al.*, American Geophysical Union, "Deciphering Mysteries of Past Climate from Antarctic Ice Cores," *Earth in Space*, 8(3), 1995.

ACIAR, "Impacts of a Warming Arctic," Arctic Climate Impact Assessment Report, Cambridge University Press, 2004.

Agrawala, S., *et al.*, "Climate Change and Natural Resource Management: Key Themes from Case Studies," *Bridge over Troubled Water: Linking Climate Change and Development*, OECD, 2005.

Aguado and Burt, *Understanding Weather and Climate*, 3rd Edition, Prentice Hall, New Jersey, 2004.

Ahern, M., Kovats, R., Wilkinson, P., Few, R., Matthies, F., "Global Health Impacts of Floods: Epidemiological Evidence," *Epidemiology Review*, 27, 36–45, 2005.

Ahmad, N., *A Framework for Estimating Carbon Dioxide Emissions Embodied in International Trade of Goods*, OECD, Paris, 2003.

Air Liquide, *Properties of Gases*, Searchable product database, 2007, at: http://www.airliquide.com/

Alliance for CO_2 solutions, Published on euractiv, June 26, 2007.

Ambus, P., *et al.*, "Sources of Nitrous Oxide Emitted from European Forest Soils," *Biogeosciences*, 3, 135–145, 2006.

American Lung Association, *Trends in Asthma Morbidity and Mortality*, Epidemiology and Statistics Unit, Research Program Services, 2005.

Anderson, J., "Ice Sheet Stability and Sea Level Rise," *Science,* 315, 180–184, 2007.

Andraea, M., "Aerosols before Pollution," *Science*, 315, 50–51, 2007.

Andraea, M., *et al.*, "Strong Present-Day Aerosol Cooling Implies a Hot Future", *Nature*, 435, 1187–1190, 2005.

Archer, C., and Jacobson, M., "Evaluation of Global Windpower," *Journal of Geophysical Research Atmospheres*, 110, DOI 10:1029, 2005.

Archer, D., "How Long Will Global Warming Last?" www.realclimate.org, 2005.

Arnell, N., "Climate Change and Water Resources: A Global Perspective," *Avoiding Dangerous Climate Change*, Cambridge University Press, 2006.

Arrhenius, S., "On the Influence of Carbonic Acid in the Air upon the Temperature of the Ground," *Philosophical Magazine*, Series 5, 41(251): 237–276, 1896.

Association of British Insurers (ABI), "The Financial Risks of Climate Change: Technical Annexes," 2005.

Australian Federal Department of Environment and Heritage, *Climate Change Impacts & Risk Management: A Guide for Business and Government*, 2005.

Australian Greenhouse Office (AGO), "AGO Factors and Methods Workbook – For Use in Australian Greenhouse Emissions Reporting," Canberra, 2006.

AGO, "Climate Change Impacts & Risk Management: A Guide for Business and Government," Canberra, 2006.

Australian Industry Group, "Environmental Sustainability and Industry: Findings of the National Survey of Environmental Sustainable Practices," AIG, September 2007.

Azar, C., and Schneider, S., "Are the Economic Costs of Stabilizing the Atmosphere Prohibitive," *Ecological Economics*, 1–8, 2002.

Baer, P., "The Worth of an Ice Sheet," www.ecoequity.org/docs/WorthofanIcesheet, 2007.

Barker, T., *et al.*, "Avoiding Dangerous Climate Change by Inducing Technological Progress," *Avoiding Dangerous Climate Change*, Cambridge University Press, 361–371, 2006.

Barnett, T., *et al.*, "Potential Impact of a Warming Climate on Water Availability in Snow Dominated Regions," *Nature*, 438, 303–309, 2005.

Barnett, T. D., *et al.*, "Penetration of Human-Induced Warming into the World's Oceans," *Science*, 309, 284–287, 2005.

Behrenfeld, *et al.*, "Biospheric Primary Production during an ENSO Transition," *Science*, 291, 2594–2597, 2001.

Benitez, P., *et al.*, "Global Potential for Carbon Sequestration: Geographical Distribution, Country Risk and Policy Implications," *Ecological Economics*, 3, 572–583, 2005.

Bernes, C., *A Warmer World: The Greenhouse Effect and Climate Change*, Swedish Environmental Agency, 2003.

Betts, R., "Offset of the Potential Carbon Sink from Boreal Reforestation by Decreases in Surface Albedo," *Nature*, 408, 187–190, 2000.

Birkett, W., "Value Creation," University of New South Wales, Unpublished White Paper, 2001.

Bony, S., *et al.*, "How Well Do We Understand and Evaluate Climate Change Feedback Processes?" *Journal of Climate*, 19, 3445–3482, 2006.

Bowers, B., "A Rising Market," *Best's Review*, 75, 2001.

Brahic, C. "Waiting for the Monsoon," *Science*, 313, 608–609, 2006.

Breon, F., "How Do Aerosols Affect Cloudiness and Climate?" *Science*, 313, 623–624, 2006.

British Antarctic Survey Dome C ice core 2006, http://news.bbc.co.uk/1/hi/sci/tech

Brown, C., and Lall, U., "Water and Economic Development: The Role of Interannual Variability and a Framework for Resilience," International Research Institute for Climate Protection, Columbia University, 2006.

Bruvoll, A., and Larsen, B., "Greenhouse Gas Emissions in Norway: Do Carbon Taxes Work?" Statistics Norway Discussion Paper 337, 2002.

Bryden, H., *et al.*, "Slowing of the Atlantic Meridional Overturning Circulation at 25°N," *Nature*, 438, 655–657, 2005.

Buesseler, K., *et al.*, "The Effects of Iron Fertilization on Carbon Sequestration in the Southern Ocean," *Science*, 304, 2004.

Buffett, B., and Archer, D., "Global Inventory of Methane Clathrate: Sensitivity to Changes in the Deep Ocean," *Earth and Planetary Science Letters*, 227, 2004.

Burtraw, D., "Cost Savings sans Allowance Trades: Evaluating the SO_2 Emissions Trading Program to Date," *Resources for the Future*, 1996.

Buseck, P., and Mihaly, P., "Airborne Minerals and Related Aerosol Particles: Effects on Climate and the Environment," *Proceedings of the National Academy of Sciences, USA*, 96(7): 3372–3379, 1999.

Cairncross, S., and Alvarinho, M., "The Mozambique Floods of 2000. Health Impact and Response," in *Flood Hazards and Health. Responding to Present and Future Risks*, ed. Few and Matthies, Earthscan, 111–127, 2006.

Caisse des Dépôts, Climat et projets domestiques CO_2: Pourquoi et comment? December 4, 2006.

Caldeira, K., and Wickett, M., " Anthropogenic Carbon and Ocean pH," *Nature*, 425, 365, 2003.

Cameron, A., and Jones, J., "Photovoltaics: Raising Objectives," *Renewable Energy World*, 9(2), 2006.

Canadell, J., *et al.*, "Saturation of the Terrestrial Carbon Sink," in *Terrestrial Ecosystems in a Changing World*, ed. Canadell, Pataki, and Pitelka, IGBP Series, 2007.

Carbon Dioxide Information Analysis Center, Carbon Dioxide Emissions by Source. Available at http: http://earthtrends.wri.org/pdf_library/data_tables/cli3_2005.pdf, 2005.

Carbon Disclosure Project, www.cdproject.net

CBC News, "Air Canada's Carbon Offset Program Takes Flight," May 28, 2007.

CERES, www.ceres.org/ceres

Challinor, A., *et al.*, "Assessing the Vulnerability of Crop Productivity to Climate Change Thresholds Using an Integrated Crop Model," *Avoiding Dangerous Climate Change*, Cambridge University Press, 2006a.

Challinor, A., *et al.*, "Indian Monsoon: Contribution to Chapter 4 of Stern Review," University of Reading, 2006b.

Charlson, R., *et al.*, "Reshaping the Theory of Cloud Formation," *Science*, 292, 2025–2026, 2001.

Chen, J., *et al.*, "Satellite Gravity Measurements Confirm Accelerating Melting of the Greenland Ice Sheet," *Science Express*, August 2006.

Church, J., and Gregory, J., "Changes in Sea Level," *Climate Change 2001, The Scientific Basis*, Cambridge University Press, 639–693, 2001.

Ciais, P., *et al.*, "Europe Wide Reduction in Primary Productivity Caused by Heat and Drought in 2003," *Nature*, 437, 529–533, 2005.

CleanEdge, "Clean Energy Trends," 2007.

Cleantech Venture Network LLC, Bloomberg Markets, 2006.

Cleary, D., "Wiring up Europe's Coastline," *Science*, 315, 788, 2006.

Cleveland, C., and Townsend, A. R., "Nutrient Additions to Tropical Forests Drive Substantial Soil Carbon Dioxide Losses to the Atmosphere," National Academy of Sciences, 2006.

Coale, H., *et al.*, "A Massive Phytoplankton Bloom Induced by an Ecosystem-Scale Iron Fertilization Experiment in the Equatorial Pacific Ocean," *Nature*, 383, 495–501, 1996.

Cochrane, M., "Fire Science for Rainforests," *Nature*, 421, 913–919, 2003.

Cogger, H. and Zweifel, R., "Reptiles and Amphibians," Southmark Publishers, New York, 1992.

Cohen, J., "Human Population Grows up," *Scientific American*, September, 2005.

CommoditiesNow, Vol. 10/1, 2006.

Conant, R., *et al.*, "Grassland Management and Conversion into Grassland: Effects on Soil Carbon," *Ecological Applications*, 11(2), 2001.

Convention of Biological Diversity, "Report of the Ad Hoc Technical Experts Group on Biological Diversity and Climate Change," CBD, Montreal, 2003.

Cook, *et al.*, "Retreating Glacier Fronts on the Antarctic Peninsula over the past half-century," *Science*, 308, 541–544, 2005.

Cox, P., *et al.*, "Acceleration of Global Warming due to Carbon Cycle Feedback in a Coupled Climate Model," *Nature*, 408, 184–187, 2000.

Cox, P., NOVA PBS WGBH, "Dimming the Sun," http://www.pbs.org/wgbh/nova/sun, 2006.

Crutzen, P., Mosier, A., *et al.*, "N_2O Release from Agro-biofuel Production Negates Global Warming Reduction by Replacing Fossil Fuels," *Atmospheric Chemistry and Physics Discussions*, 7, 11191–11205, 2007.

Curran, M., *et al.*, "Ice Core Evidence for Antarctic Sea Ice Decline since the 1950s," *Science*, 302, 1203–1206, 2003.

Curry, R., *et al.*, "A Change in the Freshwater Balance of the Atlantic Ocean over the Past Four Decades," *Nature*, 426, 826–829, 2003.

Davies, K., "The Inconvenient Truth about the Carbon Offset Industry," *The Guardian*, London, June 16, 2007.

Davis, C. H., *et al.*, "Snowfall Driven Growth in East Antarctic Ice Sheet over the Past Half Century," *Science*, 308, 1898–1901, 2005.

De Vries, E., "Trouble Spots: Gearbox Failures and Design Solutions," *Renewable Energy World*, 9(2), 2006.

Deutch, J., and Moinz, E., "The Nuclear Option," *Scientific American*, 76–83, September 2006.

Dickens, G., *et al.*, "Direct Measurements of In Situ Methane Quantities in a Large Gas-Hydrate Reservoir," *Nature*, 385, 426–428, 1997.

Doney, S., *et al.*, 'Marine Biogeochemical Modelling: Recent Advances and Future Challenges," *Oceanography*, 14(4), 93–107, 2001.

Douglas, E., "Gone with the Wind," *New Scientist*, 37–39, July 8, 2006.

Ducklow, H., *et al.*, "Upper Ocean Carbon Export and the Biological Pump," *Oceanography*, 14, 50–58, 2001.

Dukes, J., "Burning Buried Sunshine: Human Consumption of Ancient Solar Energy," *Climate Change*, 61, 31–44, 2003.

Ebi, K., "Floods and Human Health," in *Climate Change and Adaptation Strategies for Human Health*, ed. Menne, B. and Ebi, K., World Health Organization, 2006.

Eckhart, M., and Wierich, T., "Moving Renewable Energy into the Mainstream," World Power, 2006.

Ekins, P., and Barker, T., "Carbon Taxes and Carbon Emissions Trading," *Journal of Economic Surveys*, 15(3), 325–376, 2001.

Elgin, B., "Another Inconvenient Truth," *Business Week*, March 26, 2007.

Ellerman, A., and Pontero, J., "The Efficiency and Robustness of Allowance Banking in the US Acid Rain Program," Working Paper 0505, MIT Center for Energy and Environmental Policy Research, 2005.

Emanuel, K., "Increasing Destructiveness of Tropical Cyclones over the Past 30 Years," *Nature*, 426, 686–688, 2005.

Energy Information Administration (EIA), "International Energy Outlook 2005," Washington, 2005.

Enkvist, P., *et al.*, "A cost curve for greenhouse gas reductions," *McKinsey Quarterly*, 1, 35–45, 2007.

Epstein, P., "Climate Change and Health," *New England Journal of Medicine*, 2005.

Epstein, P., and Mills, E., "Climate Change Futures: Health, Ecological and Economic Dimensions," Harvard Medical School, 2005.

Epstein, R., *et al.*, "Biological and Physical Signs of Climate Change: Focus on Mosquito Borne Disease," *Bulletin of the American Meteorological Society*, 78, 409–417, 1998.

EuroActive, "ExxonMobil's Top Executives on Climate Change Policy," London, February 14, 2007.

European Parliament, Council. "Directive 2004/101/EC of the European Parliament and of the Council of 27 October 2004 Amending Directive 2003/87/EC Establishing a Scheme for Greenhouse Gas Emission Allowance Trading within the Community, in Respect of the Kyoto Protocol's Project Mechanisms," 27 October, 2004. Available at: http://ec.europa.eu/environment/climat/emission/linking_en.htm

European Union Directive 2006/40/ May 2006.

European Space Agency, "News: Satellites Witness Lowest Arctic Ice Coverage in History," September 14, 2007.

Faber, M., "Tomorrow's Gold," CLSA, 2002.

Feely, R., *et al.*, "Uptake and Storage of Carbon Dioxide in the Ocean: The Global CO_2 Survey," *Oceanography*, 14(4), 2001.

Financial Times, February 14, 2007.

Fiore, A., *et al.*, "Impact of Meteorology and Emissions on Methane Trends, 1990–2004," *Geophysical Research Letters*, 33, L12809, 2006.

Flanner, S., *et al.*, "Present-Day Climate Forcing and Response from Black Carbon in Snow," *Journal of Geophysical Resources*, 112, 2007.

Flannery, T., "The Weather Makers: The History and Future Impact of Climate Change," Text Publishing, Melbourne, 2005.

Fleury, M., *et al.*, "A Time Series Analysis of the Relationship of Ambient Temperature and Common Bacterial Enteric Infections in Two Canadian Provinces," *International Journal of Biometeorology*, 50, 385–391, 2006.

Freeman, P., and Scott, K., "Comparative Analysis of Large Scale Catastrophic Compensation Schemes," *OECD Catastrophic Risks and Insurance*, 8, 2005.

Folkestad, T., *et al.*, "Evidence and Implications of Dangerous Climate Change in the Arctic," *Avoiding Dangerous Climate Change*, Cambridge University Press, 2006.

Food and Agriculture Organisation of the United Nations (FAO), Global Forest Resource Assessment, 2001.

FAO, Global Forest Resource Assessment, 2006.

Friedlingstein, P., *et al.*, "Climate–Carbon Cycle Feedback Analysis: Results from the C4MIP Model Inter-comparison," *Journal of Climate*, 19, 3337–3353, 2006.

Frondel, M., "Energy Conservation, the Rebound Effect and Future Energy and Transport Technologies: An Introduction to Energy Conservation and the Rebound Effect," *International Journal of Energy Technology and Policy*, 2(3), 2004.

Fudeyasu, H., *et al.*, "Impact of ENSO on Landfall Characteristics of Tropical Cyclones over the Western North Pacific during the Summer Monsoon Season," *Geophysical Research Letters*, 33, 2006.

Fung, N., "China as a Global Energy Player," *World Power*, 37–40, 2006.

Gauss, M., *et al.*, "Radiative Forcing since Preindustrial Times due to Ozone Change in the Troposphere and the Lower Stratosphere," *Atmospheric Chemistry and Physics*, 6, 575–599, 2006.

Geresdi, *et al.*, "The Effect of Chemical Composition and Size Distribution of Aerosol Particles on Droplet Formation and Albedo of Stratocumulus Clouds," *Atmospheric Environment*, 40, 1845–1855, 2006.

Glantz, M. H., *Currents of Change: Impacts of El Niño and La Niña on Climate and Society*, Cambridge: Cambridge University Press, 2001.

Global Commons Institute, www.gci.org.uk

Global Environment Facility, "Status Report on the Climate Change Funds," 2008.

GLOBE-Net, "Searching for True Carbon Offsets," *GLOBE-Net News*, July 4, 2007.

Good, *et al.*, "Influence of Anthropogenic Aerosol on Cloud Optical Depth and Albedo Shown by Satellite Measurements and Climate Transport Modeling," *Proceedings of the National Academy of Sciences, USA*, 99(4): 1784–1789, 2002.

Good-Kiehl, J., "Solving the Aerosol Puzzle," *Science*, 283, 1273–1275, 1999.

Goswami, B., *et al.*, "Increasing Trend of Extreme Rain Events Over India in a Warming Environment," *Science*, 314, 1442–1445, 2006.

Greene, C., and Pershing, A., "Climate Drives Sea Change," *Nature*, 315, 1084–1085, 2007.

Greenhouse Challenge, www.greenhouse.gov.au/challenge

Gregory, J., *et al.*, " Threatened Loss of the Greenland Ice Sheet," *Nature*, 428, 616, 2004.

Grey, D., and Sadoff, C., "Water and Development," IV World Water Forum, Mexico City, World Bank, 2006.

Grieg-Gran, M., "The Cost of Avoiding Deforestation," International Institute of Environment and Development, 2006.

Gross, L., "As the Antarctic Pack Ice Recedes a Fragile Ecosystem Hangs in the Balance," *Plos Biology*, 3, 557–561, 2005.

Grubb, M., and Nuehoff, K., "Allocation and Competitiveness in the EU Emissions Trading Scheme: Policy Overview," *Cimate Policy*, 6, 7–30, 2006.

Gruber, N., *et al.*, "The Vulnerability of the Carbon Cycle in the 21st Century and Assessment of Global Climate–Human Interactions," *Global Carbon Cycle – Integrating Human, Climate and the Natural World*, 45–76, Island Press, 2004.

Guardian, August 11, 2005.

Gurney, K., *et al.*, "Towards Robust Regional Estimates of CO_2 Sources and Sinks Using Atmospheric Transport Models," *Science*, 415, 626–630, 2002.

Hadley Centre Climate Report, "Climate Change: Observations and Prediction," UK Met Office, 2003.

Hales, S., *et al.*, "Potential Effect of Population and Climate Changes on Global Distribution of Dengue Fever: An Empirical Model," *Lancet*, 360, 830–834, 2002.

Hansell, D., and Carlson, C., "Marine Dissolved Organic Matter and the Carbon Cycle," *Oceanography*, 14(4), 43–49, 2001.

Hansen, J., and Nazarenko, M., "Soot Climate Forcing via Snow and Ice Albedos," *Proceedings of the National Academy of Sciences*, 2003.

Hansen, J., *et al.*, "Global Temperature Change," *Proceedings of the National Academy*, 103, 2006.

Hardy, J., *Climate Change: Causes, Effects, and Solutions*, John Wiley & Sons, 2003.

Hare, B., "Relationship between Increases in Global Mean Temperature and Impacts on Ecosystems, Food Production, Water and Socio-Economic Systems," *Avoiding Dangerous Climate Change*, Cambridge University Press, 2006.

Harnisch, J., *et al.*, "Report #44. Primary Aluminium production: Climate policy, emissions and costs," Massachusetts Institute of Technology, Joint Program on the Science and Policy of Global Change, 1998.

Hart, S., and Milstein, M., "Creating Sustainable Value," *Academy of Management Executive*, 17(2), 56–59, 2003.

Hassol, S. J., and Corell, R. W., "Arctic Climate Impact Assessment," *Avoiding Dangerous Climate Change*, Cambridge University Press, 2006.

Hayhoe, K., *et al.*, "Regional Assessment of Climate Impacts on California under Alternative Emission Scenarios," *Avoiding Dangerous Climate Change*, 2006.

Hays, J. *et al.*, "Variations in the Earth's Orbit: Pacemaker of the Ice Ages," *Science*, 1976.

Hays, K., "Global Wind Soars to New Heights," *World Power*, 137–139, 2006.

Heath, J., *et al.*, "Rising Atmopheric CO_2 Reduces Sequestration of Root Derived Soil Carbon," *Science*, 309, 1711–1712, 2005.

Hendersen, C., "Paradise Lost," *New Scientist*, August, 2006.

Hepburn, C., "Regulating by Prices, Quantities or Both: An Update and an Overview," *Oxford Review of Economic Policy*, 22(2): 226–247, 2006.

Hoeppe, P., and Gurenko, E., "Scientific and Economic Rationales for Innovative Climate Insurance Solutions," *Climate Policy*, 6, 607–620, 2006.

Holden, C., "Going Under," Editorial, AAAS, *Science*, 316(5821): 27, 2007.

Holland, M., *et al.*, "Future Abrupt Reductions in Summer Arctic Sea Ice," *Geophysical Research Letters*, 33, 23, 2006.

Holper, "The Greenhouse Effect Information Sheet," Australian Commonwealth Scientific and Industrial Research Organisation (CSIRO), Marine and Atmospheric Research, 2001.

Holtsmark, B., "Are Global per Capita CO_2 emissions likely to remain stable?" *Energy and Environment*, 17(2), 2006.

Hooijer, A., *et al.*, "PEAT-CO_2, Assessment of CO_2 emissions from drained peatlands in SE Asia," Delft Hydraulics Report Q3943, 2006.

Hooper, R., "Something Nasty in the Air," *New Scientist*, January 21, 2006.

Hornbach, M., *et al.*, "Critically Pressured Free-Gas Reservoirs below Gas-Hydrate Provinces," *Nature*, 427, 142–144, 2004.

Houghton, R., "Revised Estimates of Annual Net Flux of Carbon to the Atmosphere from Changes in Land Use and Land Management 1850–2000," *Tellus Series B-Chemical and Physical Meteorology*, 2003.

Houghton, R., "Tropical Deforestation as a Source of Greenhouse Gas Emissions," in *Tropical Deforestation and Climate Change*, ed. Mountinho, P. and Schwartzman, S., IPAM, Instituto de Perquisa Ambiental de Amazônia; Belém, Pará (Brazil). Environmental Defense, Washington, DC (USA), pp. 13–21, 2005.

Hoyos, C., *et al.*, "Deconvolution of the Factors Contributing to the Increase in Global Hurricane Intensity," *Science*, 311, 94–97, 2006.

Hubbard Preston, H., "Hopes Are High, as are the Hurdles, for Alternative Fuel," *The International Herald Tribune*, March 15, 2008.

Hughes, T., *et al.*, "Climate Change, Human Impacts and the Resilience of Coral Reefs," *Science*, 301, 929–933, 2003.

Intergovernmental Panel on Climate Change (IPCC), First Assessment Report, 1990.

IPCC, Second Assessment Report, 1996.

IPCC, "Guidelines for National Greenhouse Gas Inventories," Tokyo, 1997.

IPCC, "Special Report: Land use, Land-use change and Forestry," IPCC, 2000a.

IPCC, "Special Report on Emission Scenarios," Intergovernmental Panel on Climate Change, 2000b.

IPCC, Third Assessment Report, 2001.

IPCC, "Climate Change and Biodiversity," IPCC Technical Paper V, IPCC, 2002.

IPCC, "Special Report: Carbon Dioxide Capture and Storage," IPCC, 2005.

IPCC, "Special Report: Safeguarding the Ozone Layer and the Global Climate System: Issues Related to Hydrofluorocarbons and Perfluorocarbons," IPCC Secretariat, Geneva, 2006a.

IPCC, "Guidelines for National Greenhouse Gas Inventories," Tokyo, 2006b.

IPCC Fourth Assessment Report, 2007.

IPCC, http://www.ipcc.ch/pub/reports.htm

IPCC, http://www.ipcc.ch/pub/techrep.htm

International Development Association, "IDA and Climate Change: Making Climate Action Work for Development," Sustainable Development Network, 2007.

International Energy Agency (IEA), "Things That Go Blip in the Night," Paris, 2001.

IEA, "Cool Appliances: Policy Strategies for Energy Efficient Homes," Paris, 2003.

IEA, "Prospects for CO_2 Capture and Storage, Paris, 2004.

IEA, "2006 World Energy Outlook," Paris, 2006a.

IEA, "Energy Technology Perspectives," Paris, 2006b.

IEA, "Light's Labour's Lost," Paris 2006c.

International Federation of the Red Cross and Red Crescent Societies, World Disasters Report 2002, IFRC, Geneva, 2002.

International Institute for Sustainable Development (IISD), "Realising the Development Dividend: Making the CDM Work for Developing Countries," 2006.

IISD, "10th Session of Working Group I of the Intergovernmental Panel on Climate Change," 12(319), 2007.

International Monetary Fund, "Fund Assistance for Countries Facing External Shocks," IMF, Washington, 2003.

INCR (Investor Network on Climate Risk), http://www.incr.com/

International Organisation for Standardisation (ISO), "ISO 14064-1 Part 1 Specification with Guidance at the Organisation Level for Quantification, Monitoring and Reporting of Greenhouse Gas Emission Reductions or Removal Enhancement," 2006.

International Petroleum Industry Environmental Conservation Association (IPIECA), "Petroleum Industry Guidelines for Reporting Greenhouse Gas Emissions," 2003.

Isherwood, G., "Biofuels: Curbing Carbon and Securing Sustainability?" *Commodities Now*, 45–48, March 2006.

Izrael, Y., and Semenov, S., "Critical Levels of Greenhouse Gases, Stabilization Scenarios, and Implications for Global Decisions," *Avoiding Dangerous Climate Change*, Cambridge University Press, 2006.

Jones, C., *et al.*, "Impact of Climate–Carbon Feedback Effects on Emission Scenarios to Achieve Stabilisation," *Avoiding Dangerous Climate Change*, Cambridge University Press, 2006.

Jones, P., *et al.*, "The Evolution of Climate over the Last Millennium," *Science*, 292, 662–667, 2001.

Joskow, P., and Schmalanese, R., "The Political Economy of Market-Based Environmental Policy: The US Acid Rain Program," *Journal of Law and Economics*, 41, 37–83, 1998.

Kallbekkan, S., and Rive, N., "Why Delaying Emission Cuts is a Gamble," *Avoiding Dangerous Climate Change*, Cambridge University Press, 311–315, 2006.

Kammen, D., "The Rise of Renewable Energy," *Scientific American*, 84–93, September 2006.

Kane, R., "El Niño Effects on Rainfall in South America: Comparison with Rainfalls in India and Other Parts of the World," *Advances in Geosciences*, 6, 35–41, 2006.

Kasting, J., "The Carbon Cycle, Climate, and the Long-Term Effects of Fossil Fuel Burning," *Consequences: The Nature and Implication of Environmental Change,* 4(1), 1998.

Katz, M., *et al.*, "The Source and Fate of Massive Carbon Input during the Latest Paleocene Thermal Maximum," *Science*, 286, 1531–1533, 1999.

Kaufman, Y., and Koren, I., "Smoke and Pollution Aerosol Effect on Cloud Cover," *Science*, 313, 655–658, 2006.

Keen, R., "Infrared Radiation and the Greenhouse Effect," Department of Atmospheric and Oceanic Studies, University of Colorado, Boulder, 2006.

Keppler, F., *et al.*, "Methane Emissions from Terrestrial Plants under Aerobic Conditions," *Nature*, 439, 2006, 187–191.

Kerr, R., "Could Mother Nature Give the Arctic a Reprieve?" *Science*, 315, 36, 2007.

Kiehl, J. and Trenberth, K., "Earth's Annual Global Mean Energy Budget," *Bulletin of the American Meteorological Society*, 78, 197–208, 1997.

Klein Goldewijk, C., "Estimating Global Land Use Change over the Past 300 Years," *Global Biogeochemistry*, 15, 415–434, 2001.

Kloor, K., "Restoration Ecology: Returning America's Forests to Their 'Natural' Roots," *Science*, 287, 573–575, 2000.

Kohler, J., "Making Waves: Simulating Long-Run Technical Change," in *Technical Knowledge and the Firm: Implications for Strategy and Industrial Change*, Edward Elgar, 404–426, 2005.

Kovats, R., and Jendritzky, G., "Heat Waves and Human Health," in *Climate Change and Adaptation Strategies for Human Health*, ed. Menne, B. and Ebi, K., World Health Organization, 2006.

Krieger, K., "Sea Cucumber Explosion," *Science Now*, 318, 2004.

Kuik, O., *et al.*, "Methodological Aspects of Recent Climate Change Damage Cost Studies," Hamburg University and Centre for Marine and Atmospheric Science, Working Paper FNU-122, 2006.

Kumar, K., *et al.*, "Unraveling the Mystery of Indian Monsoon Failure During El Nino," *Science*, 314, 115–119, 2006.

Kvenvolden, K., "A Primer on the Geological Occurrence of Gas Hydrates," The London Geological Society, Special Publication 137, 1998.

Labeyrie, L., "Glacial Climate Instability," *Science*, New Series, 290(5498): 1905–1907, 2000.

Lanchbery, J., "Climate Change Induced Ecosystem Loss and Its Implications for Greenhouse Gas Concentration Stabilisation," *Avoiding Dangerous Climate Change*, Cambridge University Press, 2006.

Landsea, C., *et al.*, "Can We Detect Trends in Extreme Tropical Cyclones?" *Science*, 313, 452–454, 2006.

Leemans, R., and Eickhout, B., "Another Reason for Concern: Regional and Local Impacts on Ecosystems for Different Levels of Climate Change," *Global Environmental Change*, 14, 210–228, 2004.

Lenton, T., *et al.*, "Climate Change on a Millennial Timescale", Tyndall Centre for Climate Research, Technical Report 41, 2006.

Le Quere, C., *et al.*, "Saturation of the Southern Ocean CO_2 Sink Due to Recent Climate Change," *Science*, 316, 1735–1738, 2007.

Levitus, S., *et al.*, "Warming of the World Ocean," *Science*, 287, 2225–2229, 2000.

Levitus, S., *et al.*, "Warming of the World's Oceans 1955–2003," *Geophysical Research Letters*, 32, 2005.

Lewis, N., "Toward Cost Effective Solar Energy Use," *Science*, 315, 798–801, 2006.

Lewis, S., *et al.*, "Tropical Forests and Atmospheric Carbon Dioxide: Current Conditions and Future Scenarios," *Avoiding Dangerous Climate Change*, Cambridge University Press, 2006.

Liepert, B. G., "Recent Changes in Solar Irradiation under Cloudy Conditions in Germany," *International Journal of Climatology*, 17, 1581–1593, 1997.

Liepert, B. G., "Observed Reductions in Surface Solar Radiation in the United States and Worldwide from 1961 to 1990," *Geophysical Research Letters*, 2002.

Lindzen, R., "Global Warming: The Origin and Nature of the Alleged Scientific Consensus," *Regulation*, 15(2), 1992.

Lindzen, R., "Climate Fear: Global Alarmists Intimidate Dissenting Scientists into Silence," *Wall Street Journal*, April 12, 2006.

Linnerooth-Bayer, J., and Mechler, R., "Insurance for Assisting Adaptation to Climate Change in Developing Countries: A Proposed Strategy," *Climate Policy*, 6, 621–636, 2006.

Long, S., *et al.*, "Food for Thought: Lower than Expected Crop Yield Stimulation with Rising CO_2 Concentrations," *Science*, 312, 1918–1921, 2006.

Lou, Y., *et al.*, "Progressive Nitrogen Limitation of Ecosystem Responses to Rising Atmospheric Carbon Dioxide," *Bioscience*, 54, 2004.

Lovins, A., "More Profit with Less Carbon," *Scientific American*, 74–82, September 2005a.

Lovins, A., "Nuclear Power: Economics and Climate Protection Potential," Rocky Mountains Institute, 2005b.

Lowe, J., *et al.*, "The Role of Sea Level Rise and the Greenland Ice Sheet in Dangerous Climate Change: Implications for the Stabilisation of Climate," *Avoiding Dangerous Climate Change*, Cambridge University Press, 29–35, 2006.

Luthcke, S., *et al.*, "Recent Greenland Ice Mass Loss by Drainage System from Satellite Gravity Observations," *Science*, 24, 1286–1289, 2006.

Lynas, M., *Six Degrees: Our Future on a Hotter Planet*, Harper Collins, 2007.

McCarthy, M., "Disaster at Sea: Global Warming Hits UK Birds," *The Independent*, July 30, 2004.

MacDonald, G., "Role of Methane Clathrates in Past and Future Climate," *Climate Change*, 16, 247–281, 1990.

McDonald, G., *et al.*, "Rapid Early Development of Circumartic Peatlands and Atmospheric CH_4 and CO_2 Variations," *Science*, 314, 285–288, 2006.

McGraw, T., *Prophet of Innovation*, Belknap Harvard, 2007.

McKinsey and Company, "Taking Stock of the World's Capital Markets," 2005.

McMichael, A., *et al.*, "Human Health and Climate Change in Oceania: Risk Assessment 2002," Department of Health and Ageing, Canberra, Commonwealth of Australia, 2003.

McMichael, A., "Comparative Quantification of Health Risks: Global and Regional Burden of Disease due to Selected Major Risk Factors," vol. 2, ed. Ezzati, M., Lopez, A., Rodgers, A., and Murray, C., World Health Organization, Geneva, 1543–1649, 2004.

McPhaden, M., *et al.*, "ENSO as an Integrating Concept in Earth Science," *Science*, 314, 1740–1745, 2006.

Magoya, E., *et al.*, "Young Organic Matter as a Source of Carbon Dioxide Outgassing from Amazonian Rivers," *Nature*, 436, 538–541, 2005.

Malthe-Sorensson, A., *et al.*, "Release of Methane from a Volcanic Basin as a Mechanism for Initial Eocene Global Warming," *Nature*, 420, 2004.

Manne, A. and Richels, R., "The impact of learning by doing on the timing and costs of CO_2 abatement', *Energy Economics*, 26, 603–619, 2004.

Manuel, J., "In Katrina's wake," *Environmental Health Perspectives*, 114, A32–A39, 2006.

Marsh, N., and Svensmark, H., "Cosmic Rays, Clouds and Climate," *Space Science Review*, 94, 215–230, 2000.

Mastrandrea, M., and Schneider, S., "Probabilistic Integrated Assessment of Dangerous Climate Change," *Science*, 304, 571–575, 2004.

Matear, R., and Hirst, A., "Climate Change Feedback on the Future Oceanic Uptake," *Tellus*, 51B, 722–733, 2001.

Meehl, G., *et al.*, "How Much More Global Warming and Sea Level Rise," *Science*, 307, 1769–1771, 2005

Meinhausen, M., "What Does a 2°C Target Mean for Greenhouse Gas Concentrations?: A Brief Analysis Based on Multi-Gas Emission Pathways and Several Climate Sensitivity Uncertainty Estimates?" *Avoiding Dangerous Climate Change*, Cambridge University Press, 2006.

Ménot, B., *et al.*, "Early Reactivation of European Rivers during the Last Deglaciation," *Science*, 313, 1623–1625, 2006.

Merton, R., "Financial Innovation and Economic Performance," *Journal of Applied Corporate Finance*, 4, 12–22, 1992.

Metz, B., and van Vuuren, D., "How, and at What Costs, can Low-Level Stabilisation Targets be Achieved?" *Avoiding Dangerous Climate Change*, 337–345, Cambridge University Press, 2006,

Meyer, S., "Weeds Shall Inherit the Earth," *New Scientist*, September 2006.

Michaelowa, A., and Ourohit, P., "Additionality Determination of Indian CDM Projects," University of Zurich, 2007.

Monaghan, D., *et al.*, "Insignificant Change in Antarctic Snowfall since the International Geophysical Year," *Science*, 313, 827–831, 2006.

Montzka, S., "Scientific Assessment of Ozone Depletion: 2002," World Meteorological Organisation, Geneva, 2002.

Morton, O., "Solar Energy: A New Day Dawning," *Nature*, 443, 19–22, 2006.

Muir-Wood, R., *et al.*, "The Search for Trends in a Global Catalogue of Normalized Weather-Related Catastrophe Losses," Proceedings of the Climate Change and Disaster Losses Workshop, Munich Re, 2006.

Murphy, J., *et al.*, "Quantification of Modelling Uncertainties in a Large Ensemble of Climate Change Simulations," *Nature*, 430, 768–772, 2004.

NASA, Ocean Surface Topography from Space, http://sealevel.jpl.nasa.gov/

Nature News, 435, 2005.

NewScientist, "It is all over for Nuclear Power," April 12, 2006.

Neuhoff, K., Keats, K., and Sato, M., "Allocation, Incentives and Distortions: The Impact of the EU-ETS Emission Allowance Allocations to the Electricity Sector," *Climate Policy*, 6, 71–89, 2006.

Niasse, M., "Climate Induced Water Conflict Issues in West Africa: Recognising and Coping With Increasing Climate Impacts on Shared Watercourses," www.gechs.org/activities/holmen/NiassePDF, 2005.

Nicholls, R., and Lowe, J., "Climate Stabilisation and Impacts of Sea Level Rise," *Avoiding Dangerous Climate Change*, Cambridge University Press, 2006.

Nicholls, R., *et al.*, "Increasing Flood Risk and Wetland Losses due to Sea Level Rise: Regional and Global Analyses," *Global Environmental Change – Human and Policy Dimensions*, 9, 69–87, 1999.

Nisbet, E., "Climate Change and Methane," *Nature*, 347, 1990.

Nisbet, E., "Sources of Atmospheric CH_4 in Early Postglacial Time," *Journal of Geophysical Research*, 97D12, 1992.

NOAA, US National Oceanic and Atmospheric Administration 2007, http://www.esrl.noaa.gov/gmd/aggi/

NOAA, Carbon Dioxide Information Analysis Center (CDIAC), http://cdiac.ornl.gov/; Global Monitoring Division, http://www.cmdl.noaa.gov/ Dec 2007

Nordhaus, W., "The Economics of Hurricanes in the United States," http://nordhaus.econ.yale.edu/hurricanes.pdf, 2006

Nordhaus, W., "The Stern Review and the Economics of Climate Change: Opposite Ends of the Globe," www.econ.yale.edu/nordhaus/homepage/SternReview, 2007

Nordhaus, W., and Boyer, J., "Warming the World: The Economics of the Greenhouse Effect," MIT Press, 2000.

Normile, D., "Eureka Moment Puts Sliced Solar Cells on Track," *Science*, 315, 785, 2007.

NOVA PBS WGBH, "Dimming the Sun," http://www.pbs.org/wgbh/nova/sun, 2006.

Nowak, R., "Bring on the Second Generation of Biofuels," *NewScientist*, 12, February 2, 2008.

Nyong, A., and Niang-Diop, I., "Impacts of Climate Change in the Tropics: The African Experience," *Avoiding Dangerous Climate Change*, Cambridge University Press, 2006.

Nuclear Energy Agency, "Projected Costs of Nuclear Electricity, 2005 Update," OECD, 2005.

Oberthur, S., and Ott, H., *The Kyoto Protocol: International Climate Policy for the 21st Century*, Springer-Verlag, 1999.

Ogden, J., "High Hopes for Hydrogen," *Scientific American*, 94–101, September 2006.

Olah, G., *et al.*, *Beyond Oil and Gas: The Methanol Economy*, Wiley, 2006.

Oldenborgh, G., *et al.*, "En Nino in a Changing Climate: A Multi-model Study," *Ocean Science, Discussions*, 2, 267–298, 2005.

Olivier, J. and Bakker, J., "Historical Global Emission Trends of the Kyoto Gases HFCs, PFCs and SF_6," Presentation at IPCC expert meeting, 2003.

Orr, J., *et al.*, "Anthropogenic Ocean Acidification over the Twenty-First Century and Its Impact on Calcifying Organisms," *Nature*, 437, 681–686, 2005.

Ourbak, T., *et al.*, "ENSO and interdecadal climate variability over the last century documented by geochemical records of two coral cores from the South West Pacific," *Advances in Geosciences*, 6, 23–27, 2006.

Oxford Dictionary of Finance, Oxford Press, 1993.

Pachauri, R., Presentation to the Exeter Conference, February 2005.

Pagani, M., *et al.*, "An Ancient Carbon Mystery," *Science*, 314, 1156–1157, 2006.

Page, E., *et al.*, "The Amount of Carbon Released from Peat and Forest Fires in Indonesia in 1997," *Nature*, 61–65, 420, 2002.

Pantic, N., *New English Translation, Canon of Insolation and the Ice Age Problem*, Alven Global, 1998.

Parmeson, C., and Yohe, G., "A Globally Coherent Fingerprint of Climate Change across Natural Ecosystems," *Nature*, 421, 37–42, 2003.

Parry, M., *et al.*, "Effects of Climate Change on Global Food Production under SRES Emissions and Socio-economic Scenarios," *Global Environmental Change*, 14, 53–67, 2004.

Pascala, R., "Stabilisation Wedges: An Elaboration of the Concept," in *Avoiding Dangerous Climate Change*, Cambridge University Press, 2006.

Pascala, S., and Socolow, R., "Stabilisation Wedges: Solving the Climate Problem for the Next 50 Years with Current Technologies," Carbon Mitigation Initiative, Princeton University, www.princeton.edu, 2004.

Pearce, F., "Fuels Gold," *New Scientist*, 36–41, September 23, 2006.

Pearce, F., *The Last Generation: How Nature Will Take Revenge for Man-made Climate Change*, Eden Books, 2006.

Peperzak, L., "Future Increase in Harmful Algal Blooms in the North Sea due to Climate Change," *Water, Science and Technology*, 51, 2005.

Peterson, B., *et al.*, "Trajectory Shifts in the Arctic and Sub-arctic Freshwater Cycle," *Science*, 313, 1061–1066, 2006.

Pew Center on Global Climate Change, *Getting Ahead of the Curve: Corporate Strategies that Address Climate Change*, Washington DC, 2006.

Pew Center on Global Climate Change, "Land Use & Global Climate Change: Forests, Land Management, and the Kyoto Protocol," Washington DC, 2007.

Phoenix, G., *et al.*, "Atmospheric nitrogen deposition in world biodiversity hotspots: the need for a greater global perspective in assessing N deposition impacts," *Global Change Biology*, 12, 470–476, 2006.

Pidwirny, M., *Fundamentals of Physical Geography*, 2nd Edition, http://www.physicalgeography.net/, UBC Unit of Biology & Physical Geography, 2006.

Popisil, R., "US Wind Finance," *Renewable Energy World*, 9(2), 2006.

Portner, H., and Knust, R., "Climate Change Affects Marine Fishes Through the Oxygen Limitation of Thermal Tolerance," *Science*, 315, 95–97, 2007.

Prentice, I., *et al.*, "The Carbon Cycle and Atmospheric Carbon dioxide, in Climate Change 2001: The Scientific Basis," Intergovernmental Panel on Climate Change Third Assessment Report Vol. 1, Cambridge University Press, 2001.

Rahmstorf, S., "A Semi-empirical Approach to Projecting Future Sea Level Rise," *Science*, 315, 2007.

Ramaswamy, R., *et al.*, "A model study of the effect of Pinatubo volcanic aerosols on the stratospheric temperatures," in *Frontiers of Climate Modelling*, Cambridge University Press, 152–178, 2006.

Randerson, J., *et al.*, "The Impact of Boreal Forest Fire on Climate Warming," *Science*, 314, 2006.

Raoux, S., "Implementing Technologies for Reducing PFC Emissions," *Solid State Technology*, January, 2007.

Rapley, C., "The Antarctic Ice Sheet and Sea Level Rise," *Avoiding Dangerous Climate Change*, Cambridge University Press, 2006.

Raupach, M., and Canadell, J., "Observing a Vulnerable Carbon Cycle," in *Observing the Continental Scale Greenhouse Gas Balance of Europe*, ed. Dolman, A. J., Valentini, R., and Freibauer, A., Springer Books, 2007.

Raupach, M., *et al.*, "Global and Regional Drivers of CO_2 Emissions," *Proceedings of the National Academy of Sciences*, 2007.

Renewables 2005, Global Status Report, Renewable Energy Policy Network for the 21st Century, Worldwatch Institute, 2005.

Reo Research Report, "In the Front Line: The Insurance Industry's Response to Climate Change," F&C Investments. London, 2007.

Reusch, W., "Introduction to Spectroscopy," Michigan State University, Department of Chemistry, 2004.

Ricklefs, R., *The Economy of Nature*, 3rd Edition. Freeman and Co., New York, 1993.

Righelato, R., and Spracklen, D., "Carbon Mitigation by Biofuels or by Saving and Restoring Forests?" *Science*, 317, 902, 2007.

Rignot, E., and Kanagaratnam, P., "Changes in the Velocity and Structure of the Greenland Ice Sheet," *Science*, 311, 986–990, 2006.

Rignot, E., *et al.*, "Contribution of the Patagonia Icefields of South America to Sea Level Rise," *Science*, 302, 434–437, 2003.

Roderick, M., and Farquhar, G., "The Cause of Decreased Pan Evaporation over the Past 50 years," *Science*, 298, 1410–1411, 2002.

Rogers, J., "Strategy, Value and Risk – The Real Options Approach," Palgrave, 2002.

Roine, K., and Tvinnereim, E., "The Global Carbon Market 2007," Greenhouse Gas Market 2007, International Emissions Trading Association.

Rosenzweig, C., and Hillel, D., *Climate Change and the Global Harvest: The Potential Impacts of the Greenhouse Effect on Agriculture*, Oxford University Press, 1998.

Rosenzweig, C., *et al.* "Climate Change and Extreme Weather Events: Implications for Food Production, Plant Diseases and Pests," *Global Change and Human Health*, 2, 2001.

Salzmann, U., and Hoelzmann, P., "The Dahomey Gap: An Abrupt Climatically Induced Rainforest Fragmentation in West Africa during the Late Holocene," *Halocene*, 15(2), 190–199, 2005.

Sathaye, J., *et al.*, "GHG Mitigation Potential, Costs and Benefits in Global Forests: A Dynamic Partial Equilibrium Approach," *The Energy Journal*, Greenhouse Gas Mitigation and Climate Policy Special Issue, 127–161, 2006.

Schaefer, D., *et al.*, "Estimating Future Emissions and Potential Reductions of HFCs, PFCs and SF_6," *The Energy Journal*, Climate Change Special Issue, 2007.

Scheehle, E., and Kruger, D., "Global Anthropogenic Methane and Nitrous Oxide Emissions," *The Energy Journal*, 34–62, 2007.

Schellnhube, H. *et al.* (eds), *Avoiding Dangerous Climate Change*, Cambridge University Press, 2006.

Schiermeier, Q., "The Power of Katrina," *Nature*, 437, 174, 2005.

Schipper, E., "Climate Change Adaptation and Development: Exploring the Linkages," Tyndall Centre Working Paper 107, 2007.

Schlamadinger, B., *et al.*, "A Synopsis of Land Use, Land-Use Change and Forestry (LULUCF) under the Kyoto Protocol and Marrakech Accords," *ScienceDirect, Environmental Science and Policy*, 10, 271–282, 2007.

Schlesinger, J., *et al.*, "Assessing the Risks of a Collapse of the Atlantic Thermohaline Circulation," *Avoiding Dangerous Climate Change*, Cambridge University Press, 2006.

Schneider, L., "Is the CDM Fulfilling Its Environmental and Sustainable Development Objectives? An Evaluation and Options for Improvement," Oeko Institute for Applied Ecology, Berlin, November 5, 2007.

Schröter, D., *et al.*, "Ecosystem Service Supply and Vulnerability to Global Change in Europe," *Science*, 310, 1333–1337, 2005.

SciDev net, June 2006.

Seaman, G., "The One-Ton Challenge," Eartheasy, www.eartheasy.com, Vancouver, 2007.

Service, R., "As the West Goes Dry," *Science*, 303, 1124–1127, 2003.

Shepherd, A., *et al.,* "Larsen Ice Shelf Has Progressively Thinned," *Science*, 302, 856–858, 2003.

Sheppard, C., "Predicted Recurrences of Mass Coral Mortality in the Indian Ocean," *Nature*, 425, 294–297, 2003.

Shindell, D. T, *et al.*, "Impact of climate change on the emissions of methane from wetlands," *Geophysical Research Letters*, 31, 2004.

Shindell, D. T., *et al.*, "An Emissions Based View of Climate Forcing by Methane and Tropospheric Ozone," *Geophysical Research Letters*, 22, 2005.

Singer, F., and Seitz, F., "Hot Talk Cold Science: Global Warming Unfinished Debate," The Independent Institute, 1998.

Shukman, D., "Carbon Dioxide Continues to Rise," *BBC*, http://news.bbc.co.uk, 2005.

Simms, A., and Reid, H., "Up in Smoke? Latin America and the Caribbean: The Threat of Climate Change to the Environment and Human Development," Working Group on Climate Change and Development, New Economics Foundation, London, 2006.

Sinha, G. "Soccer Goes Green. At the World Cup, a New Way to Offset Carbon Emissions," *Scientific American*, July 24, 2006.

Sitch, P., *et al.*, "Indirect Radiative Forcing of Climate Change through Ozone Effects on the Land-Carbon Sink," *Nature*, 448, 2007.

Sluijs, A., *et al.*, "Subtropical Arctic Ocean Temperatures during the Palaeocene/Ecocene Thermal Maximum," *Nature*, 441, 610–613, 2006.

Small, C., and Nicholls, R. J., "A Global Analysis of Human Settlement in Coastal Zones," *Journal of Coastal Resources*, 19, 584–599, 2003.

Smithers, J., and Smit, B., "Human Adaptation to Climate Variability and Change," *Global Environmental Change*, 7(2), 129–146, 1997.

Soden, B., and Held, M., "An assessment of Climate Feedbacks in Coupled Ocean Atmosphere Models," *Journal of Climate*, 19, 2006.

Sohngen, B., and Sedjo, R., "Carbon Sequestration in Global Forests under Different Carbon Price Regimes," *The Energy Journal*, Greenhouse Gas Mitigation and Climate Policy Special Issue, 109–126, 2006.

Sorrenssen, A., *et al.*, "Release of Methane from a Volcanic Basin as a Mechanism for Initial Ecocene Global Warming," *Nature*, 429, 513–515, 2004.

Stainforth, D., *et al.*, "Uncertainty in Predictions of the Climate Response to Rising Levels of Greenhouse Gases," *Nature*, 433, 403–406, 2005.

Stanhill, G., "Global Dimming: A New Dimension on Climate Change," *Weather*, 60, 11–14, 2005.

Stanhill, G., and Morshet, S., "Global Radiation Change in Israel," *Climate Change*, 22, 121–138, 1992.

Stanhill, G., and Cohen, S., "Global Dimming: A Review of the Evidence for a Widespread and Significant Reduction in Global Radiation with Discussion of Its Probable Causes and Possible Agricultural Consequences," *Agricultural and Forest Meteorology*, 107, 2001.

Steams, P., "The Industrial Revolution in World History," Westview Press, 2007.

Steffen, W., *et al.*, "Approaches to Defining Dangerous Climate Change: An Australian Perspective," *Avoiding Dangerous Climate Change*, Cambridge University Press, 2006.

Stephens, B., *et al.*, "Weak Northern and Strong Tropical Land Carbon Uptake from Vertical Profiles of Atmospheric CO_2," *Science*, 316, 1732–1735, 2007.

Stern Report, "The Economics of Climate Change," United Kingdom Government, 2006.

Stirzaker, M., "Solar Thermal Power: Mirrored Sunshine," *Renewable Energy World*, 9(2), 2006.

Stocker, T., and Schmitter, A., "Influence of CO_2 Emission Rates on the Stability of the Thermohaline," *Nature*, 388, 862–865, 1997.

Stokstad, E., "Global Survey Documents Puzzling Decline of Amphibians," *Science*, 306, 391, 2004.

Stott, P., *et al.*, "Human Contribution to the European Heatwave of 2003," *Nature*, 432, 610–614, 2004.

Syri, S., *et al.*, "Low-CO_2 Energy Pathways and Regional Air Pollution in Europe," *Energy Policy*, 29, 871–884, 2001.

Teller, J., *et al.*, "Freshwater Outbursts to the Ocean from Glacial Lake Agassiz and their role in Climate Change in the Last Deglaciation," *Quaternary Science Reviews*, 21, 879–887, 2002.

The Economist, February 20, 1999.

The Gold Standard, http://www.cdmgoldstandard.org/index.php, 2007.

The Norwegian Nobel Institute, "The Nobel Peace Prize for 2007," Oslo, 2007.

Thomas, C., *et al.*, "Extinction Risk from Climate Change," *Nature*, 427, 145–148, 2004.

Thomas, R., *et al.*, "Accelerated Sea Level Rise from West Antarctica," *Science*, 306, 255–258, 2004.

Thompson, D., *et al.*, "The North Atlantic Oscillation: Climatic Significance and Environmental Impact," Geophysical Monograph 134, American Geophysical Union, Washington, DC, 81–112, 2003.

Tietenberg, T., "Emissions Trading Principles and Practice," 2nd Edition, RFF Press, Washington, DC, 2006.

Tilman, D., *et al.*, "Carbon-Negative Biofuels from Low-Input High-Diversity Grassland Biomass," *Science*, 314, 1598–1600, 2006.

Tol, R., "Estimates of Damage Costs of Climate Change," *Environment and Resource Economics*, 21, 47–73 and 135–160, 2002.

Tol, R., "Is the Uncertainty about Climate Change too Large for Expected Costs Benefit Analysis," *Climate Change*, 56, 265–289, 2003.

Tol, R., and Yohe, G., "Of Dangerous Climate Change and Dangerous Emission Reduction," *Avoiding Dangerous Climate Change*, Cambridge University Press, 2006.

Tol, R., *et al.*, "Distributional Aspects of Climate Change Impacts," *Global Environmental Change*, 14, 259–272, 2004.

Travis, D., *et al.*, "Contrails Reduce Daily Temperature Range," *Nature*, 418, 601, 2002.

Tschumi, J., and Stauffer, B., "Reconstructing Past Atmospheric CO_2 Concentrations Based on Ice-Core Analyses: Open Questions due to In Situ Production of CO_2 in Ice," *Journal of Glaciology*, 46, 45–53, 2000.

Tudhope, A., *et al.*, "Variability in the El Nino-Southern Oscillation through a glacial-interglacial cycle," *Science*, 296, 1511–1517, 2001.

Tung, K., and Camp, C., "Solar-Cycle Warming at the Earth's Surface and an Observational Determination of Climate Sensitivity," University of Washington, 2007.

Turnball, M, "Indonesian Forest Carbon Partnership to Reduce Greenhouse Gas Emissions," Australian Government, September 2007.

Union of Concerned Scientists, "Recognising forests' role in climate change," Policy statement online at http://www.ucsusa.org/global_warming, 2007.

United Nations and United Nations Framework Convention on Climate Change, "Kyoto Protocol," Informal published version, FCCC/Informal/104, GE.05-64000, November 2005.

UN Public Web site, http://www.un.org/geninfo/ir/index.asp?id=150

United Nations Development Programme (UNDP), "The Clean Development Mechanism: An Assessment of Progress," UNDP, New York, 2006a.

UNDP, "Malaysia's Peat Swamp Forests Conservation and Sustainable Use," ISBN 983-40995-5-X, 2006b.

UNDP Human Development Report 2007, United Nations Development Programme, New York, 2007.

UNEP, "Handbook for the Montreal Protocol on Substances that Deplete the Ozone Layer," 7th Edition, Ozone Secretariat, 2006.

UNEP, "Renewable Energy," UNEP Homepage, July 2008a.

UNEP, "Insuring for Sustainability", UNEP Finance Initiative, 2007, available at www.unepfi.org/index.html

UNEP Risoe JI Pipeline Analysis and Database, www.cdmpipeline.org, 2008.

UNEP/IUCN, "Developing International Payments for Ecosystem Services," public awareness brochure available at http://www.unep.ch/etb/areas/ipes.php, 2007.

UNEP/UNFCCC, "Understanding Climate Change: A Beginner's Guide to the UN Framework Convention and Its Kyoto Protocol," 2002.

UNEP/WMO, "IPCC 16 Years of Scientific Assessment in Support of the Climate Convention," 2004.

UNFCCC Handbook, Climate Change Secretariat, Bonn, 2006

UNFCCC/SBI, "Compilation and Synthesis Report on Third National Communications," Bonn, 2003.

UNFCCC CDM Web site, www.cdm.unfccc

UNFCCC, "National Greenhouse Gas Inventory Data for the Period 1990–2005," FCCC/SBI/2007/30, October 24, 2007a.

UNFCCC, "National Reports," http://unfccc.int/national_reports/items/1408.php, 2007b.

UNFCCC, Online ghg data at http://unfccc.int/ghg_emissions_data/ghg_data_from_unfccc/items/4146.php

UNFCCC, "Report of the Conference of the Parties," 2005.

UNFCCC JI Home page, http://ji.unfccc.int/JI_Parties

UNFCCC, Press release, March 27, 2007c, at http://ji.unfccc.int/JI_News

UNFCCC Ad hoc Group on the Berlin Mandate at http://unfccc.int/resource/docs/1997/agbm/misc01a03.pdf

United Nations "United Nations Framework Convention on Climate Change", FCCC/ INFORMAL/84 GE.05-62220 (E) 200705, 1992.

United Nations Convention on Biodiversity (UNCBD) Secretariat, "Emerging Issues for Biodiversity Conservation in a Changing Climate", Technical Paper, 2003

UN General Assembly Resolution 55/5.

UN Population Division, Department of Economic and Social Affairs, 2005.

US Department of Energy (USDOE), "Climate Change Technology Program Strategic Plan," Washington, 2006.

US Environmental Protection Agency (US EPA), "Inventory of US Greenhouse Gas Emissions and Sinks 1990–2004," 2006.

US Environmental Protection Authority, PFC Reduction/Climate Partnership for the Semiconductor Industry, http://www.epa.gov/semiconductor-pcf/, March 25, 2008.

US EPA, "Ground level ozone," 2007

US Geological Survey, www.usgs.gov

US Geological Survey, "World Petroleum Assessment 2000: Description and Results," USGS, Washington, 2000.

US Office of the Press Secretary, "President Bush Discusses Global Climate Change," June 11, 2001.

University of Colorado, http://sealevel.colorado.edu/

Van Vliet, A., and Leemans, R., "Rapid Species Responses to Changes in Climate Require Stringent Climate Protection Targets," *Avoiding Dangerous Climate Change*, 2006.

Van Vuuren, D. P., *et al.*, "Exploring the Ancillary Benefits of the Kyoto Protocol for Air Pollution in Europe," *Energy Policy*, 34, 444–460, 2006.

Walter, K., *et al.*, "Methane Bubbling from Siberian Thaw Lakes as a Positive Feedback to Climate Change," *Nature*, 443, 71–75, 2006.

Wang, X., and Wiser, G., "The Implementation and Compliance Regimes under the Climate Change Convention and the Kyoto Protocol," *RECIEL*, 11, 181–199, 2002.

Wang, Y., *et al.*, "Modelling the Sun's Magnetic Field and Irradiance since 1713," *Journal of Astrophysics*, 625, 522–538, 2005.

Warren, R., "Impacts of Global Climate Change at Different Annual Mean Global Temperature Increases," *Avoiding Dangerous Climate Change*, Cambridge University Press, 2006.

Warren, R., *et al.*, Understanding the Regional Impacts of Climate Change," Tyndall Centre Working Paper 90, 2006.

Wayne, P., *et al.*, "The Production of Allergenic Pollen by Ragweed in CO_2 Enriched Atmospheres," *Annals of Allergy, Asthma and Immunology*, 88, 279–282, 2002.

Webster, P., *et al.*, "Changes in Tropical Cyclone Number, Duration, and Intensity in a Warming Environment," *Science*, 309, 1844–1846, 2005.

Weijers, J., *et al.*, "Coupled Thermal and Hydrological Evolution of Tropical Africa over the Last Deglaciation," *Science*, 315, 1701–1704, 2007.

Weldeab, S., *et al.*, "155,000 years of West African Monsoon and Ocean Thermal Evolution," *Science*, 316, 1303–1307, 2007.

Wellington, F., and Sauer, A., "Climate Risk in Portfolio Management," World Resources Institute, 2005.

Westerling, A., *et al.*, "Warming and Earlier Spring Increase Eastern US Forest Wildfire Activity," *Science*, 313, 940–943, 2006.

Wetlands International, "Tropical Peatswamp Destruction Fuels Climate Change," Factsheet, Ministry of Foreign Affairs, The Netherlands, 2007.

Weyant, J., *et al.*, "Overview of EMF-21: Multigas Mitigation and Climate Policy," *The Energy Journal*, 1–32, 2007.

Wheeler, T., *et al.*, "Growth and Yield of Winter Wheat (Triticum aestivum) Crops in Response to CO_2 and Temperature," *Journal of Agricultural Science*, 127, 37–48, 1996.

Whetton, P., "Approaches to Defining Dangerous Climate Change: An Australian Perspective," *Avoiding Dangerous Climate Change*, Cambridge University Press, 2006.

Wielicki, B., *et al.*, "Changes in Earth's Albedo Measured by Satellite," *Science*, 308, 825, 2005.

Wigley, T., "The Climate Change Commitment," *Science*, 307, 1766–1769, 2005.

Wilcox, J., "Europe's Nuclear Options," *CommoditiesNow*, 10, 53–58, December 2006.

Wild, M, *et al.*, "From Dimming to Brightening: Decadal Changes in Solar Radiation at Earth's Surface," *Science*, 308, 847–850, 2005.

Williams, S., *et al.*, "Climate Change in Australian Tropical Rainforests: An Impending Environmental Catastrophe," *Proceedings of the Royal Society of London*, 270, 1887–1892, 2003.

Wilson, E., and Perlman, D., "Conserving Earth's Diversity," Island Press, 2000.

Winton, M., "Does Arctic Sea Ice Have a Trigger Point?" *Geophysical Research Letters*, 33, 23, 2006.

Wood, E., "Wind: Getting the Facts Straight," *Renewable Energy World*, 9(2), 2006.

Wood, R., *et al.*, "Toward a Risk Assessment for Shutdown of the Atlantic Thermohaline Circulation," *Avoiding Dangerous Climate Change*, Cambridge University Press, 2006.

Wood, W., *et al.*, "Decreased Stability of Methane Hydrates in Marine Sediments Owing to Phase-Boundary Roughness," *Nature*, 420, 656–660, 2002.

Woodford, J., "Great? Barrier Reef," *Australian Geographic*, 76, 37–55, 2004.

World Bank, "Natural Resources and Violent Conflict," World Bank, Washington DC, 2003.

World Bank, "Stock-take on the Market and Compliance Gap," 2005.

World Bank, "World Development Indicators," 2006a.

World Bank, "Managing Water Resources to Maximize Sustainable Growth: A Country Water Resources Assistance Strategy for Ethiopia," World Bank, 2006b.

World Bank, "Forest Carbon Partnership Facility Launched at Bali Climate Meeting," News release 2008/142/SDN, December 11, 2007a.

World Bank, "State and Trends of the Carbon Market 2007," 2007b.

World Bank, "State and Trends of the Carbon Market 2008," 2008.

World Business Council for Sustainable Development and World Resources Institute (WBCSD/WRI), *The Greenhouse Gas Protocol: A Corporate Accounting and Reporting Standard*, Revised Edition, Geneva, Switzerland and Washington, USA, March 2004. ISBN 1-56973-568-9. Available at: http://www.ghgprotocol.org

WBCSD, "Issue Brief: Biofuels," ISBN 978-3-940388-15-5, 2007.

World Health Organization (WHO), "Africa Malaria Report 2003," Geneva, 2003.

WHO, "Air Quality Guidelines Global Update," 2005a.

WHO, "Malnutrition: Quantifying the Health Impact at National and Local Levels," *Environmental Burden of Disease Series*, 2005b.

World Resources Institute (WRI), "Navigating the Numbers: Greenhouse Gas Data and International Climate Change Policy," Washington, 2005.

World Semiconductor Council, "Joint Statement on the 8th Meeting of World Semiconductor Council," May 13, 2004.

World Wind Energy Association, 2007, www.wwindea.org

WRI Climate Analysis Indicators Tool (CAIT) on line database, http://cait.wri.org, 2007a.

WRI, "The Hidden Benefits of Climate Policy: Reducing Fossil Fuel Use Saves Lives," 2007b.

Yamin, F., and Depledge, J., "The International Climate Change Regime: A Guide to Rules, Institutions and Procedures," Cambridge University Press, 2004.

Yamin, F., *et al.,* "Perspectives on Dangerous Climate Change; or How to Operationalize Article 2 of the UN Framework Convention on Climate Change," *Avoiding Dangerous Climate Change*, Cambridge University Press, 2006.

Zachos, C., *et al.*, "A Transient Rise in Tropical Sea Surface Temperatures during the Paleocene–Ecocene Thermal Maximum," *Science*, 302, 1551–1554, 2003.

Zhou, G., *et al.*, "Old Growth Forest Can Accumulate Carbon in Soils," *Science*, 314, 1417, 2006.

Zimov, S., *et al.*, "Permafrost and the Global Carbon Budget," *Science*, 312, 1612–1613, 2006.

Index

Preliminary note: page numbers in **bold** refer to tables or figures